Engineering Innovations in Agriculture

Engineering Innovations in Agriculture

Editors

Vadim Bolshev
Vladimir Panchenko
Alexey Sibirev

MDPI • Basel • Beijing • Wuhan • Barcelona • Belgrade • Manchester • Tokyo • Cluj • Tianjin

Editors
Vadim Bolshev
Laboratory of Power and Heat Supply,
Federal Scientific Agroengineering Center VIM,
Moscow, Russia

Vladimir Panchenko
Department of Theoretical and Applied Mechanics,
Russian University of Transport,
Moscow, Russia

Alexey Sibirev
Head of Laboratory of Machine Technologies for Cultivation and Harvesting of Vegetable Crops in Open Ground,
Federal Scientific Agroengineering Center VIM,
Moscow, Russia

Editorial Office
MDPI
St. Alban-Anlage 66
4052 Basel, Switzerland

This is a reprint of articles from the Special Issue published online in the open access journal *Agriculture* (ISSN 2077-0472) (available at: https://www.mdpi.com/journal/agriculture/special_issues/B5Y1FH6WG7).

For citation purposes, cite each article independently as indicated on the article page online and as indicated below:

LastName, A.A.; LastName, B.B.; LastName, C.C. Article Title. *Journal Name* **Year**, *Volume Number*, Page Range.

ISBN 978-3-0365-8160-6 (Hbk)
ISBN 978-3-0365-8161-3 (PDF)

Contents

About the Editors

Vadim Bolshev

Vadim Bolshev received an M.S. degree in engineering from Orel State Agrarian University in 2012, and was a candidate for a technical science degree from Federal Scientific Agroengineering Center VIM in 2020. Vadim gained experience in the industry as an electrician, an electrical engineer, and a chief engineer in 2010–2018, and since 2018, he has been working as a researcher in the Laboratory of Power and Heat Supply with the Federal Scientific Agroengineering Center VIM. He has led the laboratory of Intelligent Electric Agricultural Machines and Complexes at Don State Technical University since 2021. His field of scientific activity is to develop technical and economic mechanisms to enhance the efficiency of power supply systems and electrical machines and to address other issues related to creating new engineering technologies for the agroindustrial complex.

Vladimir Panchenko

Vladimir Panchenko is an Associate Professor of the Department of Theoretical and Applied Mechanics at the Russian University of Transport, and a Senior Researcher of the Laboratory of Non-traditional Energy Systems at the Federal Scientific Agroengineering Center VIM. He graduated from the Bauman Moscow State Technical University in 2009 with an engineering qualification. His Ph.D. thesis, "Power plants based on renewable energy," was defended in 2013. He has been an Assistant Editor since 2019 at the International Journal of Energy Optimization and Engineering, and a Guest Editor since 2019 for these Special Issues of the MDPI journal (Switzerland): Applied Sciences, Energies, and Agriculture. He has also coedited the conference proceedings in Lecture Notes in Networks and Systems (Springer), and served as an Editor of IGI GLOBAL (USA) and Nova Science Publisher (USA). He has participated in more than 100 exhibitions and conferences at various levels, and has published more than 300 scientific papers, including 14 patents, 1 international patent, 6 educational publications, and 4 monographs.

https://www.researchgate.net/profile/Vladimir-Panchenko-2.

Alexey Sibirev

Alexey Sibirev received an M.S. degree in engineering and a degree of candidate of technical science from Penza State Agrarian University in 2011 and 2014, respectively. He also has a Ph.D. in technical science from the Federal Scientific Agroengineering Center VIM, which he gained in 2020. Alexey gained experience in the industry as a mechanic and chief engineer in 2010-14, and since 2015 has been working as a researcher in the laboratory of machine technologies in vegetable growing at the Federal Scientific Agroengineering Center VIM, and has been head of the laboratory since 2017. His research covers the development of technical and economic mechanisms for increasing the production of vegetables and potatoes and the efficiency of these machines, and addresses other issues related to the creation of new engineering technologies for the agroindustrial complex.

Preface to "Engineering Innovations in Agriculture"

Nowadays, the expansion of people into intact primary areas has been observed alongside an increase in land devoted to crops, pastures, etc., which has led to the destruction of natural diversity. A solution to avoid disaster is increasing agricultural production efficiency to expand crop harvest and livestock productivity, without deteriorating quality. This approach requires introducing innovative engineering technologies to agriculture. Fortunately, technology is developing rapidly, and new driving scientific forces are emerging in various fields of agriculture, which has allowed significant inroads to be made in the development of all fields of agroengineering science.

This Special Issue, titled "Engineering Innovations in Agriculture," consists of high-quality articles from academics and industry-related researchers in the field of agricultural engineering, covering the following topics: harvesting and planting crops, livestock production, livestock and agrifood waste utilization, seed treatment and transportation, agricultural robotic applications, solutions for digital and precision agriculture, energy efficiency and conservation in agriculture, agriculture mechanization and electrification, harvest technologies, postharvest technologies, renewable energy technologies, robotics, etc. The studies were carried out by scientists and researchers from around the world, and have been held in high regard by peer reviewers in their respective fields of knowledge.

The presented works contribute towards solving issues regarding the development, research, and optimization of engineering innovations in agriculture, where their results will interest specialists, researchers, and scientists in the considered areas. Of course, only a small area of the issues that are relevant today are touched upon in this Special Issue; however, the authors have carried out important work, which will significantly contribute to solving issues related to engineering innovation in agriculture in the future.

Vadim Bolshev, Vladimir Panchenko, and Alexey Sibirev
Editors

agriculture MDPI

Editorial

Engineering Innovations in Agriculture

Vadim Bolshev [1], Vladimir Panchenko [2,*] and Alexey Sibirev [3]

1 Laboratory of Power Supply and Heat Supply, Federal Scientific Agroengineering Center VIM, 109456 Moscow, Russia; vadimbolshev@gmail.com
2 Department of Theoretical and Applied Mechanics, Russian University of Transport, 127994 Moscow, Russia
3 Laboratory of Machine Technologies for Cultivation and Harvesting of Vegetable Crops in Open Ground, Federal Scientific Agroengineering Center VIM, 109456 Moscow, Russia; sibirev2011@yandex.ru
* Correspondence: pancheska@mail.ru

Nowadays, the expansion of people into intact primary areas has been observed alongside an increase in the area of land devoted to crops, pastures, etc., which has led to the destruction of natural diversity. A solution to avoid disaster is to increase agricultural production efficiency in order to expand crop harvest and livestock productivity without deteriorating their quality. This approach requires introducing innovative engineering technologies to agriculture. Fortunately, technology is developing rapidly, and new driving scientific forces are emerging in various fields of agriculture, which has allowed significant inroads to be made in the development of all fields of agroengineering science.

This Special Issue, titled "Engineering Innovations in Agriculture", consists of high-quality articles from academics and industry-related researchers in the field of agricultural engineering, covering the following topics: harvesting and planting crops, livestock production, livestock and agrifood waste utilization, seed treatment and transportation, agricultural robotic applications, solutions for digital and precision agriculture, energy efficiency and conservation in agriculture, agriculture mechanization and electrification, harvest technologies, postharvest technologies, renewable energy technologies, robotics, etc. The presented studies were carried out by scientists and researchers from around the world and have been held in high regard by peer reviewers in their respective fields of knowledge; the main objectives of the published works can be found below.

The first study [1] showed that, during the heating period, the application of the ceiling fans helped to raise the air temperature and to reduce relative air humidity in the areas where young stock is located, in accordance with the normative indicators. In [2], soybean seed particles of three varieties with different sphericities were taken as the research objects. Through the simulation analysis of repose angle and self-flow screening, it was shown that the above two parameters needed to be accurately calibrated. In [3], the authors investigated the effect of impulse (frequency 1000 Hz and duty cycle 67%), scanning (the principle of running lights), and constant 16 h and 24 h modes of operation of white light LED irradiators on the physiological, biochemical, and morphometric parameters of lettuce with red and green leaves, in order to optimize the parameters of the mechanism, reduce the pressure loss and improve the efficiency of pneumatic utilization. In [4], the principles and types of pneumatic loss in different areas were defined, and the key parameters, including the diameter of the horizontal air pipe, the angle of the air pipe, and the diameter of the negative pressure aperture, which affect the pressure loss, were analyzed. The article in [5] presents a methodology for conducting research to assess the influence of working bodies on the magnitude of the force impact on potato tubers and soil clods during harvesting, using a methodology for assessing the dynamic destruction of soil clods. In [6], the authors synthesized Mn^{4+}-doped $CaSb_2O_6$ phosphors using the conventional solid-state reaction method for plant growth lighting applications. A counting method for red jujube based on the improved YOLOv5s was proposed in [7], which realized the fast and accurate detection of red jujubes and reduced the model scale and estimation error. In [8], the effects of the

Citation: Bolshev, V.; Panchenko, V.; Sibirev, A. Engineering Innovations in Agriculture. *Agriculture* **2023**, *13*, 1328. https://doi.org/10.3390/agriculture13071328

Received: 14 June 2023
Revised: 20 June 2023
Accepted: 28 June 2023
Published: 29 June 2023

working pressure and aspect ratio (L/D) of circular and non-circular nozzles (diamond and ellipse) on water distribution and droplet kinetic energy intensity were investigated. In [9], the authors first presented the "cut-and-paste" method for synthetically augmenting the available dataset by generating additional annotated training images. In [10], the authors investigated the optimal design for an agricultural mobile robot's suspension system based on a double wishbone suspension structure. In the work presented in [11], the authors investigated the use of pyrolysis to recycle amaranth inflorescence wastes. The results indicate that the application of pyrolysis will allow the efficient conversion of amaranth inflorescence wastes into value-added products. The article in [12] presents the results of developing and testing the air-cooling system of a combined climate control unit used in pig farming. The authors found a water-evaporative system to be the most efficient for cooling the air supply. The experimental studies presented in [13] using a developed harvester with an experimental separating system made it possible to ensure the high-quality harvesting of potato and onion tubers. In [14], a method for determining the degree of infected seeds with Fusarium was developed. In [15], carbon emissions, customer satisfaction, customer value, and cost were considered, and an optimization algorithm was established to solve the time-dependent vehicle routing problem in urban cold chain logistics. In [16], each biomass sample of the redroot pigweed (leaves, inflorescences, and stems) was pyrolyzed in a lab-scale furnace, in a nitrogen atmosphere, under non-isothermal conditions. The results will help us to better understand the thermal behavior of redroot pigweed biomass and its utilization for fuels or chemicals. In [17], the authors propose a method to reduce the feed rate by decreasing the feed length of the stalk and design a double-cutter bar combine header. The purpose of the study presented in [18] is to identify and evaluate the effect of high-voltage electrical pulses on the irreversible damage to the intracellular structures of the plant tissue of weeds and unwanted grasses during electric weed control, characterizing and evaluating the parameters and modes associated with such processing. The article presented in [19] is devoted to questions of increasing the efficiency and attractiveness of solar photovoltaic equipment for the purpose of ensuring the appropriate conditions for widespread power supply to agricultural consumers through the use of solar energy.

Thus, the presented works contribute towards solving issues regarding the development, research, and optimization of engineering innovations in agriculture, where their results will be of interest to specialists, researchers, and scientists in the considered areas. Of course, only a small area of the issues that are relevant today are touched upon in this Special Issue, titled "Engineering Innovations in Agriculture"; however, the authors have carried out important work, which will make a significant contribution to solving issues related to engineering innovation in agriculture in the future.

Author Contributions: Conceptualization, V.B., V.P. and A.S.; formal analysis, V.P. and V.B.; resources, V.P.; writing—original draft preparation, V.P.; writing—review and editing, V.B.; supervision, V.B. and A.S.; project administration, V.B. and V.P. All authors have read and agreed to the published version of the manuscript.

Funding: The studies cited in this review were funded by the National Natural Science Foundation of China (No. 52130001); by the Ministry of Science and Higher Education of the Russian Federation for large scientific projects in priority areas of scientific and technological development, contract no. 075-15-2020-774; by the National Natural Science Foundation of China (Grant No. 52175260), the China Agriculture Research System (CARS-03), the Key Research and development plan of Hainan Province (Grant No. ZDYF2022XDNY184), Project of Scientific Research in Colleges and Universities of Hainan Province (Grant No. Hnky2022-93), and Research Start-up Fund Project of Hainan University (Grant No. KYQD(ZR)-22036); by the Russian Science Foundation Grant No. 22-76-10002; by the Talent start-up Project of Zhejiang A&F University Scientific Research Development Foundation (2021LFR066) and the National Natural Science Foundation of China (C0043619, C0043628); by the Postgraduate Research & Practice Innovation Program of Jiangsu Province, grant number SJCX22_1870, the Jiangsu Province and Education Ministry Co-sponsored Synergistic Innovation Center of Modern Agricultural Equipment, grant number XTCX2018, the Changzhou Key Research and Development Program, grant number CE20222024, Zhenjiang Key

Research and Development Program, grant number CN2022003, and the Youth Talent Development Program of Jiangsu University; by the National Natural Science Foundation of China (61871061); by the Project of Advanced Scientific Research Institute of CQUPT under Grant E011A2022329; by the USDA National Institute of Food and Agriculture (Hatch Project number ME0-22021 to YJZ) through the Maine Agricultural & Forest Experiment Station; by an intra-university grant from the Russian State Agrarian University—Moscow Timiryazev Agricultural Academy "Food Sovereignty" section for research projects in the field of import substitution within the framework of the "Priority 2030" strategic academic leadership program; by the Russian Science Foundation of the 2022 contest "Conducting research by scientific groups led by young scientists" of the Presidential Program of research projects implemented by leading scientists, including young scientists No. 22-76-10002; by the China Education Ministry of Humanities and Social Science Research Youth Fund project (No. 18YJCZH192), the Special Project of National Characteristic Freshwater Fish Industrial Technology System for Construction of Modern Agricultural Industrial Technology System (No. CARS-46), and the major project of the National Social Science Fund "Research on the development strategy of China's deep blue fishery under the background of accelerating the construction of a marine power" (No. 21 & ZD100); and by China Agriculture Research System (CARS-03), Henan Province Science and Technology Projects (222102110235), and Henan Province Science and Technology Projects (232102110271).

Acknowledgments: The authors of the presented studies are grateful to the Postgraduate Research & Practice Innovation Program of Jiangsu Province (SJCX22_1870), the Jiangsu Province and Education Ministry Co-sponsored Synergistic Innovation Center of Modern Agricultural Equipment (XTCX2018), the Changzhou Key Research and Development Program (No. CE20222024), Zhenjiang Key Research and Development Program (No. CN2022003), the Youth Talent Development Program of Jiangsu University, and Zhengzhou Zhonglian Harvesting Machinery Co., Ltd.

Conflicts of Interest: The authors declare no conflict of interest.

References

1. Khimenko, A.; Tikhomirov, D.; Trunov, S.; Kuzmichev, A.; Bolshev, V.; Shepovalova, O. Electric Heating System with Thermal Storage Units and Ceiling Fans for Cattle-Breeding Farms. *Agriculture* **2022**, *12*, 1753. [CrossRef]
2. Yan, D.; Yu, J.; Wang, Y.; Sun, K.; Zhou, L.; Tian, Y.; Zhang, N. Measurement and Calibration of DEM Parameters of Soybean Seed Particles. *Agriculture* **2022**, *12*, 1825. [CrossRef]
3. Smirnov, A.A.; Semenova, N.A.; Dorokhov, A.S.; Proshkin, Y.A.; Godyaeva, M.M.; Vodeneev, V.; Sukhov, V.; Panchenko, V.; Chilingaryan, N.O. Influence of Pulsed, Scanning and Constant (16- and 24-h) Modes of LED Irradiation on the Physiological, Biochemical and Morphometric Parameters of Lettuce Plants (*Lactuca sativa* L.) while Cultivated in Vertical Farms. *Agriculture* **2022**, *12*, 1988. [CrossRef]
4. Zhang, X.; Wen, Z.; Wang, Q.; Li, H.; Zhang, Z.; Liu, J. Research on Characteristics of Airway Pressure Loss in Seeding-Wheel-Type Pneumatic Seeder. *Agriculture* **2022**, *12*, 2021. [CrossRef]
5. Dorokhov, A.; Didmanidze, O.; Aksenov, A.; Sibirev, A.; Sazonov, N.; Mosyakov, M.; Godyaeva, M. The Results of Studies on the Assessment of the Destruction of Soil Clods during Combine Harvesting of Potatoes. *Agriculture* **2022**, *12*, 2024 [CrossRef]
6. Bharat, L.K.; Patnam, H.; Sokolov, A.; Gudkov, S.V.; Yu, J.S. Red Light Emitting Transition Metal Ion Doped Calcium Antimony Oxide for Plant Growth Lighting Applications. *Agriculture* **2022**, *12*, 2066. [CrossRef]
7. Qiao, Y.; Hu, Y.; Zheng, Z.; Yang, H.; Zhang, K.; Hou, J.; Guo, J. A Counting Method of Red Jujube Based on Improved YOLOv5s. *Agriculture* **2022**, *12*, 2071. [CrossRef]
8. Wang, Z.; Jiang, Y.; Liu, J.; Li, H.; Li, H. Experimental Study on Water Distribution and Droplet Kinetic Energy Intensity from Non-Circular Nozzles with Different Aspect Ratios. *Agriculture* **2022**, *12*, 2133. [CrossRef]
9. Obsie, E.Y.; Qu, H.; Zhang, Y.-J.; Annis, S.; Drummond, F. Yolov5s-CA: An Improved Yolov5 Based on the Attention Mechanism for Mummy Berry Disease Detection. *Agriculture* **2023**, *13*, 78. [CrossRef]
10. Qu, Z.; Zhang, P.; Hu, Y.; Yang, H.; Guo, T.; Zhang, K.; Zhang, J. Optimal Design of Agricultural Mobile Robot Suspension System Based on NSGA-III and TOPSIS. *Agriculture* **2023**, *13*, 207. [CrossRef]
11. Karaeva, J.; Timofeeva, S.; Islamova, S.; Bulygina, K.; Aliev, F.; Panchenko, V.; Bolshev, V. Pyrolysis of Amaranth Inflorescence Wastes: Bioenergy Potential, Biochar and Hydrocarbon Rich Bio-Oil Production. *Agriculture* **2023**, *13*, 260. [CrossRef]
12. Ignatkin, I.; Kazantsev, S.; Shevkun, N.; Skorokhodov, D.; Serov, N.; Alipichev, A.; Panchenko, V. Developing and Testing the Air Cooling System of a Combined Climate Control Unit Used in Pig Farming. *Agriculture* **2023**, *13*, 334. [CrossRef]
13. Dorokhov, A.; Aksenov, A.; Sibirev, A.; Mosyakov, M.; Sazonov, N.; Godyaeva, M. Evaluation of Comparative Field Studies for Root and Onion Harvester with Variable Angle Conveyor. *Agriculture* **2023**, *13*, 572. [CrossRef]
14. Moskovskiy, M.N.; Belyakov, M.V.; Dorokhov, A.S.; Boyko, A.A.; Belousov, S.V.; Noy, O.V.; Gulyaev, A.A.; Akulov, S.I.; Povolotskaya, A.; Efremenkov, I.Y. Design of Device for Optical Luminescent Diagnostic of the Seeds Infected by Fusarium. *Agriculture* **2023**, *13*, 619. [CrossRef]

15. Wu, D.; Li, J.; Cui, J.; Hu, D. Research on the Time-Dependent Vehicle Routing Problem for Fresh Agricultural Products Based on Customer Value. *Agriculture* **2023**, *13*, 681. [CrossRef]
16. Karaeva, J.; Timofeeva, S.; Gilfanov, M.; Slobozhaninova, M.; Sidorkina, O.; Luchkina, E.; Panchenko, V.; Bolshev, V. Exploring the Prospective of Weed Amaranthus retroflexus for Biofuel Production through Pyrolysis. *Agriculture* **2023**, *13*, 687. [CrossRef]
17. Yuan, L.; Lan, M.; He, X.; Wei, W.; Wang, W.; Qu, Z. Design and Experiments of a Double-Cutterbar Combine Header Used in Wheat Combine Harvesters. *Agriculture* **2023**, *13*, 817. [CrossRef]
18. Yudaev, I.; Daus, Y.; Panchenko, V.; Bolshev, V. Influence of Factors Determining Weeds' Plant Tissuereaction to the Electric Pulse Damage Impact. *Agriculture* **2023**, *13*, 1099. [CrossRef]
19. Shepovalova, O.; Izmailov, A.; Lobachevsky, Y.; Dorokhov, A. High-Efficiency Photovoltaic Equipment for Agriculture Power Supply. *Agriculture* **2023**, *13*, 1234. [CrossRef]

agriculture

MDPI

Article

Electric Heating System with Thermal Storage Units and Ceiling Fans for Cattle-Breeding Farms

Aleksei Khimenko *, Dmitry Tikhomirov, Stanislav Trunov, Aleksey Kuzmichev, Vadim Bolshev and Olga Shepovalova

Laboratory of Power and Heat Supply, Federal Scientific Agroengineering Center VIM, 1-St Institutskiy Proezd, 5, 109428 Moscow, Russia

* Correspondence: avmkh87@gmail.com; Tel.: +7-499-174-8595

Abstract: A combined energy–saving heat supply system was proposed that included a combined ETS unit and a ceiling fan, and provided the normative air parameters in a livestock room, with an air temperature of −17 °C and air relative humidity (ARH) of −75%. A heat supply system of a preventive maintenance premises for calves was chosen as the subject of the study. Comparative analysis of the temperature and ARH distribution with height in the preventive maintenance premises, was carried out, with and without a ceiling fan. The study showed that, during the heating period, application of the ceiling fans helped to raise the air temperature and to reduce ARH, in the areas where young stock is located, in accordance with the normative indicators. The energy-saving effect was achieved by supplying warmer ventilation air, which accumulated in the upper zone of the premises from the ceiling fan to the locations of the animals. At the same time, there was a decrease in the consumption of electric energy for the heat supply system of up to 14%.

Keywords: microclimate; ceiling fans; electric thermal storage unit; energy–saving; heat supply system

Citation: Khimenko, A.; Tikhomirov, D.; Trunov, S.; Kuzmichev, A.; Bolshev, V.; Shepovalova, O. Electric Heating System with Thermal Storage Units and Ceiling Fans for Cattle-Breeding Farms. *Agriculture* **2022**, *12*, 1753. https://doi.org/10.3390/agriculture12111753

Academic Editor: Claudia Arcidiacono

Received: 15 September 2022
Accepted: 21 October 2022
Published: 23 October 2022

1. Introduction

In the case of group housing for calves, animals must be kept in a rather limited space, which may cause critically bad environment conditions, leading to the loss of livestock that, in some circumstances, attains 30% to 40%. Besides, it is associated with productivity reduction of 15% and with the growth of specific fodder consumption by 10% to 15% per product unit [1,2]. This is why, in conditions of intensive animal housing, major attention must be paid to maintaining the optimal microclimate (first of all, temperature and ARH).

Technical–economic studies show that, in case of group housing for calves (i.e., in boxes, preventive maintenance premises, cages, etc.), it is advisable to use storage-type electric units for air heating to serve as internal heat sources designed for small premises operating on the principle of convective heat exchange [3].

Applications of electric thermal storage (ETS) units in combination with energy generating installations in a net-zero multi-energy system, including those designed for farms, belong to the most effective solutions, as well. Such installations must be based mainly on various renewable energy sources (RES), first of all, solar- and wind-power installations for heat and water supply [4–8].

Simulation and tests on an electric thermal storage heating system with solid-state heat storage materials (SS-ETSHSM) using electric energy generated by coal combined heat and power (CHP) units [9] and wind, and solar power stations have been carried out [10,11]. ETS unit are charged during the minimum load period of the energy system. Then, this energy is transferred to the heat-carrier into the water-heating system with the help of 'air-water' heat-exchangers. Sun Y. et al. [9] investigated the characteristics of thermal processes and the regularities of temperature changes in different parts of the heating system occurring in the modes of charging and heat emission ETS under quantitative and qualitative regulation. Zhao H. et al. [10] investigated the heat transfer characteristics of SS-ETSHSM. The thermal calculation method for key parameters of the electric heating system was proposed. Based on the experimental data, the amount of accumulated heat was calculated along with the heat-exchange coefficient, in air channels of various geometries. Amounts of thermal energy transferred from the heat-storage unit to the heated air were estimated as well. The perational effectiveness of ETS unit installations can be improved by a reasonable selection of the air channels number in ETS units, as well as by optimizing their shape and dimensions. Correctly defining the optimal values of air velocity, in channels, and location of air inlets and outlets is also important [11]. Xu G. et al. [12] developed a static-type ETS unit with a heat-storage phase change material (PCM) in which heat-exchange occurs in air channels, owing to natural convection. The proposed device is charged by the off-peak electricity. The dynamic process of the thermal state under a fully charging/discharging cycle of the device was tested. The amount of heat transferred in the air channels to the heated air and the heat given off by the ETS casing by natural convection and radiation were determined as a percentage.

Klymchuk et al. [13] considered various schemes of arrangement of tubular heating elements for the charging of heat storage cells. The temperature distribution over the cross-section of the thermal storage battery at the key points as a result of mathematical simulations was obtained. The obtained simulation results were tested on an experimental installation consisting of four heat storage cells during the charging and discharging of the thermal storage unit. The most effective arrangement of tubular heating elements was determined, allowing for maximum use of the volume of heat storage material. Dependences for determining the index and averaging coefficient of the heat flux were also found. An algorithm for defining the design parameters of the thermal storage depending on the heat supply system operating conditions was described by Klymchuk et al. [14] which made it possible to calculate the optimal position of the heating elements in the heat-storage core, their number, the thickness of the heat-storage layer, dimensions of the heat-storage core and the specific heat flux upon the heated surface. Beknazarian et al. [15] developed a method for selecting and optimizing the high-temperature thermal insulation that can be applied in ETS units.

Theoretical studies of the heat-storage core thermal state of the dynamic-type ETS units were performed by Janssen et al. [16], for both heat charging and heat emission operation modes. Three-dimension models for the heat-exchange processes were developed and the calculations for the non-stationary temperature field were carried out using the finite element method.

When selecting a reasonable method of heat and air distribution in premises, one must consider the requirements for providing the specified air–thermal conditions in work areas where calves are kept, including those of cages.

Various design options for inflow-exhaust ventilation installations and systems, can be implemented enabling solutions for the following assigned tasks, to a significant degree: providing natural ventilation, conventional air distribution equipment (air-ducts for uniform air distribution); sputtering heads, inflow exhaust installations of AHU type), ejection-axifugal air terminal units, conventional air distribution units, etc. [17]. Also, in order to improve the energy efficiency of heat supply systems when using the heat of low-potential energy sources and environmental protection, it is possible to use heat pumps with an eco-friendly refrigerants such as CO_2 [18,19].

However, in such systems, non-homogeneous air temperature distribution with the height takes place. Warm air, having lower volumetric density, tends to concentrate directly under the ceiling, thus forming an airlock. On the other hand, cold and excessively wet air is accumulated in the lower layers where young stock is located, creating unfavorable conditions for animals. Air velocity in the area where calves are located should not exceed 0.3 m/s, during the cold and period of the year [20].

Based on the results of experimental studies, Bodrov [21] developed a method for defining the required value of the heat transfer resistance, for unheated livestock premises depending on the individual biological characteristics of the particular animal type. Methods for calculating the natural ventilation, on an annual basis, were developed as well.

We assume that the maintenance and energy-related parameters of heat- and air-distribution installations and systems can be improved substantially by application of ceiling fans. Such fans, installed in addition to the conventional heating equipment, ensure effective air circulation (within the permissible air velocity range, in areas where animals are located). As a result, the air temperature distribution becomes more uniform with the height, thus reducing the heat energy loss through the enclosing structures and increasing the heat-exchange coefficient of heating devices. The ceiling fan speed control function makes it possible to maintain its specified performance and, consequently, the air flow velocity. Fans can be mounted in any area of premises making it possible to produce their effect on the animals locally depending on their age and physical status.

A ceiling fan installed in the upper area of the premises draws in air from above and directs the fan-twisted air jets having an internal vortex core towards the floor. These air jets reach the ceiling in form of an overlapping flux that gets spread over the premises and moves towards the walls, penetrating the areas where animals are kept.

In these conditions, the required air mobility is provided, and warm air is supplied to the space where animals are located, including in zones under the cages, thus insuring temperature equalization along the vertical axis. This technical solution makes it possible to reduce the required heating system capacity and to save considerable amounts of energy.

Shah et al. [22] reported the results of a technical-economic effectiveness evaluation, for various ceiling fan options, having optimized fan blade design and for wider application of DC electric motors. Babich et al. [23] developed and verified a three-dimensional time-dependent implicit model of the standard ceiling fan by simulation results comparison with the experimental data. Present et al. [24] analyzed the results of the ceiling fans practical application, in commercial premises. Liu et al. [25] and Raftery et al. [26] carried out a series of experimental studies of air velocity fields formed with the use of the ceiling fans, in enclosed areas. Based on the results of experimental data, parameters were specified that, to a major extent, defined the air flow rate in premises, for the case when the ceiling fans operate. Li W. et al. [27] have carried out numerical simulations and experimental studies of air conditioning system and ceiling fans combined operation. Application of a ceiling fan made it possible to reduce the density of airborne participates by more than 20% in the area of the human breathing zone, owing to a better dispersion of airborne participates over the premises. Omrani et al. [28] have performed an incisive analysis of the ceiling fan effects on the microclimate, i.e., air flow rate, thermal acceptability and air quality and energy consumption. It was pointed out that the thermal acceptability and air the flow rate belong to the most extensively studied parameters. Major attention was paid to the air quality and electric energy consumption. Besides, that analysis has brought to light the gap in our knowledge concerning the specifics of natural ventilation and ceiling fans combined operation including their influence on the air quality, in premises, which is vitally important during pandemic periods.

2. Materials and Methods

In this research area, previous works were devoted to simulations of the heat storage cells thermal state of the ETS unit [29] and calculation of the ETS unit thermal characteristics [30,31]. The results of experimental studies of the thermal state of heat-storage cells in both heat-charging and heat-emission modes of ETS unit have been reported [32,33]. Also, an energy-saving infrared heater (IR) for calves was developed for preventive maintenance premises of cattle-breeding farm with an adjustable heat flux depending on the position of the animal and an evaluation of the parameters of its operation was carried out [34].

The heat supply system for the preventive maintenance of an 8 m long, 4 m wide and 4 m high premises, for housing calves, was chosen as the subject of research. The recommended air temperature value in the premises was 17 °C, while the maximum permissible relative humidity (ARH) of the air was 75% [20].

In our research, the air heating combined type ETS unit [31] designed for operation under the electric energy time of use price plan was chosen. A ceiling fan of the type MR-1 [35] was applied for temperature equalization and uniform air flow distribution in the premises.

As part of this work, field studies of the thermal and humidity parameters of the air were carried out in two similar preventive maintenance premises of cattle-breeding farm. Two variants of the operation of the heat supply system were investigated. In the first room, air heating is carried out from the combined ETS unit without the use of ceiling fans, and in the second, with their use. Experimental studies of the proposed heat supply system were carried out in winter from mid-December to the end of March. The temperature and ARH were measured along the height of the preventive maintenance premises at several characteristic control points.

During the experimental studies, the consumption of electrical energy for heat supply of the preventive maintenance premises was recorded for two variants of the system operation.

The results of measurements of the thermal and humidity parameters of the air in the preventive maintenance were processed using the probabilistic-statistical method. A calculation method was also used to justify the parameters and selection of the ceiling fan.

3. Calculating and Selecting the Ceiling Fan Parameters

At the present time, the most commonly applied air heating systems in cattle-breeding farms are those with heated air distribution into the upper area of premises. In this case, the air temperature becomes uniform, in the upper zone, within a short period of time (there is no sensible temperature gradient) owing to the intensive turbulent air mixing. At the same, air flows do not practically reach the lower ('sluggish') zone. It happens because the temperature of air exhausted from the outlet into an unlimited area differs from that of the ambient air in premises with anisothermal air jets. Therefore, air jet parameters, as well as their trajectory, depend on not only inertial forces but also on those of gravity [33].

The following factors have an essential effect on the circulation of air flows produced by the ceiling fans: the height of the premises, the length of suspension, the number fans and their installation points, etc.

The reasonable length of fan suspension h_f (see Figure 1) insuring the maximum air flow rate in the operating area, can be defined from the following equation:

$$h_f = \frac{H_p - L}{4}, \quad (1)$$

where H_p is distance between the floor and the ceiling (m), L is that between the floor and the animal shoulder (m).

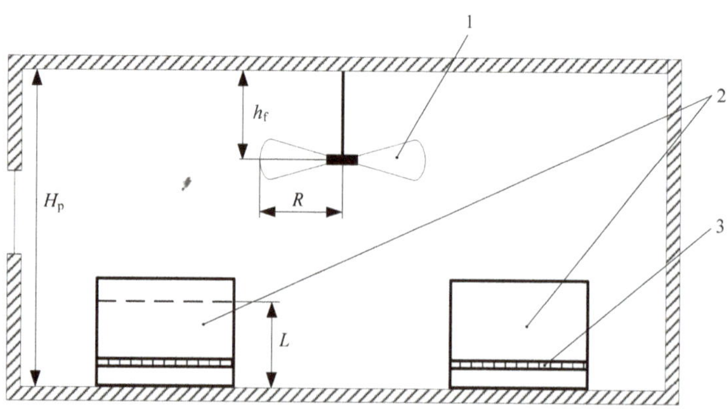

Figure 1. Diagram for optimal ceiling fan installation: 1—ceiling fan; 2—cages for calves; 3—slatted floor of the cage.

In this case, fan blades diameter ($D = 2R$) shall not exceed 1/3 of the premises width [3]. As applied to up-to-date premises designed for calves housing, fans must be mounted along the longer premises axis.

The number of fans N_f capable to ensure the required values of air velocity and flow rate, for reasonable fan suspension length, can be found from the following expression deduced using experimental methods [3]:

$$N_f = c_f P^{2.91} H_p^{-2.14} \omega_{av}^{4.43}, \tag{2}$$

where P is room perimeter (m), ω_{av} is average air flow rate, in operation area, i.e., standing or rest zone, (m/s); c_f is coefficient depending on the ceiling fan design and its operation conditions.

Since the air velocity field has practically uniform profile, we can assume $\omega_{ave} = 0.5\ \omega_0$ where ω_0 is initial value of air velocity (m/s).

It has been found out that coefficient $c_f = 0.07$, for premises having 4×8 m, in plain view, and 4 m in height, for ceiling fans type MR-1. By substituting the variables in formula (2) with their known values we obtained $N_f = 1.3$. One 60 W fan unit type MR-1 with controlled air flow performance rate was chosen, having three 620 mm long blades [35].

4. Calculating Basic Thermal Characteristics of the Electric Thermal Storage Unit

Combined type ETS unit comprises two independently operating heaters, i.e., heat storage core and convector heater (see Figure 2). The heat storage core is designed to be charged with thermal energy during the lower electric power price rate period in order to supply heat during periods of higher electric power price rates. The electric convector-type heater serves as a direct heating energy source supplying heat mainly during the period when the heat storage core is being charged. Technical parameters of the installation are presented in Table 1.

Figure 2. Electric thermal storage unit.

Table 1. Technical characteristics of ETS unit.

Input voltage (V)	380/220
Heat storage capacity (kW)	4.8
Electric convector heater capacity (kW)	2.4
Minimum heat charging period (h)	4
Heat emission period (h)	48
Weight (kg)	200

The required volume of ETS unit heat storage core can be calculated using the following expression:

$$V_{HSC} = \frac{3600 Q_{ch\_ave} \tau_{he}}{\rho_{HSM} c_{HSM} (T_{HSC,max} - T_{HSC,min})}, \quad (3)$$

where Q_{ch_ave} is average heat capacity of ETS unit (W); τ_{he} is duration of ETS unit heat emission period (h); ρ_{HSM} is volumetric density of the heat storage core material (kg/m^3); c_{HSM} is heat capacity ratio of the core material (kJ·kg^{-1}·K^{-1}); $T_{HSC,max}$ and $T_{HSC,min}$ are temperature values of ETS unit heat storage core, in the initial (600 °C to 650 °C) and the final (50 °C to 100 °C) moments of the heat charging period, respectively [36,37].

The maximum temperature of the casing outer surface, by the end moment of the ETS unit charging mode is $T_{sh,max}$ = 60 °C to 70 °C, while at the end moment of the heat emission period it is in the range of $T_{sh,min}$ = 25 °C to 30 °C which is in the reasonable compliance with the requirements for ETS unit casing outer surface temperature specified in [36].

In view of the fact that there is, practically, no heat loss, in the beginning of the core charging period, the average value of thermal energy Q_{loss} emitted from the surface of the ETS unit casing into surrounding space, in the course of charging, should be calculated for the casing surface temperature value $T_{sh} = (T_{sh,min} + T_{sh,max})/2$, using the following formula:

$$Q_{loss} = k_{rad} \alpha_{sh\_ave} F_{ETS} (T_{sh} - T_{ap}), \quad (4)$$

where α_{sh_ave} is average value of the coefficient of heat-exchange between the electric thermal storage outer surface and ambient air (premises), (W·m^{-2}·°K^{-1}); T_{ap} is ambient air temperature in premises (°C); F_{ETS} is surface area of the ETS casing (m^2); k_{rad} is coefficient taking into account the heat loss by radiation from the surface of the ETS casing (value is 1.25).

When calculating thickness δ_{ins} of the thermal insulation layer, it should be considered as a single-layer plain wall. The required thickness of ETS unit thermal insulation layer insuring compliance of the heat loss from the thermal storage unit with the specified Q_{loss} value can be defined from the following equation:

$$\delta_{ins} = \frac{\lambda_{ins} F_{ave} (T_{ins,int} - T_{sh,max})}{Q_{loss}}, \quad (5)$$

where λ_{ins} is thermal-conductivity coefficient of the thermal insulation material (W·m^{-1}·°K^{-1}); $T_{ins,int}$ is temperature of the thermal insulation internal layer, by the end of the charging period (°C).

$$F_{ave} = \frac{F_{ins} + F_{ETH}}{2}, \quad (6)$$

where F_{ins} is surface area of the thermal insulation (m^2).

Electric heater capacity W_{unit} of ETS unit is defined from the following equation:

$$W_{unit} = k_r Q_{ch\_ave} \frac{\tau_{he}}{\tau_{ch}} + Q_{loss}, \quad (7)$$

where τ_{ch} is duration of ETS unit charging period (h); k_r is power reserve coefficient that takes into account the aging of electric heating elements and changes in the supply voltage (value is 1.2).

Quantity of the accumulated heat can be calculated from the following expression:

$$\begin{gathered} Q_{st} = Q_{ins} + Q_{HSM} = \\ = c_{ins} \rho_{ins} V_{ins} (T_{ins,max} - T_{ins,min}) + c_{HSM} \rho_{HSM} V_{HSM} (T_{HSC,max} - T_{HSC,min}) \end{gathered} \quad (8)$$

where c_{ins} is heat capacity ratio of the thermal insulation (kJ·kg^{-1}·°K^{-1}); ρ_{ins} is its volumetric density (kg/m^3); $T_{ins,max}$ and $T_{ins,min}$ are temperature values of the thermal insulation in the end moments of ETS unit charging and heat emission periods, respectively.

Quantity of heat Q_{st} accumulated in ETS unit makes it possible to define time period τ_{warm} required for its heating to temperature value $T_{HSC,max}$:

$$\tau_{warm} = \frac{Q_{st}}{3600(0.8W_{unit?} - Q_{loss,max})}, \quad (9)$$

where $Q_{loss,max}$ is heat loss, for the maximum temperature of the casing surface $T_{sh,max}$ of ETS unit, at the end moment of the heat charging period (W).

In paper by [31] the method of the thermal and aerodynamic calculation of the basic thermal characteristics has been described, for dynamic type ETS unit.

When calculating the average value of heat-exchange coefficient α_{sh_ave}, for the outer ETS unit casing surface, similarity criterion Gr should be defined from Formula (10), following which similarity criterion Nu is calculated in accordance with expression (11). After that, α_{sh_ave} value can be determined [30]:

$$\mathrm{Gr} = \frac{\beta h^3 g(T_{sh} - T_{ap})}{v^2}, \quad (10)$$

where $\beta = \frac{1}{273+T_{ap}}$ (K^{-1}); h—is ETS unit height (m); g is gravity factor (m/s^2); ν is air kinematic viscosity (m^2/s).

$$\overline{\mathrm{Nu}} = 0.695 Gr_{dliq}^{0.25}. \quad (11)$$

In order to define the average value of the heat-exchange coefficient α_{ch_ave}, in air channels of ETS unit heat storage cells, similarity criterion Nu has to be calculated using Equation (12), with the account of the temperature difference $\theta_{ch} = T_{wch}/T_{air\_ch}$ [38–40]:

$$\overline{\mathrm{Nu}} = 0.023\mathrm{Pr}^{0.4}\mathrm{Re}^{0.8}\left(\frac{T_{wch}}{T_{air\_wch}}\right)^{-0.55}. \quad (12)$$

Thermal-physical properties of solid heat storage materials (HSM) have been analyzed. Formulas for calculating temperature dependent values of the thermal-conductivity coefficient and heat capacity ratio are presented in Table 2 [41].

Table 2. Thermal-physical properties of solid-state HSM.

Material	c (kJ·kg^{-1}·°K^{-1})	λ (W·m^{-1}·°K^{-1})	ρ_{ave} (kg/m^3)
Magnesium oxide	$1.05 + 0.29 \cdot 10^{-3}\ T_{HSM}$	$4.7–1.7 \cdot 10^{-3}\ T_{HSM}$	3000
Chamotte	$0.88 + 0.23\ 10^{-3}\ T_{HSM}$	$0{,}84 + 0.58 \cdot 10^{-3}\ T_{HSM}$	2200
Corundum	$0.79 + 0.42 \cdot 10^{-3}\ T_{HSM}$	$2{,}1 + 1.9 \cdot 10^{-3}\ T_{HSM}$	3300
Dinas	$0.837 + 0.25 \cdot 10^{-3}\ T_{HSM}$	$0.93 + 0.69 \cdot 10^{-3}\ T_{HSM}$	2200

Figures 3 and 4 shows the change of the thermal-conductivity coefficient and heat capacity ratio, in temperature range from 50 °C to 650 °C. These are the minimum and the maximum temperatures of the heat storage elements, in ETS unit heat emission and charging modes, respectively.

Analysis of the obtained results provides the reason to conclude that chamotte, corundum and dinas feature linear temperature dependence of the thermal-conductivity coefficient (see Figure 3). Magnesium oxide has the highest values of the thermal-conductivity coefficient and heat capacity ratio in the course of heating and cooling processes, compared to other materials in the study (see Figures 3 and 4). It should also be noted that the thermal-conductivity coefficient of magnesium oxide is inversely proportional to the temperature [42].

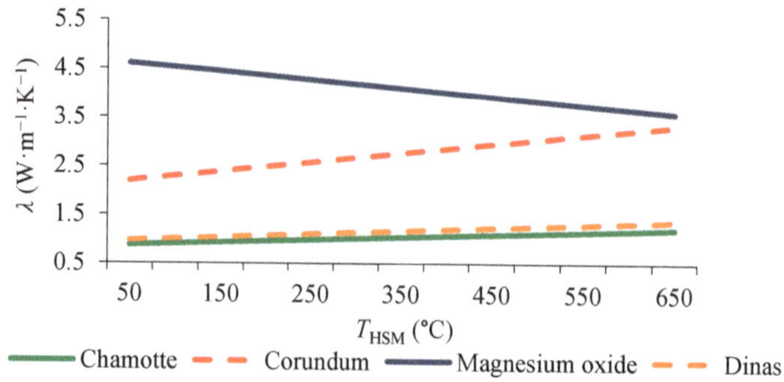

Figure 3. Dependence of thermal-conductivity coefficient on temperature T_{HSM}.

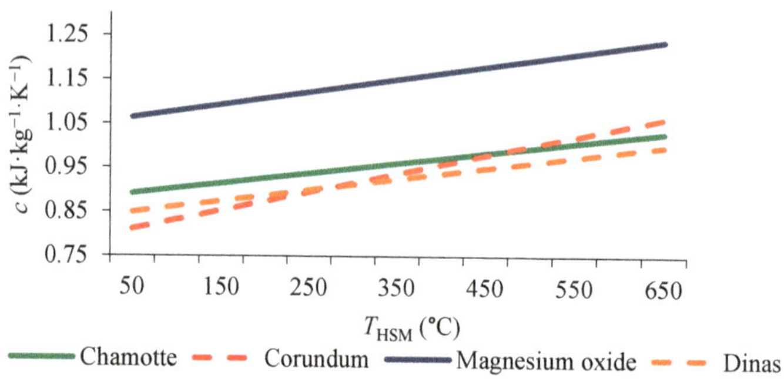

Figure 4. Dependence of heat capacity ratio on temperature T_{HSM}.

5. Discussion

A comparative analysis of temperature and ARH distribution with height of the preventive maintenance premises was made for the ceiling fan switching on and off operation modes in this study. An initial evaluation of the energy efficiency for the system with ceiling fans was also carried out.

5.1. Experimental Studies of Thermal and Humidity Parameters of Air

Two system operation modes applied to preventive maintenance premises were studied. One of these operation modes involves inflow ventilation with heating internal air from electric thermal storage units thus implementing the combined air heating method (without the use of ceiling fans). In this case, there exists a temperature drop in the lower areas and under the cage, i.e., directly under the slatted floor, and 'no flow areas' occur. Besides, minimum required air temperature of 17 °C is not achieved in areas where animals are located (area under the cage [20]). This means that either a heating installation of higher capacity should be used during or the period of its operation should be extended.

In the other operation mode, air flows driven with the use of the ceiling fans circulate with permissible velocity limited in accordance with the recommendations for technological production design (not exceeding 0.3 m/s) [20]. Avoiding forming a heat cushion directly under the ceiling helps to inject a certain amount of heat into the work area, thus maintaining a required temperature environment. It also contributes to reduce the heat loss through the enclosing structures, first of all through the ceiling.

Recommendations for technological production design [20] stipulate the optimal air temperature for housing calves that must be maintained within the cage and under the slatted floor.

Temperature conditions and the aggregate thermal energy consumption for heating depend essentially on the temperature distribution over the space of the premises.

Figure 5 shows vertical air temperature profile, in preventive maintenance premises for calves.

It should be noted that operating ceiling fans ensure ARH reduction by up to 5% in areas where animals are located because of the warmer air supplied into these areas (see Figure 6).

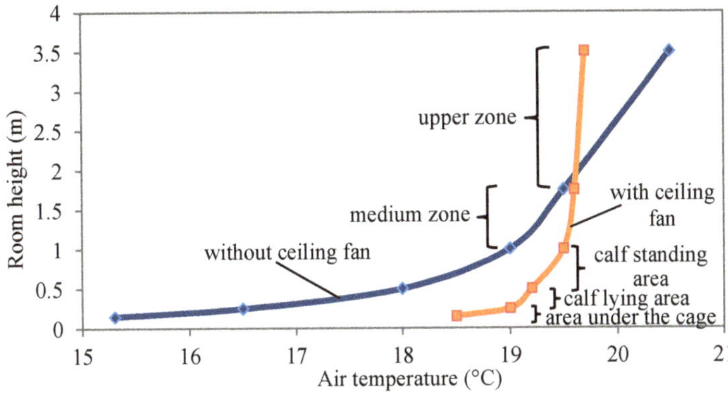

Figure 5. Air temperature distribution, in preventive maintenance premises for calves.

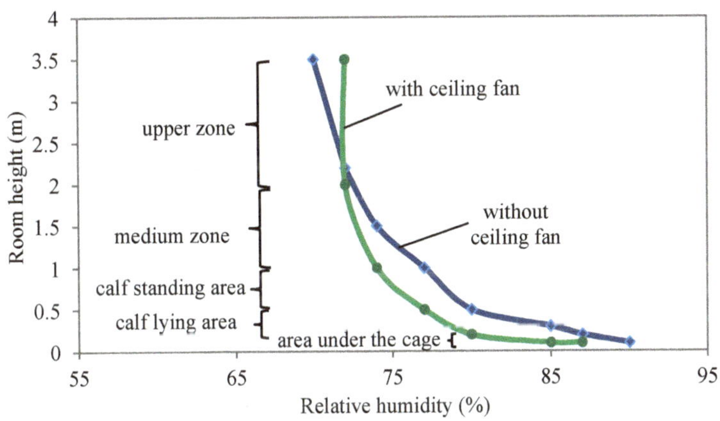

Figure 6. ARH distribution, in preventive maintenance premises for calves.

5.2. Evaluation of the Energy Efficiency of the Combined Heat Supply System

Normally, internal air temperature is kept within a specified range with the use of the on-off action control.

In case that heating-ventilation system is not equipped with ceiling fans the time period within which air gets heated to a specified temperature equals to [43]:

$$\tau_{p1} = \frac{c_{air} G \Delta T}{Q_p}, \quad (13)$$

where c_{air} is air heat capacity ratio (kJ·kg^{-1}·°K^{-1}), G is mass of air (kg), ΔT is specified air temperature control interval in work zone (°C), Q_p is thermal energy production by the heating installation (W).

In case of the ceiling fans application, time period τ_{p2} within which air gets heated to a specified temperature will be equal to:

$$\tau_{p2} = \frac{c_{air} G \Delta T}{k_{te} Q_p}, \quad (14)$$

where k_{te} is coefficient that takes account of additional thermal energy incoming as a result of the ceiling fans operation.

It was experimentally found out that the values of k_f fall into the interval between 1.2 and 1.25.

Then it is follows from Expressions (13) and (14) that $\tau_{p2} = 0.8\tau_{p1}$.

Let us consider the effect of the ceiling fans operation on the air-cooling rate in premises. Evidently, air cooling time period is defined by the heat dissipation rate:

$$Q_{out} = \alpha_{es}\left(T_{ap} - T_{es}\right)F_{es} + Q_{air}, \tag{15}$$

where α_{es} is heat-exchange coefficient of the enclosing structure internal surface (W·m^{-2}·°K^{-1}), T_{ap} is ambient air temperature in premises (°C), T_{es} and F_{es} are, respectively, temperature (°C) and area (m^2) of the enclosing structure internal surface, Q_{air} is thermal energy loss with exhausted air (W).

Heat-exchange coefficient αfen comprises both convective α_{con} and radiant α_{rad} energy components:

$$\alpha_{es} = \alpha_{con} + \alpha_{rad}, \tag{16}$$

where the surfaces of the premises are blown by air flows there the forced and combined convection operation mode takes place in which case the following expression is valid, with regard to [44]:

$$\alpha_{con} = 3.38\left(\frac{\omega}{l}\right)^{-0.5}, \tag{17}$$

where ω is air flow velocity (m/s), l is distance between the floor and arbitrary cross-section (m).

Studies carried out in [3] have shown that, in case of ceiling fans application, velocity of air flows blowing onto the surfaces increases by the average rate from 0.15 m/s to 0.18 m/s, i.e., increases by a factor of 1.2. If we assume l = 1 m (area where animals are located), then the coefficient of heat convective exchange is 1.1 times higher, for ceiling fans active state, compared to their idle status. Coefficient α_{rad} = const since it does not depend on the air flow velocity and it can be assigned a value in the range of 4 W·m^{-2}·K^{-1} to 4.5 W·m^{-2}·K^{-1}, for animal-housing premises [44].

Having performed the relevant transformations, we can deduce the relationship between the time of air cooling, in premises, in case of ceiling fans application τ_{o2} and without ceiling fans τ_{o1} (in the assumption that Q_{air} = 0):

$$\tau_{o2} = \tau_{o1}\frac{\alpha_{con} + \alpha_{rad}}{1.1\alpha_{con} + \alpha_{rad}}. \tag{18}$$

With the account of optic and thermal-technical parameters of standard buildings [44] designed for young stock housing, Expression (18) can be reduced to the following form $\tau_{o2} = 0.95\tau_{o1}$.

Therefore, in the case when no ceiling fan is applied, the period of self-sustained oscillating process of maintaining a required air temperature in areas where animals are kept is equal to:

$$T_1 = \tau_{p1} + \tau_{o1}. \tag{19}$$

With operating ceiling fans, this parameter can be calculated as follows:

$$T_2 = 0.8\tau_{p1} + 0.95\tau_{o1}. \tag{20}$$

The relative switching frequency of heaters without ceiling fans defined from Expressions (19) and (20) equals to:

$$n_1 = \frac{\tau_{p1}}{\tau_{p1} + \tau_{o1}}. \tag{21}$$

The same parameter defined for the case when ceiling fans are applied has the following form:

$$n_2 = \frac{0.8\tau_{p1}}{0.8\tau_{p1} + 0.95\tau_{o1}}. \tag{22}$$

It is clear from Formulas (21) and (22) that $n_1 > n_2$. Consequently, the average heat energy income, for $Q_{p1} = Q_{pn1}$, exceeds that, for $Q_{p2} = Q_{pn2}$. It means that the thermal energy consumption is greater, in the first case.

Electric power consumption by ceiling fans is insignificant (its maximum input power is just 0.06 kW). Besides, thermal energy dissipated by the electric motor remains in premises thus contributing to the positive component of the thermal balance.

Results of a number of tests and initial technical-economic evaluation of the newly designed heat supply system of preventive maintenance premises for calves have shown that electric power consumption is 1760 kWh for a month-long heating period without ceiling fans, while in case of ceiling fans application it is 1520 kWh. Therefore, electricity consumption can be reduced by 14%.

6. Conclusions

The combined energy–saving heat supply system, which includes a combined ETS unit and a ceiling fan, allows the provision of normative air parameters in the livestock premises: air temperature and ARH.

Experimental studies of the heating–ventilation system carried out for preventive maintenance premises of cattle-breeding farmin the winter period have shown that the application of ceiling fans makes it possible to reduce the heating and cooling times of air in the premises. At the same time, there was a decrease in the consumption of electric energy for the heat supply system by up to 14%. The electric capacity of the ETS unit or the overall time of its operation can be also reduced. The energy-saving effect is achieved by using the heat of the air that accumulates in upper zone of the premises, when it is supplied by a ceiling fan, to the locations of the animals.

Based on the results of tests, a comparative analysis of temperature and ARH distribution with height of the preventive maintenance premises was made for the ceiling fan switching on and off operation modes. An initial evaluation of the energy efficiency for the system with ceiling fans was performed.

Application of combined type ETS unit for air heating makes it possible to reduce the current end user's annual expenditures on electricity by up to 30%, provided that the time of use price plan for electricity is adhered to.

Moreover, it ensures more uniform daily electric power load schedules in power networks, reduction of the equipment installed capacity, as well as that of nighttime electric energy loss. Besides, a mass-scale implementation of ETS units in farming will not require putting into operation considerable additional power-generating capacities.

Author Contributions: Conceptualization, D.T. and A.K. (Aleksei Khimenko); methodology, S.T. and A.K. (Aleksei Khimenko); validation, D.T., S.T. and A.K. (Aleksey Kuzmichev); formal analysis, A.K. (Aleksei Khimenko), A.K. (Aleksey Kuzmichev) and V.B.; investigation, S.T. and D.T.; resources, A.K. (Aleksei Khimenko); data curation, D.T. and A.K. (Aleksei Khimenko); writing—original draft preparation, D.T., S.T. and A.K. (Aleksei Khimenko); writing—review and editing, A.K. (Aleksei Khimenko) and V.B.; visualization, A.K. (Aleksei Khimenko) and D.T.; supervision, D.T. and O.S. All authors have read and agreed to the published version of the manuscript.

Funding: This research received no external funding.

Institutional Review Board Statement: Not applicable.

Informed Consent Statement: Not applicable.

Data Availability Statement: Not applicable.

Conflicts of Interest: The authors declare no conflict of interest.

References

1. Roland, L.; Drillich, M.; Klein-Jobstl, D.; Iwersen, M. Invited review: Influence of climatic conditions on the development, performance, and health of calves. *J. Dairy Sci.* **2016**, *99*, 2438–2452. [CrossRef]
2. López-Bellido, L.; Wery, J.; López-Bellido, R.J. Energy crops: Prospects in the context of sustainable agriculture. *Eur. J. Agron.* **2014**, *60*, 1–12. [CrossRef]
3. Rastimeshin, S.A.; Trunov, S.S. *Energy-Saving Systems and Technical Means for Heating and Ventilation in Cattle-Breeding Premises*; FGBNU VIESH: Moscow, Russia, 2016.
4. Li, J.; Qi, W.; Yang, J.; He, Y.; Luo, J.; Guo, S. The capacity optimization of wind-photovoltaic-thermal energy storage hybrid power system. *E3S Web Conf.* **2019**, *118*, 02054. [CrossRef]
5. Sokolnikova, P.; Lombardi, P.; Arendarski, B.; Suslov, K.; Pantaleo, A.M.; Kranhold, M.; Komarnicki, P. Net-zero multi-energy systems for Siberian rural communities: A methodology to size thermal and electric storage units. *Renew. Energy* **2020**, *155*, 979–989. [CrossRef]
6. Bruno, S.; Dicorato, M.; La Scala, M.; Sbrizzai, R.; Lombardi, P.A.; Arendarski, B. Optimal Sizing and Operation of Electric and Thermal Storage in a Net Zero Multi Energy System. *Energies* **2019**, *12*, 3389. [CrossRef]
7. Dorjiev, S.S.; Bazarova, E.G.; Pimenov, S.V.; Rozenblum, M.I. Development of wind power installations with the accelerator of an air stream for areas with a low speed of wind. *J. Phys. Conf. Ser.* **2018**, *1111*, 012053. [CrossRef]
8. Dorzhiev, S.S.; Bazarova, S.V.; Pimenov, S.V.; Dorzhiev, S.S. Application of renewable energy sources for water extraction from atmospheric air. *Energy Rep.* **2021**, *7S5*, 343–357. [CrossRef]
9. Sun, Y.; Feng, Y.; Ma, Y.; Wang, Y.; Shi, Y. Thermal Storage and Release Features of Electric Thermal Storage Heating Systems with Solid Storage Material. *Int. J. Heat Technol.* **2019**, *37*, 1089–1098. [CrossRef]
10. Zhao, H.; Yan, N.; Xing, Z.; Chen, L.; Jiang, L. Thermal Calculation and Experimental Investigation of Electric Heating and Solid Thermal Storage System. *Energies* **2020**, *13*, 5241. [CrossRef]
11. Xing, Z.; Fu, Q.; Chen, L.; Xu, T. Research on multi-physical field coupling of solid electrothermal storage unit. *Energy Rep.* **2020**, *6*, 775–791. [CrossRef]
12. Xu, G.; Hu, X.; Liao, Z.; Xu, C.; Yang, C.; Deng, Z. Experimental and Numerical Study of an Electrical Thermal Storage Device for Space Heating. *Energies* **2018**, *11*, 2180. [CrossRef]
13. Klymchuk, O.; Denysova, A.; Shramenko, A.; Borysenko, K.; Ivanova, L. Theoretical and experimental investigation of the efficiency of the use of heat-accumulating material for heat supply systems. *EUREKA Phys. Eng.* **2019**, *3*, 32–40. [CrossRef]
14. Klymchuk, A.A.; Luzhanska, G.V.; Shramenko, A.N. Construction modernization of heat accumulators based on solid materials for electricity night tariffs operation. *Refrig. Eng. Technol.* **2017**, *53*, 44–48. [CrossRef]
15. Beknazarian, D.V.; Kanevets, G.E.; Strogonov, K.V. Methodological bases of optimization of thermal insulation structures of glass furnaces. *J. Phys. Conf. Ser.* **2020**, *1683*, 052027. [CrossRef]
16. Janssen, N.T.; Peterson, R.A.; Wies, R.W. Generalized heat flow model of a forced air electric thermal storage heater core. *J. Therm. Sci. Eng. Appl.* **2017**, *9*, 041008. [CrossRef]
17. Pavlukhin, L.V.; Teterevnikov, V.N. *Industrial Microclimate, Ventilation and Air Conditioning*; Stroyizdat: Moscow, Russia, 1993.
18. Qin, X.; Wang, D.; Jin, Z.; Wang, J.; Zhang, G.; Li, H. A comprehensive investigation on the effect of internal heat exchanger based on a novel evaluation method in the transcritical CO_2 heat pump system. *Renew. Energy* **2021**, *178*, 574–586. [CrossRef]
19. Qin, X.; Zhang, Y.; Wang, D.; Chen, J. System development and simulation investigation on a novel compression/ejection transcritical CO_2 heat pump system for simultaneous cooling and heating. *Energy Convers. Manag.* **2022**, *259*, 115579. [CrossRef]
20. *RD-APK 1.10.01.01-18*; Methodological Recommendations on Technological Design of Cattle-Breeding Farms and Complexes. Minselkhoz RF: Moscow, Russia, 2018.
21. Bodrov, M.V. Justification, Selection and Calculation of Year-round Systems of Natural Ventilation in Livestock Premises. *Sci. Her. Voronezh State Univ. Archit. Civ. Eng. Constr. Archit.* **2011**, *9*, 1.
22. Shah, N.; Sathaye, N.; Phadke, A.; Letschert, V. Efficiency improvement opportunities for ceiling fans. *Energy Effic.* **2015**, *8*, 37–50. [CrossRef]
23. Babich, F.; Cook, M.; Loveday, D.; Rawal, R.; Shukla, Y. Transient three-dimensional CFD modelling of ceiling fans. *Build. Environ.* **2017**, *123*, 37–49. [CrossRef]
24. Present, E.; Raftery, P.; Brager, G.; Graham, L.T. Ceiling fans in commercial buildings: In situ airspeeds & practitioner experience. *Build. Environ.* **2019**, *147*, 241–257. [CrossRef]
25. Liu, S.; Lipczynska, A.; Schiavon, S.; Arens, E. Detailed experimental investigation of air speed field induced by ceiling fans. *Build. Environ.* **2018**, *142*, 342–360. [CrossRef]
26. Raftery, P.; Fizer, J.; Chen, W.; He, Y.; Zhang, H.; Arens, E.; Schiavon, S.; Paliaga, G. Ceiling fans: Predicting indoor air speeds based on full scale laboratory measurements. *Build. Environ.* **2019**, *155*, 210–223. [CrossRef]
27. Li, W.; Chong, A.; Hasama, T.; Xu, L.; Lasternas, B.; Tham, K.W.; Lam, K.P. Effects of ceiling fans on airborne transmission in an air-conditioned space. *Build. Environ.* **2021**, *198*, 107887. [CrossRef]
28. Omrani, S.; Matour, S.; Bamdad, K.; Izadyar, N. Ceiling fans as ventilation assisting devices in buildings: A critical review. *Build. Environ.* **2021**, *201*, 108010. [CrossRef]
29. Khimenko, A.V.; Tikhomirov, D.A.; Vasilyev, A.N.; Samarin, G.N.; Shepovalova, O.V. Numerical simulation of the thermal state and selecting the shape of air channels in heat-storage cells of electric-thermal storage. *Energy Rep.* **2022**, *8*, 1450–1463. [CrossRef]

30. Khimenko, A.V.; Tikhomirov, D.A.; Kuzmichev, A.V.; Trunov, S.S.; Shepovalova, O.V. Thermal characteristics and operation efficiency of solid-state electro thermal storage. *Energy Rep.* **2021**, *7*, 219–231. [CrossRef]
31. Tikhomirov, D.; Dudin, S.; Trunov, S.; Rastimeshin, S.; Tikhomirov, A.; Kuzmichov, A. Combined electric accumulation unit for air heating. *Rev. De La Univ. Del Zulia* **2019**, *10*, 168–183.
32. Khimenko, A.V.; Tikhomirov, D.A.; Kuzmichev, A.V. Experimental studies of the thermal characteristics of electro-thermal storage. *IOP Conf. Ser. Earth Environ. Sci.* **2021**, *839*, 052058. [CrossRef]
33. Khimenko, A.V.; Tikhomirov, D.A.; Kuzmichev, A.V.; Trunov, S.S. Experimental studies of the thermal state of heat storage units of electro thermal storage. *IOP Conf. Ser. Earth Environ. Sci.* **2021**, *866*, 012015. [CrossRef]
34. Tikhomirov, D.A.; Khimenko, A.V.; Trunov, S.S.; Kuzmichev, A.V. Energy-saving automated IR heater for calves. *IOP Conf. Ser. Earth Environ. Sci.* **2021**, *848*, 012088. [CrossRef]
35. Product Catalogue by Comp. Group 'Euromash'. Ceiling Fans MR-1. Available online: https://www.evromash.ru/catalog/venti/pov/mr1/ (accessed on 16 August 2022).
36. Shepelev, I.A. *Aerodynamics of Air Flows in Premises*; Stroyizdat: Moscow, Russia, 1978.
37. Simonov, A.A. Household electric heating with heat storage is the most important factor in the efficient use of electricity. *Energ. I Elektrif.* **1992**, *1*, 26–30.
38. Vysotskii, L.I. *Electric Heaters of Storage Type*; Informelectro: Moscow, Russia, 1971.
39. Isachenko, V.P.; Osipova, V.A.; Sukomel, A.S. *Heat Transfer*, 4th ed.; Energoizdat: Moscow, Russia, 1981.
40. Kutateladze, S.S.; Borishanskii, V.M. *Handbook of Heat Transfer*; Gosenergoizdat: Moscow, Russia, 1958.
41. Kazantsev, E.I. *Industrial Furnaces: Reference Guide for Calculations and Designing*, 2nd ed.; Metallurgy: Moscow, Russia, 1975.
42. Khimenko, A.V.; Prantsuz, O.S.; Glebova, I.A.; Ponomarev, A.K. Evaluation of the efficiency of heat storage by a solid-state electric thermal storage. *J. Phys. Conf. Ser.* **2020**, *1560*, 012049. [CrossRef]
43. Klimenko, A.V.; Zorin, V.M. *Thermal Energy Industry and Heat Engineering*, 3rd ed.; MPEI Publishing: Moscow, Russia, 2004; Volume 4.
44. Bogoslovskii, V.N. *Thermal-Physics in Buildings (Thermal-Physical Principles of Heating, Ventilation and Air Conditioning): Guide-Book for Higher Education Institutes*, 2nd ed.; Vysshaia shkola: Moscow, Russia, 1982.

agriculture

MDPI

Article

Measurement and Calibration of DEM Parameters of Soybean Seed Particles

Dongxu Yan [1], Jianqun Yu [2], Yang Wang [2], Kai Sun [2], Long Zhou [3], Ye Tian [4] and Na Zhang [2,*]

1 Hua Lookeng Honors College, Changzhou University, Changzhou 213164, China
2 School of Biological and Agricultural Engineering, Jilin University, Changchun 130022, China
3 School of Agricultural Engineering and Food Science, Shandong University of Technology, Zibo 255022, China
4 Center of Industry and Technology, Hebei University of Technology Petroleum, Chengde 067000, China
* Correspondence: zna18@mails.jlu.edu.cn; Tel.: +86-138-449-41571

Abstract: In discrete element method (DEM) simulations, accurate simulation parameters are very important. For ellipsoidal soybean seed particles, the rolling friction coefficient between seed particles (RFCP-P) and the rolling friction coefficients between seed particle and boundary (RFCP-B) are difficult to measure experimentally and therefore need to be calibrated. In this paper, soybean seed particles of three varieties with different sphericities were taken as the research objects. Through the simulation analysis of repose angle and self-flow screening, it was shown that the above two parameters needed to be accurately calibrated. In addition, the sensitivity of the RFCP-P and RFCP-B to the angle of repose was analyzed by simulating the repose angle test. The results showed that the RFCP-P had a significant effect on the test results of the repose angle, and the RFCP-B had little effect on the test results of the repose angle. Therefore, the RFCP-P was calibrated using a single-factor test of repose angle, and the RFCP-B was calibrated using the repose angle test with soybean particles mixed with organic glass spheres. The accuracy of the calibration parameters was verified by rotating cylinder test and self-flow screening test.

Keywords: discrete element method; soybean seed; ellipsoidal shape; parameter calibration; rolling friction coefficient

Citation: Yan, D.; Yu, J.; Wang, Y.; Sun, K.; Zhou, L.; Tian, Y.; Zhang, N. Measurement and Calibration of DEM Parameters of Soybean Seed Particles. *Agriculture* **2022**, *12*, 1825. https://doi.org/10.3390/agriculture12111825

Academic Editors: Vadim Bolshev, Vladimir Panchenko and Alexey Sibirev

Received: 14 September 2022
Accepted: 28 October 2022
Published: 1 November 2022

1. Introduction

The DEM is widely used in the field of agricultural engineering, and its parameters are crucial to the simulation results [1–10]. For soybean seed particles, parameters play a crucial role in the simulation results when DEM is used to analyze the movement of the particles, the contact interaction between the particles, and between the particles and the boundary.

Nguyen, et al. [11] studied soybean seed particles of one variety, which were approximated as being spherical. The physical properties of the soybean (particle size distribution and weight properties) and the static friction coefficient between the particles and the material surface were determined by test. The rest of the DEM simulation parameters were calibrated by different particle flow tests. In fact, with such a calibration method, there are multiple combinations of parameters that meet the requirements. The applicability of such parameter results needs further analysis when the calibration is performed without determining whether the calibration test is sensitive to only one of the parameters.

Bhupendra et al. [12] used a spherical soybean seed particle as their study object. A set of calibration results were obtained by stacking tests, as follows: the restitution coefficient, static friction coefficient, and rolling friction coefficient between the particles, and the restitution coefficient, static friction coefficient, and rolling friction coefficient between the particles and the boundary. However, the restitution coefficient and static friction coefficient between the particles obtained by calibration were quite different from the actual values. In order to make the simulation closer to the test and accurately analyze the particle population movement, the restitution coefficient and static friction coefficient between the particles can be determined by the test method.

Some scholars [13,14] believe that non-spherical particles do not need to consider the rolling friction coefficient, and only need to calibrate the corresponding sliding friction coefficient to meet testing requirements. However, others believe that the effect of rolling friction coefficient on the test results is significant even for non-spherical particles [15–17]. For soybean seed particles of ellipsoidal shape [18,19], how much the rolling friction coefficient affects the particle population motion and whether the rolling friction coefficient needs to be accurately calibrated requires further study.

Long Zhou [15] demonstrated that the rolling friction coefficient has a significant effect on test results using sensitivity tests during the modeling of different shapes of corn seeds, and further calibrated the rolling friction coefficients between particles, and between particles and boundaries through a piling angle test. For soybean seed particles, the sensitivity of the rolling friction coefficient to the test results needs to be determined further, and the calibration method of the sensitivity parameter also needs to be studied in-depth.

Based on the above problems, in this paper, we verified the importance of the parameters for the first time, and a method for calibrating the above simulation parameters was proposed and verified by tests. Three representative soybean varieties, SN42, JD17, and ZD39, were used in this study. Some parameters of soybean seed particles were measured by test methods. The effect of RFCP-P and RFCP-B on powder motion was analyzed by a repose angle test and self-flow screening simulation. After analyzing the sensitivity of parameters, the test method of parameter calibration was determined and parameter values were calibrated. The accuracy of the parameters was verified using a rotating cylinder test and self-flow screening test. This paper provides some reference for the calibration of parameters for soybean seed particles.

2. Measurement of Soybean Seed Particle Parameters

In this section, the moisture content, triaxial dimensions, particle density, stiffness coefficient, elasticity modulus, restitution coefficient, and static friction coefficient of soybean seed particles are measured by test measurements; Poisson's ratio is taken to be 0.4, according to reference [20].

2.1. The Moisture Contents of Soybean Seed Particles

The moisture content of soybean seed particles was measured using XY-102MW type halogen moisture meter (accuracy 0.001), and the test was repeated 5 times for each variety, The moisture content was 10.31, 8.08, and 11.1% for SN42, JD 17, and ZD39, respectively.

2.2. Particle Density of Soybean Seed Particles

In this paper, the particle density of soybean seeds were measured by the pycnometer method, and the calculation formula is as follows:

$$\rho_0 = m_0 / V_0 \tag{1}$$

where ρ_0 is the density of soybean seed particles, g/cm^3; m_0 is the mass of soybean seeds particles, g; V_0 is the volume of soybean seed particles, cm^3. The formula for the volume of soybean particles is as follows:

$$V_0 = \frac{(m_2 - m_1)(m_3 - m_1 - m_0)}{\rho_w} \tag{2}$$

where m_1 is the mass of the dry specific gravity bottle, g; m_2 is the mass of the specific gravity bottle filled with water, g; m_3 is the mass of soybean seed particles, water, and specific gravity bottle, g; and ρ_w is the density of water, g/cm^3.

2.3. *Stiffness Coefficients of Soybean Seed Particles*

The stiffness coefficients of the three varieties were measured using the compression test method [21,22]. As soybean seed particles are ellipsoidal particles with three unequal axes, their stiffness coefficients are different in all directions. Test measurements with three different placement methods (horizontal, lateral, and vertical) are shown in Figure 1. The average value was obtained and used as its final stiffness coefficient.

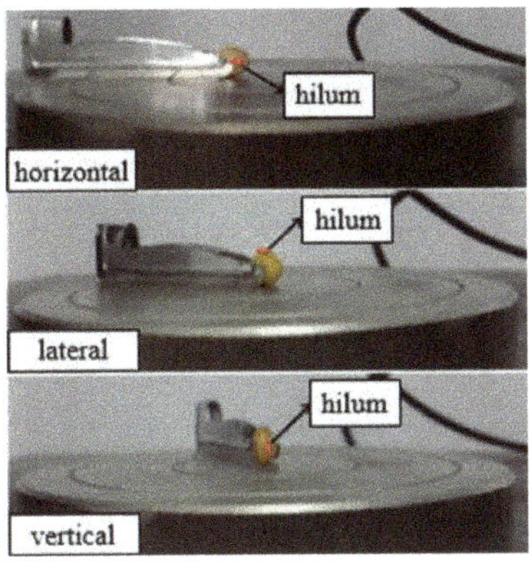

Figure 1. Placement of soybean seed particles.

Taking SN42 as an example, during the loading process, the test force was gradually increased with an increase in deformation before the abrupt change of the test force. The curve is divided into three sections for discussion. In the first section, the deformation (0–0.01 mm) is very small, the test force is small and the growth trend is not obvious; in the second section, the deformation (0.01–0.05 mm) is small, and the test force increases slowly in this range; in the third section, the deformation (0.05–0.4 mm)is larger, and the test force basically grows linearly, as shown in Figure 2.

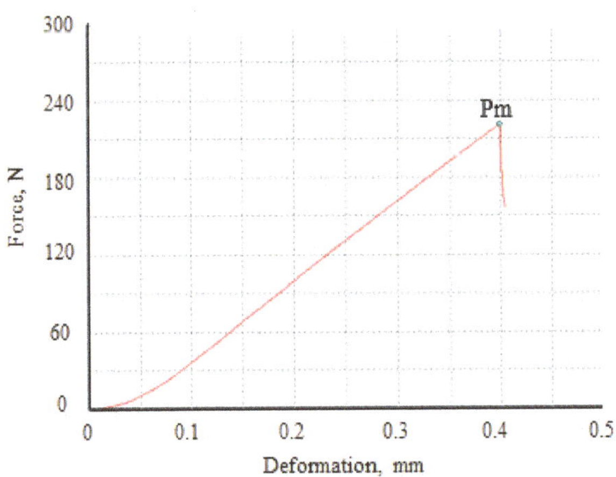

Figure 2. The force–deformation relationship of SN42 soybean seeds.

In our study, the deformation caused by collision during particle movement were within a small range of changes (0.01–0.05 mm) [21]. Therefore, the second section of the curve in the range of smaller deformations was analyzed and processed in this paper. This segment of data was processed in an Excel sheet and a straight line was fitted, as shown in Figure 3. The slope of this straight line is the stiffness coefficient of the soybean seed particles.

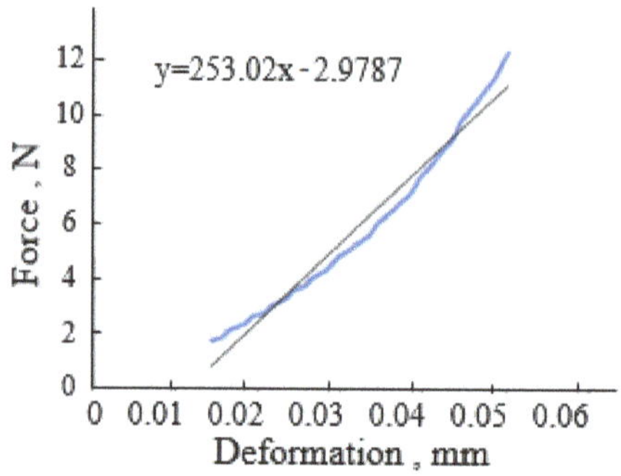

Figure 3. The force–deformation relationship of SN42 soybean seeds when the deformation is small.

2.4. Elastic Modulus of Soybean Seed Particles

The elasticity modulus of soybean seed particles was measured by compression tests [23]. Soybean seed particles are approximately ellipsoidal in shape, so the radii of curvature of soybean seed particles in contact with the upper and lower surfaces of the platen are the same. According to the standard ASAE S368.4 DEC2000 (R2008) [20], the elasticity modulus was calculated as follows:

$$E = \frac{0.338F(1-\mu^2)}{D^{3/2}}\left[2K_U\left(\frac{1}{R}+\frac{1}{R'}\right)^{1/3}\right]^{3/2} \quad (3)$$

where E is the modulus of elasticity of soybean seed particles, Pa; D is the amount of deformation, mm, which is the middle value of the deformation corresponding to the previous measurement of the stiffness coefficient; F is the test force corresponding to the current deformation, N, which can be directly found in the Excel database; μ is the Poisson's ratio, with a value of 0.4; R and R' are the primary and secondary curvature radii when soybean seed particles come in contact with the surface of the plate, m; and K_U is the coefficient.

2.5. Restitution Coefficient of Soybean Seed Particles

2.5.1. Restitution Coefficient between Soybean Seed Particles and Boundary

A drop test [24,25] was used to measure the restitution coefficient between the soybean seed particle and boundary. Using SN42 as an example, the soybean seed particle is placed vertically at the vacuum nozzle along its length and marked directly in front of the seed particle, as shown in Figure 4a–c. A high speed camera is used to record the above test procedure, while ensuring that the soybean seed particle is moving in a vertical direction. The software accompanying the high-speed camera was used to analyse the experimental results. Neglecting air resistance, the restitution coefficient was calculated as follows:

$$e = \sqrt{\frac{h_1}{h_0}} \quad (4)$$

where h_0 is the distance between the initial position and coordinate origin of soybean seed particles, as shown in Figure 4b, and h_1 is the distance between the highest point and coordinate origin of the first vertical rebound of soybean seed particles, as shown in Figure 4c.

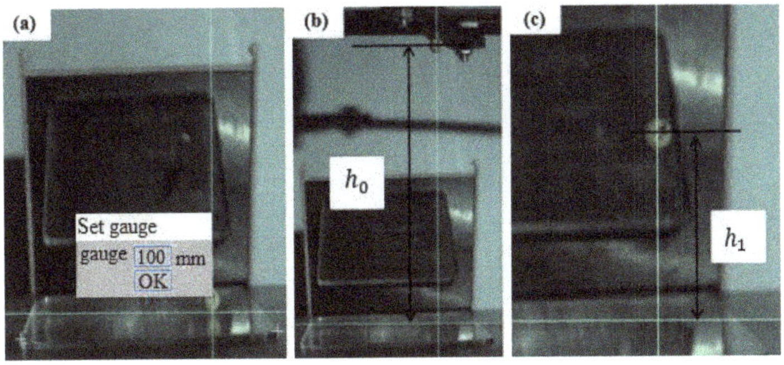

Figure 4. PCC image analysis process: (**a**) coordinate origin selection and distance calibration, (**b**) the distance between the initial position and coordinate origin of soybean seed particles, and (**c**) the distance between the highest point and coordinate origin of the first vertical rebound of soybean seed particles.

2.5.2. Restitution Coefficient between Soybean Seed Particles

A single pendulum collision test [26,27] was used to measure the restitution coefficient between soybean seed particles. A soybean was lifted with a straightedge to the position shown in Figure 5a–c. The software that accompanied the high-speed camera was used to analyse the experimental results. The equation for the restitution coefficient, ignoring air resistance, was as follows:

$$e = (\sqrt{h_1} - \sqrt{h_2}) / \sqrt{h_0} \tag{5}$$

where h_0 is the vertical distance between the initial position of soybean seed particles and coordinate origin, as shown in Figure 5b, and h_1 and h_2 are the distances between the highest point of the vertical rebound of the two soya beans and the origin of the coordinate after the first collision, respectively, as shown in Figure 5c.

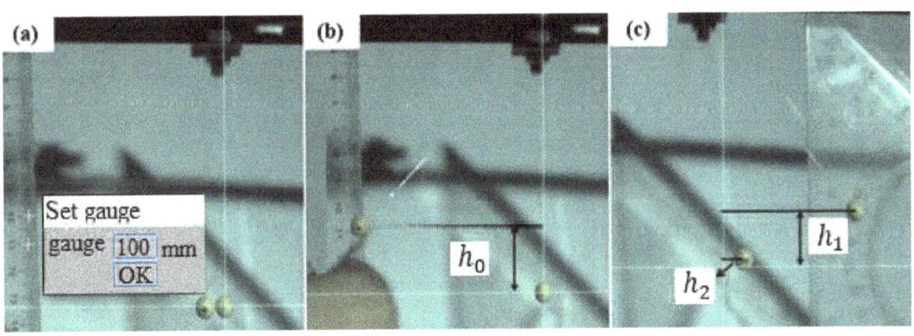

Figure 5. PCC image analysis process: (**a**) coordinate origin selection and distance calibration, (**b**) the vertical distance between the initial position of soybean seed particles and coordinate origin, (**c**) the distances between the highest point of the vertical rebound of the two soya beans and the origin of the coordinate after the first collision.

2.6. Static Friction Coefficient of Soybean Seed Particles

2.6.1. Static Friction Coefficient between Soybean Seed Particles and the Boundary

The method for measuring the static friction coefficient between the particles and boundary was the slope method [28,29]. Three soybean seed particles with an intact appearance were fixed on a small square glass plate with glue. The boundary material and inclinometer were fixed to the slope meter and the test material was placed on the boundary material, as shown in Figure 6. For the trials, three test specimens of each variety were used to produce three replicate experiments for each specimen. The specimens were made with the soybean in a random orientation with no fixed direction. The formula for the coefficient of static friction between the soya seed pellet and the boundary is shown below:

$$\mu_{SP-B} = \tan\alpha \tag{6}$$

where μ_{SP-B} is the static friction coefficient between the soybean seed particle and boundary and α is the indication of the inclinometer when the test specimen is just sliding, rad.

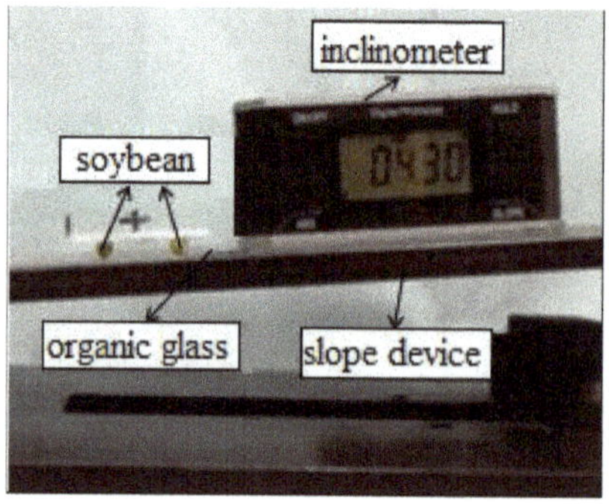

Figure 6. Static friction coefficient between soybean seed particle and boundary measured by the slope method.

2.6.2. Static Friction Coefficient between Soybean Seed Particles

The slope method was used to measure the static friction coefficient between soybean seed particles. The test specimens were made by fixing three intact appearing soybean seed particles to a small square glass piece with glue. We fixed one specimen on the inclinometer and placed the other on top, to ensure that the two test soybean particles were in exact vertex contact, as shown in Figure 7. The equation for calculating the static friction coefficient between soybean seed particles is as follows:

$$\mu_{SP-P} = \tan\beta \tag{7}$$

where μ_{SP-P} is the static friction coefficient between soybean seed particles and β is the indication of the inclinometer when the test specimen is just sliding, rad.

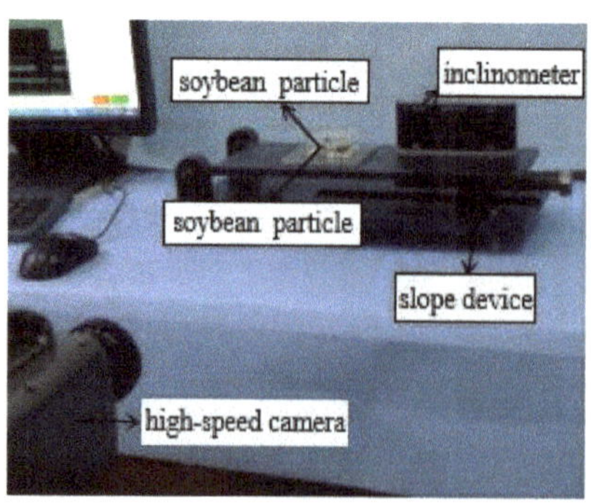

Figure 7. Static friction coefficient between soybean seed particles measured by the slope method.

2.7. Soybean Seed Particle Model and Simulation Parameters

It was clear from our previous work that geometric models of soybean seed particles with different sphericities can be built by 5-, 9-, and 13-sphere models [19]. In this paper, two hundred grains of each variety were selected and measured to obtain the average triaxial dimensions, as well as the sphericity of soybean seed particles, as shown in Table 1.

Table 1. Triaxial dimensions of soybean seed particles of three varieties.

Variety	Length, mm	Width, mm	Thickness, mm	Sphericity, %
SN42	7.44	7.24	6.51	94.78
JD17	6.95	6.2	5.11	86.86
ZD39	7.36	6	4.73	80.6

The 13-sphere model was chosen for the simulation, as shown in Figure 8, with powder generation performed according to volume normal distribution; the simulation parameters are shown in Table 2.

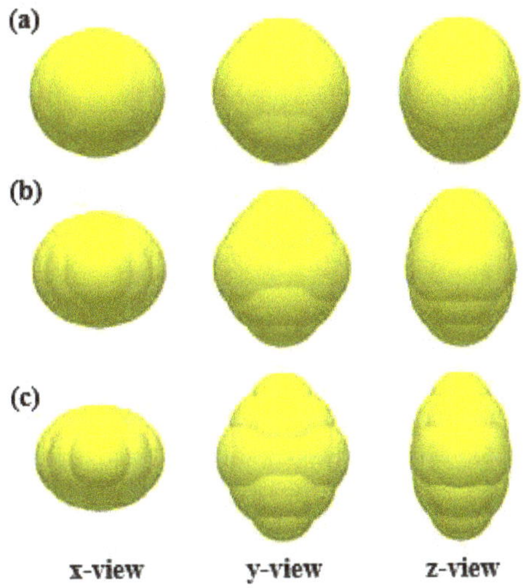

Figure 8. The 13-sphere model of (**a**) SN42, (**b**) JD17, and (**c**) ZD39.

Table 2. Parameters used in the simulation.

Parameters	Symbol	SN 42		JD 17		ZD 39	
		Soybean Seeds	Organic Glass	Soybean Seeds	Organic Glass	Soybean Seeds	Organic Glass
Density, kg/m^3	ρ	1257	1800	1213	1800	1192	1800
Poisson's ratio	ν	0.4	0.25	0.4	0.25	0.4	0.25
Elastic modulus, Pa	E	7.59×10^8	1.30×10^8	6.07×10^8	1.30×10^8	2.55×10^8	1.30×10^8
Static friction coefficient	μ_S	0.205	0.228	0.211	0.228	0.207	0.235
Restitution coefficient	e	0.627	0.542	0.562	0.642	0.607	0.705

3. Analysis of the Influence of the Rolling Friction Coefficient

This section analyses the influence of the RFCP-P on the angle of repose by means of a repose angle test and the influence of the RFCP-B on the percentage passing by means of a self-flow screening test. The results confirmed that accurate calibration of the above two parameters is necessary.

3.1. Repose Angle Simulation

The RFCP-B was set to 0.025 and the RFCP-P was set to nine values of 0, 0.025, 0.05, 0.075, 0.1, 0.125, 0.15, 0.175, and 0.2. The single factor test was conducted to analyze the effect of the RFCP-P on the angle of repose.

The test procedures were same as in previous studies; the dimensions of the device were 220*48 mm [19]. The EDEM (Version, 2018, School of Biological and Agricultural Engineering, Jilin University, Changchun, Jilin, China) software with the parameters in Table 1 was used to simulate the repose angle test. First, 4000 particles were generated in the particle factory. After two seconds, the insert plate was pulled out and the seed particles flowed out of the loading box; at the same time, the angle of repose was formed in the loading box. The repose angle results were analyzed using image processing software, as shown in Figure 9. The simulation tests were repeated five times for SN42, JD17, and ZD39, respectively, according to the above steps.

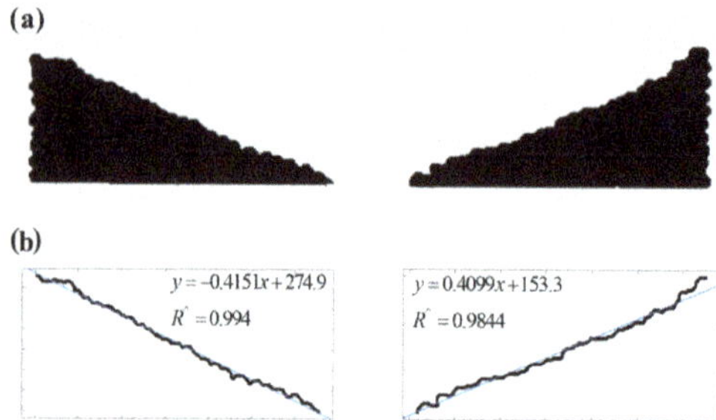

Figure 9. Image of (**a**) repose angle simulation results and (**b**) processing diagram.

3.2. Self-Flow Screening Simulation

The RFCP-P was set to 0.025 and the RFCP-B was set to nine level values of 0, 0.025, 0.05, 0.075, 0.1, 0.125, 0.15, 0.175, and 0.2, respectively. A single factor test was conducted to analyze the effect of the RFCP-B on the percentage passing.

The test procedures were the same as in previous studies [19].The inclination angle and aperture sizes of the three varieties of self-flow screening simulation were 11° and 8 mm, respectively. The EDEM software with the parameters in Table 1 was used to simulate the self-flow screening test. In the simulation, 1000 particles were generated in the particle factory. After the particles were stable, we pulled out the insert plate, and the soybean particles moved downward along the sieve. At the end of the movement, the particles were present in the receiving area below or on the sieve deck. The numbers of soybean particles in the corresponding areas were counted and numbered as 1–5, as shown in Figure 10.

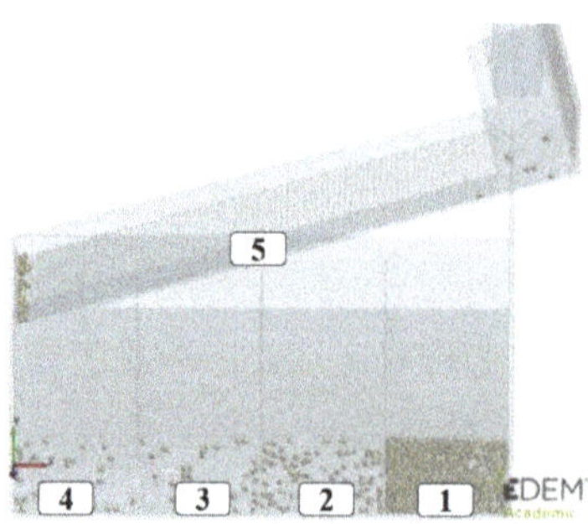

Figure 10. The statistical areas of self-flow screening simulation.

3.3. Analysis of Simulation Results

The relationship between the angle of repose and RFCP-P is shown in Figure 11. With the increase in RFCP-P from 0 to 0.2, the increasing trend of the angle of repose of the three varieties was obvious. For SN42, the angle of repose gradually increased from 19.33 to 37.57°; for JD17, the angle of repose gradually increased from 24.8 to 37.02°; and for ZD39, the angle of repose gradually increased from 27.31 to 39.1°. Analysis of the results shows that the RFCP-P had a significant effect on the angle of repose. Therefore, an accurate RFCP-P needs to be calibrated through calibration tests.

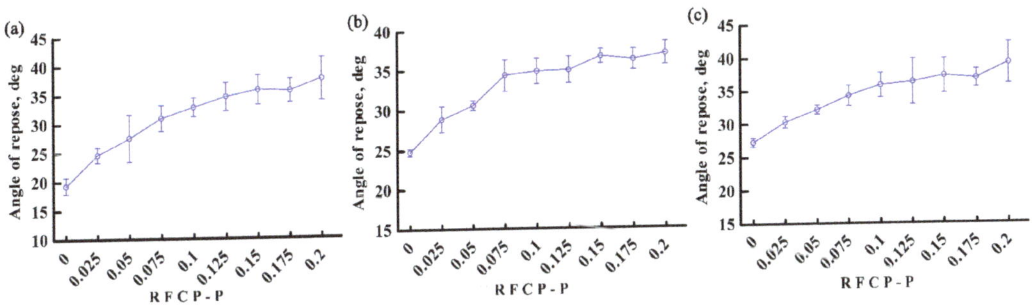

Figure 11. The relationship between angle of repose and RFCP-P for (**a**) SN42, (**b**) JD17, and (**c**) ZD39.

The relationship between percentage passing and RFCP-B is shown in Figure 12. For SN42, due to its high sphericity, the percentage passing did not change much when the RFCP-B gradually increased from 0 to 0.075, whereas the percentage passing tended to significantly decrease when the RFCP-B gradually increased from 0.075 to 0.2, with a range of 94.53 to 29.23%. For JD17 and ZD39, the percentage passing was significantly reduced as the RFCP-B increased, with a range of 95.43 to 26.4% and 86.97 to 23.53%, respectively. The analysis showed that the RFCP-B had a significant effect on the percentage passing. Therefore, an accurate RFCP-B needs to be calibrated by calibration tests.

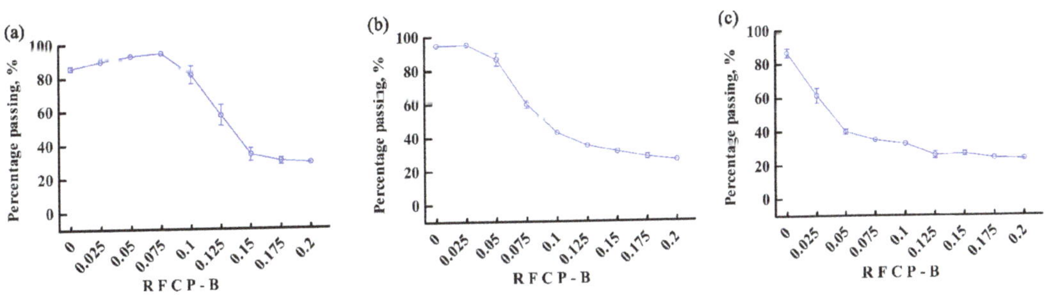

Figure 12. The relationship between percentage passing and RFCP-B for (**a**) SN42, (**b**) JD17, and (**c**) ZD39.

4. Study on the Sensitivity of RFCP-P and RFCP-B

4.1. Comprehensive Simulation Test of Sensitivity Analysis

As mentioned before, the RFCP-P and RFCP-B need to be calibrated. If two parameters are simultaneously calibrated, there are multiple sets of solutions. Therefore, the sensitivity of RFCP-P and RFCP-B to the angle of repose was first analyzed by simulating pitch angle tests. Seven levels of 0.01–0.07 were taken for both RFCP-P and RFCP-B, and a full 7*7 simulation was performed.

4.2. Analysis of the Results

The results of the effect of RFCP-P and RFCP-B on the angle of repose for the three varieties of soybean seed particles are shown in Figure 13. For the three varieties, the effect of RFCP-P on the angle of repose was highly significant, whereas the effect of RFCP-B on the angle of repose was insignificant.

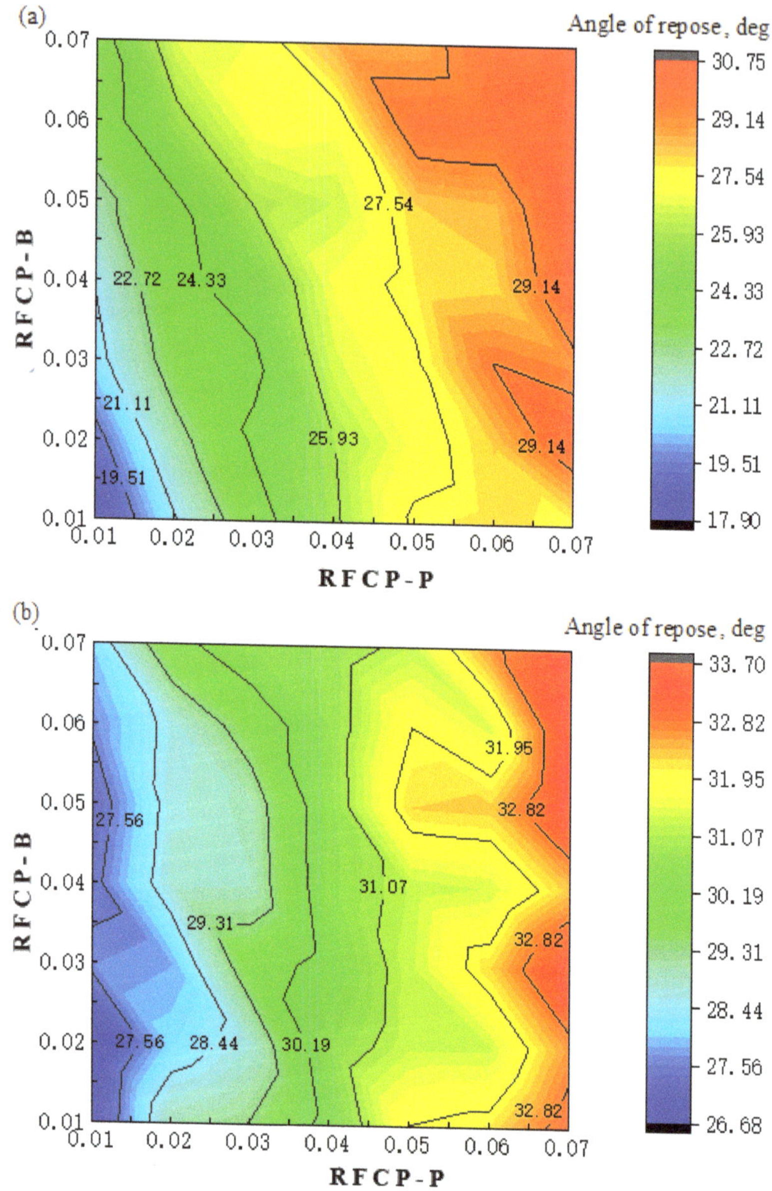

Figure 13. *Cont.*

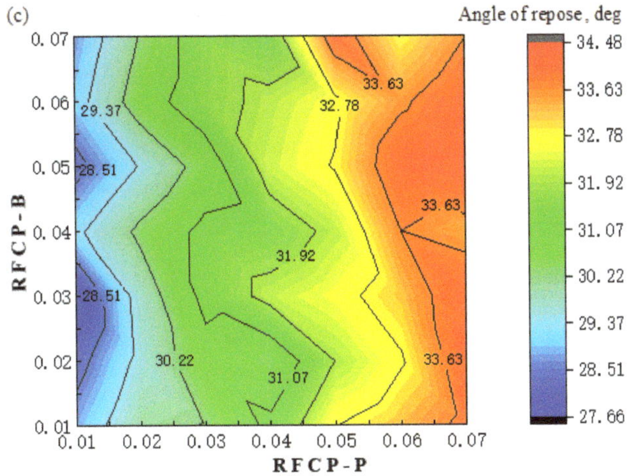

Figure 13. Sensitivity analysis of RFCP-P and RFCP-B to angle of repose for (**a**) SN42, (**b**) JD17, and (**c**) ZD39.

According to the above analysis, the RFCP-P can be calibrated by a single factor test of the repose angle.

5. Calibration of the Rolling Friction Coefficient

5.1. Calibration of the RFCP-P

From the previous analysis, the angle of repose was only sensitive to the RFCP-P, so the RFCP-P was calibrated by simulation of the repose angle. The RFCP-P was taken as 0.01, 0.02, 0.03, 0.04, and 0.05, and the RFCP-B was taken as 0.02 for the single factor test of the repose angle. The RFCP-P was calibrated by comparing the simulation and test results.

5.2. Calibration of the RFCP-B

After the calibration of the RFCP-P, the repose angle test was also used to calibrate the RFCP-B. In this paper, spheres with the same boundary material were processed, with a radius of 5 mm. The soybean seed particles and organic glass balls were 350 g each. They were uniformly mixed and poured into the loading box for testing, and the repose angle was simulated and analyzed, as shown in Figure 14. The rolling friction coefficient between the soybean seed particles and organic glass spheres obtained from the calibration was the RFCP-B.

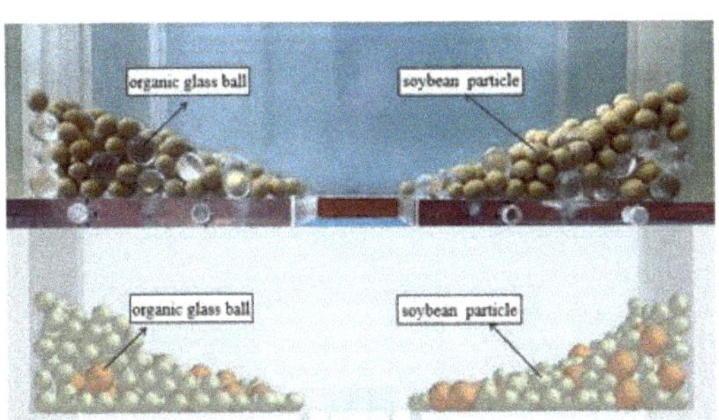

Figure 14. Test and simulation diagram of the repose angle of soybean seed particles mixed with organic glass spheres.

The static friction coefficient and rolling friction coefficient between the organic glass spheres and boundary involved in the simulation were measured by the slope method, and the restitution coefficient was measured by the drop test.

5.3. Analysis of Results

5.3.1. Calibration Results of the RFCP-P

The relationship between the angle of repose and the RFCP-P for the three varieties are shown in Figure 15.

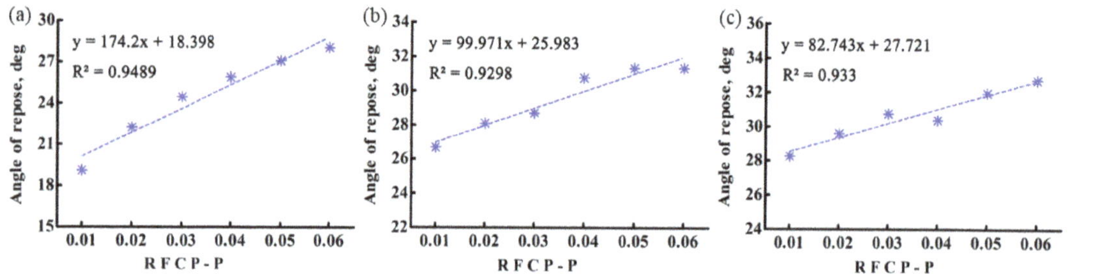

Figure 15. The relationship between the angle of repose and the RFCP-P for (**a**) SN42, (**b**) JD17, and (**c**) ZD39.

For the three varieties, the angle of repose tended to increase as the RFCP-P increased. The repose angle results for the three varieties (SN42, 23.86°; JD17, 27.78°; and ZD39, 38.97°.) were obtained by measuring the repose angle test of the soybean seed particles. Taking SN42 as an example, the relationship between the angle of repose and RFCP-P was obtained by liner fitting, and the formula is shown as follows.

$$y = 174.2x + 18.398\ (R^2 = 0.9298) \tag{8}$$

The result (23.86°) of the repose angle test for SN42 was entered into Formula 8, and the RFCP-P was calculated to be 0.031. The same method was used to calculate the RFCP-P, which was 0.018 and 0.136 for JD17 and ZD39, respectively.

5.3.2. Calibration Results for RFCP-B

The relationship between the angle of repose formed by mixing the soybean seed particles with the organic glass spheres and rolling friction coefficient between the soybean seed particles and boundary (organic glass spheres) is shown in Figure 16.

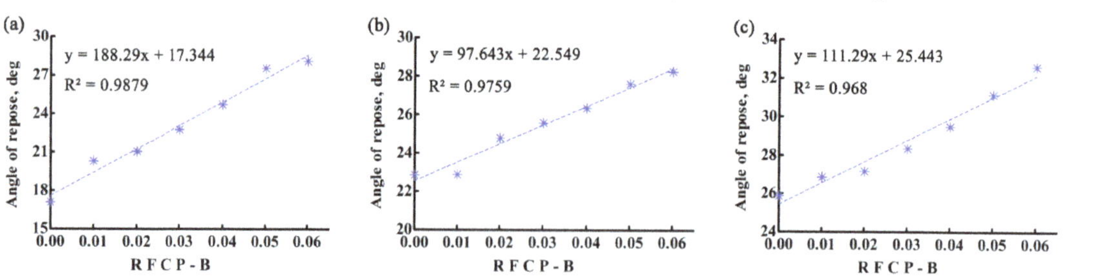

Figure 16. The relationship between the angle of repose and the RFCP-B for (**a**) SN42, (**b**) JD17, and (**c**) ZD39.

For the three varieties, the angle of repose tended to increase as the RFCP-B increased. The repose angle results for the three varieties (SN42, 18.72°; JD17, 26.16°; and ZD39, 28.56°.) were obtained by measuring the repose angle test of the soybean–glass ball mixture.

Taking SN42 as an example, the relationship between the angle of repose and RFCP-B was obtained by liner fitting, and the formula is shown as follows.

$$y = 188.29x + 17.344\ (R^2 = 0.9879) \tag{9}$$

The result (18.72°) of the repose angle test for SN42 was entered into the Formula 9, and the RFCP-B was calculated to be 0.008. The same method was used to calculate the RFCP-B, which was 0.037 and 0.028 for JD17 and ZD39, respectively.

6. Test Verification of Calibration Parameters

The accuracy of the calibration parameters was verified by rotating cylinder test and self-flow screening test. At the same time, the difference between the test results and simulation results when the rolling friction coefficient was ignored was also analyzed.

6.1. Rotating Cylinder Test

The rotating cylinder test apparatus is shown in Figure 17a, where the inner diameter of the cylinder is 160 mm and the height is 40 mm. Taking SN42 as an example, the soybean seed particles were first filled through the inlet, and when the height of the soybean seed particle filling in the cylinder reached 40–50 mm, it met the particle filling requirements of the rotating cylinder test, as shown in Figure 17b.

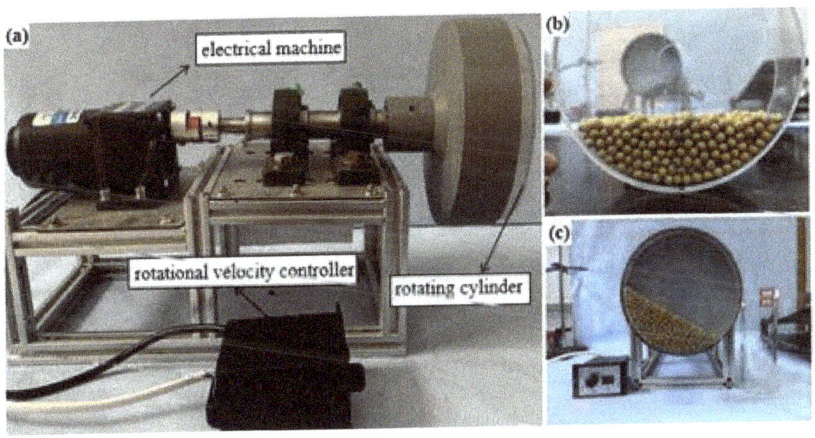

Figure 17. (**a**) The rotating cylinder test apparatus and (**b**,**c**) the screenshot of test process.

After the test device was installed, the power supply was connected, and the speed controller was adjusted to make the drum speed 7.5 rpm. After the speed was stable, the soybean seed particles in the drum formed a dynamic repose angle, such as Figure 17c, recording the dynamic piling process for 30 s. The cylinder speed was adjusted to 11.5 and 15.5 rpm, respectively, and the dynamic stacking process was recorded for 30 s. At the end of the test, the soybean seed particles were weighed for mass. Three replicate tests were conducted for each variety.

For the simulated rotating cylinder test, the soybean seed particle model was equal to the actual soybean seed particle mass. Taking SN42 as an example, the drum speed was 7.5 rpm. Figure 18a,b shows the test photo and binarization image, respectively. Figure 18c,d shows the simulation screenshot using the calibration parameters and binarization image, respectively. Figure 18e,f shows the simulation screenshot without considering the rolling friction coefficient and binarization image, respectively.

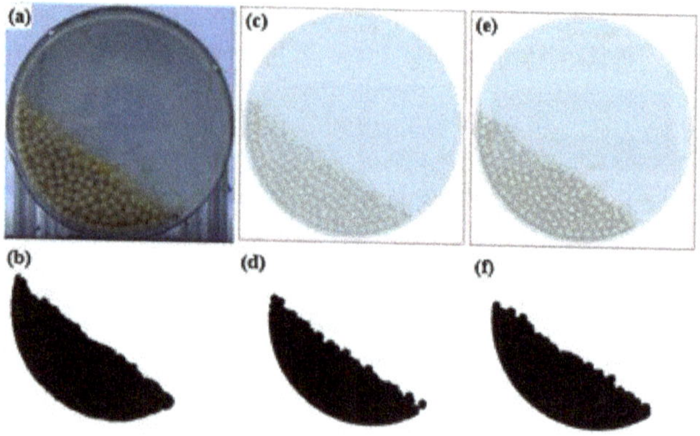

Figure 18. (**a**) The test photo of the rotating cylinder and (**b**) its binarization image; (**c**) simulation result snapshot using calibration parameters and (**d**) its binarization image; and (**e**) simulation result snapshot without considering rolling friction coefficient and (**f**) its binarization image.

Figure 19 shows the dynamic angle of repose versus cylinder speed for each variety. For SN42, the relative errors between the simulation results using the calibration parameters and experimental results were 2.28, 6.93, and 4.78% for rotational speeds of 7.5, 11.5, and 15.5 rpm, respectively, whereas the relative errors between the simulation results ignoring the rolling friction coefficient and experimental results were 13.85, 17.08, and 12.83% for rotational speeds of 7.5, 11.5, and 15.5 rpm, respectively, as shown in Figure 19a.

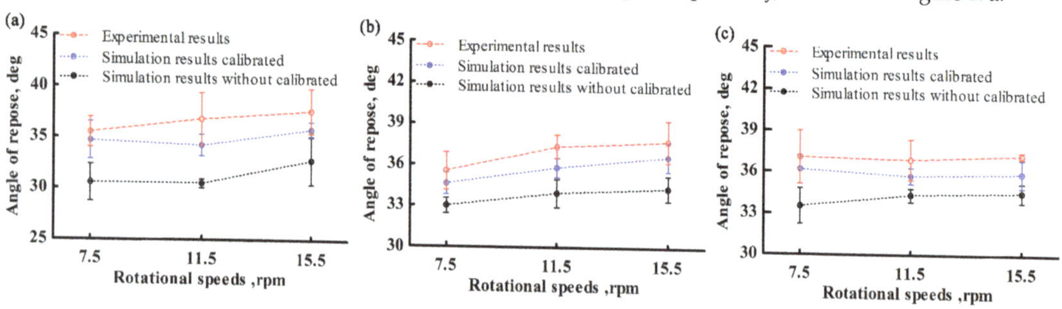

Figure 19. The relationship between dynamic angle of repose and rotational speed for (**a**) SN42, (**b**) JD17, and (**c**) ZD39.

For JD17, the relative errors between the simulation results using the calibration parameters and experimental results were 2.56, 4.1, and 2.84% for rotational speeds of 7.5, 11.5, and 15.5 rpm, respectively, whereas the relative errors between the simulation results ignoring the rolling friction coefficient and experimental results were 7.15, 9.11, and 9.11% for rotational speeds of 7.5, 11.5, and 15.5 rpm respectively, as shown in Figure 19b.

For ZD39, the relative errors between the simulation results using the calibration parameters and experimental results were 2.43, 3.25, and 3.62% for rotational speeds of 7.5, 11.5, and 15.5 rpm, respectively, whereas the relative errors between the simulation results ignoring the rolling friction coefficient and experimental results were 9.55, 6.94, and 7.28% for rotational speeds of 7.5, 11.5, and 15.5 rpm, respectively, as shown in Figure 19c.

The analysis shows that the relative error between the simulation results using calibration parameters and experimental results was small and within the error range of the experimental results. The simulation results without taking the rolling friction coefficient into account were much smaller than the experimental results, and were not within the experimental error range.

6.2. Self-Flow Screening Tests

According to previous studies, the inclination angles of the self-flow screening devices were 7, 11, and 15° for SN42, JD17, and ZD39, respectively. The material of the device was organic glass and the aperture size was 8 mm. Figure 20a–c shows the test photo for self-flow screening, the simulation screenshot using the calibrated parameters, and the simulation screenshot without considering the rolling friction coefficient, respectively.

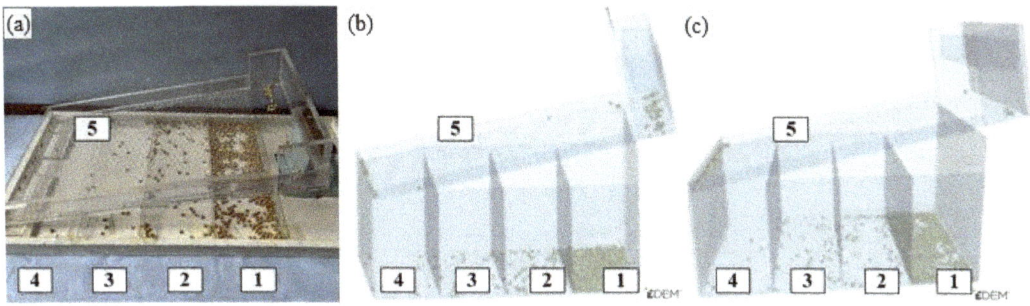

Figure 20. (**a**) The test photo, (**b**) the simulation screenshot using the calibrated parameters, and (**c**) the simulation screenshot without considering the rolling friction coefficient for self-flow screening test.

Figure 21a–c shows comparisons of the simulation results with experimental results of the percentage passing into the five statistic areas for the three varieties. The analysis shows that the simulation results using the calibrated parameters, simulation results without considering the rolling friction coefficient, and experimental results have similar trends.

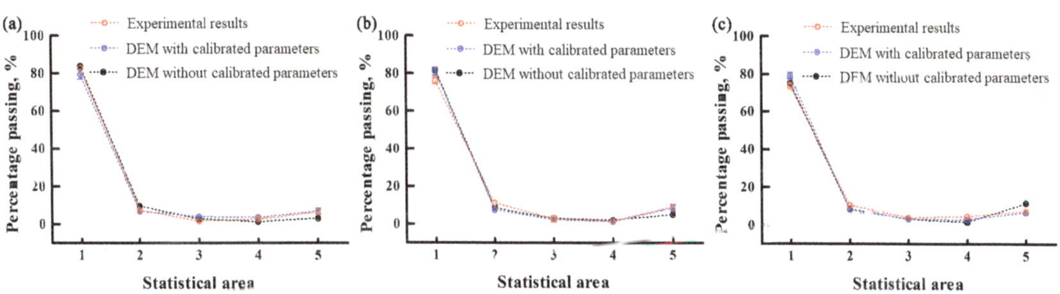

Figure 21. Comparisons of the simulation results with the test results of the percentage passing from different statistic areas in the "self-flow screening" for (**a**) SN42, (**b**) JD17, and (**c**) ZD39.

Further analysis of the percentage passing of the simulation and experimental results for the three varieties is shown in Figure 22.

For the SN42, the simulation results using the calibrated parameter was slightly smaller than the experimental results, with a relative error of 0.7%. The difference between the simulation results ignoring the rolling friction coefficient and experimental results was larger, with a relative error of 3.3%.

For the JD17, the simulation results using the calibrated parameter was slightly smaller than the experimental results, with a relative error of 0.1%. The difference between the simulation results ignoring the rolling friction coefficient and experimental results was larger, with a relative error of 4.9%.

For the ZD39, the simulation results using the calibrated parameter was slightly larger than the experimental results, with a relative error of 0.27%. The difference between the simulation results ignoring the rolling friction coefficient and experimental results was larger, with a relative error of 5.23%.

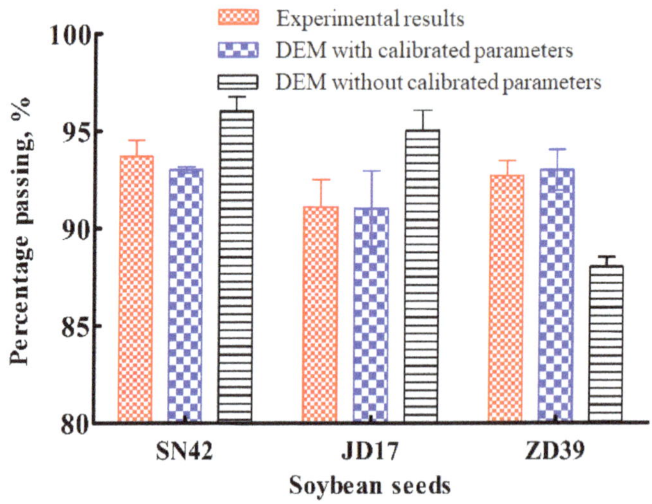

Figure 22. The percentage passing of the simulation and experimental results for three varieties.

Therefore, for the self-flow screening test, the simulation results using the calibrated parameters were closer to the experimental results.

Through the comprehensive analysis of the simulated and experimental results of the rotating cylinder test and self-flow screening test, our results showed that, for the three varieties of soybean seed particles, the simulation results using calibrated parameters were closer to experimental values than simulation results without the rolling friction coefficient. At the same time, the simulation results using the calibrated parameters were within the error range of the experimental results. Therefore, the results of the parameter calibrations in this paper were high in accuracy.

7. Conclusions

In this paper, the physical parameters of soybean seed particles of different varieties were tested, and the rolling friction coefficients, which could not be measured by test, were determined by calibration methods. The accuracy of the calibration parameters was verified by rotating cylinder test and self-flow screening test. The conclusions are as follows:

(1) The simulation of the repose angle demonstrated that the RFCP-P had a large effect on the angle of repose. The simulation of self-flow screening demonstrated that the RFCP-B had a large effect on the percentage passing. It showed that RFCP-P and RFCP-B needed to be accurately calibrated.

(2) A comprehensive test of the repose angle demonstrated that the RFCP-P had a significant effect on the angle of repose, whereas the RFCP-B did not have a significant effect on the angle of repose.

(3) The RFCP-P was calibrated using a repose angle test. By mixing organic glass spheres with soybean seed particles for the repose angle test, the RFCP-B was calibrated further. The calibrated parameters were verified by means of a rotating cylinder test and self-flow screening test. The results showed that the calibrated parameters were accurate and valid. The RFCP-P and RFCP-B should be considered in the simulation of soybean seed particles.

In summary, parameters are important for simulation. It is particularly important to select and calibrate the relevant parameters accurately. Though the experimental apparatus in this paper is not necessarily applicable to other research subjects, a similar research methodology could be adopted. In addition, there are various methods for modelling particles when the object of study is a different shape. Due to time constraints, only the multi-sphere method was used to model particles in this paper; therefore, other methods for modelling particles and more in-depth analytical studies on the calibration of parameters should be considered for the next step of research. This paper could provide some reference for relevant areas of inquiry.

Author Contributions: Conceptualization, D.Y.; methodology, D.Y.; validation, D.Y. and Y.T.; investigation, K.S. and resources, N.Z. and K.S.; writing—original draft preparation D.Y.; writing—review and editing, Y.W.; supervision, L.Z.; project administration, J.Y.; funding acquisition, D.Y. All authors have read and agreed to the published version of the manuscript.

Funding: The authors are grateful to the National Natural Science Foundation of China (No. 52130001) for the financial support of this work.

Institutional Review Board Statement: Not applicable.

Informed Consent Statement: Not applicable.

Data Availability Statement: Not applicable.

Conflicts of Interest: The authors declare no conflict of interest.

References

1. Yu, Y.J.; Fu, H.; Yu, J.Q. DEM-based simulation of the corn threshing process. *Adv. Powder Technol.* **2015**, *26*, 1400–1409. [CrossRef]
2. Kafashan, J.; Wiacek, J.; Ramon, H.; Mouazen, A.M. Modelling and simulation of fruit drop tests by discrete element method. *Biosyst. Eng.* **2021**, *212*, 228–240. [CrossRef]
3. Yan, D.X.; Yu, J.Q.; Wang, Y.; Zhou, L.; Sun, K.; Tian, Y. A Review of the Application of Discrete Element Method in Agricultural Engineering: A Case Study of Soybean. *Processes* **2022**, *10*, 1305.
4. Wang, Y.; Yu, J.Q.; Yu, Y.J. Validation of a coupled model of discrete element method with multibody kinematics to simulate the screening process of a swing-bar sieve. *Powder Technol.* **2019**, *346*, 193–202. [CrossRef]
5. Yan, D.X.; Yu, J.Q.; Wang, Y.; Zhou, L.; Tian, Y.; Zhang, N. Soil Particle Modeling and Parameter Calibration Based on Discrete Element Method. *Agriculture* **2022**, *12*, 1421. [CrossRef]
6. Yan, D.X.; Xu, T.Y.; Yu, J.Q.; Wang, Y.; Guan, W.; Tian, Y.; Zhang, N. Test and Simulation Analysis of the Working Process of Soybean Seeding Monomer. *Agriculture* **2022**, *12*, 1464.
7. Kafashan, J.; Van Zeebroeck, M.; Ramon, H.; Tijskens, B. A novel approach to a realistic discrete element modelling (DEM) in 3D. *Commun. Agric. Appl. Biol. Sci.* **2007**, *72*, 205–208.
8. Kafash, J.; Van Liedekerke, P.; Ramon, H.; Tijskens, B. An Approach to represent realistic particles of bulk assembly in three dimensional-DEM simulations and applications. *Commun. Agric. Appl. Biol. Sci.* **2011**, *76*, 33–36.
9. Kafashan, J.; Tijskens, B.; Ramon, H. Shape modelling of fruit by image processing. *Commun. Agric. Appl. Biol. Sci.* **2005**, *70*, 161–164.
10. Wiacek, J.; Stasiak, M.; Kafashan, J. Structural and micromechanical properties of ternary granular packings: Effect of particle size ratio and number fraction of particle size classes. *Materials* **2020**, *13*, 339.
11. Nguyen, T.X.; Le, L.M.; Nguyen, T.C.; Nguyena, N.T.H.; Le, T.; Pham, B.T.; Le, V.M.; Ly, H. Characterization of soybeans and calibration of their DEM input parameters. *Part. Sci. Technol.* **2021**, *39*, 530–548. [CrossRef]
12. Bhupendra, M.G.; Manish, P.; Namdeo, R. Calibration of discrete element model parameters, soybeans. *Comput. Part. Mech.* **2019**, *6*, 3–10.
13. Darius, M.; Álvaro, R.; Rimantas, K.; Zdancevičius, E. Maize grain shape approaches for DEM modeling. *Comput. Electron. Agric.* **2015**, *118*, 247–258.
14. Pasha, M.; Hare, M.C.; Ghadiri, M.; Gunadi, A.; Piccione, P.M. Effect of particle shape on flow in discrete element method simulation of a rotary batch seed coater. *Powder Technol.* **2016**, *296*, 29–36. [CrossRef]
15. Zhou, L.; Yu, J.Q.; Liang, L.S. DEM Parameter Calibration of Maize Seeds and the Effect of Rolling Friction. *Processes* **2021**, *9*, 914. [CrossRef]
16. Wang, L.; Li, R.; Wu, B.; Wu, Z.; Ding, Z. Determination of the coefficient of rolling friction of an irregularly shaped maize particle group using physical experiment and simulations. *Particuology* **2018**, *38*, 185–195. [CrossRef]
17. Roessler, T.; Richter, C.; Katterfeld, A.; Will, F. Development of a standard calibration procedure for the DEM parameters of cohesionless bulk materials—Part I, Solving the problem of ambiguous parameter combinations. *Powder Technol.* **2019**, *343*, 803–812. [CrossRef]
18. Yan, D.X.; Yu, J.Q.; Wang, Y.; Long, Z.; Yu, Y.J. A general modelling method for soybean seeds based on the discrete element method. *Powder Technol.* **2020**, *372*, 212–226. [CrossRef]
19. Yan, D.X.; Yu, J.Q.; Zhang, N.; Tian, Y.; Wang, L. Test and Simulation Analysis of Soybean Seed Throwing Process. *Processes* **2022**, *10*, 1731. [CrossRef]

20. Yan, D.X.; Yu, J.Q.; Liang, L.S.; Wang, Y.; Yu, Y.; Zhou, L.; Sun, K.; Liang, P. A Comparative Study on the Modelling of Soybean Particles Based on the Discrete Element Method. *Processes* **2021**, *9*, 286. [CrossRef]
21. Wang, Y. *Simulation Analysis and Experimental Study of the Screening Process of a Swing-Bar Sieve Based on the Coupling of DEM with MBK*; Jilin University: Changchun, China, 2018.
22. Zhang, X.; Vu-Quoc, L. A method to extract the mechanical properties of particles in collision based on a new elasto-plastic normal force–displacement model. *Mech. Mater.* **2002**, *34*, 779–794.
23. Gao, L.X.; Jiao, W.P. Effect of moisture content on mechanical properties of soybean seed under static pressure. *Trans. Chin. Soc. Agric. Eng.* **2012**, *28*, 40–44.
24. Yang, Y.; Schrock, M.D. Analysis of grain kernel rebound motion. *Trans. ASAE* **1994**, *37*, 27–31. [CrossRef]
25. Locurto, G.; Zhang, X.; Zakirov, V.; Bucklin, R.; Vu-Quoc, L.; Hanes, D.; Walton, O. Soybean impacts: Experiments and dynamic simulations. *Trans. ASAE* **1997**, *40*, 789–794. [CrossRef]
26. Wong, C.X.; Daniel, M.C.; Rongong, J.A. Energy dissipation prediction of particle dampers. *J. Sound Vib.* **2009**, *319*, 91–118.
27. González-Montellano, C.; Fuentes, J.M.; Ayuga-Téllez, E.; Ayuga, F. Determination of the mechanical properties of maize grains and olives required for use in DEM simulations. *J. Food Eng.* **2012**, *111*, 553–562. [CrossRef]
28. Fraser, B.M.; Verma, S.S.; Muir, W.E. Some Physical Properties of Fababeans. *J. Agric. Eng. Res.* **1978**, *23*, 53–57.
29. Seifi, M.R. Moisture-Dependent Physical Properties of Sunflower Seed. *J. Food Eng.* **2010**, *4*, 135–144.

agriculture

MDPI

Article

Influence of Pulsed, Scanning and Constant (16- and 24-h) Modes of LED Irradiation on the Physiological, Biochemical and Morphometric Parameters of Lettuce Plants (*Lactuca sativa* L.) while Cultivated in Vertical Farms

Alexandr A. Smirnov [1,*], Natalya A. Semenova [1], Alexey S. Dorokhov [1], Yuri A. Proshkin [1], Maria M. Godyaeva [1], Vladimir Vodeneev [2], Vladimir Sukhov [2], Vladimir Panchenko [3] and Narek O. Chilingaryan [1]

1 Department of Agricultural, Federal State Budgetary Scientific Institution "Federal Scientific Agroengineering Center VIM" (FSAC VIM), 109428 Moscow, Russia
2 Department of Biophysics, N.I. Lobachevsky State University of Nizhny Novgorod, 603950 Nizhny Novgorod, Russia
3 Department of Mechanics, Russian University of Transport, 127994 Moscow, Russia
* Correspondence: alexander8484@inbox.ru; Tel.: +7-905-723-6219

Abstract: In city farming, when growing green crops, a significant part of the production cost is the cost of electricity for lighting. The physiology, biochemistry, morphology and productivity of plants can be affected by changing irradiation modes and these changes reduce electricity costs. However, the results of studies in the literature are contradictory. In this work, we investigated the effect of impulse (frequency 1000 Hz and duty cycle 67%), scanning (the principle of running lights) and constant 16 h and 24 h modes of operation of white light LED irradiators on the physiological, biochemical and morphometric parameters of lettuce with red and green leaves. The daytime integral of light in all variants remained unchanged ~15.6 mol m^{-2} day^{-1}. Daily electricity consumption also did not differ significantly. Plants were grown on racks in a climatic chamber up to 35 days of age. For lettuce with red leaves, the most optimal for biomass accumulation and synthesis of anthocyanins was the impulse illumination mode, while for lettuce with green leaves, no statistically significant differences in biomass were observed under different irradiation modes. For red-leaved lettuce, it was found that the highest concentration of carotenoids in the leaf was observed under constant (24 h) and scanning irradiation, which is associated with a more active reaction of the photosynthetic system to prolonged irradiation and increased intensity during scanning irradiation. Also, increased photosynthetic activity was found in both varieties of lettuce at 16 h of operation of LED irradiators, which, however, did not affect their final productivity. The results may be useful for the development of LED illuminators for use in rack growing.

Keywords: pulsed LED light; continuous LED; scanning LED light; energy saving; lettuce; vertical farms; growth; cultivation

Citation: Smirnov, A.A.; Semenova, N.A.; Dorokhov, A.S.; Proshkin, Y.A.; Godyaeva, M.M.; Vodeneev, V.; Sukhov, V.; Panchenko, V.; Chilingaryan, N.O. Influence of Pulsed, Scanning and Constant (16- and 24-h) Modes of LED Irradiation on the Physiological, Biochemical and Morphometric Parameters of Lettuce Plants (*Lactuca sativa* L.) while Cultivated in Vertical Farms. *Agriculture* **2022**, *12*, 1988. https://doi.org/10.3390/agriculture12121988

Academic Editor: Athanasios Koukounaras

Received: 9 November 2022
Accepted: 19 November 2022
Published: 23 November 2022

1. Introduction

During the cultivation of plants in vertical farms, the cost of electrical energy can be quite high due to the many hours of operation of the irradiation system, which can be a limiting factor for the development of technologies and the profitability of the farm. One of the ways to reduce the cost of electricity for plant irradiation is the use of radiation sources with an impulse mode, which saves up to 12% of electricity compared to the constant mode without reducing plant productivity by varying the duration of working cycles and the ratio of light and dark (l/d) periods in each cycle [1,2].

Energy savings are also achieved due to a slight increase in the light output of LEDs with impulse power. For example, Cree LEDs of the X-Lamp XP-C series, when powered by an impulse current with a frequency of 1 kHz with a duty ratio (D) of 5%, showed a

10–15% higher light output compared to continuous mode [3]. Also, impulse irradiation has an influence on plant physiology, biochemistry, morphology and productivity. Thus, in basil plants, impulse irradiation caused a significant increase in biomass by 47% compared to continuous light [4], while the overall rate of photosynthesis (Pn) did not significantly decrease [5]. Interesting results were obtained while wheat was irradiated with impulse light. Instead of a photoprotective reaction, an increase in photosynthetic carbon assimilation occurred [6]. The use of intermittent illumination with a frequency of 1 kHz and D = 50%, 70% and 80% and with the same intensity on wheat plants showed that at D = 80% illumination, the plants showed the highest Pn, while the concentration of chlorophyll, antioxidant capacity, yield index and the weight of a thousand grains corresponded to the variant with constant illumination [7].

The etiochloroplasts defined from wheat leaves acquired green colour with repeated flashes of light at intervals of 15 min in the dark. These plastids exhibited light-induced decolourization of chlorophyll and carotenoids, and pigment stabilization occurred after additional repeated flashes with short intervals of darkness (5 s). However, when comparing plastids with similar photochemical activity, stabilization was much lower than under constant illumination [8]. Experiments carried out on pepper plants show that the addition of impulse radiation to the flux of fluorescent lamps affects the functional state of the photosynthetic apparatus—the efficiency of electron transport by open photosystem II (PSII) reaction centres and the quantum efficiency of PSII increased. Additional impulse irradiation contributed to an increase in the biomass and content of chlorophyll in the tissues of the pepper leaf blade [9].

A study of the response of tomato leaves to impulse irradiation showed that all light generated a photosynthetic photon flux (PPF) equivalent of 50 μmol m^{-2} s^{-1} from 5000 μmol m^{-2} s^{-1} pulses for 1.5 μs followed by periods of dark 148.5 μs (D = 1%, frequency 6.6 kHz), did not affect photosynthesis compared to continuous illumination of 50 μmol m^{-2} s^{-1}. When l/d pulses were extended to 2 ms of light and 198 ms of dark (D = 1%, frequency 5 kHz), net photosynthesis decreased by a factor of two [10]. In this work, the theory was also confirmed that the pigments of the xanthophyll cycle were not affected by any of the impulse light modes.

There are quite a lot of works on the study of the effect of impulse light sources on the physiology, biochemistry, and morphology of various varieties and species of plants [11–16], but an all-round study of the effect on plant growth and their productivity has not been previously conducted. The parameters of irradiation modes that allow to reduce the consumption of electrical energy without loss in the quality of the grown products were not shown.

Another way to reduce energy costs during the artificial irradiation of plants is the use of scanning phytoemitters (Light-Mover). This is a device that combines a phyto-illuminator mounted on a rail attached to the ceiling of a greenhouse or growing room, along which the lamp is constantly moving electrically. The most popular and widespread are the scanning phytoirradiators moved linearly. There are also systems with a circular motion of the irradiators. To cover large areas, it is possible to use several parallel guides. The advantage of scanning phytoirradiators is a more uniform distribution of radiation over the growing area and a significant expansion of the irradiation zone of each phytoirradiator.

Manufacturers claim that such devices reduce lighting costs (for any type of lamp) or increase yields for equal energy costs. It has been established that shade-loving plants can quickly adapt to changing periods of diffuse and direct radiation under natural conditions [17]. Relatively few experiments have been carried out on photophilous plants that have shown the species-specific response of plants to scanning irradiation [18,19]. The results of studies on the use of impulse and scanning radiation systems in crop production are controversial and require further study. The effectiveness of such systems depends on many factors, such as plant species and variety, planting frequency, intensity and spectral composition of radiation. The purpose of this work is to study the time modes of LED irradiation (constant, impulse and scanning) of lettuce grown on racks.

The purpose of this work is to study the effect of impulse, scanning, and constant 16- and 24-h LED irradiation modes on the physiological, biochemical, and morphometric parameters of lettuce (*Lactuca sativa* L.) grown in vertical farms. In contrast to earlier studies, for the scanning mode of irradiation, we used LED irradiators, which have several advantages over gas discharge lamps. They can be placed closer to plants without causing leaf burns from excess radiation, have a longer lifespan, and are promising for vertical farms.

2. Materials and Methods

2.1. Characteristics of Lettuce Varieties

For our experiments, two varieties of lettuce were chosen—a semi-head green-leaved variety called 'Azart' and a red-leaved variety called 'Lollo Rossa'. These varieties belong to different groups in terms of colour and structure, in addition, they are grown well in hydroponic culture and have been tested by us in previous experiments [20]. The 'Azart' lettuce variety ("Prestige" company)—mid-season semi-head lettuce. The rosette of leaves is semi-erect, 25–27 cm high, and 31–32 cm in diameter. The texture of the leaf tissue is tender and semi-crispy. The mass of a head of cabbage reaches 230 g 'Lollo Rossa' lettuce variety is a mid-season variety of leaf lettuce. The rosette of leaves is semi-erect, 20 cm high, 30 cm in diameter, weighing 200–350 g. Productivity is 3 kg/m^2.

2.2. Cultivation Conditions

The research was conducted in a climate chamber for growing green crops. Plants were grown in plastic trays on racks equipped with periodic flood hydroponic systems (Figure 1a). Seeds were sown in cubes with a mineral wool substrate. To avoid the negative effect of far-red radiation on seed germination, plant lighting was turned off before germination in the climatic room. After the appearance of the cotyledons, the lighting was turned on again. After seedlings were found and the first true leaves appeared, three plants were left in each cube. For growing plants, seven light-insulated racks of three tiers each were used. The cultivation area for each light treatment was 0.68 m^2. The planting area was 50 plants per 1 m^2. The microclimate in the room was maintained by an automatic system. The day/night air temperature was 25/22 °C with a relative humidity of 75%. The ventilation system maintained the concentration of carbon dioxide and corresponded to atmospheric values of 400–450 ppm. Additional CO_2 was not used. Nutrient solutions were prepared using Flora Series® hydroponic fertilizer complex (Terra Aquatica, Fleurance, France). The electrical conductivity of the nutrient solution was maintained within 1500–1600 µS/cm.

2.3. Irradiation Conditions

Irradiation of plants was carried out by irradiators based on white light LEDs with a colour temperature of 4000 K Samsung LM281b+ 2835 (Seoul, South Korea). The specific power consumption per one shelf of the rack 0.68 m^2 was 90 W m^{-2}. The daytime integral of light (DLI) in all variants of the experiments was ~15.6 mol m^{-2} day^{-1}. Measurements of the photon flux density and the spectral composition of irradiation were carried out using an MK350D Compact Spectrometer (UPRtek Corp. Miaoli County, Taiwan). The spectral composition of the radiation is presented in Figure 1b.

On the first rack, round-the-clock irradiation of plants was implemented 24 h a day. This mode was chosen because some studies on growing plants with a long photoperiod show some potential advantages of a 24-h irradiation regimen [21,22]. The PAR irradiance was ~180 µmol s^{-1} m^{-2}. Daily electricity consumption was 5.7 kWh.

On the second rack, impulse irradiation of plants with a frequency of 1 kHz and D = 67% was implemented since early studies showed the maximum efficiency of light use in this mode of lettuce illumination [23]. The PAR irradiance (average value) was ~180 µmol s^{-1} m^{-2}. Daily electricity consumption was 6.1 kWh.

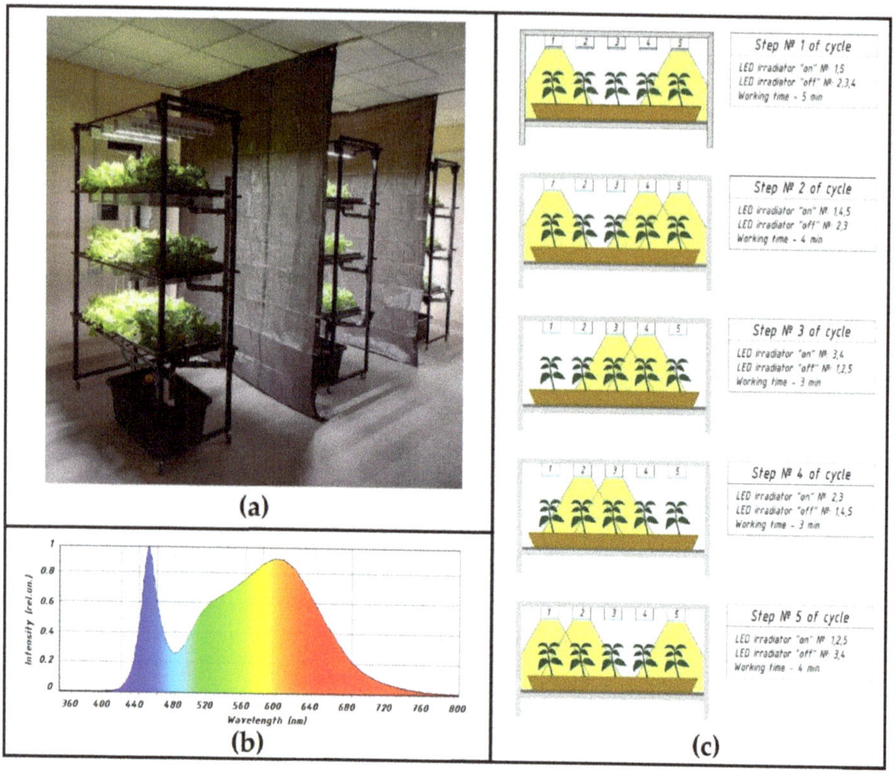

Figure 1. Climatic chamber with plants (**a**), the spectral composition of phytoirradiators radiation (**b**), operating parameters of scanning irradiation (**c**).

On the third rack, scanning irradiation of plants was implemented with five irradiators, which were switched on in pairs according to the "running lights" principle. The work cycle was divided into five steps (Figure 1c). The duration of each step was adjusted so that DLI for each plant on the shelf was ~15.6 mol m^{-2} day^{-1}. The PAR irradiation at the plant level varied from ~45 µmol s^{-1} m^{-2} to ~285 µmol s^{-1} m^{-2}. Daily electricity consumption was 5.8 kWh.

The lettuce of 'Azart' and 'Lollo Rossa' varieties were irradiated 16 h a day on the fourth rack (control variant). The PAR irradiance was ~275 µmol s^{-1} m^{-2}. Daily electricity consumption was 6.0 kWh.

2.4. Biometric Measurements of Lettuce Plants

On the 25th and 35th days of cultivation, the fresh and dry mass of the aerial parts of plants was weighed using a Sartorius LA230S balance (Laboratory Scale, Göttingen, Germany). To obtain dry matter, the selected plants were dried to constant weight in an oven at a temperature of 105 °C. The leaf area was determined on an LI-COR LI-3100 AREA METER photoplanimeter (LI-COR, Inc., Lincoln, NE, USA). The nitrate concentration was measured by an ionometric method using the ion-meter "Itan" (Tom'analit, Tomsk, Russia).

2.5. Spectrophotometric Measurements of the Pigment Composition of Leaves

Quantitative pigments were analysed by extracting them from plant tissues with solvents. The optical density of the pigment extract was determined on a SPECS SSP-705 (Moscow, Russia) spectrophotometer at wavelengths 662, 644, and 440 nm using cuvettes with an absorbing layer thickness of 10 mm. The concentration of chlorophylls a, b and carotenoids was calculated using the Holm–Wettstein formula for 100% acetone [24]. The measurements were carried out on the 25th and 35th days of cultivation.

2.6. Reflection Measurements in Lettuce Leaves

Leaf reflectance spectra were measured using a portable PolyPen RP 410 UVIS system (Photon Systems Instruments, Drásov, Czech Republic). Leaf reflectance measurements were taken 35 days after the start of cultivation. The spectra of ten plants were measured under each illumination and at each cultivation period. Three spectral measurements were carried out on different leaves of each plant.

Using the PolyPen RP 410 UVIS software, the main reflectance indices were automatically calculated. In this case, we analysed, first of all, the normalized difference vegetation index (NDVI) [25], which is a widely used indicator of the photosynthetic biomass of plants, the Zarco–Tejada–Miller index (ZMI) [26] and Gitelson–Merzlyak Indices 1 and 2 (GM1 and GM2) [27], which are highly sensitive to the content of chlorophylls, carotenoid reflectance indices 1 and 2 (CRI1 and CRI2) [28], which are sensitive to the content of carotenoids, and anthocyanin reflectance index (ARI) [29], which is sensitive to the content of anthocyanins.

2.7. Measurement of Chlorophyll Fluorescence Parameters

To measure the activity of the light stage of photosynthesis, a portable fluorimeter FluorPen FP 110/S (Photon Systems Instruments, Drásov, Czech Republic) was used to detect active chlorophyll fluorescence and its further analysis using the PAM method or the OJIP test. FluorPen FP 110/S includes a detector (PIN photodiode with a narrow band filter, working optical range from 667 to 750 nm) and a blue LED emitter (maximum about 455 nm), an ambient light sensor. To assess Fv/Fm, the leaf was preliminarily dark-adapted for at least 20 min [30].

2.8. Raman Spectroscopy of Lettuce Leaves

Analysis of the Raman spectra (RS) was carried out on the 25th day of cultivation, on undisturbed samples of living tissues of lettuce plants. The measurements were carried out on a Senterra II confocal Raman microscope (Bruker, San Jose, CA, USA) using a red (785 nm) laser. Additional processing of the samples was not carried out, the accumulation time and the number of additions did not change, and corrective substrates (SERS) were not used in the measurements. The resolution was 4 cm^{-1}. The measurements were carried out in quadruple repetition.

2.9. Statistical Data Processing

All experiments were carried out threefold. Statistical processing of measurement results and plotting were performed in Python 3.9 and MS Excel. To estimate the statistical significance by the considered parameters the Independent two-sample *T*-test at $p< 0.05$ significant levels.

3. Results and Discussion

3.1. Morphological Parameters of Lettuce Plants

According to biometric indicators, the studied lettuce varieties had different responses to irradiation regimes (Table 1). The indicator of the accumulation of a fresh mass of plants for the semi-head 'Azart' variety had the highest value with continuous illumination for 24 h, and for the 'Lollo Rossa' leaf variety—when using impulse irradiation; at the same time, the differences between these lighting options remained unreliable. The obtained results are consistent with similar studies carried out on red leaf lettuce (*Lactuca sativa* L. cv. 'Sunmang'), the biomass of which did not differ significantly from the control with continuous illumination at a duty cycle of 1 kHz (D = 75%) [2]. In general, the fresh mass of both lettuce cultivars was somewhat less under scanning irradiation, as was the case in studies conducted on rose and bluebells plants [18,19].

Table 1. Morphological parameters of lettuce plant of 'Azart' and 'Lollo Rossa' varieties on the 35th day of cultivation. Values represent mean SEM (n = 10). Letters indicate significant differences among treatment and control samples ($p < 0.05$).

Irradiation Mode	Fresh Mass, g	Dry Mass, %	Leaf Surface Area, sm^2	Stem Height, sm
		'Azart'		
Constant (24 h)	42.17 (a) ± 11.46	6.31 (a) ± 1.05	954.74 (a) ± 188.3	3.65 (ab) ± 1.28
Impulse	36.22 (a) ± 9.17	6.56 (a) ± 0.65	816.66 (a) ± 174.68	2.75 (b) ± 0.69
Scanning	35.06 (a) ± 7.07	4.87 (b) ± 0.47	869.36 (a) ± 146.38	5.69 (a) ± 1.62
Constant (16/8 h) (control)	36.77 (a) ± 7.61	5.12 (ab) ± 0.62	893.38 (a) ± 165.66	3.19 (b) ± 0.85
		'Lollo Rossa'		
Constant (24 h)	31.42 (ab) ± 7.22	6.48 (a) ± 0.55	937.51 (ab) ± 163.05	2.8 (a) ± 0.54
Impulse	40.39 (a) ± 10.2	6.53 (a) ± 0.56	1073.66 (a) ± 241.45	2.47 (a) ± 0.68
Scanning	25.76 (b) ± 5.33	5.9 (b) ± 0.33	747.13 (b) ± 145.44	1.53 (b) ± 0.29
Constant (16/8 h) (control)	32.6 (ab) ± 5.84	6.04 (ab) ± 0.55	945.78 (ab) ± 158.25	1.36 (b) ± 0.44

According to the accumulation of dry mass, both varieties showed a similar reaction depending on the lighting regime. In the control variant (16/8) and the variant with the use of scanning irradiation, the ratio of dry to fresh mass did not have statistically significant differences, however, in the variant with continuous illumination for 24 h and using the impulse irradiation mode, the accumulation of dry matter of the 'Azart' variety was significantly increased by 23% and 28%, respectively, and for 'Lollo Rossa' the increase in dry weight was not statistically significant. The studied lighting regimes had a different effect on the development of the leaf apparatus of plants of the 'Lollo Rossa' variety and did not have a significant effect on changes in the area of the leaf surface of the 'Azart' variety. The greatest increase in the leaf surface area of the 'Lollo Rossa' variety (13.5%) was observed in the variant with impulse irradiation, while scanning irradiation significantly reduced this indicator compared to the control variant of illumination (by 21%). The head lettuce variety 'Azart' in the variant with the scanning mode of lighting was significantly extended, while the height of the stem increased by 1.8 times and the rosettes of the leaves were non-compact, the plants acquired an unmarketable appearance. Studies of scanning irradiation modes on flower crops showed a different effect: when growing a rose, the height of the stem remained unchanged [18], and when growing a bell, it even significantly decreased [19]. This difference is due to the fact that for these flower crops, scanning irradiation was used as an addition to the main one, and in our case, it was the main one. A significant elongation of the stem of the 'Lollo Rossa' lettuce variety was observed in the variants with impulse and constant 24-h irradiation by 1.8 and 2 times, respectively.

3.2. Biochemical Parameters of Lettuce Plants

In the phase of active growth on the 25th day of cultivation, for 'Lollo Rossa' leaf lettuce in terms of concentration and ratio of pigments, the options with constant illumination 16/8 and impulse were optimal. The plants grown under these light modes had the highest content of total chlorophyll—1.2 mg per 1 g of fresh mass (Figure 2a). The ratio of total chlorophyll to carotenoids was more than four (Figure S1b) which corresponds to the period of active vegetation. As for the semi-head 'Azart' variety, there were no significant differences in the ratio of chlorophylls to carotenoids (Figure S1a,b), and the highest concentration of pigments was observed in the control variant.

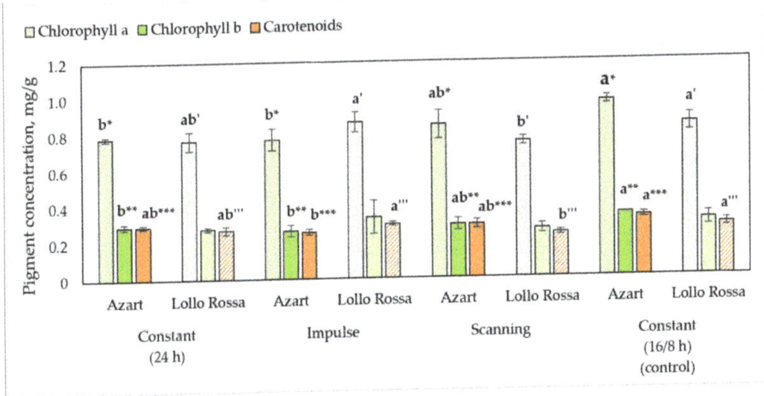

(a)

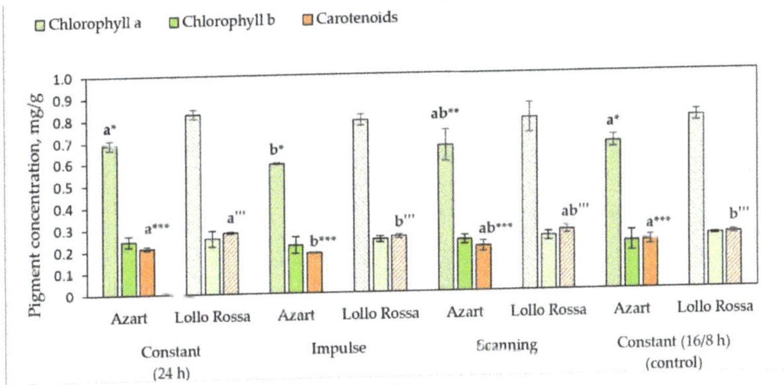

(b)

Figure 2. The concentration of pigments on the 25th (**a**) and 35th (**b**) day of cultivation. The solid colour indicates the columns of the diagram related to the 'Azart' variety; the shading columns of the diagram indicate data related to the 'Lollo Rossa' variety. Values represent mean SEM (n = 10). Letters indicate significant differences among treatment and control samples ($p < 0.05$). Asterisk sign near letters is used for the 'Azart' variety and apostrophe - for the 'Lollo Rossa' variety. One sign is used for Chlorophyll *a*, two signs – for Chlorophyll *b* and three signs – for Carotenoids.

On the 35th day of cultivation, the 'Lollo Rossa' leaf lettuce was distinguished by a more intense accumulation of pigments (Figure 2b). This may be due to the screening effect of anthocyanins [31]. It also accumulated more carotenoids in relation to the total content of chlorophylls in the variants with Constant 24 and Scanning irradiation (Figure S1d), what indicates increased stress level since carotenoids protect the photosynthetic apparatus from negative light exposure [32]. Plants of the 'Azart' variety in the variant with the use of impulse irradiation were distinguished by a reduced concentration of photosynthetic pigments. The ratio of chlorophyll a to chlorophyll b (Figure S1c) in the half-head 'Azart' variety in all variants, except for the Constant 24, was below three, which indicates a slight lack of illumination, and the total chlorophyll content ratio to the carotenoid content was the highest under Impulse and Scanning irradiation modes. Thus, judging by the pigment composition, both constant 16/8 and impulse irradiation modes are suitable for 'Lollo Rossa' leaf lettuce and Constant 24 for the 'Azart' variety.

It is known that the prolongation of the photoperiod can increase the fresh weight and chlorophyll content, meanwhile, a decrease in the photoperiod helps to reduce the content of nitrates in lettuce [33]. In this regard, we checked the content of nitrates in plants at the time of green mass harvesting. The concentration of nitrates was minimal in the control variant and the variant with the use of impulse irradiation (Figure 3). When using the scanning and constant 24 modes, the concentration of nitrates was slightly higher but did not exceed the maximum allowable concentrations for green products. For these lighting options, the absence of a night period contributes to a greater accumulation of nitrates in the leaves of crops. The observed results are a consequence of the influence of the light environment on plant metabolism and carbon and nitrogen exchange processes. The data presented in this article correlates with previous studies [33,34].

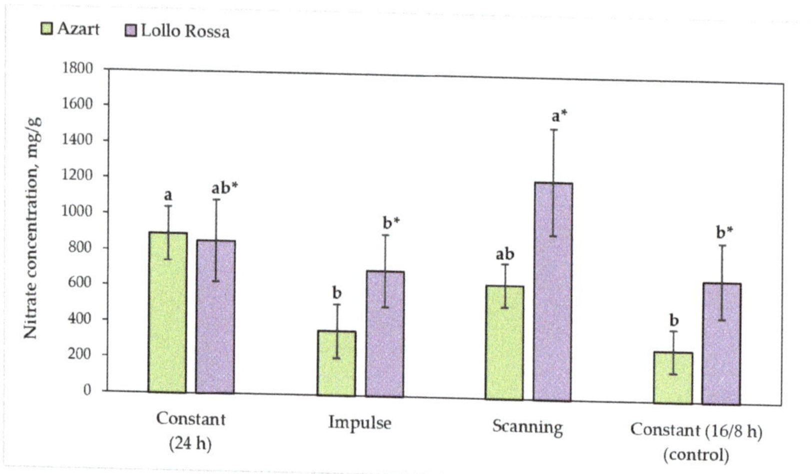

Figure 3. The concentration of nitrates on the 35th day of cultivation. Values represent mean SEM (n = 5). Letters indicate significant differences among treatment and control samples of 'Azart' variety, letters with asterisk sign–of 'Lollo Rossa' variety ($p < 0.05$).

3.3. *Chlorophyll Fluorescence Parameters*

At the next stage of the study, the influence of the illumination mode on the magnitude of the photosynthetic electron flux through photosystem II (ETR) and non-photochemical fluorescence quenching (NPQ) was studied.

It was shown (Figure 4a) that the lighting regime did not cause significant changes in the 'Azart' lettuce variety; at the same time, there was a trend towards an increase in ETR with increasing actinic light intensity in plants cultivated under continuous light conditions (control). In the 'Lollo Rossa' variety, this increase in the ETR of plants grown under continuous illumination was significant (Figure 4b). NPQ values at intermediate light intensities were reduced in both 'Azart' leaves (Figure 4c) and 'Lollo Rossa' leaves (Figure 4d); under conditions of low or high intensity of actinic light, the effect was absent. At the same time, there were practically no differences between the studied photosynthetic parameters in plants grown under other lighting conditions that were used in the work. These results are consistent with data obtained from red leaf lettuce, where there was no significant difference in the total rate of leaf photosynthesis under impulse illumination with Kf = 75% compared to continuous illumination with LEDs [2], and on green lettuce, where impulse irradiation did not significantly affect the concentration of chlorophyll and fluorescence parameters, except for the lowest frequency (0.2 Hz) [16].

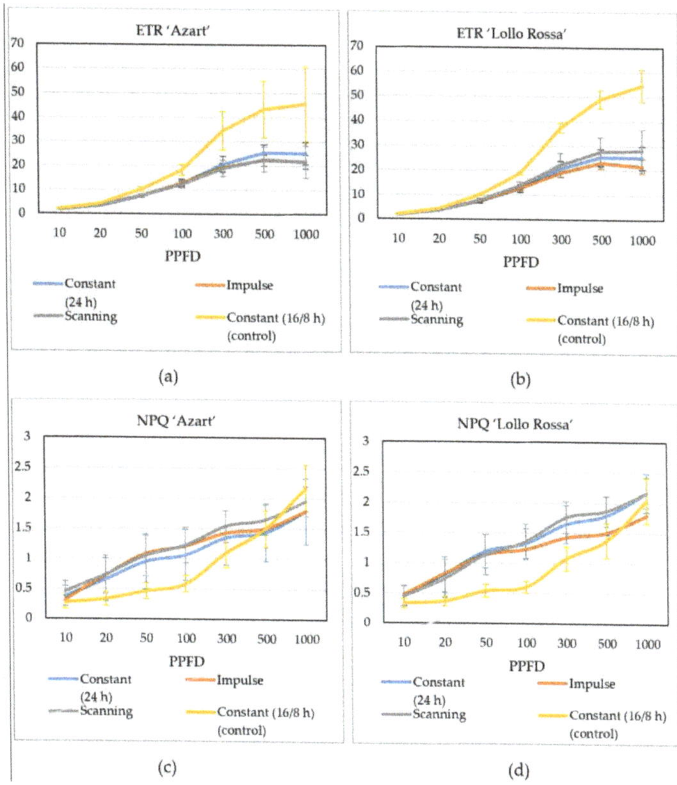

Figure 4. Dependence of the rate of electron transport (**a**,**b**) and non-photochemical (**c**,**d**) quenching of chlorophyll fluorescence on PPFD of lettuce plants on the 35th day of cultivation. Values represent mean SEM (n = 6).

The obtained results suggest that it is the high ETR value and low NPQ that can be the reason for the high productivity of lettuce plants under continuous illumination (Table 1); however, they do not explain the high efficiency of impulse lighting, since the photosynthetic parameters when grown under such conditions do not differ from the options with scanning lighting or simulated daylight hours (l/d = 16/8 h).

3.4. Study of the Influence of the Light Regime during Cultivation on Leaf Reflectance Indices

Analysis of the studied reflectance indices (NDVI, ZMI, GM1, GM2, CRI1 and CRI2) (Figure S2a,b) showed that their values were somewhat lower in the leaves of the 'Lollo Rossa' lettuce variety compared to the leaves of the 'Azart' variety. This result was in good agreement with the lower values of the chlorophyll index and the net productivity of photosynthesis in the 'Lollo Rossa' variety (Table 1 and Figure 2) since the studied indices correlate with the content of photosynthetic pigments.

Comparison of NDVI, ZMI, GM1, GM2, CRI1 and CRI2 indices in lettuce leaves of the same variety grown under different lighting conditions did not show statistically significant differences between the studied groups. At the same time, with the use of impulse and scanning irradiation, the 'Azart' variety showed a tendency to decrease in several indices sensitive to the concentration of chlorophyll in plants (NDVI, ZMI, GM1 and GM2), which is consistent with a decrease in the content of chlorophyll (Figure 2). Similar trends were also observed for carotenoid-sensitive indices CRI1 and CRI2 (Figure S2b), which correlate with changes in carotenoid concentration (Figure 2).

Even though the changes in the studied reflectance indices were not statistically significant, the observed trends corresponded well to changes in the content of photosynthetic pigments under different lighting conditions: a decrease in the content of chlorophyll a was accompanied by a decrease in the reflectance indices NDVI [25], ZMI [26], GM1 and GM2 [27], while a decrease in the content of carotenoids led to a decrease in CRI1 and CRI2 [28]. At the same time, small values and low significance of changes showed a relatively weak effect of the lighting regime on the reflectance spectrum of the sheet; Given the more pronounced differences in indices between the studied lettuce varieties (Figure S2a,b), the variety factor seems to be more significant for reflectance indices than the light regime during cultivation. On the other hand, Table 1 and Figure 2 show that under different growing regimes, differences in morphometric parameters of plants were observed, which suggests that such a difference may be associated with a change in the anatomical parameters of the leaf. Such features can significantly depend on the illumination parameters [35,36] and modify the reflectivity of the sheet [37] by decreasing or increasing changes in the reflectance indices.

For the "Lollo Rossa" lettuce variety, which has a high content of anthocyanins, the anthocyanin content index (ARI) was additionally measured (Figure 5). It was shown that the impulse lighting mode led to the increase in this index and has a positive effect on the anthocyanin concentration in lettuce leaves.

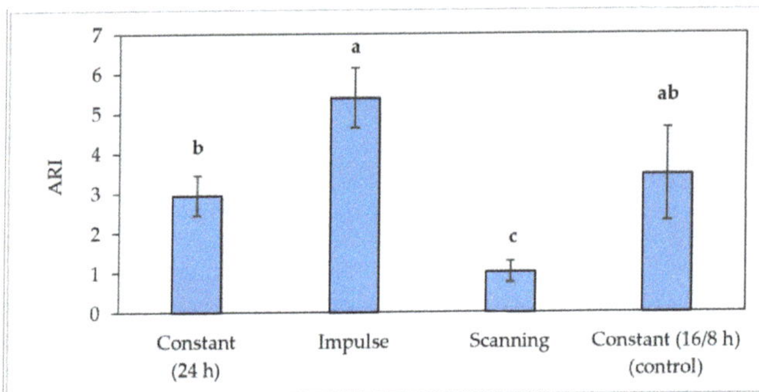

Figure 5. Anthocyanin content index (ARI) of lettuce plants on the 35th day of cultivation. Values represent mean SEM (n = 10). Letters indicate significant differences among treatment and control samples ($p < 0.05$).

3.5. Raman Spectroscopy of Lettuce Leaves

Usually, as a plant culture in Raman spectroscopy, *Lactuca sativa* is used to determine toxicants, plastics and heavy metals [38–40]. To identify the optimum points for collecting the Raman spectrum for both varieties, spectral data were collected from the inner space of the cell, the centre of the open stomata, and its valve. Based on the data obtained, the most informative were the spectra of the intracellular space (Figure 6).

Most often, peaks associated with pigments, especially with carotenoids and chlorophylls, can be detected in the spectra of plant samples. The main pigment peaks are observed in the range of 1500–1550 and 1150–1170 cm^{-1}, this part of the range is associated with in-phase stretching oscillations C=C and C-C of the polyene chain. In addition, swing patterns in the plane of CH_3 groups attached to the polyene chain can be identified as peaks of average intensity in the range of 1000–1020 cm [41]. Assessing the average spectra of 'Azart' and 'Lola Rossa' lettuce varieties, it should be noted that the overall colour and content of pigments significantly affected the data: the leaves of plants of the 'Lola Rossa' variety showed a higher and noisier spectrum, with isolated peaks characteristic of photosynthetic pigments and organic acids (Figure S3). In general, the intensity of the red-leaf 'Lola Rossa' lettuce variety detected by CR was more than two times higher than that of the green-leaf 'Azart' variety, which is associated with an increased content of carotenoids and anthocyanins (Figure 2), which have the properties to raise the overall intensity of the spectrum. A change in the illumination mode had little effect on the cell biochemistry reflected in the spectral data. This was more pronounced in the leaves of 'Lola Rossa' variety lettuce, which is explained by a change in the concentration of anthocyanins with a change in daylight hours (Figure 7).

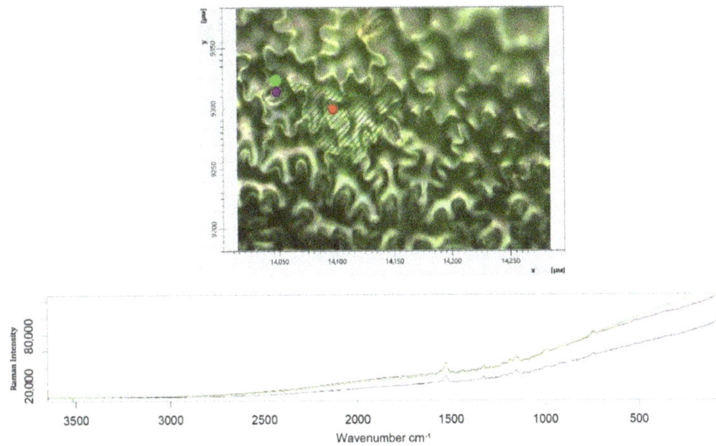

Figure 6. Measurement and selection of points on a lettuce leaf of the 'Azart' variety on the 35th day of cultivation, where red is the intracellular space, purple is the centre of the stomata, and green is the stomatal valve.

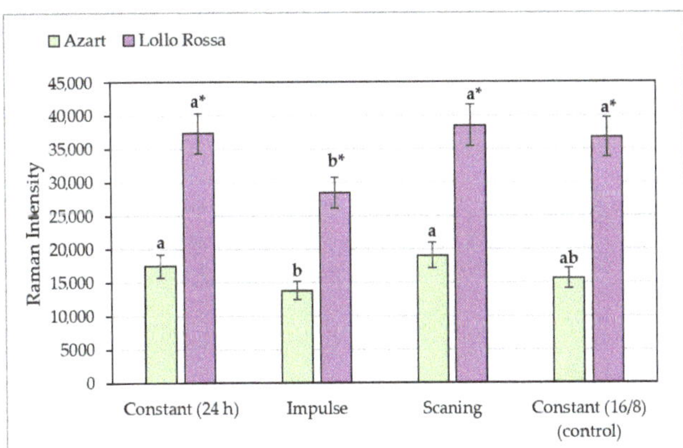

Figure 7. Varieties and variations in average intensity of significant peaks of the 'Azart' variety and 'Lola Rossa' on the 35th day of cultivation. Values represent mean SEM (n = 5). Letters indicate significant differences among treatment and control samples of 'Azart' variety, letters with asterisk sign–of 'Lollo Rossa' variety ($p < 0.05$).

As a result of the analysis of various types of lighting, it was found that the most pronounced spectra are plants grown under the scanning and constant type of lighting, which indicates a greater overall biochemical activity, expressed in an increase in the concentrations of pigments, sugars and organic acids in plant cells. The spectral data for both varieties of lettuce correlate with each other (Figure 7).

4. Conclusions

For the semi-head 'Azart' lettuce variety, the largest accumulation of the fresh mass of the plant was recorded on the stand with constant irradiation (24 h), and the smallest under scanning illumination. For the red-leaved 'Lollo Rossa' lettuce variety, the accumulation of fresh mass under constant (24 h), impulse and constant (16/8 h) were approximately the same values with a slight advantage in impulse irradiation, and the smallest value was also recorded for the scanning variant. From the point of view of qualitative analysis, the highest concentrations of carotenoids in the 'Lollo Rossa' lettuce variety were found for constant (24 h) and scanning irradiation, since carotenoids protect the photosynthetic apparatus from negative and intense light exposure. The highest value of chlorophyll content was found for plant variants under constant (16/8 h) and impulse irradiation. It was shown that impulse lighting during plant cultivation led to a statistically significant increase in the anthocyanin content index (ARI), which indicates a positive effect of impulse lighting on the concentration of anthocyanins in lettuce leaves. For the 'Azart' lettuce variety, the minimum value of the pigment concentration was found for impulse irradiation, for constant (24 h), scanning and constant (16/8) irradiation, the obtained values did not differ significantly. Thus, judging by the accumulation of fresh mass and the pigment composition, constant (16/8 h) and impulse irradiation is suitable for the 'Lollo Rossa' lettuce variety.

For the semi-head 'Azart' variety, the options with constant (24 h) and impulse irradiation are optimal. Further research will be aimed at researching and evaluating the modes and methods for introducing impulse irradiation for growing leafy crops in greenhouses.

Supplementary Materials: The following supporting information can be downloaded at: https://www.mdpi.com/article/10.3390/agriculture12121988/s1, Figure S1: The ratio of the concentrations of chlorophyll a to chlorophyll b on the 25th day (a) and the 35th day (c) of 'Azart' and 'Lollo Rossa' lettuce variety cultivation; the ratio of the concentrations of total chlorophyll to carotenoids on the 25th day (b) and the 35th day of cultivation (d) using different LED irradiation time modes; Figure S2: Vegetation indices of lettuce leaves NDVI, ZMI, GM1, GM2 (a) and CRI1, CRI2 (b) of 'Azart' and 'Lollo Rossa' varieties on the 35th day of cultivation using different LED irradiation time modes. Values represent mean SEM (n = 10); Figure S3: Comparison of normalized Raman spectra of 'Azart' (a) and 'Lola Rossa' (b).

Author Contributions: Conceptualization, N.O.C., A.A.S. and N.A.S.; methodology, A.A.S., A.S.D. and N.O.C.; formal analysis, Y.A.P. and V.S.; investigation, A.A.S., N.A.S. and M.M.G.; resources, A.A.S., A.S.D. and N.O.C.; writing—original draft preparation, A.A.S. and N.A.S.; writing—review and editing, N.O.C., V.V. and V.S.; visualization, Y.A.P.; supervision, V.P.; project administration, A.S.D. All authors have read and agreed to the published version of the manuscript.

Funding: This investigation was funded by the Ministry of Science and Higher Education of the Russian Federation for large scientific projects in priority areas of scientific and technological development, contract no. 075-15-2020-774.

Institutional Review Board Statement: Not applicable.

Informed Consent Statement: Not applicable.

Data Availability Statement: The data presented in this study are available upon request from the corresponding author.

Conflicts of Interest: The authors declare no conflict of interest. The funders had no role in the design of the study; in the collection, analyses, or interpretation of data; in the writing of the manuscript; or in the decision to publish the results.

References

1. Olvera-Gonzalez, E.; Escalante-Garcia, N.; Myers, D.; Ampim, P.; Obeng, E.; Alaniz-Lumbreras, D.; Castaño, V. Pulsed LED-Lighting as an Alternative Energy Savings Technique for Vertical Farms and Plant Factories. *Energies* **2021**, *14*, 1603. [CrossRef]
2. Son, K.H.; Jeon, Y.M.; Oh, M.M. Application of supplementary white and impulse light-emitting diodes to lettuce grown in a plant factory with artificial lighting. *Hortic. Environ. Biotechnol.* **2016**, *57*, 560–572. [CrossRef]
3. Skripnichenko, A. Pulsed power supply of Cree XLamp LEDs with increased current. *Semicond. Lighting Eng.* **2011**, *1*, 16–19.

4. Avgoustaki, D.D.; Bartzanas, T.; Xydis, G. Minimising the energy footprint of indoor food production while maintaining a high growth rate: Introducing disruptive cultivation protocols. *Food Control* **2021**, *130*, 108290. [CrossRef]
5. Avgoustaki, D.D.; Li, J.Y.; Xydis, G. Basil plants grown under intermittent light stress in a small-scale indoor environment: Introducing energy demand reduction intelligent technologies. *Food Control* **2020**, *118*, 107389. [CrossRef]
6. Li, Y.T.; Yang, C.; Zhang, Z.S.; Zhao, S.J.; Gao, H.Y. Photosynthetic acclimation strategies in response to intermittent exposure to high light intensity in wheat (*Triticum aestivum* L.). *Environ. Exp. Bot.* **2021**, *181*, 104275. [CrossRef]
7. Dong, C.; Shao, L.Z.; Liu, G.H.; Wang, M.J.; Liu, H.; Xie, B.Z.; Li, B.W.; Fu, Y.M.; Liu, H. Photosynthetic characteristics, antioxidant capacity and biomass yield of wheat exposed to intermittent light irradiation with millisecond-scale periods. *J. Plant Physiol.* **2015**, *184*, 28–36. [CrossRef] [PubMed]
8. Mathieu, Y. Photochemical activities of etiochloroplasts isolated from plants greened under intermittent light.2. Photobleaching of pigments. *Physiologia Plantarum* **1976**, *37*, 55–61. [CrossRef]
9. Kotik, M.I.; Andriychuk, V.A.; Kostik, L.N.; Gerts, N.V.; Gerts, A.I. Pulse light stimulation of pepper sprouts cultivation. *Light Eng.* **2019**, *27*, 84–91.
10. Tennessen, D.J.; Bula, R.J.; Sharkey, T.D. The efficiency of photosynthesis in continuous and impulse light emitting diode irradiation. *Photosynth. Res.* **1995**, *44*, 261–269. [CrossRef]
11. Kurata, H.; Mochizuki, A.; Okuda, N.; Seki, M.; Furusaki, S. Intermittent light irradiation with second- or hour-scale periods controls anthocyanin production by strawberry cells. *Enzyme Microb. Technol.* **2000**, *26*, 621–629. [CrossRef]
12. Pinker, I.M. Chopper-light for shoot cultures. *Acta Hortic.* **2000**, *520*, 195–202. [CrossRef]
13. Chow, W.S.; Funk, C.; Hope, A.B. Greening of intermittent-light-grown bean plants in continuous light: Thylakoid components in relation to photosynthetic performance and capacity for photoprotection. *Ind. J. Biochem. Biophys.* **2000**, *37*, 395–404.
14. Grobbelaar, J.U.; Nedbal, L.; Tichý, V. Influence of high frequency light/dark fluctuations on photosynthetic characteristics of microalgae photoacclimated to different light intensities and implications for mass algal cultivation. *J. Appl. Phycol.* **1996**, *8*, 335–343. [CrossRef]
15. Jao, R.C.; Fang, W. Effects of frequency and duty ratio on the growth of potato plantlets in vitro using light-emitting diodes. *Hortsci.* **2004**, *39*, 375–379. [CrossRef]
16. Kanechi, M.; Maekawa, A.; Nishida, Y.; Miyashita, E. Effects of impulse lighting based light-emitting diodes on the growth and photosynthesis of lettuce leaves. *Acta Hort.* **2016**, *1134*, 207–214. [CrossRef]
17. Chazdon, R.L.; Pearcy, R.W. The Importance of Sunflecks for Forest Understory. *Plants. BioScience.* **1991**, *41*, 760–766. [CrossRef]
18. Marissen, N.; Snel, J.; Elings, A.; Warmenhoven, M. Mobile light in roses. *Acta Hortic.* **2006**, *711*, 189–194. [CrossRef]
19. Blom, T.J.; Zheng, Y.B. The effect of moving high intensity lights on potted Campanula haylodgensis. *Acta Hortic.* **2006**, *711*, 157–164. [CrossRef]
20. Semenova, N.A.; Smirnov, A.A.; Grishin, A.A.; Pishchalnikov, R.Y.; Chesalin, D.D.; Gudkov, S.V.; Chilingaryan, N.O.; Skorokhodova, A.N.; Dorokhov, A.S.; Izmailov, A.Y. The effect of plant growth compensation by adding silicon-containing fertilizer under light stress conditions. *Plants* **2021**, *10*, 1287. [CrossRef]
21. Velez-Ramirez, A.I.; van Ieperen, W.; Vreugdenhil, D.; Millenaar, F.F. Plants under continuous light. *Trends Plant Sci.* **2011**, *16*, 310–318. [CrossRef] [PubMed]
22. Proietti, S.; Moscatello, S.; Riccio, F.; Downey, P.; Battistelli, A. Continuous Lighting Promotes Plant Growth, Light Conversion Efficiency, and Nutritional Quality of *Eruca vesicaria* (L.) Cav. in Controlled Environment with Minor Effects Due to Light Quality. *Front. Plant Sci.* **2021**, *12*, 730119. [CrossRef] [PubMed]
23. Son, K.H.; Lee, S.R.; Oh, M.M. Comparison of Lettuce Growth under Continuous and Impulse Irradiation Using Light-Emitting Diodes. *Hortic. Sci. Technol.* **2018**, *36*, 542–551. [CrossRef]
24. Wettstein, D. Chlorophyll-letale und der submikroskopische Formwechsel der Plastiden. *Exp. Cell Res.* **1957**, *12*, 427–434. [CrossRef]
25. Rouse, J.W.; Haas, H.R.; Deering, D.W.; Schell, J.A.; Harlan, J.C. *Monitoring the Vernal Advancement and Retrogradation (Green Wave Effect) of Natural Vegetation*; NASA/GSFC Type III Final Rep.; The Goddard Space Flight Center: Greenbelt, MD, USA, 1974; 371p.
26. Zarco-Tejada, P.J.; Berjón, A.; López-Lozano, R.; Miller, J.R.; Martín, P.; Cachorro, V.; González, M.R.; de Frutos, A. Assessing vineyard condition with hyperspectral indices: Leaf and canopy reflectance simulation in a row-structured discontinuous canopy. *Remote Sens. Environ.* **2005**, *99*, 271–287. [CrossRef]
27. Gitelson, A.A.; Merzlyak, M.N. Remote estimation of chlorophyll content in higher plant leaves. *Int. J. Remote Sens.* **1997**, *18*, 2691–2697. [CrossRef]
28. Gitelson, A.A.; Zur, Y.; Chivkunova, O.B.; Merzlyak, M.N. Assessing carotenoid content in plant leaves with reflectance spectroscopy. *Photochem. Photobiol.* **2002**, *75*, 272–281. [CrossRef]
29. Gitelson, A.A.; Merzlyak, M.N.; Chivkunova, O.B. Optical properties and nondestructive estimation of anthocyanin content in plant leaves. *Photochem. Photobiol.* **2001**, *74*, 38–45. [CrossRef]
30. Yudina, L.; Sukhova, E.; Mudrilov, M.; Nerush, V.; Pecherina, A.; Smirnov, A.A.; Dorokhov, A.S.; Chilingaryan, N.O.; Vodeneev, V.; Sukhov, V. Ratio of Intensities of Blue and Red Light at Cultivation Influences Photosynthetic Light Reactions, Respiration, Growth, and Reflectance Indices in Lettuce. *Biology* **2022**, *11*, 60. [CrossRef]

31. Proshkin, Y.A.; Smirnov, A.A.; Semenova, N.A.; Dorokhov, A.S.; Burynin, D.A.; Ivanitskikh, A.S.; Panchenko, V.A. Assessment of Ultraviolet Impact on Main Pigment Content in Purple Basil (*Ocimum basilicum* L.) by the Spectrometric Method and Hyperspectral Images Analysis. *Appl. Sci.* **2021**, *11*, 8804. [CrossRef]
32. Chivkunova, O.; Dalke, I.; Dymova, O.; Elsakov, V.; Fiedor, L.; Fujita, M.; Garmash, E.; Hasanuzzaman, M.; Jedynak, P.; Jemiola-Rzeminska, M.; et al. *Photosynthetic Pigments: Chemical Structure, Biological Function and Ecology*; Komi Scientific Centre of the Ural Branch of the Russian Academy of Sciences: Syktyvkar, Russia, 2014.
33. Song, J.; Huang, H.; Song, S.; Zhang, Y.; Su, W.; Liu, H. Effects of Photoperiod Interacted with Nutrient Solution Concentration on Nutritional Quality and Antioxidant and Mineral Content in Lettuce. *Agronomy* **2020**, *10*, 920. [CrossRef]
34. Shao, M.J.; Liu, W.K.; Zhou, C.B.; Wang, Q.; Li, B.S. Alternation of temporally overlapped red and blue light under continuous irradiation affected yield, antioxidant capacity and nutritional quality of purple-leaf lettuce. *Sci. Hortic.* **2022**, *295*, 110864. [CrossRef]
35. Hogewoning, S.W.; Trouwborst, G.; Maljaars, H.; Poorter, H.; van Ieperen, W.; Harbinson, J. Blue light dose-responses of leaf photosynthesis, morphology, and chemical composition of Cucumis sativus grown under different combinations of red and blue light. *J. Exp. Bot.* **2010**, *61*, 3107–3117. [CrossRef] [PubMed]
36. Hogewoning, S.W.; Wientjes, E.; Douwstra, P.; Trouwborst, G.; van Ieperen, W.; Croce, R.; Harbinson, J. Photosynthetic quantum yield dynamics: From photosystems to leaves. *Plant Cell.* **2012**, *24*, 1921–1935. [CrossRef] [PubMed]
37. Kior, A.; Sukhov, V.; Sukhova, E. Application of Reflectance Indices for Remote Sensing of Plants and Revealing Actions of Stressors. *Photonics* **2021**, *8*, 582. [CrossRef]
38. De Matos, Y.M.L.S.; Vasconcelos, D.L.M.; Barreto, A.C.H.; Rocha, J.E.; De Araújo-Neto, J.B.; Campina, F.F.; Da Silva, M.M.C.; Al Yafawi, T.T.; Sobral-Souza, C.E.; Pinheiro, J.C.A.; et al. Protection against the Phytotoxic Effect of Mercury Chloride by Catechin and Quercetin. *J. Chem.* **2022**, *2022*, 3770935. [CrossRef]
39. De Paiva Pinheiro, S.K.; Rangel Miguel, T.B.A.; Chaves, M.D.M.; Barros, F.C.D.F.; Farias, C.P.; De Moura, T.A.; Ferreira, O.P.; Paschoal, A.R.; Souza Filho, A.G.; De Castro Miguel, E. Silver nanoparticles (AgNPs) internalization and passage through the Lactuca sativa (Asteraceae) outer cell wall. *Fun. Plant Biol.* **2021**, *48*, 1113–1123. [CrossRef]
40. Uzu, G.; Sobanska, S.; Sarret, G.; Muñoz, M.; Dumat, C. Foliar Lead uptake by lettuce exposed to atmospheric fallouts. *Environ. Sci. Technol.* **2010**, *44*, 1036–1042. [CrossRef]
41. Baranska, M.; Schulz, H. Application of infrared and Raman spectroscopy for analysis of selected medicinal and spice plants: A review. *J. Med. Spice Plants* **2006**, *11*, 72–80.

agriculture

MDPI

Article

Research on Characteristics of Airway Pressure Loss in Seeding-Wheel-Type Pneumatic Seeder

Xirui Zhang [1,2], Zhentuo Wen [1], Qingjie Wang [3], Hongwen Li [3], Zhifu Zhang [1] and Junxiao Liu [1,2,*]

1 School of Mechanical and Electrical Engineering, Hainan University, Haikou 570228, China
2 Sanya Nanfan Research Instituteof Hainan University, Sanya 572025, China
3 College of Engineering, China Agricultural University, Beijing 100083, China
* Correspondence: 995229@hainanu.edu.cn

Abstract: In order to optimize the parameters of the mechanism, reduce the pressure loss and improve the efficiency of pneumatic utilization, the principles and types of pneumatic loss in different areas were defined, and the key parameters, including the diameter of the horizontal air pipe, the angle of the air pipe, and the diameter of the negative pressure aperture, which affect the pressure loss, were analyzed. In addition, an orthogonal simulation experiment was carried out using Fluent software, and determined the best parameter combination, which is as follows: the diameter of the horizontal air pipe must be 15 mm, the angle of the air pipe must be 105°, and the diameter of the negative pressure aperture must be 34 mm. It can be concluded from the simulation results that the average airflow velocity of the seed-sucking hole is 102.59 $m \cdot s^{-1}$, the minimum airflow velocity of the seed-sucking hole is 101.58 $m \cdot s^{-1}$, and the airflow velocity standard deviation of the seed-sucking hole is 0.54 $m \cdot s^{-1}$. The trend of the test results was consistent with the simulation results, which verified the reliability of the numerical analysis. These results will provide theoretical guidance for research on the obstruction effect of complex airways on airflow.

Keywords: pneumatic seeder; pressure loss characteristics; complex airway; airway optimization

Citation: Zhang, X.; Wen, Z.; Wang, Q.; Li, H.; Zhang, Z.; Liu, J. Research on Characteristics of Airway Pressure Loss in Seeding-Wheel-Type Pneumatic Seeder. *Agriculture* **2022**, *12*, 2021. https://doi.org/10.3390/agriculture12122021

Academic Editors: Vadim Bolshev, Vladimir Panchenko and Alexey Sibirev

Received: 1 November 2022
Accepted: 23 November 2022
Published: 26 November 2022

1. Introduction

Single seed precision seeding refers to a type of seeding technology that separates crop seeds one by one and controls their distribution position precisely on the seedbed. This technology encounters difficulties in the separation of seeds and in preventing the formed uniform seed flow from being re-dispersed on the inner wall of the seed pipe and seedbed [1–3]. To solve the above problems, a seeding-wheel-type pneumatic seeder that works via pneumatic single-grain separation and orderly and stable casting in a low position was designed. However, its internal airway structure is complex with many branches, has a dense arrangement, variable cross-sectional areas, and displays long and narrow bending, which will inevitably cause pressure losses, and lead to further problems, such as low pneumatic utilization efficiency and energy waste [4].

Scholars have conducted extensive research on the airflow field associated with the pneumatic seed-metering mechanism [5–8]. For example, Yazgi et al. [9,10] used the CFD method to study the effect of a vacuum plate with different numbers of holes on the uniformity of seed spacing. The highest performance was reported when 26 and 36 holes were used for cotton and corn, respectively. Ghafori et al. [11] analyzed the pressure drop function during the suction and transport of corn and barley. The lowest pressure drop for corn and barley occurred at the air velocity of 20 $m \cdot s^{-1}$ and 15 $m \cdot s^{-1}$, respectively. Yang Shandong et al. [12–14] used the large eddy simulation method with Fluent software to analyze the flow field of the positive-pressure air chamber inside a seed-metering device, and studied the variation law of the internal flow field of the seed-metering device under different parameters. It was shown that the seeding performance was better when the plate speed was lower than 18 $r \cdot min^{-1}$. Zhao Zhan et al. [15–17] used Fluent software to calculate the force of seeds in the real seed suction airflow field and high-speed photography technology to obtain the seed transient adsorption and falling trajectory. The results showed that the average seed-spacing interval error reached its minimum value at 1.5 kPa pressure and a −5° angle and that the error increased almost linearly with increasing cylinder rotational speed. Xiaolong Lei et al. [18,19] carried out a numerical simulation of seed motion in the distribution head and tracked the seed migration trajectory and distribution behavior. The simulation results showed that, when the streamlined angle increased from 10° to 50°, the variation coefficient of seed distribution decreased initially and then increased for an inlet diameter of 20 mm and an airflow velocity of 20 $m \cdot s^{-1}$. Song shi et al. [20] calculated the gas–solid two-phase flow of a seed-metering device by introducing the pressure gradient force model into the coupled analysis of EDEM-CFD. The results were given as follows: the curvature coefficient of the seed guide groove curve was 0.265, the depth of the seed guide groove was 2.57 mm, and the slope angle of the seed guide groove was 15.33°.

The above scholars all used the computational fluid dynamics method to simulate the internal flow field that is difficult to measure, and the simulation results were similar to the experimental results, which shows that it is an effective method to study the pneumatic seed-metering mechanism. However, the above research mainly focuses on the large-diameter pneumatic and intensive conveying air duct or the cylindrical chamber inside the seed-metering device, which is characterized by a spacious air duct and simple structure. There is no relevant research on long, narrow and complex air ducts with variable diameters.

Therefore, we focused on the fluid domain of seeding-wheel-type pneumatic seeders and used the computational fluid dynamics method to explore the types and principles of the pressure loss in each area. The key parameters that affect the pressure loss were identified and the airflow field distribution and the airflow motion law in the seed-placing device were defined to determine the best structure and working parameters. The results show that it is possible to weaken the obstruction of the complex airway structure to the airflow, reduce the pressure loss, and improve pneumatic utilization and the effect of sowing. The results will also provide theoretical guidance for pressure drop analysis of complex airways.

2. Structure and Fluid Analysis

2.1. The Structure and Working Principle of Seeding-Wheel-Type Pneumatic Seeder

Seeding wheels are the main carriers used for picking seeds, transporting seeds, and casting seeds. Their structure and installation are shown in Figure 1 [21]. Several seed-sucking holes are evenly distributed along the edge of the seeding wheel and are connected to the internal pneumatic distribution center through a long and narrow airway with variable diameters. The upper-end face of the pneumatic distribution center is connected to a rubber sealing ring fixed with a pneumatic distribution cover, and they are press-fitted and sealed. The inner air cavity of the rubber sealing ring is connected to the negative pressure aperture to provide negative pressure force for the seed-sucking hole. During operation, the pneumatic distribution cover remains stationary, the seeding wheel rotates with the ground wheel, and the two are connected by thin-walled bearings. A spring in the locking mechanism is used to press the seed wheel and air distribution cover together. The outer wall of the seeding wheel, which is installed on both sides of the edge of the seeding wheel, flattens the soil with a certain force, which makes a difference to the depth limit and level of flattening. The seed-sucking boss can extend to the lower soil layer to complete the casting of seeds at a suitable position.

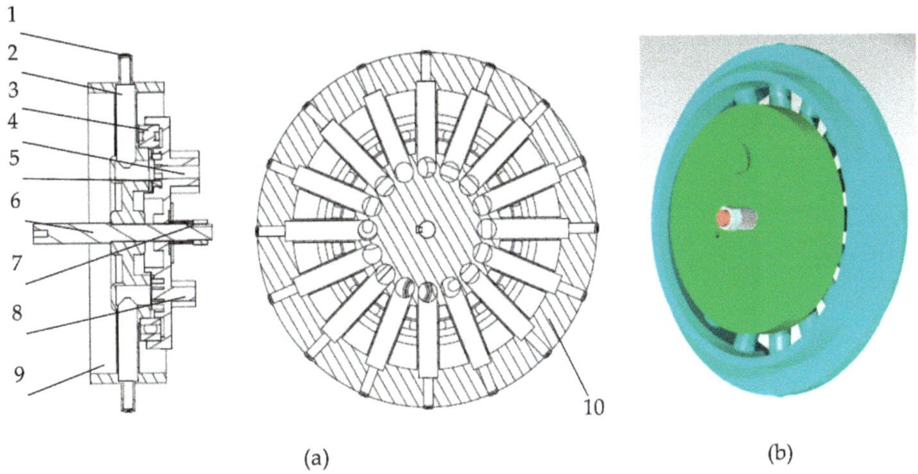

Figure 1. Schematic diagram and three-dimensional modeling of seeding wheel structure. (**a**) Schematic diagram of seeding wheel structure: (1) seed-sucking hole; (2) airway; (3) thin-walled bearing; (4) negative pressure aperture; (5) rubber sealing ring; (6) seeding axle; (7) locking mechanism; (8) positive pressure aperture; (9) outer wall; (10) seed-sucking boss; (**b**) three-dimensional modeling of seeding wheel structure.

2.2. Fluid Domain Modeling

The function of the airway inside the seeding wheel is to transport the fan's pneumatic supply to the seed-sucking hole, and try to reduce the pressure loss during conveying [16,17,22], so that the seed-sucking hole has enough airflow velocity and negative pressure to perform its function. In addition, the airway should demonstrate good pneumatic distribution, making the difference between negative pressure and airflow velocity in each seed-sucking hole small, which is not significantly related to their location.

The negative pressure fluid domain inside the seeding device is the inner space enclosed by the seeding wheel, rubber sealing ring, and pneumatic distribution cover. The seeding wheel and the pneumatic distribution cover will rotate during operation, thereby dividing the entire negative pressure fluid domain into two parts, the static part and the moving part. The static part is the upper fluid domain surrounded by the pneumatic distribution cover, which is the green part shown in Figure 1b. The moving part is the under-fluid domain surrounded by the seeding wheel, which is the blue part shown in Figure 1b. The seeding wheel has 16 sub-airways that are not connected, as shown in Figure 1a. Because the inner pneumatic distribution cover is divided into areas with air and areas without air, the seed-sucking hole at the end of the sub-airway has negative pressure only when the sub-airway is connected to areas with air. These two parts will rotate during operation, which contributes to the airflow's on–off control by switching the corresponding position. Using Space Claim software to extract the fluid domain, a simplified 3D model of the seeding device was imported into the workbench of ANSYS, as shown in the central part of Figure 2.

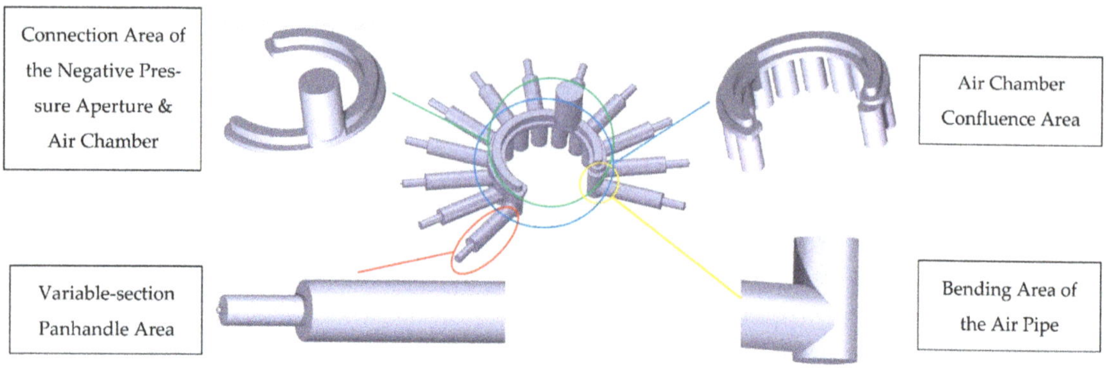

Figure 2. Fluid domain division.

According to Figure 2, there are 12 lower sub-air channels connected to the upper air chamber, and there is a lower sub-air channel at the edge of the domain (only about half of this is connected to the upper air chamber at the top). The analysis below focuses on the airway in the connected position only. The structure of the negative pressure fluid domain is complex, involving various forms of airway structural changes, and it is difficult to conduct an overall analysis. Therefore, the negative pressure fluid domain is divided into the following four areas according to the different types of aerodynamic changes: the variable-section panhandle area, the bending area of the air pipe, the air chamber confluence area, and the connection area of the negative pressure aperture and air chamber. These are shown in Figure 2.

2.3. *Analysis of Pressure Loss in Airway*

2.3.1. Aerodynamic Analysis of Variable-Section Panhandle Area

The energy loss of the fluid can be divided into the along-path loss and the local loss [23] according to the difference in the airway geometrical boundary that causes the fluid energy loss. The airflow velocity inside the airway is much less than the sound velocity, so air can be regarded as an incompressible fluid. The pneumatic input is negative pressure. The entire pneumatic system's inlet is the 12 seed-sucking holes, and the outlet is a negative pressure aperture. The analyses of the four areas were conducted using the airflow inlet. The cross-sectional view of the variable-section panhandle area is shown in Figure 3.

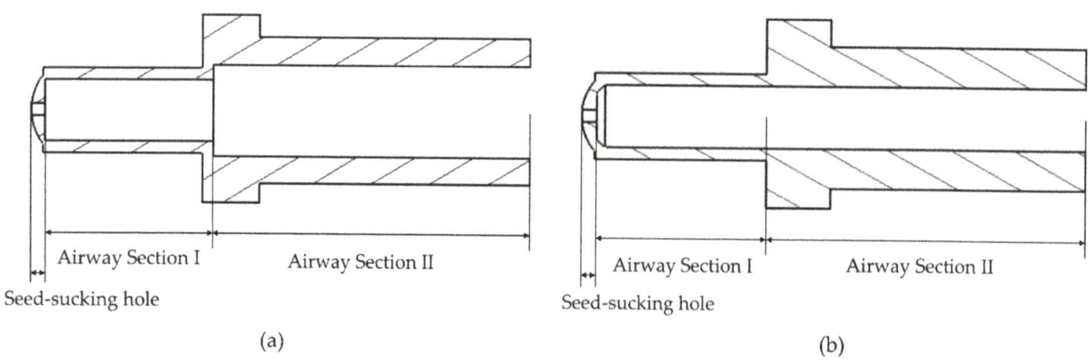

Figure 3. Sectional view of the variable-section panhandle area: (**a**) the section with an unequal diameter; (**b**) the section with an equal diameter.

The airway structure in this area is shown in Figure 3a. There are three cylindrical cavities arranged from small to large. The first cavity is the seed-sucking hole, which is also the executed structure for the terminal, and its cross-sectional area is mainly related to the performance of seed absorption. Its structure is not studied in this paper. The second cavity is the airway section I, which is contained in the seed-sucking boss. Its cross-sectional area is limited by the size of the seed-sucking boss and has a maximum size of 10 mm. The third cavity is the airway section II, and its cross-sectional area is adjustable. When the diameters of airway section I and airway section II are equal, the airway structure shown in Figure 3b can be observed.

The variable-section panhandle area mainly involves two forms of pressure loss. When the air passes from the seed-sucking hole to the airway section I and from the airway section I to the airway section II (unequal diameter), the type of aerodynamic change that occurs is the sudden expansion of the pipe. It is mainly caused by the local pressure loss; while in the airway section I and II, it is caused by the pressure loss along the path.

2.3.2. Aerodynamic Analysis of Variable Diameter Transition

The type of aerodynamic change in the variable diameter is caused by diffusion flow [24], which can be divided into the following two forms: sudden and gradual expansion. The space with a variable diameter is narrow, and it is difficult to make a tapered airway with a small angle and a long lead. Therefore, this is regarded as the sudden expansion type. After the cross-sectional area of the sudden expansion pipe changes, the fluid will move away from the wall and form a vortex, resulting in the loss of airflow energy. This part of the local energy loss can be calculated as follows [25]:

$$p_{j1} = (1 - \frac{A_1}{A_2})\frac{\rho v^2}{2} \tag{1}$$

where p_{j1} is the energy loss per unit volume of the gas in the form of sudden expansion (pressure loss), Pa; ρ is the medium density, $kg \cdot m^{-3}$; A_1 is the inlet cross-sectional area, m^2; A_2 is the outlet cross-sectional area, m^2; v is the inlet airflow velocity, $m \cdot s^{-1}$.

It can be observed from Equation (1) that the pressure loss here is mainly related to the cross-section ratio and the velocity in the inlet. Under the condition of a given inlet with negative pressure, the velocity in the inlet can be regarded as constant. Moreover, because the diameter of the seed-sucking hole and the cross-sectional area of the airway section I are not variable, A_1 can be regarded as invariable. Therefore, the pressure loss of this part is mainly related to the outlet cross-sectional area A_2 (the cross-sectional area of the airway section II). When the cross-sectional areas of airway section I and section II are equal, local pressure loss occurs once. Otherwise, it will occur twice.

2.3.3. Aerodynamic Analysis of Long and Narrow Air Pipe

The size, shape, and direction of the flow cross-sections inside airway section I and section II remain unchanged, so the fluid encounters only the frictional resistance provided by the airway wall, which is the loss that occurs along the path [26,27]. The resistance loss coefficient along the path (λ) is related to the fluid flow state and often is calculated using the Reynolds number (*Re*). The Reynolds number calculation formula [28] is as follows:

$$\text{Re} = \frac{\rho \overline{v} d_0}{\mu} \tag{2}$$

where μ is the dynamic viscosity of the fluid, Pa·s; d_0 is the characteristic length of the airflow section, m; $\overline{v}$ is the average flow velocity of the airway section, $m \cdot s^{-1}$.

Air is the fluid object studied in this paper and by taking the parameters under normal pressure at 20 °C, ρ can be calculated as $1.205\ kg \cdot m^{-3}$ and μ as 1.79×10^{-5} Pa·s. The characteristic length of the airflow section, which is also the diameter of the air pipe, is 0.02 m. The average velocity of the airway section II after stabilization is calculated by pre-simulation to be about $2\ m \cdot s^{-1}$. The Reynolds number (*Re*) is solved at 2692, which is greater than the critical value of 2300. Therefore, at this time, the fluid flow state in the airway is turbulent, but the value is less than 4000, indicating that the fluid in the airway is in the transition state between laminar flow and turbulent flow [29,30]. Therefore, the resistance loss coefficient (λ) along the path in the area increases with the increase in Reynolds number (Re) and has a weak relationship with the relative roughness of the wall [31].

$$p_{f1} = \frac{l_g \rho^{\frac{3}{4}} Q^{\frac{7}{3}}}{800 d_g^3 \mu^{\frac{1}{3}}} \tag{3}$$

where l_g is the length of the pipe, mm; Q is the flow rate, $m^3 \cdot s^{-1}$; d_g is the diameter of the pipe, mm. p_{f1} is the pressure loss along the path of the long and narrow airway part.

It can be observed from the formula that the pressure loss (p_{f1}) in the airway part is mainly related to the length of the pipe (l_g), the diameter of the pipe (d_g), and the flow rate (Q). The length of the pipe (l_g) is the pneumatic conveying distance, which is determined by the size of the seeding wheel. The flow rate (Q) is mainly related to the given negative pressure. Therefore, in order to reduce air resistance in design, one must adjust the diameter of the pipe (d_g). The larger the diameter of the pipe, the smaller the pressure loss along the path in the long and narrow airway.

Therefore, according to the above research, there are three kinds of design schemes in this area, which are as follows:

Scheme 1. The diameter of airway section II is equal to the diameter of section I, that is, the diameter of airway section II is 10 mm, so there is only one instance of local pressure loss in this area, but the pressure loss along the path is higher.

Scheme 2. The diameter of the airway section II is unequal to that of section I, so this will result in twice the local pressure loss. However, due to the larger diameter of the airway section II, the pressure loss along the path will be lower.

Scheme 3. The optimal value can be found between Scheme 1 and Scheme 2. However, based on theoretical analysis alone, it is impossible to determine which scheme is better, and numerical analysis and experimental methods are still needed for research.

2.3.4. Aerodynamic Analysis of Bending Area of Air Pipe

The direction of the airway shifts from horizontal to longitudinal in the bending area of the air pipe, as shown in Figure 4.

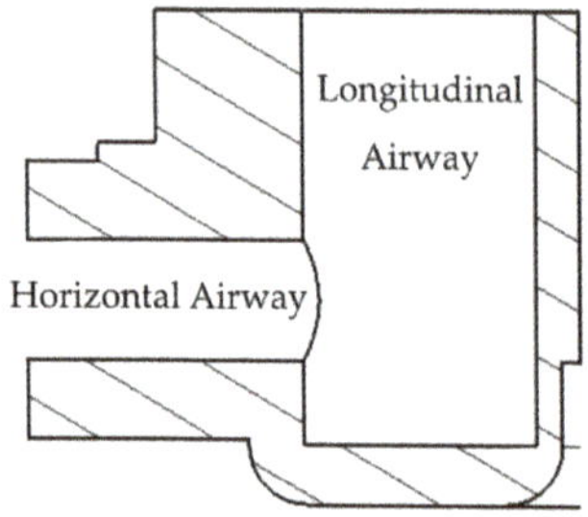

Figure 4. Bending area of the air pipe.

When the air goes flows through the connection, there is a velocity difference between the inner and outer airflows, which generates eddy currents, resulting in a large local energy loss. Due to space constraints, this is in the form of a sharp bend. The pressure loss in the bending area of the air pipe [32] can be calculated as follows:

$$p_w = [0.946 \sin^2(\frac{\theta}{2}) + 2.05 \sin^4(\frac{\theta}{2})] \frac{\rho \overline{v}^2}{2} \tag{4}$$

where p_w is the local pressure loss in the bending area of the air pipe, Pa; θ is the angle between the two air pipes, (°).

It can be observed from the formula that the pressure loss in this part is related to the angle (θ) between the two air pipes. The larger the angle, the smaller the pressure loss. However, when the position of the air-passing aperture is fixed, the larger the angle is and the larger the horizontal length of the inclined section is. The relationship of the angle (θ) between the horizontal length and the two air pipes is as follows:

$$\theta = 180^\circ - \arctan(\frac{h_f}{l_c - R}) \tag{5}$$

where h_f is the vertical distance from the center of the outermost air pipe to the plane where the air-passing aperture is located, mm; l_c is the horizontal length of the distribution center, mm. R is the distance between the center of the longitudinal airway and the center of the sowing wheel, mm.

Restricted by the structure and thin-walled bearings, the horizontal length of the distribution center should not be greater than 150 mm. The vertical distance should be less than the width of the outer wall of the seeding wheel (100 mm). It can be solved by formula 5 that the angle (θ) between the two air pipes should be less than 122.62°, meaning that the horizontal value range of the air pipes' angle is 90–120°.

2.3.5. Aerodynamic Analysis of Air Chamber Confluence Area

The air chamber confluence area is shown in Figure 5.

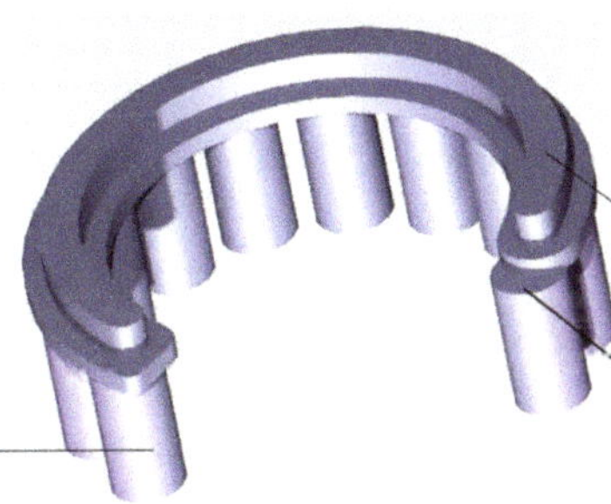

Figure 5. Air chamber confluence area.

This area is the air chamber space formed by the air afflux from the longitudinal air pipes of each lower sub-airway to the rubber sealing ring in the pneumatic distribution cover [33], which is essentially the flow form of multiple small-section pipes that transform into large-section cavities. The energy loss that occurs is mainly local pressure loss, which mainly depends on the cross-sectional area on both sides, but it is difficult to change, and therefore optimize.

2.3.6. Aerodynamic Analysis of Connection Area of Negative Pressure Aperture and Air Chamber

The connection area of the negative pressure aperture and air chamber is shown in Figure 6. The geometry of the airway in this area is transformed from an arc-shaped air chamber to a cylindrical negative pressure aperture. The aerodynamic change that occurs is in the form of a sudden reduction in the cross-sectional area, and the reported energy loss is mainly local pressure loss [34].

$$p_s = 0.5(1 - \frac{A_2}{A_1})\frac{\rho v^2}{2} \tag{6}$$

where p_s is the local pressure loss in the connection area of the negative pressure aperture and air chamber, Pa.

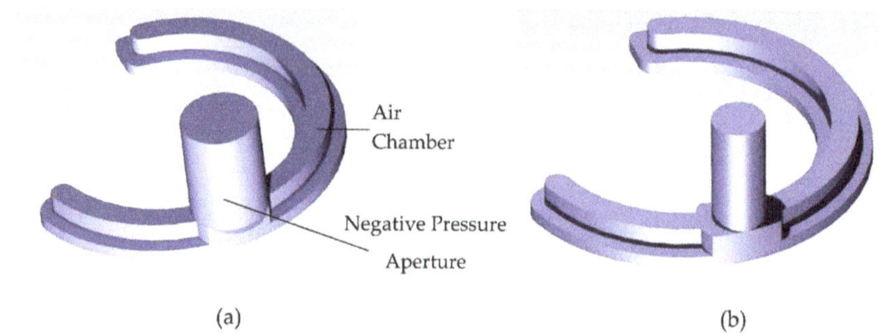

Figure 6. Connection area of negative pressure aperture and air chamber. (**a**) Diameter of negative pressure aperture is 34 mm; (**b**) diameter of negative pressure aperture is 21.5 mm.

It can be observed from the formula that the local pressure loss in this area is mainly related to the cross-sectional area of the negative pressure aperture, and the larger the negative pressure aperture, the lower the pressure loss. However, the space of the pneumatic distribution cover is limited, and it is necessary to leave space for the connection with the frame, so the maximum inner diameter of the negative pressure aperture is 34 mm. In addition, the maximum diameter of the fluid domain of the negative pressure aperture in order to maintain a complete circle is 21.5 mm under the structural limitations, as shown in Figure 6b. If it increases continually, the rubber sealing rings will block the two sides, and the bottom section of the negative pressure aperture will change from a circle to a rectangle-like shape, as shown in Figure 6a. It is difficult to determine the optimal inner diameter of the negative pressure aperture only by theoretical analysis, and further experimental research is needed.

3. Simulation Analysis

3.1. Flow Field Static Simulation Analysis of Seeding Device

The fluid domain was imported into ANSYS FLUENT, and the hexahedral meshing method was used to mesh the nine groups of the fluid domain in the experiment scheme. Since there was no need to solve the viscous bottom layer, we used the wall function method to solve the simulation [35–37]. The divided fluid domain contained about 1 million meshes under the simulation parameter of five boundary layers and a 1.2 mesh growth coefficient. The mesh quality was good, with a minimum orthogonal quality is above 0.5 and a maximum aspect ratio of less than 10. The fluid domain is shown in Figure 7.

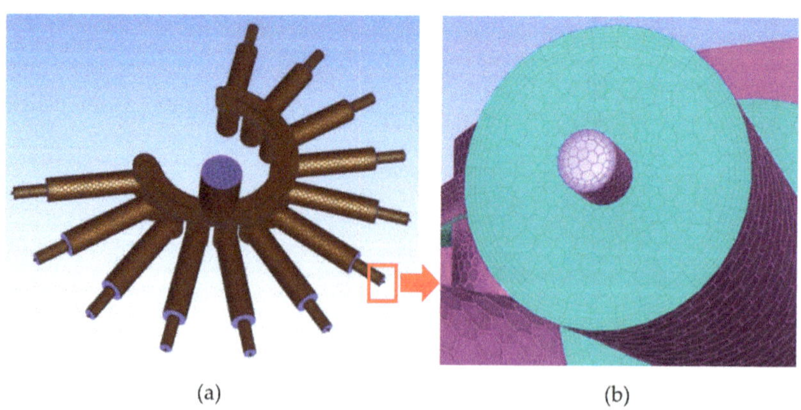

Figure 7. Fluid domain meshing: (**a**) overall meshing diagram; (**b**) meshing diagram for seed-sucking hole.

According to the above theoretical analysis, it can be concluded that the diameter (d_h) of the horizontal air pipe, the angle (θ) of the air pipe, and the diameter (d_z) of the negative pressure aperture have a significant influence on the pressure loss in the airway. Taking the above three parameters as the experimental factors, a three-factor three-level orthogonal simulation experiment was designed. The experiment factors and levels are shown in Table 1.

Table 1. Experiment factors and levels.

Level	Factor		
	A Diameter of Horizontal Air Pipe d_h/mm	B Angle of Air Pipe θ	C Diameter of Negative Pressure Aperture d_z/mm
1	10	90	21.5
2	15	105	27.75
3	20	120	34

The airflow velocity at the seed-sucking hole was used as the core index to measure the performance of the seed sucking, so the seed-sucking holes' average airflow velocity, the minimum airflow velocity, and the airflow velocity standard deviation were selected as the experiment indicators. There were 12 sets of data on seed-sucking holes for each set of simulations. The airflow velocity at the center of the seed-sucking hole was taken as the experiment results. The standard three-factor three-level orthogonal table was selected and simulations were carried out. The method of steady-state calculation and the model of standard k-epsilon turbulence in Fluent were used. The pressure inlet boundary was set at −7 kPa, the outlet boundary condition was the pressure outlet boundary, and the pressure was 0 Pa. The solution method used was the SIMPLE method. The simulation results are shown in Table 2.

Table 2. Simulation experiment results.

Number	Factor			Average Airflow Velocity of Seed-Sucking Hole /(m·s^{-1})	Minimum Airflow Velocity of Seed-Sucking Hole /(m·s^{-1})	Airflow Velocity Standard Deviation of Seed-Sucking Hole /(m·s^{-1})
	A	B	C			
1	1	1	1	101.46	99.66	0.86
2	1	2	2	102.42	101.17	0.68
3	1	3	3	102.31	101.05	0.89
4	2	1	2	102.02	100.39	0.78
5	2	2	3	102.59	101.58	0.54
6	2	3	1	101.62	100.29	0.73
7	3	1	3	102.12	100.45	0.72
8	3	2	1	101.75	100.78	0.67
9	3	3	2	102.02	100.99	0.77

3.2. Analysis of Simulation Results

3.2.1. Analysis of Variance and Range

The range analysis of the simulation experiment results [21] is shown in Table 3. The selection of the optimal parameter combination is based on the following information: the larger the average airflow velocity and the minimum airflow velocity of the seed-sucking hole, the lower the airway resistance and pressure loss and airflow velocity standard deviation of the seed-sucking hole; the smaller the difference in seed sucking performance, the better the sucking stability. According to this information, it was concluded that the optimal parameter combination of the airway structure parameters for the three experiment indicators is $A_2B_2C_3$, which is the fifth level of the orthogonal simulation experiment. The simulation results show that the average airflow velocity of the seed-sucking hole is 102.59 $m \cdot s^{-1}$, the minimum airflow velocity of the seed-sucking hole is 101.58 $m \cdot s^{-1}$, and the airflow velocity standard deviation of the seed-sucking hole is 0.54 $m \cdot s^{-1}$, which are the best values in every group. Therefore, the optimal airway structural parameters can be determined as follows: the diameter of the horizontal air pipe must be 15 mm, the angle of the air pipe must be 105°, and the diameter of the negative pressure aperture must be 34 mm.

Table 3. Range analysis.

Items	Average Airflow Velocity of Seed-Sucking Hole /(m·s^{-1})			Minimum Airflow Velocity of Seed-Sucking Hole /(m·s^{-1})			Airflow Velocity Standard Deviation of Seed-Sucking Hole /(m·s^{-1})		
	A	B	C	A	B	C	A	B	C
K_1	102.06	101.87	101.61	100.63	100.17	100.24	0.81	0.79	0.75
K_2	102.08	102.25	102.15	100.75	101.18	100.85	0.68	0.63	0.74
K_3	101.96	101.87	102.34	100.74	100.78	101.03	0.72	0.80	0.72
R	0.12	0.38	0.73	0.12	1.01	0.79	0.13	0.17	0.03
Optimal level combination		$A_2B_2C_3$			$A_2B_2C_3$			$A_2B_2C_3$	

Further variance analysis was conducted on the experimental data to determine the significance of each factor. The variance analysis is shown in Table 4.

Table 4. Variance analysis.

Source of Difference		SS	df	MS	F	Significance
Average airflow velocity of seed-sucking hole	A	0.02	2	0.01	11.75	
	B	0.23	2	0.17	118.91	**
	C	0.87	2	0.43	442.65	**
	Error	0.00	2	0.00		
	Sum	1.13	8			
Minimum airflow velocity of seed-sucking hole	A	0.03	2	0.01	1.37	
	B	1.55	2	0.78	72.45	*
	C	1.01	2	0.5	47.05	*
	Error	0.02	2	0.01		
	Sum	2.61	8			
Airflow velocity standard deviation of seed-sucking hole	A	0.03	2	0.01	2.9	
	B	0.05	2	0.03	5.87	
	C	0.00	2	0.00	0.28	
	Error	0.01	2	0.00		
	Sum	0.09	8			

Note: In the table, the critical value of significance judgment is F0.001 (2, 2) = 999, F0.01 (2, 2) = 99, and F0.05 (2, 2) = 19. ** indicates that the influence is very significant. * indicates that the influence is significant. The data in the table were rounded up, but in order to ensure accuracy, the calculation was based on the original value.

From the analysis of variance results, it can be observed that the angle of the air pipe and the diameter of the negative pressure aperture have a significant influence on the average airflow velocity index of the seed-sucking hole. The diameter of the horizontal air pipe does not influence the average airflow velocity of the seed-sucking hole. For the minimum airflow velocity of the seed-sucking hole, the angle of the air pipe and the diameter of the negative pressure aperture have less significant influence, while the diameter of the horizontal air pipe has no influence on the experiment results, and the degree of influence on the average airflow velocity of the seed-sucking hole is ranked as follows: angle of air pipe > diameter of negative pressure aperture > diameter of horizontal air pipe. From the results of the variance analysis for the airflow velocity standard deviation index of the seed-sucking hole, it can be observed that three factors do not influence this index. By analyzing the simulation experiment data, it was found that there is little difference in airflow velocity between the 12 seed-sucking holes, resulting in a small standard deviation value, indicating that the distribution of airflow in the air chamber is very uniform, which is beneficial for the collection of seeds.

3.2.2. Analysis of Airflow Velocity Difference of Various Seed-Sucking Holes

Taking the optimal parameter combination as the research object, the airflow velocity at the center of the 12 seed-sucking holes with negative pressure was calculated, as shown in Figure 8. It can be found from the figure that the position of the seed-sucking hole has no significant influence on the airflow velocity of the seed-sucking hole. The deviation in the airflow velocity value of the seed-sucking hole compared with the average value is less than 1%, which shows that the air pressure stability of the seeding-wheel-type seed precision placing device is higher.

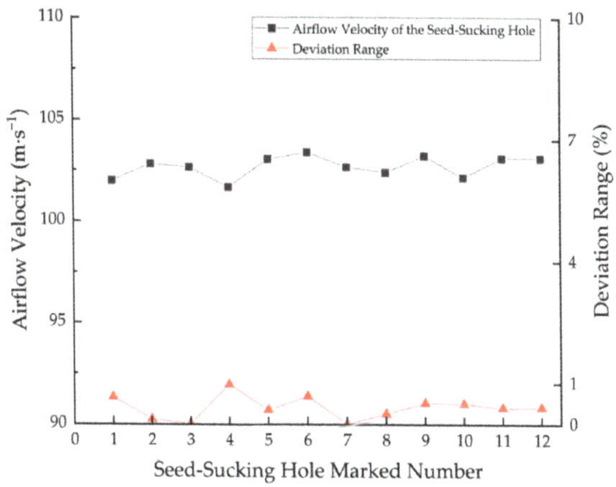

Figure 8. The line chart of the airflow velocity of seed-sucking holes with the optimal parameter combination.

3.2.3. Airflow Trajectory Analysis

Taking the airflow inlet (seed-sucking hole) as the starting point, it is possible to study the airflow trajectory of air that enters the negative pressure fluid domain and flows out the airflow outlet (negative pressure port). A streamlined diagram of the fluid domain is shown in Figure 9.

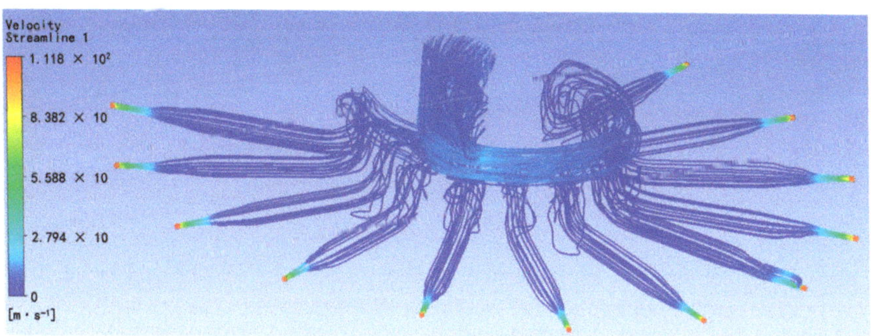

Figure 9. Streamline diagram of fluid domain.

It can be observed from the streamlined diagram that after the airflow flows in from the seed-sucking hole, the flow velocity gradually decreases after diffusion occurs twice. When entering the vertical air pipe, most of the airflow flows into the air chamber after a 90° turn, but a vortex composed of a small amount of airflow is formed near the wall, and the streamline is chaotic, resulting in a large local pressure loss. After entering the arc-shaped air chamber, the airflow flows to the negative pressure aperture. In this trajectory, a clear light blue area to the right of the negative pressure aperture can be observed, indicating that the airflow in this area accelerates. It is speculated that the reason for this phenomenon is that there were more seed-sucking holes contained on the right side, resulting in a large airflow, and a longer flow distance than on the left side. When all the airflow converges at the negative pressure aperture, the airflow swirls upward, generating a cyclone, and also causes pressure loss, but a larger space means fewer collisions; therefore, the optimal level was recorded at the upper part of the negative pressure aperture.

3.3. Study on Negative Pressure Characteristics of Locally Connected Seed-Sucking Holes

During operation, a seed-sucking hole that is locally connected can be observed, but its airflow velocity does not significantly differ from other seed-sucking holes. Therefore, it can be determined that when more than half of the air-passing aperture of the longitudinal air pipe is connected to the air chamber, the seed-sucking hole's performance will not be affected. However, this phenomenon will lead to a certain deviation in the seeding position compared with the design value, making it impossible to cast the seed precisely at the designated position. To compensate for this deviation, it is necessary to further determine when the air-passing aperture of the longitudinal air pipe and the air chamber are in any of the connection states, as the airflow velocity of the seed-sucking hole will drop significantly, thus triggering seeding.

A dynamic simulation was carried out that gradually reducing the connected part of the air-passing aperture, and the time when the negative pressure of the seed-sucking hole dropped significantly and sharply was determined as the trigger time of seed casting. According to the actual working situation, we simulated the rotation of the seeding device, set the fluid domain of the upper air chamber to remain static, and the lower fluid domain to rotate, and studied the variation law of the negative pressure of the seed-sucking hole in the lower sub-fluid domain. Separate fluid domain extraction was performed for each sub-airway, obtaining 14 fluid domain units, as shown in Figure 10a.

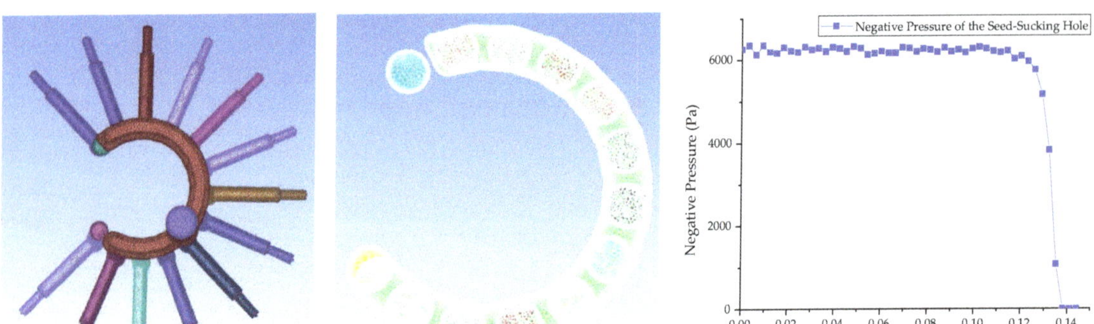

(a) (b) (c)

Figure 10. Diagram of simulation process and results: (**a**) fluid domain flow diagram; (**b**) interface setting; (**c**) dot–line graph of experiment results.

To observe the spatial position changes in rotating objects in real time, the sliding mesh method [38–41] was adopted. The surface of the air-passing aperture of the lower sub-airway and the lower surface of the fluid domain of the upper air chamber were set as the interfaces, and data exchange could be performed, as shown in Figure 10b. The rotational speed of the lower sub-airway was set to 2.31 rad $\cdot$ s^{-1}, and the time step was 1.5×10^{-4} s. The simulation time was 0.141 s, which could cover the complete process of lower sub-airway separation. We plotted the experimental data as a dot–line graph, as shown in Figure 10c. We recorded the significant negative pressure changes in the seed-sucking holes for groups 41–48, and calculated the corresponding opening angles of the air-passing aperture according to the structural parameters. The results are shown in Table 5.

Table 5. Sliding mesh simulation results for groups 41–48.

Group	Simulation Time (s)	Opening Angle of Air-Passing Aperture (°)	Negative Pressure of Seed-Sucking Hole (Pa)
41	0.12	2.518	−6101.03
42	0.123	2.121	−5974.7
43	0.126	1.723	−5768.42
44	0.129	1.326	−5179.11
45	0.132	0.929	−3831.63
46	0.135	0.532	−1072.76
47	0.138	0.135	−5.13
48	0.141	0	0

It can be observed from Figure 10c and Table 5 that the air pressure of the seed-sucking hole remains stable for most of the time, fluctuating only slightly. However, there is a certain decline after 0.12 s, but the range is not large, and after 0.129 s (the corresponding opening angle of the air-passing aperture is 1.326°), it falls sharply, and the negative pressure of the seed-sucking hole is only −5.13 Pa until 0.138 s. The negative pressure distribution of groups 41–48 is shown in Figure 11.

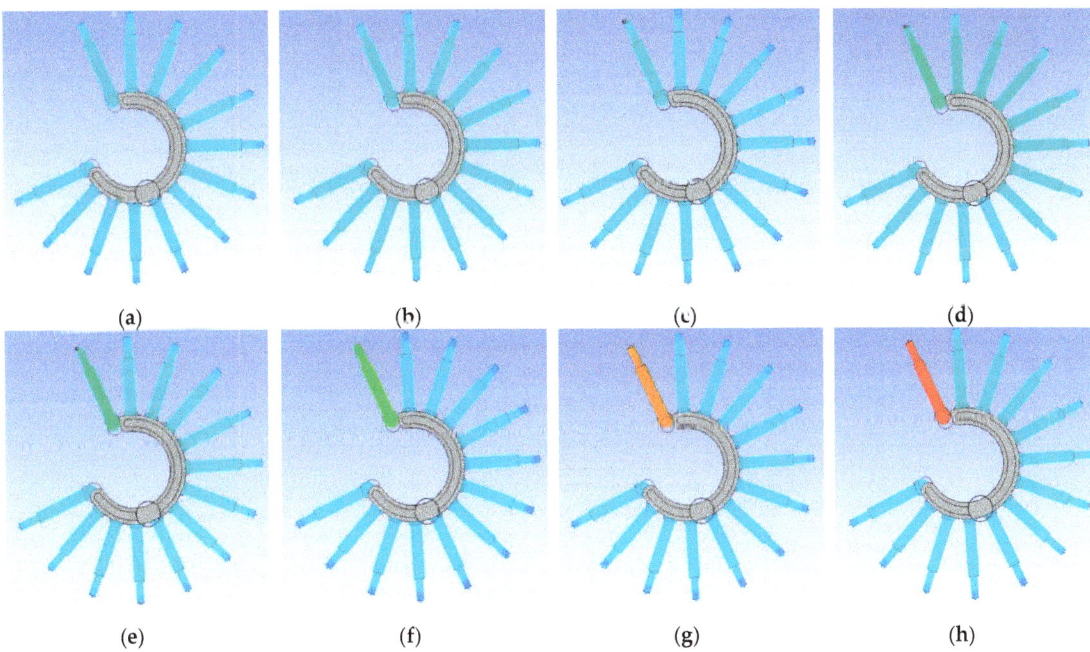

Figure 11. Negative pressure distribution chart of groups 41–48: (**a**) group 41; (**b**) group 42; (**c**) group 43; (**d**) group 44; (**e**) group 45; (**f**) group 46; (**g**) group 47; (**h**) group 48.

On the whole, when only a small part of the air-passing aperture is connected to the upper air chamber, the seed-sucking hole still retains the ability to maintain seed adsorption. The trigger of seed casting may be associated with the period when the negative pressure of the seed-sucking hole drops sharply, but it is difficult to judge its exact time. However, the period in which the negative pressure of the seed-sucking hole sharpy drops is very short and only lasts about 0.01 s, so the setting error is small.

4. Prototype Test Verification

4.1. Prototyping and Testing Methods

Three-dimensional printing technology was used to process complex key components, such as the seeding wheel and pneumatic distribution cover, with a Stratasys F900 3D printer (Stratasys Ltd., Rehovot, Israel) and 8000 photosensitive resin, and the rest of the components were processed using metal. The prototype is shown in Figure 12a. The measuring equipment included a DN2000 intelligent anemometer (Beijing Instrument Equipment Factory, Beijing, China), as shown in Figure 12b. Its measuring range of wind speed was 1.00–365.00 $m \cdot s^{-1}$ with an accuracy of 0.01 $m \cdot s^{-1}$, its wind pressure range was $\pm$80 kPa, with an accuracy is 0.01 kPa, and its air volume range was <999,999 $m^3 \cdot h^{-1}$.

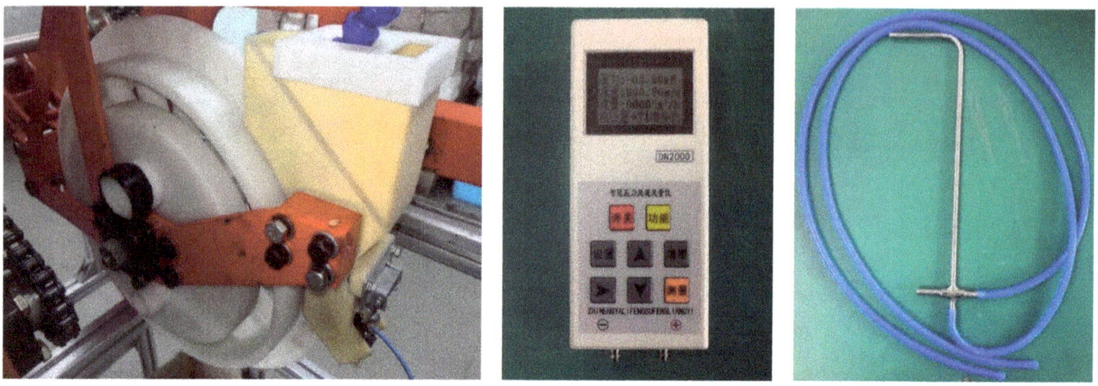

(a) (b) (c)

Figure 12. Test prototype and measuring equipment: (**a**) the prototype; (**b**) DN2000 intelligent anemometer; (**c**) the Pitot tube.

The prototype was fixed according to the relative position of the pneumatic distribution cover and the seeding wheel in the simulation experiment, and the airflow velocity of the airway at this position was measured by an anemometer. When the seeding wheel was processed, a measuring slot hole in an air pipe was created to facilitate the insertion of the Pitot tube. During the measurements, the front end of the Pitot tube was inserted into the hollow airway and then hot-melt adhesive was used to seal the measuring hole to ensure air tightness. In addition, it is necessary to ensure that the extended part of the Pitot tube is concentric with the pipe. The negative pressure fan was turned on, and the vacuum degree of negative pressure was 7 kPa during the test. After the working state was confirmed as stable, the value on the DN2000 intelligent anemometer was read, the airflow velocity of this group was recorded, the measurement was repeated 3 times, and the average value was recorded. After one measurement was completed, the seeding wheel was rotated to the next position and the steps above were repeated to derive the actual airway wind velocity at the same location as in the simulation.

However, because of the small diameter of the seed-sucking hole, it was difficult to use an anemometer to measure air velocity accurately at this position. It would also affect flow field distribution if it was too close to the seed-sucking hole. Therefore, the measurement position was adjusted by 20 mm. From the simulation results, it can be concluded that the airflow velocity trend at this position is consistent with the results for the seed-sucking hole.

4.2. Results and Analysis

Figure 13 shows the simulated and tested values of the airflow velocity in the air pipe.

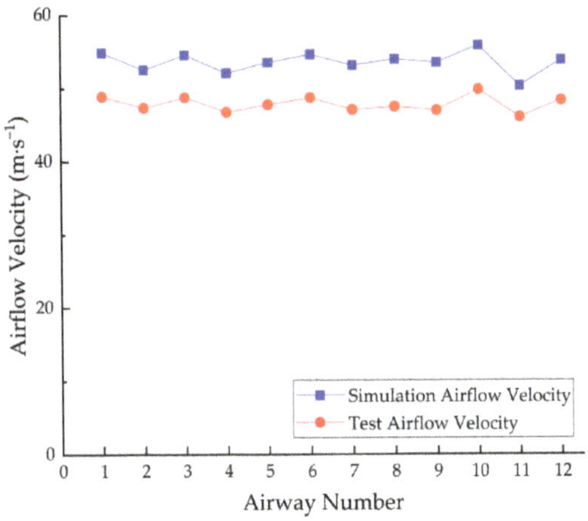

Figure 13. Dot–line graph of test results.

The data in the figure were plotted after taking the average of the three groups of tests. In fact, the results of the anemometer fluctuated in a narrow range due to the uncertainty of the flow field during the measurement, and the recorded value was taken as the fluctuation center. The analysis of the readings obtained from the three groups of tests showed that there was little difference in the values measured, so it was reasonable to characterize the values using averages. The dot–line graph shows that the test results were generally lower than the simulation results. There are three main reasons for the results. First, the prototype processing could not guarantee absolute air tightness as in the numerical analysis, resulting in additional pressure loss. Second, the detection head of the anemometer inserted into the air pipe inserted an artificial obstacle, which would affect the airflow in the airway. Third, the inserted part of the Pitot tube should be concentric with the air pipe during the measurements. However, it was hard to ensure this during the actual measurements, which would have affected the measurement results.

However, both the simulation results and the test results show a trend of narrow fluctuations around the mean value, and the airflow velocity of each airway shows little difference, which can verify the accuracy of the simulation.

5. Conclusions

In this paper, an evaluation scheme for internal complex airway pressure loss was formulated, the influence of airway structure on pressure loss under different parameter combinations was explored, the internal flow field distribution and airflow movement law of a precision seeding device were defined, the drag reduction and efficiency increase were analyzed, and a new method was provided for the research of the pneumatic characteristics of complex narrow airway groups. The specific conclusions are as follows:

1. A functional model of pressure loss in variable-section panhandle areas, the bending area of the air pipe, air chamber confluence area, and connection area of the negative pressure aperture and air chamber were established and it was concluded that the airway structural factors that may have a significant influence on the pressure loss are the diameter of the horizontal air pipe, the angle of the air pipe, and the diameter of the negative pressure aperture.
2. It was concluded that the optimal parameter combination was as follows: the diameter of the horizontal air pipe was 15 mm, the angle of the air pipe was 105°, and the diameter of the negative pressure aperture was 34 mm. Under these parameters, the average airflow velocity of the seed-sucking hole was 102.59, the minimum airflow velocity of the seed-sucking hole was 101.58, and the airflow velocity standard

deviation of the seed-sucking hole was 0.54. The position of the seed-sucking hole has no significant influence on its airflow velocity.

3. A dynamic simulation test was carried out and it was concluded that there is a sharp decline when the opening angle of the air-passing aperture is 1.326°. In addition, it triggers the casting action of seeds when the opening angle of the air-passing aperture is 0.929°.
4. Compared with the simulation results, the tested measurement results are generally lower, but the difference is not significant, which proves that the position of the seed-sucking holes has little influence on the airflow velocity, and the overall pneumatic distribution is uniform.
5. This paper mainly explores the distribution law of the internal flow field when the seed-sucking hole is open, and the interaction between the flow field and the seeds can be further explored by CFD-DEM coupling technology in the future, which will help us to study the mechanism and effect of the flow field on seeds more accurately.

Author Contributions: Conceptualization, X.Z., Q.W., H.L. and J.L.; software, Z.Z. and J.L.; validation, Z.Z., Z.W. and J.L.; investigation, Q.W., H.L. and X.Z.; resources, Q.W., H.L. and X.Z.; visualization, Z.W. and J.L. All authors have read and agreed to the published version of the manuscript.

Funding: This research was funded by the National Natural Science Foundation of China (Grant No.52175260), the China Agriculture Research System (CARS-03), the Key Research and development plan of Hainan Province (Grant No. ZDYF2022XDNY184), Project of Scientific Research in Colleges and Universities of Hainan Province (Grant No. Hnky2022-93) and Research Start-up Fund Project of Hainan University (Grant No. KYQD(ZR)-22036). The authors are grateful for their generous financial assistance.

Institutional Review Board Statement: Not applicable.

Data Availability Statement: Not applicable.

Conflicts of Interest: The authors declare that there is no conflict of interest.

References

1. Dowlati, M.; Namjoo, M. Development and Evaluation of a Pneumatic Dibble Punch Planter for Precision Planting. *AMA Agric. Mech. Asia Afr.* **2016**, *47*, 53–59.
2. Kroulík, M.; Hula, J.; Rybka, A.; Honzík, I. Seed passage speed through short vertical delivery tubes at precise seeding. *Agron. Res.* **2016**, *14*, 442–449.
3. Soza, E.L.; Agnes, D.W.; Agostinelli, F. Variability of wheat seeds distribution by one seeder on three tillage systems, impact in the establish and growth crop. *Rev. Fac. Agron.* **2010**, *30*, 123–132.
4. Liu, J.; Wang, Q.; Li, H.; He, J.; Lu, C.; Wang, C. Numerical Analysis and Experiment on Pneumatic Loss Characteristic of Pinhole-Tube Wheat Uniform Seeding Mechanism. *Trans. Chin. Soc. Agric. Mach.* **2020**, *51*, 29–37.
5. Tu, J.; Yeoh, G.; Liu, C.; Wang, X. *Computational Fluid Dynamics*, 2nd ed.; Northeastern University Press: Shenyang, China, 2014; pp. 1–17.
6. John, D.; Anderson, J.R. *Introduction to Computational Fluid Dynamics*, 1st ed.; Tsinghua University Press: Beijing, China, 2010; pp. 1–28.
7. Ben Amara, H.; Bouadila, S.; Fatnassi, H.; Arici, M.; Allah Guizani, A. Climate assessment of greenhouse equipped with south-oriented PV roofs: An experimental and computational fluid dynamics study. *Sustain. Energy Technol. Assess.* **2021**, *45*, 100–101. [CrossRef]
8. Panigrahi, S.S.; Singh, C.B.; Fielke, J. Strategies to mitigate dead-zones in on-farm stored grain silos fitted with aeration ducting modelled using computational fluid dynamics. *Biosyst. Eng.* **2021**, *205*, 93–104. [CrossRef]
9. Yazgi, A.; Degirmencioglu, A. Optimisation of the seed spacing uniformity performance of a vacuum-type precision seeder using response surface methodology. *Biosyst. Eng.* **2007**, *97*, 347–356. [CrossRef]
10. Yazgi, A.; Degirmencioglu, A. Measurement of seed spacing uniformity performance of a precision metering unit as function of the number of holes on vacuum plate. *Measurement* **2014**, *56*, 128–135. [CrossRef]
11. Ghafori, H.; Hemmat, A.; Borghaee, A.M.; Minaei, S. Physical properties and conveying characteristics of corn and barley seeds using a suction-type pneumatic conveying system. *Afr. J. Agric. Res.* **2011**, *6*, 5972–5977. [CrossRef]
12. Yang, S. Study on Maize Seed-Metering Mechanism of Side Positive Pressure and Gas Supply System. Ph.D. Thesis, China Agricultural University, Beijing, China, 28 May 2015.

13. Yang, S.; Zhang, D.; Diao, P.; Guo, Z.; Song, J.; Zhang, X. Design and experiment of side positive pressure seed metering device. *Trans. Chin. Soc. Agric. Eng.* **2015**, *31*, 8–13.
14. Yang, S.; Zhang, D.; Gao, Z.; Liu, X. Flow Field Simulation and Working Parameters Analysis of Side Positive Pressure Maize Seeding Device. *Trans. Chin. Soc. Agric. Mach.* **2014**, *45*, 35–39.
15. Zhan, Z.; Li, Y.; Chen, J.; Xu, L. Numerical analysis and laboratory testing of seed spacing uniformity performance for vacuum-cylinder preci-sion seeder. *Biosyst. Eng.* **2010**, *106*, 344–351. [CrossRef]
16. Zhan, Z.; Li, Y.; Chen, J.; Han, Z. Dynamic analysis on seeds pick-up process for vacuum-cylinder seeder. *Trans. Chin. Soc. Agric. Eng.* **2011**, *27*, 112–116.
17. Zhan, Z.; Li, Y. Numerical Analysis on the Inner Flow Field of Vacuum-cylinder Precision Seeder. In Proceedings of the 2010 International Conference on Agricultural Engineering, Shanghai, China, 18 November 2010.
18. Lei, X.L.; Liao, Y.T.; Zhang, Q.S.; Wang, L.; Liao, Q.X. Numerical simulation of seed motion characteristics of distribution head for rapeseed and wheat. *Comput. Electron. Agr.* **2018**, *150*, 98–109. [CrossRef]
19. Lei, X.; Liao, Y.; Liao, Q. Simulation of seed motion in seed feeding device with DEM-CFD coupling approach for rapeseed and wheat. *Comput. Electron. Agric.* **2016**, *131*, 29–39. [CrossRef]
20. Shi, S.; Liu, H.; Wei, G.; Zhou, J.; Jian, S.; Zhang, R. Optimization and Experiment of Pneumatic Seed Metering Device with Guided Assistant Filling Based on EDEM-CFD. *Trans. Chin. Soc. Agric. Mach.* **2020**, *51*, 54–66.
21. Liu, J.; Wang, Q.; Li, H.; He, J.; Lu, C. Design and seed suction performance of pinhole-tube wheat precision seeding device. *Trans. Chin. Soc. Agric. Eng.* **2019**, *35*, 10–18.
22. Logachev, K.I.; Ziganshin, A.M.; Popov, E.N.; Averkova, O.A.; Kryukova, O.S.; Gol'Tsov, A.B. Experiment determining pressure loss reduction using a shaped round exhaust hood. *Build. Environ.* **2021**, *190*, 107572. [CrossRef]
23. Borzenko, E.I.; Ryl'Tsev, I.A.; Schrager, G.R. Pressure Losses in Power-Law Fluid Flow through a Tube of Variable Cross-Section. *Fluid Dynam.* **2021**, *56*, 1–9. [CrossRef]
24. Mo, S.; Zuo, C.; Jia, F. Calculation of Local Pressure Loss Coefficient for Flow Distribution Baffle and Cross Flow in Steam Generator. *At. Energy Sci. Technol.* **2015**, *49*, 1039–1044.
25. Bernardi, C.; Dib, S.; Girault, V.; Hecht, F.; Murat, F.; Sayah, T. Finite element methods for Darcy's problem coupled with the heat equation. *Numer. Math.* **2018**, *139*, 315–348. [CrossRef]
26. Que, X.; Su, L.; Yang, W.; Hu, C. Research on influencing factors of fire hose's pressure loss and its calculation method. *J. Univ. Shanghai Sci.* **2006**, *28*, 561–565.
27. Li, Z.; Wang, D.; Liu, G.; Yang, M.; Wang, Z. Experimental study on sowing seeds by air-stream metering mechanism. *Trans. Chin. Soc. Agric. Eng.* **2009**, *25*, 89–93.
28. Zinurov, V.E.; Dmitriev, A.V.; Badretdinova, G.R.; Galimova, A.R.; Dmitrieva, O.S. Analysis of various mathematical models of turbulence when calculating the gas dynamics in a classifier with coaxially arranged pipes. *IOP Conf. Ser. Mater. Sci. Eng.* **2021**, *1089*, 012002. [CrossRef]
29. Guo, Y.; Yang, K.; Guo, X.; Li, J.; Fu, H. Quick evaluation of the pipeline equivalent sand-grain roughness based on the surface roughness parameters. *J. Hydraul. Eng.* **2018**, *49*, 178–185.
30. Uralov, B.; Choriev, R.; Maksudova, L.; Sapaeva, M.; Shernaev, A.; Nurmatov, P. Substantiation of the influence of the channel shape and the roughness of machine canals on the pressure loss of irrigation pumping stations. *IOP Conf. Ser. Mater. Sci. Eng.* **2021**, *1030*, 12148. [CrossRef]
31. Han, Z.; Wang, G. *Fundamentals of Engineering Fluid Dynamics*, 2nd ed.; Beijing Institute of Technology Press: Beijing, China, 2016; pp. 71–97.
32. Mao, Z.; Zhao, K.; Zhao, X.; Luo, S. Experimental study on local flow resistance at junctions of circular pipes. *J. Hydraul. Eng.* **2007**, *38*, 812–818.
33. Shen, Y.; Wang, J.; Cao, K. Loss property analysis of throttle with large aperture. *J. Air Force Eng. Univ.* **2001**, *2*, 10–12.
34. Salikov, V.; Antonyuk, S.; Heinrich, S.; Sutkar, V.S.; Deen, N.G.; Kuipers, J.A.M. Characterization and CFD-DEM modelling of a prismatic spouted bed. *Powder Technol.* **2015**, *270*, 622–636. [CrossRef]
35. Tao, Z.; Xiao-jun, Z.; Fei, P.; Shao-song, M. Analysis of effect of near-wall treatments on numerical computation of turbulent flow. *J. Nav. Univ. Eng.* **2013**, *25*, 104–108.
36. Kun, H.; Tingting, H.; Haifeng, M. *ANSYS CFD Introductory Guide Fundamentals and Applications of Computational Fluid Dynamics*, 1st ed.; China Machine PRESS: Beijing, China, 2019; pp. 47–58.
37. Richard, W.; Richard, M.L. Analysis of variance in soil research: Examining the assumptions. *Eur. J. Soil. Sci.* **2019**, *70*, 990–1000.
38. Guarella, P.; Pellerano, A.; Pascuzzi, S. Experimental and Theoretical Performance of a Vacuum Seeder Nozzle for Vegetable Seeds. *J. Agric. Eng. Res.* **1996**, *64*, 29–36. [CrossRef]
39. Zhang, J.; Li, Z.; Liu, H.; Wu, G. Mathematical modeling and validation of seeder's suction-boundary on pneumatic-roller type metering. *Trans. Chin. Soc. Agric. Eng.* **2016**, *32*, 12–20.
40. Liu, F. Discrete Element Modelling of the Wheat Particles and Short Straw in Cleaning Devices. Ph.D. Thesis, Northwest A&F University, Xi'an, China, 5 June 2018.
41. Shi, S.; Zhang, D.; Yang, L.; Cui, T.; Zhang, R.; Yin, X. Design and experiment of pneumatic maize precision seed-metering device with combined holes. *Trans. Chin. Soc. Agric. Eng.* **2014**, *30*, 10–18.

agriculture

MDPI

Article

The Results of Studies on the Assessment of the Destruction of Soil Clods during Combine Harvesting of Potatoes

Alexey Dorokhov, Otari Didmanidze, Alexander Aksenov, Alexey Sibirev *, Nikolay Sazonov, Maxim Mosyakov and Maria Godyaeva

FSBSI "Federal Scientific Agronomic and Engineering Center VIM", 109428 Moscow, Russia
* Correspondence: sibirev2011@yandex.ru; Tel.: +7-964-584-3518

Abstract: Existing potato harvesters cause damage to marketable products as a result of the interaction of potato tubers with each other, with working bodies and soil clods, given the wide variety of soil and climatic conditions in which the harvesting process takes place. In addition, under homogeneous soil and climatic conditions within the same accounting area, there is a large deviation from the average values of the main physical soil constants—moisture and hardness. Field studies were carried out to determine the fractional composition of soil clods, size-mass parameters, as well as their physical and mechanical properties with the identification of the greatest force for their destruction. The article presents a methodology for conducting research to assess the influence of working bodies on the magnitude of the force impact on potato tubers and soil clods during harvesting, a methodology for assessing the dynamic destruction of soil clods. The results of comparative studies of the force impact of the working bodies of modern potato harvesters, which affect the destruction of soil clods, causing damage to potato tubers as a result of their interaction with soil clods are presented.

Keywords: harvesting; force action; potato; working body; harvesting machines; device for assessing the suitability for harvesting

Citation: Dorokhov, A.; Didmanidze, O.; Aksenov, A.; Sibirev, A.; Sazonov, N.; Mosyakov, M.; Godyaeva, M. The Results of Studies on the Assessment of the Destruction of Soil Clods during Combine Harvesting of Potatoes. *Agriculture* **2022**, *12*, 2024. https://doi.org/10.3390/agriculture12122024

Academic Editor: Jacopo Bacenetti

Received: 3 October 2022
Accepted: 23 November 2022
Published: 26 November 2022

1. Introduction

The main factors that inflict damage to potato tubers during mechanized harvesting are the design of potato harvesters and the material [1–3] from which the working bodies of the machines and operating modes are made [4,5]. An important role is played by the physical and mechanical properties of tubers [6,7], which, in turn, depend on the variety, agricultural techniques of cultivation, soil structure, and climatic conditions. To determine the location and register the magnitude of the greatest force impact of the working bodies of potato harvesters on potato tubers, as well as recommendations for subsequent changes in the design and technological parameters of harvesting machines and the development of devices that ensure the stability of potato tubers to mechanized harvesting, experimental studies were conducted using the software tool "Electronic Potato Tuber Log" in the Moscow region of JSC "Ozery". In addition, when conducting research on the assessment of the force impact of the working bodies of harvesting machines, it was found that despite the saturation of the machine with separating working bodies, the efficiency of their work remains quite low, due to the fact that the entire layer of earth dug up by the plowshare is subjected to separation. Moreover, the mass of the layer undermined in this case increases by 1.6–1.8-times, compared with undermining with a passive plowshare, which, in turn, has a small loosening and unloading of the soil (Figure 1).

Despite the saturation of machines with separating working bodies of various designs and their intensification, the mass of soil impurities in the total volume of the tuberous heap reaches threshold values of 46–50% (Figure 2), depending on the fractional composition.

Figure 1. General view of the tuberous heap in the bunker of a potato harvester.

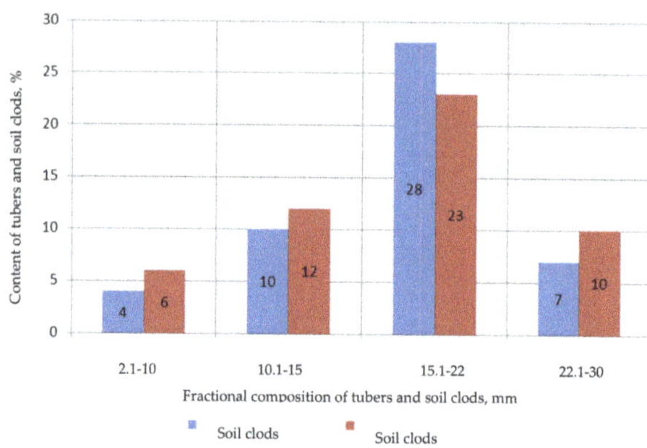

Figure 2. Fractional composition of the tuberous heap in the hopper of a potato harvester.

Therefore, studies aimed at establishing the patterns of interaction between tubers and working bodies, soil clods, and the establishment of physical and mechanical properties to determine the magnitude of the dynamic impact that contribute to the destruction of soil clods are relevant, so they will ensure the development of harvester devices, providing an exception, whether or not there is a decrease in the receipt of soil clods in the device for collecting marketable potato products.

2. Materials and Methods

The programmable tool "Electronic Potato Tuber Log" (Figure 3) includes: a data logger 1, made in shape, size, and density by a standard potato tuber, a personal 2 or tablet computer 3, with installed software for processing, registered root crop damage data and its subsequent analysis, as well as an auxiliary 4. Electronic potato equipment allows you to record the magnitude of the acquired acceleration, as well as the impulse of the impact force from its interaction with the working body. Studies to determine the places of damage on the web of the rod elevator of the potato harvesting machine were carried out at different values of the translational VEL of the speed of movement of the rod elevator [8,9]. To fix the location and time of damage to the data logger 1, an Inspector Tornado DVR with a screen resolution of 960 × 240 and a diagonal viewing angle of 150° mounted on a laboratory tripod was fixed on the rod elevator [7,10].

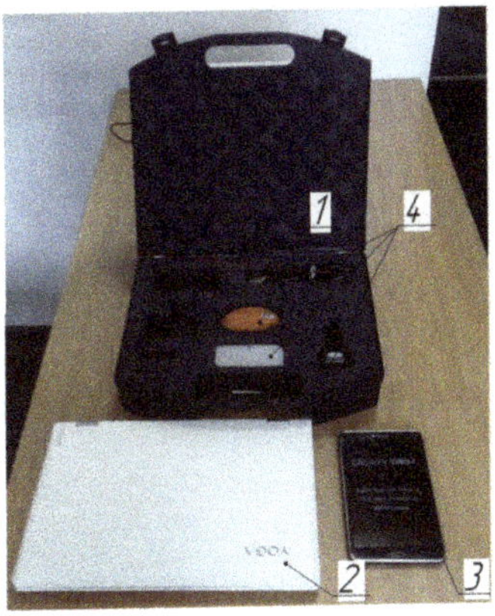

Figure 3. General view of the programmable tool "Electronic potato Tuber Log": 1—data logger; 2—computer; 3—tablet computer; 4—auxiliary equipment.

The use of video recording of the movement of the data logger on the surface of the bar elevator is due to the need to compare the time intervals obtained from the DVR with the diagrams of the personal computer of the software tool "Electronic Potato Tuber Log", followed by their superimposition to determine the location of the greatest force impact of the bar elevator on the data logger [11–14].

Experiments to assess the force effect of working bodies on potato tubers were carried out on potato harvesting machines of the following brands: AVR-Spirit-6200 combine harvester (Belgium, Roeselare), Dewulf RA-3060 combine harvester (The Netherlands, Winsum), and Bolko combine harvester (Poland, Grudziądz). The methodology of conducting experimental studies is as follows. Before conducting experimental studies on the accounting plot, the physical and mechanical properties of the soil were determined: moisture and hardness, according to the method of STO AIST 8.7–2013 "Machines for harvesting vegetable and melon crops. Methods for assessing functional indicators" (STO AIST 8.7–2013, et al.). The technological parameters of the potato harvesting machine before the research were set to optimal operating modes, in which the depth h_P of immersion of the digging ploughshare into the soil was set below the depth of potato tubers in a range of 0.12–0.18 m, which is due to the depth of planting tubers. The translational speed V_M of the potato harvester movement in the course of experimental studies was set in a range of 3–5.2 km/h. The optimal values of the translational V_{EL} of the speed of movement of the rod elevator of the potato harvesting machine under study were set in a range of values from 3.6 to 6.4 km/h, due to the prevention of unloading of the tuberous pile on the surface of the rod elevator. With the steady mode of movement of the web of the bar elevator, the DVR was turned on, and the data logger "Electronic Potato Tuber Log" was fed from container 2 for the preliminary placement of the pile. Consequently, after data logger 1 (Figure 1) passed the surface of the bar elevator, the logger was turned off, the studied factors were changed, and the experiment was repeated by the chosen research plan. In order to determine the destruction of soil clods under the dynamic action of the separating devices of the harvester, laboratory studies were carried out on a device (simulator of dynamic effects) that simulates the operation of the cleaning devices of the harvester (Figures 4 and 5). This specified device consists of a drum with rubberized rods 1, a blade 2 and an open window behind it for loading and unloading a batch of test tubers. Inside the drum, a finger slide 3 is installed above its axis, which allows you to change the angle of inclination. Fixed walls are installed at the ends of the drum, designed to keep tubers from rolling out of the drum and simulate the sides of the elevator and the sides of other working bodies of potato harvesters. The device consists of a drum with rubberized rods 1, with a blade 2 and an open window behind it for loading and unloading a batch of test tubers. Inside the drum, a finger slide 3 is installed above its axis, which allows you to change the angle of inclination. Fixed walls are installed at the ends of the drum, designed to keep tubers from rolling out of the drum and simulate the sides of the elevator and the sides of other working bodies of potato harvesters.

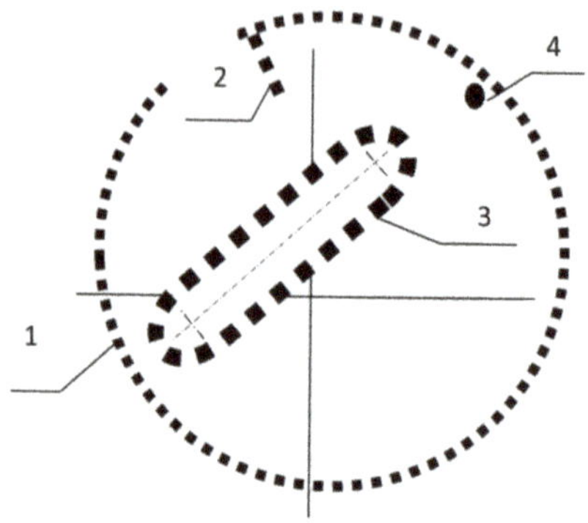

Figure 4. Design and technological scheme of the device for assessing the suitability of varieties and hybrids for mechanized harvesting: 1—rod drum; 2—blade; 3—finger slide; 4—shaker.

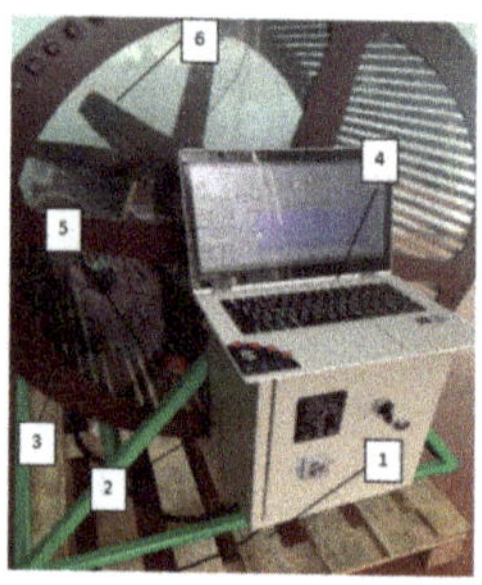

Figure 5. A general view of the laboratory installation for assessing the force effect of a mock-up sample of a device for assessing the suitability of varieties and hybrids for mechanized harvesting: 1—a support platform; 2—a control cabinet; 3—a drum; 4—a personal computer; 5—a data recorder; 6—a finger slide.

The separating devices imitate the instantaneous impact that takes place in the field by changing the rotational speed of the bar drum 3 and the translational speed of the finger slide 6.

The estimated number of cycles is determined based on the correspondence between the number of drops and the path travelled by potato tubers along the elevator of the potato harvester and the simulator. The determinant of damage to tubers works as follows: A portion of the test potato tubers is placed in the rod drum 1 through the window. When the drum rotates clockwise, blade 2 captures the tubers and lifts them to a predetermined height, which is set by changing the angle of inclination of the blade. With further rotation of the drum, the tubers fall on finger slide 3, simulating the drops and movement of tubers along the combine's hedgehog conveyors. From finger slide 3, the tubers roll down onto the rubberized rods of the drum (imitation of another drop) along which they roll until they return to blade 2, simulating the elevator of a potato harvester. When rolling over the inner surface of the drum, the tubers bump into rod 4, simulating the elevator shaker. Fixed sidewalls at the ends of the drum keep the tubers from rolling out of the drum and the finger slide while imitating the sidewalls of the elevator and the finger conveyors of potato harvesters. Then the cycle repeats. When the set number of cycles is counted, the drum is reversed and the tubers are poured out through the window into the receiving container and the movement of the drum is automatically turned off. The estimated number of cycles is determined based on the correspondence between the number of drops and the path travelled by potato tubers along the elevator of the potato harvester and the simulator. Experimentally, the number of cycles for each type of potato harvester is specified by the calibration method when harvesting potatoes in the field. The height of the differences on the simulator is limited by the design dimensions of the drum and must correspond to the differences in potato harvesters. A greater fine-tuning of the damage simulator for a specific harvesting machine is achieved by changing the angle of rotation of blade 2 and the location of the finger slide 3 of the simulator.

Experimental studies on the assessment of the force effect on potato tubers were carried out on a mock-up sample of a device for assessing the suitability of varieties and hybrids for mechanized harvesting, the design and operating principle of which were developed at the FSBI FSAC VIM (Figure 5).

The electronic potato allows you to record the magnitude of the acquired acceleration, as well as the momentum of the impact force from its interaction with the working elements of the laboratory installation.

In addition, using this device, experimental studies were carried out to determine the dynamic impact on the destruction of soil clods.

The mass of the lump was determined with an accuracy of 0.1 g on an electronic balance M-ER 122ACFJR-300.01 LCD (Figure 6).

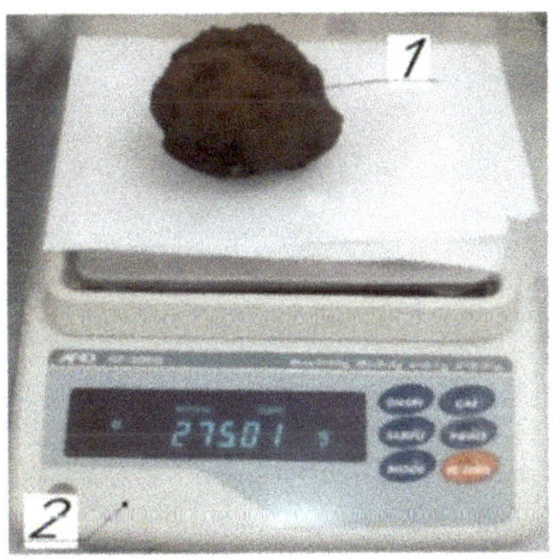

Figure 6. Weighing the soil clod: 1—soil clod; 2—electronic scales M-ER 122ACFJR-300.01 LCD.

The dimensions of the soil clods were determined with a caliper (Figure 7), which provides a measurement accuracy of 0.02 mm.

Figure 7. Determination of the dimensional characteristics of soil clods.

To determine the energy of destruction of soil clods, it is necessary to conduct studies on the interaction of soil clods with the material of the separating surface of harvesting machines in identifying the critical height of the fall of the destruction of the clod. To conduct these studies, soil samples were prepared (Figure 8), corresponding to the moisture content during harvesting and formation of a monolithic soil layer (Figure 9) with the appropriate dimensions: length (Lk = 5 cm), height (Hk = 5 cm), and thickness (Tk = 5 cm).

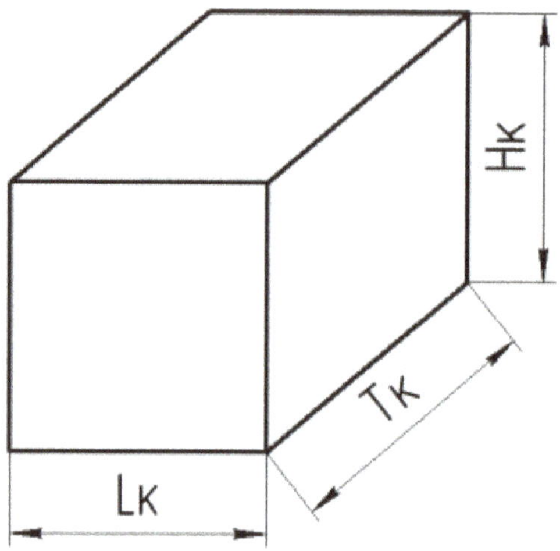

Figure 8. Soil sample diagram.

Figure 9. General view of the destruction of soil clods on the bars of the separating surface: 1—the canvas of the rod elevator; 2—reference scale; 3—soil sample.

When carrying out studies on the energy of destruction of soil clods, the moisture content of the studied soil samples varied from 10 to 24% with an interval of variation towards an increase of 1% via the method of surface moistening and its determination by the thermostatic-weight method.

3. Results and Discussion

Therefore, an assessment of the graphical dependencies (Figure 2) obtained during experimental studies on the Bolko S combine harvester is distinguished by the most "gentle" force effect of the working bodies of the harvesting machine on the tuberous pile, where throughout the entire cleaning process, there is a minimal force effect on the separated products in a range from 3 N to 6.5 N, which is 28–31% of the maximum force effect of the working bodies of the AVR-Spirit harvesting machines-6200 and "Dewulf RA-3060" (Figure 10) [14].

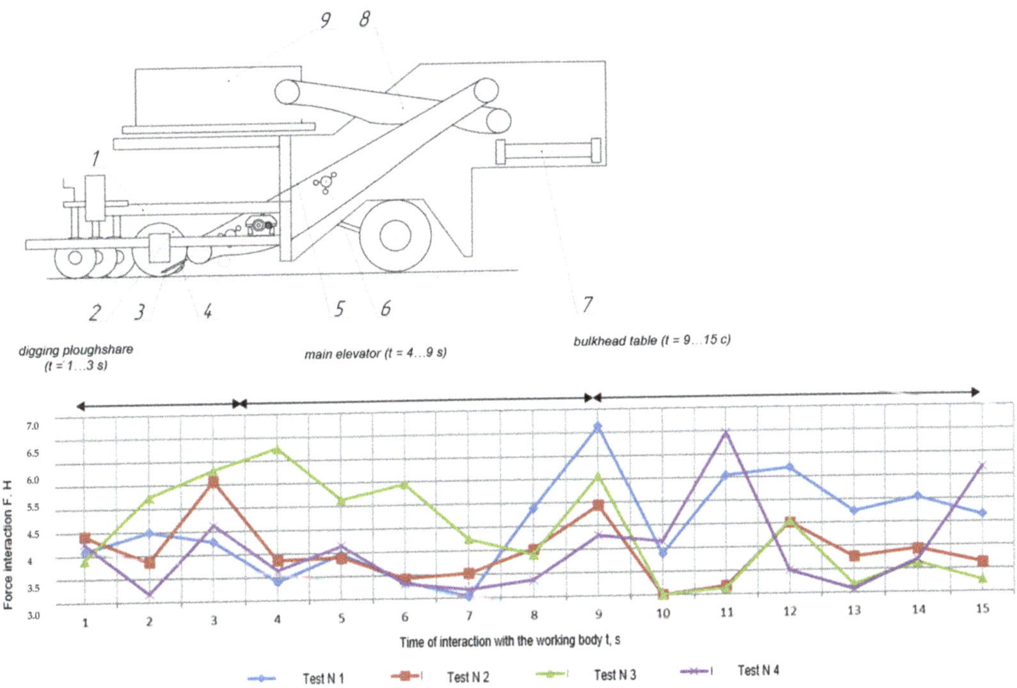

Figure 10. Structural and technological scheme of the Bolko combine harvester: 1—frame; 2—vertical discs; 3—digging ploughshare; 4—support wheel; 5—main elevator; 6—shaker; 7—transverse elevator; 8—bulkhead table; 9—hopper [15].

The conducted analysis of the experimental data presented in Figure 9 allows us to conclude that in the process of soil separation, tubers certainly interact with the active and passive working organs of the potato harvester.

Therefore, tubers fall during the passage of the shaking section on the bars of the elevator and during the transition from one elevator to another, which is reflected in the graphical dependencies presented above.

At differences from one elevator to another, tubers fall from a height of no more than 0.2 m, which corresponds to a collision speed of 1.9 m/s and satisfies the permissible collision speed of tubers with working bodies (2.2 m/s). The speed of collision of tubers thrown by the bars of the elevator from the impact of the shaker, with the vertical component of the speed of the bars of the elevator 0.7 m/s or more, will be more than 2.6 m/s, which is higher than the permissible speed of collision of tubers with the working bodies of machines. As a result, tubers are damaged in this area.

Each parameter of the size/mass characteristics was measured in triplicate, after which the average values of mass measurements were used to estimate the variation series.

At the same time, the concepts and elements generally accepted in variation statistics to characterize the variation series were used: average variation—X, standard deviation—σ, and coefficient of variation—ν. Each of the main elements was determined according to the known formulas of variation statistics.

This made it possible to determine the accuracy of the experimental data and to establish the acceptable limits within which they were sufficiently reliable.

To determine the number of intervals (K) for varying the values of the parameters of the size/mass characteristics of tubers, we used the empirical relationship [2,16].

$$K = \sqrt{n} \tag{1}$$

where n is the number of tubers, pcs.

$$K = \sqrt{100} = 10$$

The sampling range can be expressed as:

$$R = xmax - xmin, \tag{2}$$

where xmax and xmin denote the maximum and minimum values of the investigated feature.

The interval of the investigated feature is calculated as follows:

$$D = R/K \tag{3}$$

The size/mass characteristics of tubers combine the following features: the shape, size, and weight.

While planning a multifactorial experiment, it is necessary to analyze and determine the input parameters of the sorting process, which most significantly affect the quality indicators of the sorting. The selected factors should be controllable, unambiguous, compatible, independent, and the accuracy of measurements of the factor levels should be higher than the accuracy of the optimization parameter values.

When studying the process of separation of soil impurities and damage to potato tubers, factors were identified, the total number of which was initially 15, which covered the technological, design parameters of the separating devices, as well as the physical and mechanical properties of the soil.

As the results of studies of the fractional composition of soil clods show, one should take into account the fact that within one fractional group, there are also soil clods that differ in mass within one accounting plot (sampling in the harvester bunker was carried out every 15 m of movement). The mass of soil clods during the research is displayed in the diagrams in Figure 11.

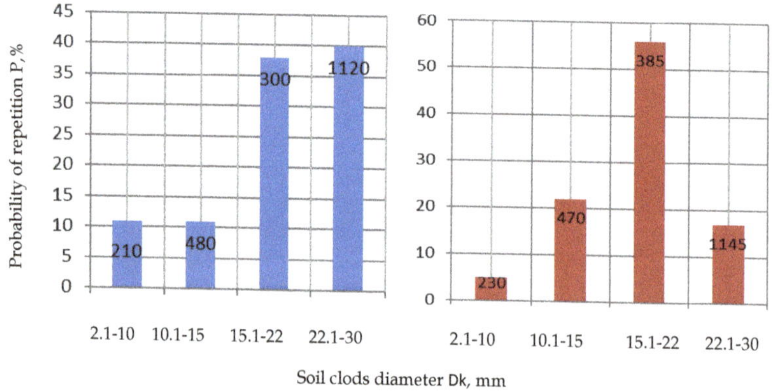

Figure 11. *Cont.*

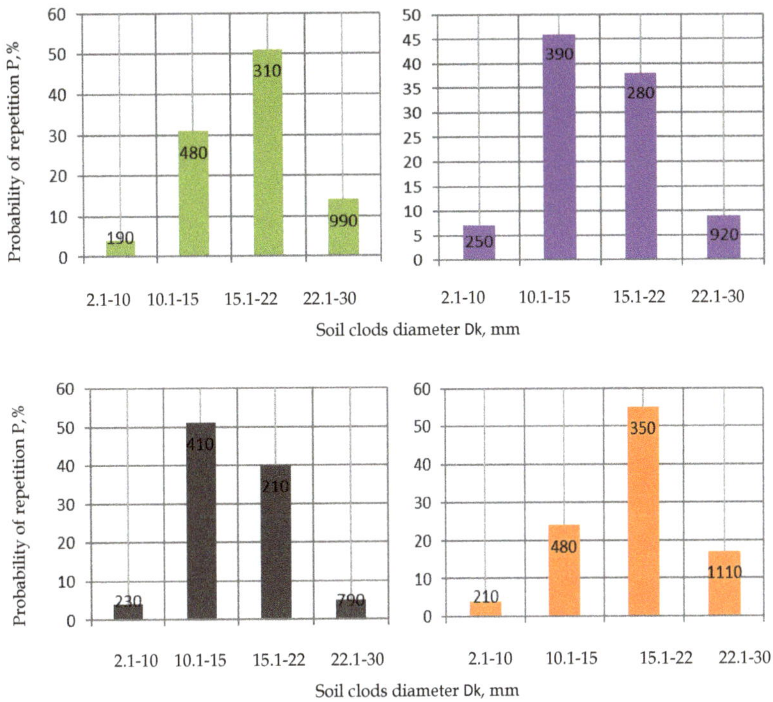

Figure 11. Fractional composition and mass of soil clods.

Analysis of the diagrams presented in Figure 9 indicates that within one fractional group, the mass of soil clods differs by an average of 10–15.4%, which should be explained by the distinctive physical and mechanical properties of soil clods, even within the same sample taken and the pattern of mass change soil clods is probabilistic in nature.

At the same time, it should be noted that the strength properties of soil clods and the boundaries of their destruction, depending on the density and moisture content, were determined under a five-fold dynamic impact, based on the condition of destruction of at least 90% of soil clods.

At the same time, the allowable loads were limited by the condition of tuber damage: the maximum allowable fall height on the metal lattice surface is not more than 0.25 m; the value of static compression is within 200–250 N. The results of the studies of the destruction of the soil layer are presented in Table 1.

Conducting research on the dynamic destruction of soil clods was carried out when determining the mass of a soil sample of clods, the probability of the appearance of a fractional composition of which is the greatest, namely, 20, 25, and 30 mm.

The results of the studies on the force impact of the working bodies of harvesters on damage to potato tubers and the destruction of soil clods indicate that the implementation of the process of destruction of soil clods on the separating devices of modern machines for harvesting root crops is not possible, due to the limiting correlation—wearing the injury rate of tubers up to 18 N, which does not allow one to ensure the destruction of the soil clod, the minimum value of the dynamic impact of which is 243.6 N, which exceeds the maximum allowable force effect exerted on the potato tuber by 2.3-times.

The correlation dependence of the statistical destruction and density of soil clods on their moisture content is represented by a system of expressions:

$$\begin{cases} P_K(W) = 326.14 - 9.54W + 0.27W^2 \\ \rho(W) = 2293.5 + 327.6W - 6.39W^2 \end{cases}, \quad (4)$$

Table 1. The results of studies on the dynamic destruction of soil clods.

Absolute Humidity of Leached Chernozem, %	Dynamic Impact, H	Soil Density, kg/m^3	Soil Clods Diameter, mm	Weight, g
18	243.6	1430	20.0	10.8
18	244.2	1540	25.0	46.6
18	242.8	1720	30.0	75.4
20	248.7	1530	20.0	12.4
20	246.2	1760	25.0	56.4
20	246.5	1840	30.0	78.7
22	253.4	1740	20.0	14.3
22	253.2	1850	25.0	58.7
22	253.6	1920	30.0	81.2
24	254.4	1830	20.0	16.2
24	254.8	1910	25.0	59.3
24	256.6	1950	30.0	81.7
26	261.5	1900	20.0	16.7
26	258.7	1930	25.0	61.2
26	262.6	1970	30.0	82.4
28	282.6	1920	20.0	17.1
28	283.8	1960	25.0	64.5
28	282.2	1980	30.0	83.2
30	287.4	1950	20.0	17.6
30	288.7	1970	25.0	67.2
30	287.6	1980	30.0	84.3

An analysis of the graphical dependence presented in Figure 12 allows us to state that the strength and density of soil clods increase, depending on the increase in the absolute moisture content of the soil, which is due to an increase in the plasticity and connectivity of soil particles among themselves and the impossibility of their destruction.

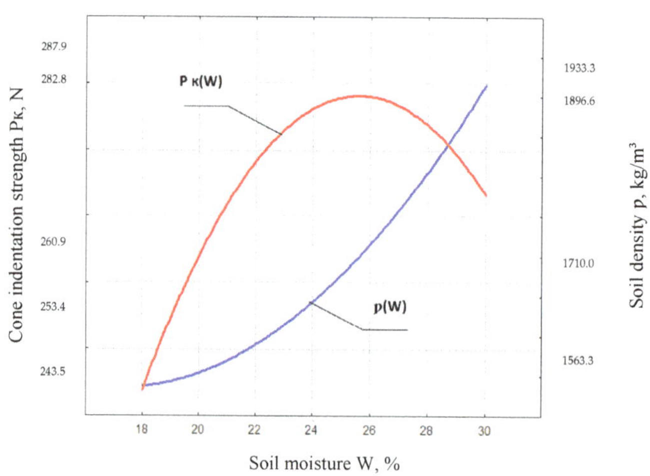

Figure 12. Dependence of the static destruction of soil clods on their density and moisture content.

The results of studies of the destruction energy of the soil layer depending on soil moisture and fall height are shown in Figure 13.

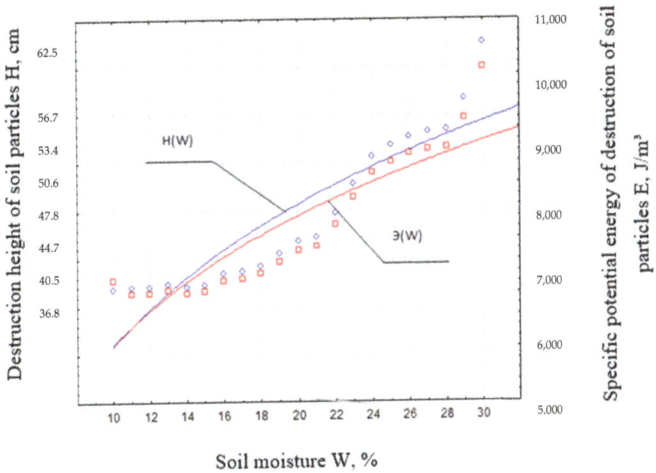

Figure 13. Dependence of the specific potential energy of the destruction of soil particles on the density, humidity, and height of fall.

The correlation dependence of the destruction energy of soil clods on their moisture content and fall height is represented by a system of expressions:

$$\begin{cases} H(W) = -18.21 + 49.12 \lg W \\ E(W) = -2684.87 + 7755.31 \lg W \end{cases}, \quad (5)$$

An analysis of the graphical dependence allows us to conclude that in a humidity range of 12–21%, the strength of the lumps does not change significantly; with a higher humidity, there is a sharp increase in the strength properties. With an increase in humidity from 21% to 25%, the destruction energy increases from 6848.21 to 8386.21 J/m^3. With soil moisture above 25%, the height corresponding to the height of destruction was in a range of 50–60 cm. These circumstances are explained by the presence of moisture, which plays the role of a damper, which leads to swelling of soil particles and a decrease in the intercolloidal distance between them and the absence of a violation of bonds between them. The height corresponding to the destruction height and belonging to the central area of the graph, with an absolute soil moisture content of 18–24%, is minimal and amounts to 37–41 cm. This should be explained by the physical maturation of the soil and an increase in the connectivity between particles.

When studying the process of mechanized harvesting of potatoes, factors were identified, the total number of which was initially 15, which covered the technological, design parameters of functioning elements, as well as the physical and mechanical properties of the soil.

In view of the fact that it is impossible to cover the influence of all factors and their interaction in studies, based on a priori information, as well as based on the specific objectives of the study, the most significant factors affecting the quality of potato harvesting were identified, which are presented in Table 2 and Figure 14.

Table 2. Factors affecting the quality of potato harvesting.

Designation	Name of Factors	Levels of Factor Variation	
		+1	−1
X_1	Angle of inclination of the bar elevator β, hail	15	5
X_2	Number of separators z2, things	6	2
X_3	Physical and mechanical properties of the soil l_C, things	5	2
X_4	Forward speed of movement of separating devices v_D, m/s	1.2	0.8

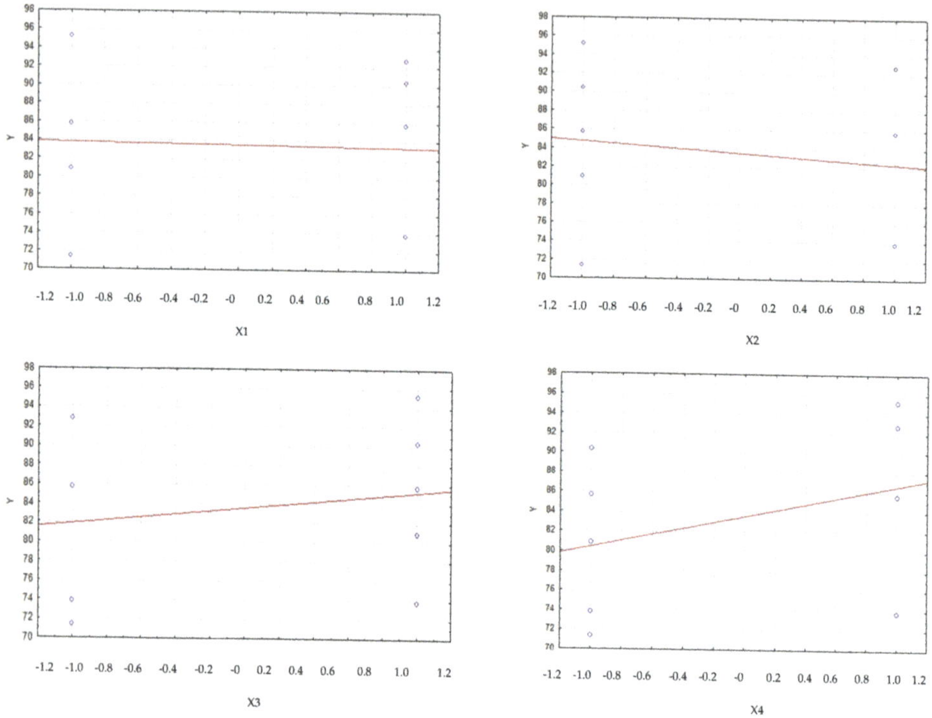

Figure 14. Scatter plots of levels of factors of the screening experiment.

To conduct a screening experiment, an experiment planning matrix was compiled (Table 3) from a full factorial experiment, the formation of two half-replicas of the type 2^{4-1}.

Table 3. Planning matrix and screening experiment results.

Experience Number	Factors				Criterion Optimization		
	X_1	X_2	X_3	X_4	Y	Y_1	Y_2
1	−	−	−	−	71.4	71.4	71.4
2	−	−	+	+	95.2	95.2	95.2
3	+	+	+	−	73.8	76.2	71.4
4	−	−	+	−	80.9	85.7	76.2
5	+	+	−	+	92.8	95.2	90.4
6	+	−	−	−	85.7	85.7	85.7
7	+	+	−	+	73.8	71.4	76.2
8	−	+	+	+	85.7	85.7	85.7
9	−	+	−	+	85.7	85.7	85.7
10	+	−	+	−	90.4	90.4	90.4

To evaluate the effects of factors, we construct a scatterplot. At the first stage, a selection of factors was chosen X_1 (angle of inclination of the bar elevator) и X_2 (number of separators).

The effects of the selected factors were assessed using a multi-entry table. The numerical values of the effects were (Table 4):

$$\begin{cases} X_7 = \frac{\overline{Y}_1+\overline{Y}_3}{2} - \frac{\overline{Y}_2+\overline{Y}_4}{2} = \frac{72.6+75.3}{2} - \frac{92.8+92.8}{2} = -18.85, \\ X_8 = \frac{\overline{Y}_1+\overline{Y}_2}{2} - \frac{\overline{Y}_3+\overline{Y}_4}{2} = \frac{72.6+92.8}{2} - \frac{75.3+92.8}{2} = -1.35 \end{cases} \quad (6)$$

Table 4. Table for calculating the effects of factors.

Factors Assessed	$+X_7$	$-X_7$
$+X_8$	73.8 71.4 - $\sum Y_1 = 145.2$ $\overline{Y}_1 = 72.6$	90.4 92.8 95.2 $\sum Y_2 = 278.4$ $\overline{Y}_2 = 92.8$
$-X_8$	71.4 73.8 80.9 $\sum Y_3 = 226.1$ $\overline{Y}_3 = 75.3$	90.4 92.8 95.2 $\sum Y_4 = 278.4$ $\overline{Y}_4 = 92.8$

The significance of the effects of the selected factors was checked by the *t*-test, the calculation results of which are summarized in Table 5, and the numerical values were $t_{X_1} = 2.28$ и $t_{X_1} = 4.95$.

Table 5. Table for calculating t—criteria.

№ Cells	$\sum Y_i$	$(\sum Y_i)^2$	$\sum Y_i^2$	n_i	$S_R{}^2 = \frac{\sum Y_i^2}{n_i-1} - \frac{(\sum Y_i)^2}{n_i(n_i-1)}$,	$\frac{S_{R^2}}{n_i}$
1	145.2	21,083.04	10,544.36	2	2.84	1.42
2	278.4	77,506.56	25,847.04	3	5.76	1.92
3	226.1	51,121.21	17,089.17	3	24.38	8.12
4	278.4	77,506.56	25,847.04	3	5.76	1.92

If the selected factor is significant, the tabular value of the t-test should be less than the calculated one. Tabular value of t (criterion) for the number of degrees of freedom f = 7 At 5%th level of significance $t_{0.05} = 2.365$, at 10% ом significance level $t_{0.1} = 1.895$.

Based on the calculation t (criterion), it can be concluded that the factor X_1 and X_2 is significant, with a confidence probability 0.95.

After highlighting the effects of X_1 and X_2, the results of the experiment were corrected.

Based on the adjusted results of the optimization parameter, scatterplots were again built, according to which the factors were visually identified X_3 and X_4. The numerical values of the effects were:

$$\begin{cases} X_3 = -12.23, \ t_{X_1} = -1.05, \\ X_4 = -6.45, t_{X_4} = 2.82 \end{cases} \quad (7)$$

Based on the calculation of the *t*-test, it can be concluded that the factor X_4 is significant with confidence probability 0.95, and the factor X_1 significant for 5%, significant for 10% the significance level of Student's tabular test, respectively $t_{0.05} = 2.447$ and $t_{0.1} = 1.943$ with the number of degrees of freedom f = 6.

The results of the quantitative assessment of the selected factors are displayed in Table 6.

Table 6. The results of the quantitative assessment of the identified factors.

Selection Stage	Factors	Importance of Factors	Estimated Value of t—Criteria
According to initial data	X_1	−18.85	2.28 ^{X}
	X_2	−1.35	4.95 ^{XX}
After adjustment	X_3	−12.23	−1.05 ^{XXX}
	X_4	−6.45	2.82 ^{X}

X—significant to the level 0.05($t_{0.05}$); XX—significant to the level 0.1($t_{0.1}$); XXX—significant to a level less than 0.1.

Insignificant factors as a result of the screening experiment were eliminated and we began to produce the search for the equation of the surface and the description of this surface.

As a result of calculating the regression coefficients, we obtained a linear equation for the factors of the screening experiment:

$$Y = 98.65 - 0.007x_1 0.015x_2 - 0.085x_3 \quad (8)$$

Fisher's calculated F_P (criterion) was obtained when determining the adequacy of the presentation of the results of the experiment using a polynomial of the first degree:

$$F_P = \frac{S_{LF}^2}{S_y^2} = \frac{0.0698}{0.022} = 3.17 \quad (9)$$

Tabular value of the Fisher criterion − F_T c 95% probability for degrees of freedom of the denominator $f_y = N \cdot (k-1) = 10 \cdot (3-1) = 20$ and numerator $f_{LF} = N - n - 1 = 10 - 4 - 1 = 5$ makes up $F_{0.05} = 3.84$.

The tabular value of the Fisher criterion F_T is greater than the calculated value F_T; therefore, the adequacy hypothesis can be accepted.

Therefore, it is necessary to ensure the development of separating working bodies of potato harvesters with fundamentally new principles of influencing the material being cleaned, the implementation of which will ensure not only the separation of potato tubers from soil clods but also reduce damage to marketable products from interaction with soil clods. The results of the studies obtained are consistent with the previously known data on the development and testing of machines in terms of cleaning marketable products of root and tuber crops. Further, the design of the separating rod elevator, which is a separation intensifier, contains within it a passive two-shoulder shaker 4 located under the upper branch of the web of the rod elevator 3 [14].

The disadvantages of the well-known design of the rod elevator include increased damage to root crops during the transition from one cascade to another, as well as the inability to disperse the pile of root crops over the entire width of the conveyor [15].

4. Conclusions

The results of the experimental studies carried out indicate the prospects of the work performed in the direction of obtaining knowledge of the physical and mechanical properties of the tuberous heap in order to create devices that ensure the separation of marketable potato products from soil clods in the development of modern machines for harvesting potatoes.

It has been established that in a humidity range of 12–21%, the strength of the lumps does not change significantly; at higher humidity, a sharp increase in strength properties occurs.

With an increase in humidity from 21% to 25%, the destruction energy increases from 270 to 288 N. With soil moisture above 25%, the height corresponding to the destruction height was in a range of 50–60 cm. These circumstances are explained by the presence of moisture, which is the role damper, which leads to swelling of soil particles and a decrease in the intercolloidal distance between them and the absence of a violation of the bonds between them.

The height corresponding to the destruction height and belonging to the central area of the graph, with an absolute soil moisture content of 21–27%, is minimal and amounts to 37–41 cm. This should be explained by the physical maturation of the soil and an increase in the connectivity between particles. In addition, the results of the studies obtained made it possible to establish that, regardless of the types of loads, the influence of density and moisture content on the strength properties of soil clods can be traced significantly.

The greatest destruction of soil clods is provided at a moisture content of 22–24%, which should be explained by the formation of the physical ripeness of the leached chernozem. In addition, a decrease in soil moisture leads to a more intensive gluing of soil particles among them and, consequently, an increase in the force effect on the soil lump for its destruction. Under dynamic impact, soil clods with a density of 1600–1700 kg/m^3 are destroyed and under static compression—1300–1400 kg/m^3 at a moisture content of soil clods of 22–24%.

This, in our opinion, is explained by the fact that vibration provides a more targeted development of cracks due to the constancy of the impact of external forces in a certain (limited) clod zone, which will ensure more efficient removal of soil clods and an increase in the purity of the tuberous heap. On heavy soils with high humidity, the use of destructive devices using the static compression method is also impractical.

The novelty of this study lies in obtaining empirical dependences of the physical and mechanical properties of soil clods on their size and mass parameters, with the possibility of changing the size of soil clods as a result of force action to the passage values of fractions through the separating devices of the harvesting machine.

The results of the research carried out will ensure the development of automated separating devices with an adaptive force effect of the cleaning executive elements on soil clods, with their physical and mechanical properties changing during the technological process of potato harvesting.

Author Contributions: Conceptualization, O.D., A.D. and A.S.; methodology, A.S. and M.G.; software, N.S., A.A. and M.M.; validation, A.D. and A.S.; investigation, A.S.; resources, A.S.; writing—original draft preparation, N.S.; writing—review and editing, A.S.; project administration, A.A.; funding acquisition, A.S. and A.D. All authors have read and agreed to the published version of the manuscript.

Funding: The study was supported by the Russian Science Foundation grant No. 22-76-10002.

Institutional Review Board Statement: Not applicable.

Informed Consent Statement: Not applicable.

Data Availability Statement: The raw data supporting the conclusions of this article will be made available by the authors, without undue reservation.

Conflicts of Interest: The authors declare that they have no known competing financial interest or personal relationships that could have appeared to influence the work reported in this paper.

References

1. Sibirev, A.V.; Aksenov, A.G.; Mosyakov, M.A. Substantiation of design and technological parameters of a separating bar conveyor with an asymmetric arrangement of shakers. *Bull. Fed. State Educ. Inst. High. Prof. Educ. Mosc. State Agroeng. Univ. V.P. Goryachkina* **2018**, *4*, 15–20.
2. Akdemir, B.; Ulger, P.; Arin, S. Mechanized panting and harvestiong of onion. *Agric. Mech. Asia* **1993**, *24*, 23–26.
3. Sibirev, A.V.; Aksenov, A.G.; Mosyakov, M.A. Experimental Laboratory Research of Separation Intensity of Onion Set Heapson Rod Elevator apparatus. *J. Eng. Appl. Sci.* **2018**, *23*, 10086–10091.
4. Sibirev, A.V.; Aksenov, A.G.; Mosyakov, M.A. Results of experimental studies of separation of a heap of onion sets on a bar elevator with asymmetrically installed shakers. *Eng. Technol. Syst.* **2019**, *1*, 91–107. [CrossRef]
5. Dorokhov, A.S.; Sibirev, A.V.; Aksenov, A.G. Results of field studies on the separation of a heap of onion sets on a bar elevator with asymmetrically installed shakers. *Eng. Technol. Syst.* **2020**, *1*, 133–149. [CrossRef]
6. Pasaman, B.; Zakharchuk, V. The determination of the parameters of a ploughshare-rotor potato digger. *Econtechmod. Int. Q. J.* **2012**, *1*, 43–47.
7. El-Rahman, A.; Magda, M. Development and performance evaluation of a simple grading machine suitable for onion sets. *J. Soil Sci. Agric. Eng. Mansoura Univ.* **2011**, *2*, 213–226. [CrossRef]

8. Khura, T.; Mani, I.; Srivastava, A. Design and development of tractor-drawn onion (Allium cepa) harvester. *Indian J. Agric. Sci.* **2011**, *6*, 528–532.
9. Indraja, D.; Ajkhilesh, J.; Vishal, P.; Amarsingh, P.; Ashutosh, D. A Review Paper Based on Design and Development of An Onion Harvesting Machine. *J. Inf. Comput. Sci.* **2019**, *9*, 333–337.
10. Bachche, S. Deliberation on Design Strategies of Automatic Harvesting Systems: A Survey. *Robotics* **2015**, *4*, 194–222. [CrossRef]
11. Dai, F.; Guo, X.H.; Zhao, W.Y.; Xin, S.L.; Liu, X.L.; Wu, Z.W. Design and experiment of canvas belt combined operation machine for potato digging and plastic film collecting. *Trans. CSAM* **2018**, *3*, 104–113.
12. Byshov, N.V.; Yakutin, N.N.; Koveshnikov, R.Y.; Rodionov, V.V.; Serzhantov, N.V.; Smirnov, P.S. Modernization of the KST-1.4 digger. *Bull. Ryazan State Agrotechnol. Univ. Named V.I. P.A. Kostychev* **2016**, *2*, 75–78.
13. Bachche, S.; Oka, K. Modeling and performance testing of end-effector for sweet pepper harvesting robot. *J. Robot. Mechatron.* **2013**, *25*, 705–717. [CrossRef]
14. Xiong, P.; Lin, X.; Wang, Y.; Shen, S.B. Simulation of southern potato grader based on adams. *INMATEH—Agric. Eng.* **2016**, *1*, 5–13.
15. Sibirev, A.; Aksenov, A.; Dorokhov, A.; Ponomarev, A. Comparative study of force action of harvester work tools on potato tubers. *CAAS Agric. Eng. J.* **2019**, *55*, 85–90. [CrossRef]
16. Erokhin, M.N.; Dorokhov, A.S.; Sibirev, A.V.; Aksenov, A.G.; Mosyakov, M.A.; Sazonov, N.V.; Godyaeva, M.M. Development and Modeling of an Onion Harvester with an Automated Separation System. *AgriEngineering* **2022**, *4*, 380–399. [CrossRef]

agriculture

MDPI

Article

Red Light Emitting Transition Metal Ion Doped Calcium Antimony Oxide for Plant Growth Lighting Applications

Lankamsetty Krishna Bharat [1,2], Harishkumarreddy Patnam [2], Alexander Sokolov [1], Sergey V. Gudkov [1,3,*] and Jae Su Yu [2,*]

1 Federal State Budgetary Scientific Institution "Federal Scientific Agroengineering Center VIM", 1st Institutsky Proezd 5, 109428 Moscow, Russia
2 Department of Electronic Engineering, Institute for Wearable Convergence Electronics, Kyung Hee University, Yongin-si 17104, Republic of Korea
3 Prokhorov General Physics Institute of the Russian Academy of Sciences, Vavilova Str. 38, 119991 Moscow, Russia
* Correspondence: s_makariy@rambler.ru (S.V.G.); jsyu@khu.ac.kr (J.S.Y.)

Abstract: In this work, we synthesized Mn^{4+}-doped $CaSb_2O_6$ phosphors using the conventional solid-state reaction method for plant growth lighting applications. The morphological, structural, and optical properties were analyzed based on the results obtained from scanning electron microscope, X-ray diffraction, and spectrophotometer. The results of the spectrophotometer illustrate that the phosphors showed a red emission band in 550–800 nm wavelength range with peak maxima at 642 nm. The red emission in these phosphors is attributed to the $^2E_g \rightarrow {}^4A_{2g}$ transition of Mn^{4+} ions. The emission intensity is increased with the doping of a charge compensator. The emission range of the phosphor covers the absorption range of photosynthetic pigments such as chlorophyll a, chlorophyll b, phytochrome P_r, and phytochrome P_{fr}. The results signify that the prepared phosphor materials are suitable candidates for application in plant growth lighting.

Keywords: transition metal; charge compensator; red emission; artificial lighting; photosynthetic pigments

Citation: Bharat, L.K.; Patnam, H.; Sokolov, A.; Gudkov, S.V.; Yu, J.S. Red Light Emitting Transition Metal Ion Doped Calcium Antimony Oxide for Plant Growth Lighting Applications. *Agriculture* **2022**, *12*, 2066. https://doi.org/10.3390/agriculture12122066

Academic Editor: Athanasios Koukounaras

Received: 20 October 2022
Accepted: 25 November 2022
Published: 1 December 2022

1. Introduction

Recently, luminescent materials have been the focus of interest of the research community due to their vast suitability for applications, namely, display, solar cells, biomedical, sensing, etc. [1–3]. The benefits of these luminous materials in a variety of applications have prompted scientists to look for new and better materials with enhanced luminescence capabilities [4]. This artificial luminescence is used for plant growth and development in greenhouses [5]. The lighting conditions in plant growth are directly related to the success of the production. Plant lighting demand and energy usage are growing in tandem with the advancement of modern agriculture. Generally, in plant production systems, traditional light sources such as fluorescent lamps, incandescent lamps, and high-pressure sodium lamps are used [6]. However, traditional light sources were primarily designed to support human activities based on the sensitivity of the human eye rather than the absorption spectra of plants. This type of traditional lamp for crop systems suffers from high energy consumption and a serious spectral mismatch between its emitting spectra and the absorption spectra of plants. Hence, light-emitting diodes (LEDs) have become an unavoidable choice in comparison to traditional light sources due to their quick response time, long lifetimes, energy savings, low cost, and reliability [7–9]. LED wavelengths can be modified using various phosphors to fit the spectral range of plant photosynthesis and photo morphogenesis, potentially affecting plant growth and development through modulating phytochrome. It is known that light impacts various developmental processes in plants, including seed germination, blooming, fruiting, and other morphogenesis, in addition to being a vital energy source for photosynthesis [10,11]. Blue (400–500 nm), red

(620–690 nm), and far-red (700–740 nm) lights are responsible for photosynthesis, phototropism, and photo morphogenesis, respectively [12]. It is important to note that plants' photoreceptor systems are particularly sensitive to red and far-red light, which are involved in the entire growth process [13,14]. As a result, the spectrum required for plant growth must correspond to that of photosensitive pigments.

Typically, Eu^{3+}- and Eu^{2+}-doped phosphors were reported for red emission. Eu^{3+}-doped materials show sharp and narrow peaks with peak maxima in the wavelength range of 610 to 620 nm, which is considerably far from deep to far-red emission [15,16]. Similarly, Eu^{2+}doped nitride phosphors are also confined due to precise and rigorous synthesis conditions and the high cost of rare-earths [17,18]. To overcome difficulties, transition metal ions such as Cr^{3+}, Mn^{4+} can be chosen as alternative dopants for plant growth applications [19–21]. In this work we studied the optical properties of Mn^{4+}doped $CaSb_2O_6$ materials. Mn^{4+} with a $3d^3$ electronic configuration generally stabilizes in an octahedron environment in most of the host materials. The optical properties of Mn^{4+} ions were significantly affected by the crystal field which is stronger in oxide- than in fluoride-based host materials [22]. The Mn^{4+} doped host materials show a broad excitation band due to the O^{2-}-Mn^{4+} charge transfer and the $^4A_{2g} \rightarrow {}^4T_{1g}$, $^4A_{2g} \rightarrow {}^2T_{2g}$, and $^4A_{2g} \rightarrow {}^4T_{2g}$ transitions. Likewise, the emission is seen in the red region (550–800 nm) due to the $^2E_g \rightarrow {}^4A_{2g}$ transition of the Mn^{4+} ion [23]. On this account, we can say that the similarity of the red or far-red emission of the Mn^{4+} doped oxide host materials and the desired spectral range of plants make the Mn^{4+} doped oxide materials an ideal choice for the LED application of plant growth.

In this context, we prepared an Mn^{4+}doped $CaSb_2O_6$ (CSO: Mn^{4+}) sample using the conventional high-temperature solid-state method. The sample was characterized to study the crystal structure, morphology, elemental composition, and luminescence properties. The emission of the CSO: Mn^{4+} sample was further compared with the absorption spectra of the photosensitive pigments of the plant to portray their use for plant growth LED application. Furthermore, the effect of the charge compensator on the luminescence properties was also studied.

2. Materials and Methods

2.1. Experimental Procedure

The Mn^{4+}doped $CaSb_2O_6$ (CSO: Mn^{4+}) phosphors with different concentrations ($CaSb_{2\text{-}x}O_6$: xMn^{4+} (x = 0.25, 0.5, 0.75, 1, 2, and 3 mol %)) were prepared by a conventional solid-state reaction method. All analytical grade chemicals, i.e., calcium carbonate ($CaCO_3$), antimony oxide (Sb_2O_3), and manganese carbonate ($MnCO_3$) were purchased from Sigma Aldrich Co, Gangnam-gu, Seoul, Korea. Raw materials were used as received without further purification. Initially, the stoichiometric ratios of the desired raw materials were weighed and put in an agate mortar for grinding with the support of the pestle. After homogeneous mixing, the powders were carefully transferred to alumina crucibles and calcined at 1000 °C for 6 h. After the completion of the calcination process, the obtained powders were ground smooth for characterizations.

2.2. Characterizations

The crystal structure of the optimized Mn^{4+}activated $CaSb_2O_6$ (CSO: $2Mn^{4+}$) phosphor sample was analyzed by X-ray diffraction (XRD) (Mac Science, M18XHF-SRA, Yokohama, Japan) with Cu kα (λ = 1.5406 Å). The morphology of the CSO: $2Mn^{4+}$ phosphor sample was studied using a field-emission scanning electron microscope (FE-SEM) (ZEISS, LEO SUPRA 55, Oberkochen, Baden-Württemberg, Germany). The elemental analysis was performed using an energy dispersive X-ray (EDX) spectrometer attached with FE-SEM. The oxidation states of the elements were found using X-ray Photoelectron spectroscopy (XPS) (K-alpha, Thermo electron, Gangnam-gu, Seoul, Republic of Korea). The photoluminescence excitation (PLE) and PL emission (PL) properties were investigated using a Sinco FluoroMate FS-2 spectrofluorometer, Gwangju, Korea. The lifetime was measured on a Photon Technology International (PTI) fluorimeter, Kyoto, Japan, which was equipped with a phosphorimeter attachment to the main system along with a 25 watt Xe-flash lamp.

3. Results and Discussion

Figure 1 shows the morphology and composition analysis of the CSO: $2Mn^{4+}$ sample. The FE-SEM image shown in Figure 1a depicts that the particles formed were of sub-micrometer size. The particles have irregular shapes with a smooth surface. The elemental analysis of the as-prepared sample was performed using the EDX attachment of FE-SEM. The EDX spectrum of the sample is shown in Figure 1b and contains all elements, i.e., Ca, Sb, O, and Mn. The Ca, O, and Mn occupy K-shell and show peaks at 3.7, 0.52, and 5.9 eV, respectively. Sb and Mn occupy the M-shell and show peaks at 0.73 and 0.64 eV, respectively, and Mn also occupies the K shell with a peak at 5.9 eV. Elemental mapping of a small area is also presented in the figure which shows the elemental distribution in the sample. From Figure 1c–f it can be said that the elements Ca, Sb, O, and Mn were uniformly distributed in the sample. A layered image of all the elements present in the sample for the measured area is shown in the inset of Figure 1b.

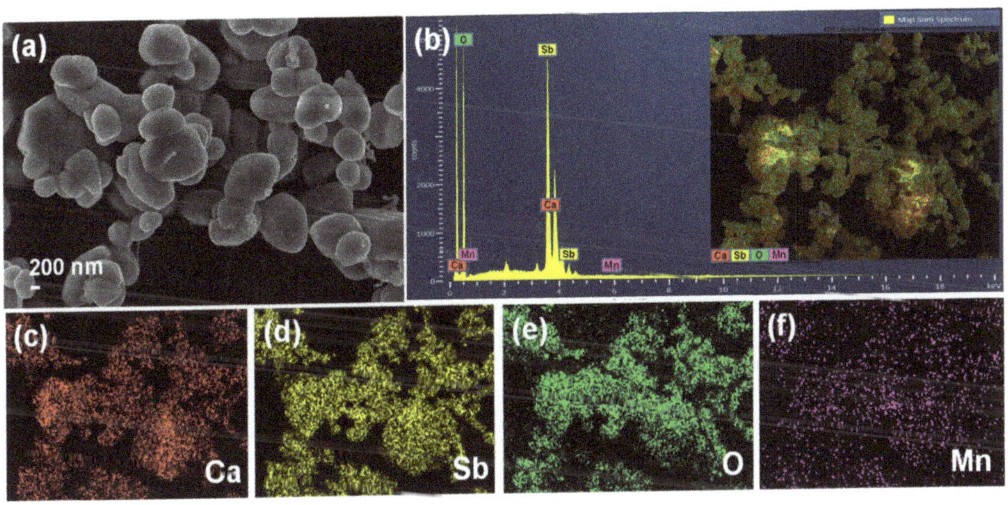

Figure 1. (**a**) FE-SEM image, (**b**) EDX spectrum, and elemental mapping of elements (**c**) Ca, (**d**) Sb, (**e**) O, (**f**) Mn. Inset of (**b**) shows a layered image of all the elements present in the sample.

The XRD pattern of the CSO: $2Mn^{4+}$ sample is shown in Figure 2a. The XRD pattern shows sharp diffraction peaks with major peaks at 2θ values of 17.62°, 19.53°, 26.44°, 34.17°, 38.71°, and 50.27° which corresponds to (001), (100), (101), (110), (2-11), and (2-12) planes, respectively, with other minor peaks at higher 2θ values. All the diffraction peaks of the sample match well with the standard JCPDS card #46-1496. The results show the sample crystallized in a hexagonal system with a space group of P -3 1 m (162). The lattice parameters of the sample are a = b (Å): 5.24, c (Å): 5.02, and V ($Å^3$): 119.42 [24].

The crystal structure of the sample was drawn using diamond software and is shown in Figure 2b. The crystal structure shown in the figure contains alternating layers of CaO_6 octahedra and SbO_6 octahedra. Each of the SbO_6 octahedra share corners with six CaO_6 octahedra, on the other hand, each CaO_6 octahedra shares corners with twelve SbO_6 octahedra. The length of all Ca–O bonds is 2.43 Å and the lengths of all Sb–O bonds are 2.02 Å. Generally, Mn^{4+} occupies and stabilizes in an octahedral site with 6-fold coordination [25] but, in this structure, both Ca and Sb form octahedra coordination, but the ionic radio of Ca is 1.00 Å and Sb is 0.6 Å, while it is 0.53 Å for Mn. Usually, the difference of ionic radii between the dopant and the host atom should not be more than 30% which suggests that Mn will be doped in the Sb site rather than the Ca site. The oxidation states of the elements present in CSO: $2Mn^{4+}$ sample were studied using the results obtained from XPS. The survey scan spectrum and high resolution Ca 2p, Sb 3d, O 1s, and Mn 2p were shown in Figure 2c. The survey scan spectrum taken in the binding energy range of 0–1300 eV shows peaks of all elements at respective binding energy values. The extrinsic hydrocarbon environment during the XPS measurement results in the visibility of C 1s peak around 280 eV in the survey scan spectrum. High resolution spectrum of Ca 2p shows two peaks at 349.5 and 346.03 which correspond to Ca $2p_{1/2}$ and Ca $2p_{3/2}$, respectively. Similarly, the high resolution spectrum of Sb 3d shows three peaks at binding energy values of 538.65, 531.89, and 529.3 eV. The peaks at 538.65 and 529.3 correspond to Sb $3d_{3/2}$ and Sb $3d_{5/2}$, respectively, and the peak at 531.89 eV belongs to O 1s. The high resolution spectrum of Mn 2p in the binding energy range of 660–630 eV shows high noise to signal ratio as the doping concentration of Mn in the host lattice is nominal.

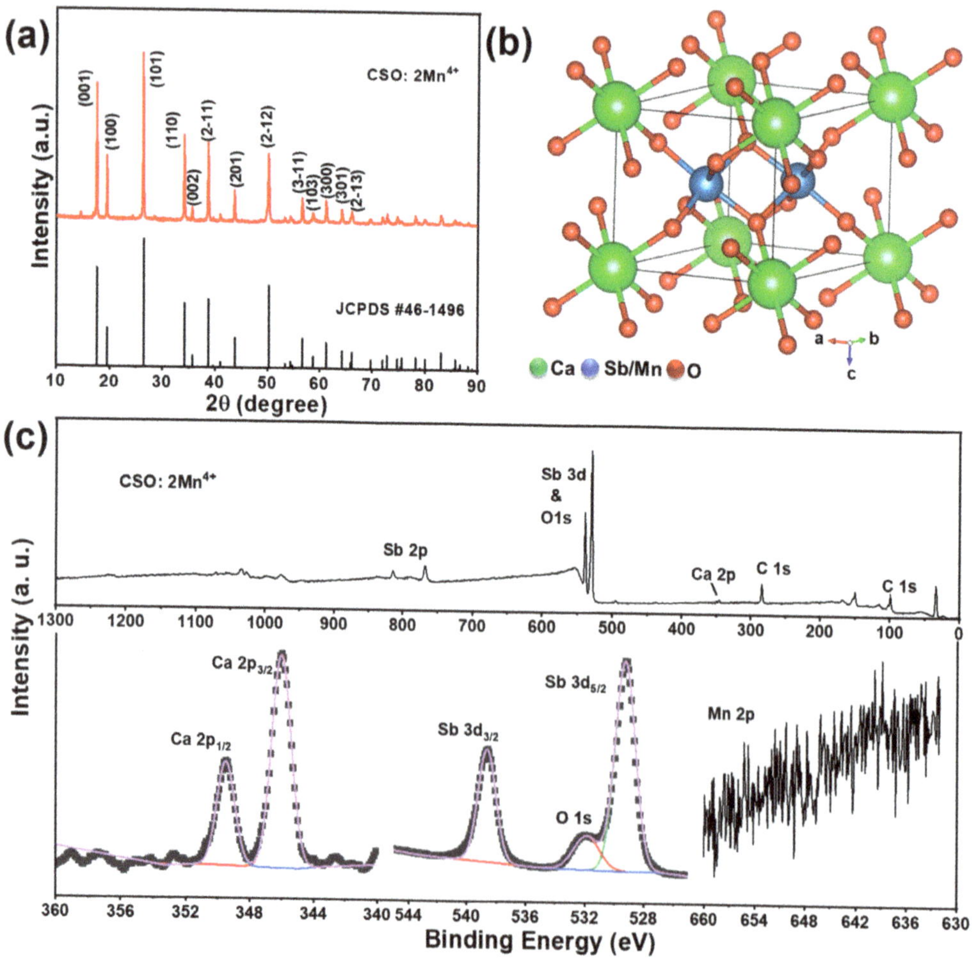

Figure 2. (**a**) XRD spectrum, (**b**) crystal structure, and (**c**) XPS results of CSO: $2Mn^{4+}$.

The room temperature photoluminescence excitation (PLE) spectrum of the CSO: $2Mn^{4+}$ is shown in Figure 3a.

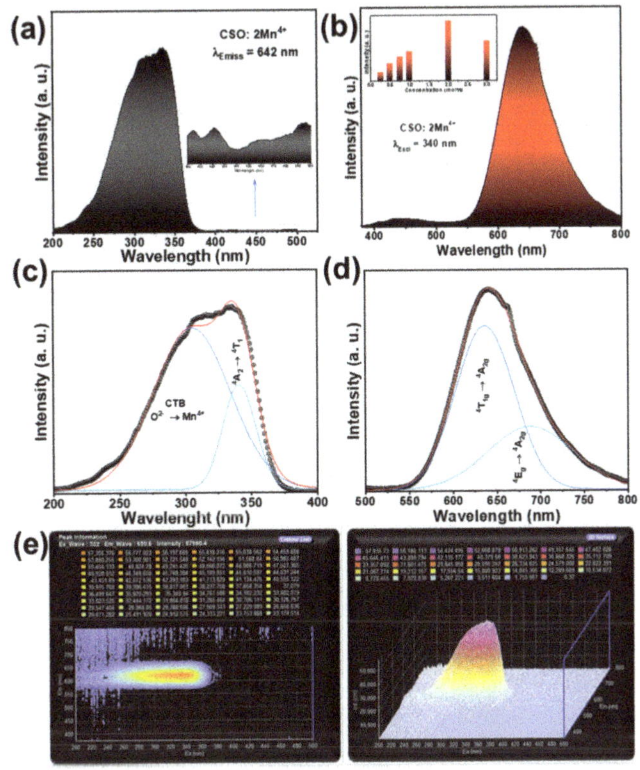

Figure 3. (**a**) PLE, (**b**) PL, (**c**,**d**) Gaussian fitting results and (**e**) 3D luminescence of CSO: $2Mn^{4+}$. Inset of (**b**) shows a graph of intensity variation with different concentrations of Mn^{4+}.

The PLE spectrum shows broadband ranging from 200 to 400 nm with peak maxima located at 340 nm. The broadband is attributed to the spin-allowed transitions which have large electron-phonon coupling, i.e., $^{4}A_{2g} \rightarrow {}^{4}T_{1g}$ and Mn^{4+}-O^{2-} charge transfer. As can be seen from the Gaussian fitting results (Figure 3c), the broad band can be de-convoluted into two peaks, of which the charge transfer band is dominant, which is centered at 305 nm. Weak bands arising from the $^{4}A_{2g} \rightarrow {}^{2}T_{2g}$ and $^{4}A_{2g} \rightarrow {}^{1}T_{2g}$ electronic transitions are observed in the longer wavelength region (magnified in Figure 3a) and the results are in accordance with previous reports [21,26]. These results confirm that the Mn is in +4 oxidation state in the host material. On the other hand, the photoluminescence (PL) spectrum of the sample is shown in Figure 3b. The Mn^{4+}-doped sample when excited at 340 nm shows a bright far-red emission band ranging from 575 to 800 nm with a peak maximum at 642 nm. The band observed is due to the spin-forbidden $^{2}E_{g} \rightarrow {}^{4}A_{2g}$ electronic transitions in Mn^{4+} ions. However, the broadband splits into two peaks when fitted with a Gaussian function (Figure 3d) with peak maxima at 636 and 688 nm corresponding to the $^{4}T_{1g} \rightarrow {}^{4}A_{2g}$ and $^{2}E_{g} \rightarrow {}^{4}A_{2g}$ electronic transitions, respectively. The results are consistent with the previous reports of Mn-doped oxide materials [26–29]. To determine the optimal doping concentration of Mn^{4+} ions, the host sample was doped with different concentrations and the PL spectra was measured. As the concentration of Mn^{4+} increases, the PL intensity increases and reaches the maximum at a concentration of 2 mol% and then decreases with further increase in concentration (inset of Figure 3b). This observed decrease in concentration is due to the quenching effect caused by the energy migration between neighboring Mn^{4+} activator ions. There are several reasons for the concentration quenching effect; as the PLE and PL spectra do not overlap, the re-absorption of radiation will not be one among them. So, to further evaluate, the critical distance (R_c) between the two Mn^{4+} ions was calculated. If the R_c value is less than 5 Å, the exchange interaction is dominant and the multipole-multipole interaction will be dominant if the value is greater than 5 Å. The ionic radii of Sb and Mn are almost similar, and very dilute concentrations were doped into the host lattice. Therefore, this does not affect the volume of the unit cell much, and the critical concentration is X_c = 0.02. The critical distance calculated was found to be 15.6 Å which depicts that the multipole-multipole interaction is dominant in the concentration quenching effect [30,31]. On the other hand, the three-dimensional (3D) spectra of the CSO: $2Mn^{4+}$ sample were obtained to confirm the broad excitation and narrow emission of the sample. The contour line and 3D surface plots were shown in Figure 3e. The 3D surface plot was measured in the wavelength range of 400–800 nm for each nanometer excitation wavelength from 200 to 500 nm. From the contour line and 3D surface plots we can say that emission is only observed with lower excitation wavelengths, and at higher excitation wavelengths, i.e., above 380 nm, there is no emission observed. Strong red emission is observed in the excitation wavelengths range of 320–350 nm.

Figure 4 shows the Tanabe-Sugano energy level diagram of Mn^{4+} which is generally used to examine the luminescence mechanism. In general, the $3d^3$ electronic configuration of Mn^{4+} ions make them sensitive to the surrounding crystal field. To explain the effect of the crystal field strength on the PL properties of the CSO: Mn^{4+} phosphors, the Racah parameters and crystal field strength (D_q) are estimated. The D_q value can be estimated with the support of the ${}^4A_{2g} \rightarrow {}^4T_{2g}$ peak energy (23,753 cm^{-1}) according to the following calculation:

$$D_q = E\frac{({}^4\mathrm{A}_{2g} \rightarrow {}^4\mathrm{T}_{2g})}{10} \tag{1}$$

Then, the Racah parameter can be evaluated by using the peak energy difference between the ${}^4A_{2g} \rightarrow {}^4T_{1g}$ and ${}^4A_{2g} \rightarrow {}^4T_{2g}$ transitions by the following equations:

$$\frac{D_q}{B} = \frac{15(x-8)}{(x^2-10x)} \tag{2}$$

$$x = E\left({}^4\mathrm{A}_{2g} \rightarrow {}^4\mathrm{T}_{1g}\right) - E\frac{({}^4\mathrm{A}_{2g} \rightarrow {}^4\mathrm{T}_{2g})}{D_q} \tag{3}$$

The Racah parameter C can be estimated by the peak energy of the ${}^2E_g \rightarrow {}^4A_{2g}$ emission transition using

$$E\frac{({}^2Eg \rightarrow {}^4\mathrm{A}_{2g})}{B} = \frac{3.05C}{B} - \frac{1.8B}{D_q} + 7.9 \tag{4}$$

According to Equations (1)–(4), the values of D_q, B, and C were estimated to be 2375.3, 502.96, and 3525.67 cm^{-1}, respectively. The D_q/B value estimation will decide the presence of Mn^{4+} in strong or weak crystal filed [32]. The calculated D_q/B was found to be 4.72. The obtained result shows that the Mn^{4+} ions are placed in a strong crystal field in CSO host lattice. Furthermore, the nephelauxetic ratio ($\beta 1$) shows a substantial effect on the position of ${}^2E_g$ energy level in a different host which can be obtained using the equation as shown below:

$$\beta 1 = \sqrt{\left(\frac{B}{B_o}\right)^2 / \left(\frac{C}{C_o}\right)^2} \tag{5}$$

B_o and C_o are the Racah parameters of Mn^{4+} free-ions, which are 1160 and 4303 cm^{-1}, respectively. The $\beta 1$ value calculated for CSO: $2Mn^{4+}$ was found to be 0.927.

Figure 5 shows the comparison of the emission spectrum of CSO: $2Mn^{4+}$ and the absorption spectra of chlorophyll a and chlorophyll b.

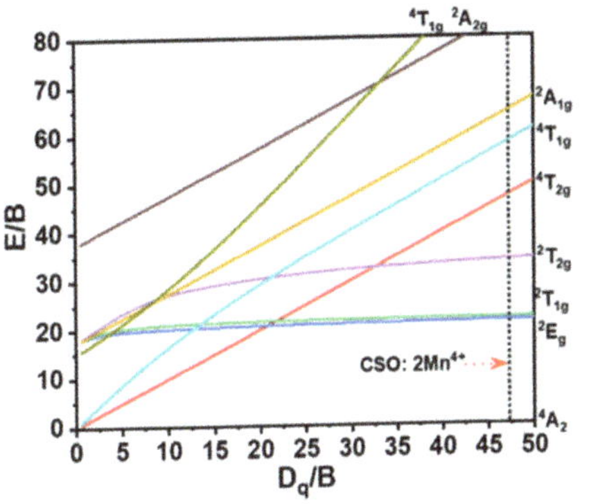

Figure 4. Tanabe-Sugano energy level diagram of Mn^{4+}.

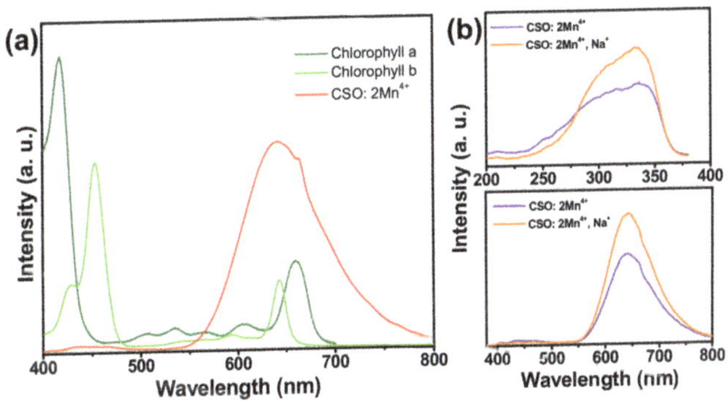

Figure 5. (**a**) Comparison of PL spectrum of CSO: $2Mn^{4+}$ and absorption spectra of chlorophyll a and chlorophyll b. (**b**) PLE and PL spectra of Na^+-doped CSO: $2Mn^{4+}$.

When analyzing the figure, we can say that there is a significant overlap between the emission spectrum of CSO: $2Mn^{4+}$ and the absorption spectra of chlorophyll a and chlorophyll b in the red region. Generally, chlorophyll is the green pigment of plants that absorbs the incident light and converts it into usable energy through a process called photosynthesis. On the other hand, the emission spectrum also overlaps with the absorption spectra of phytochrome P_r and phytochrome P_{fr}. P_r is the inactive form and converts to the active P_{fr} form on the absorption of red light and P_{fr} converts back to the inactive P_r form on absorption of far-red light or in the absence of light [33]. The results indicate that the CSO: $2Mn^{4+}$ sample, which has the potential to produce red light, can tune and accelerate the plant growth cycle and the carbohydrate yield. Furthermore, the effect of a charge compensator on the luminescence property of the CSO: Mn^{4+} sample was studied by doping Na^+ ions in the host lattice. Usually, the ionic radii of the charge compensator should be more than the dopant ion, otherwise the charge compensator easily occupies the Sb site more than the actual dopant ion and decreases the concentration of Mn^{4+} at the Sb site [34,35]. The ionic radio of Na^+ and Mn^{4+} are 1.02 and 0.53 Å, respectively. So, when Na^+ was doped along with Mn^{4+} ions at the Sb site, an increase in the intensity i.e., both in PL and PLE spectra, was observed.

Figure 6a shows the decay profiles of the CSO: $2Mn^{4+}$ and CSO: $2Mn^{4+}$, Na^+ samples for excitation and emission wavelengths of 340 and 642 nm, respectively. The Mn^{4+} single doped sample best fit with the single exponential function as shown below:

$$I(t) = I_o + Ae^{\frac{-t}{\tau}} \tag{6}$$

where I_o and $I(t)$ are initial luminescence intensity and luminescence intensity at time t, A is fitting constant and τ is decay time. The CSO: $2Mn^{4+}$ sample shows a decay time of 2111.51 µs. Similarly, CSO: $2Mn^{4+}$, Na^+ sample was fitted for a double exponential function as shown below:

$$I(t) = I_o + A_1 e^{\frac{-t}{\tau_1}} + A_2 e^{\frac{-t}{\tau_2}} \tag{7}$$

where A_1 and A_2 are fitting constants and τ_1 and τ_2 are short and long decay times. The obtained τ_1 and τ_2 values are 431.47 and 1968.78 µs, respectively. The short and long life times indicate the presence of Mn^{4+} ions near and far from the charge compensation defects, respectively [36]. Figure 6b shows the CIE chromaticity diagram where the CIE values were plotted. The CIE values were calculated from the emission data (400–800 nm) of CSO: $2Mn^{4+}$ and CSO: $2Mn^{4+}$, Na^+ samples obtained at the excitation and emission wavelengths of 340 and 642 nm, respectively. The calculated color coordinates for Mn^{4+} (0.6107, 0.3313) and Mn^{4+}- and Na^+- (0.6338, 0.341) doped samples, respectively. The inset of Figure 6b shows photographic images of powder samples under 365 nm ultraviolet light.

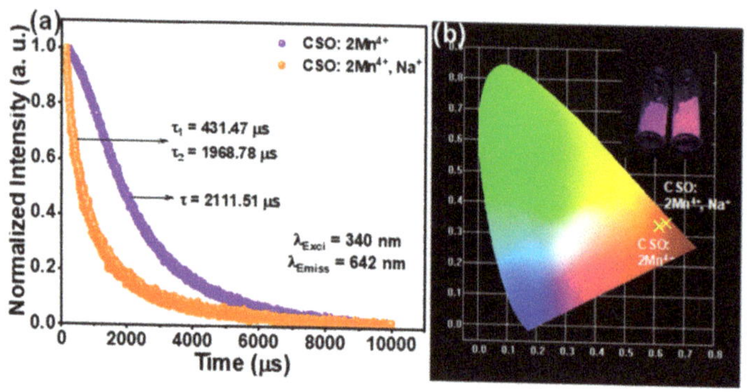

Figure 6. (**a**) Decay curves of CSO: $2Mn^{4+}$ and Na^+-doped CSO: $2Mn^{4+}$, (**b**) CIE chromaticity diagram for the prepared phosphors. The inset of (**b**) shows the powder images under 365 nm ultraviolet lamp.

4. Conclusions

In summary, the CSO: Mn^{4+}-doped phosphors were prepared using a conventional solid-state reaction method. The powders were characterized using different techniques, and the results were analyzed. The samples crystallized in a hexagonal system with a space group of P -3 1 m (162). The diffraction peaks matched well with the standard values, and no other impurities or peaks related to other phases were observed. SEM and EDX analysis showed that the particles are in the sub-micrometer range and the elements were uniformly distributed throughout the sample. The PLE spectrum displays that the sample can be effectively excited at 340 nm wavelength. The strong red emission centered at 642 nm is mainly due to the spin-forbidden $^2E_g \rightarrow {}^4A_{2g}$ electronic transitions in Mn^{4+} ions. The optimum doping concentration is found to be 2 mol%, and thereafter concentration quenching was observed, which is due to the multipole-multipole interaction. In addition, a charge compensator (Na^+) doping increased the emission intensity by 35%. The emission results match well with the absorption range of chlorophyll a, chlorophyll b, P_r, and P_{fr}, indicating that the materials are suitable candidates for plant growth LED applications.

Author Contributions: L.K.B., H.P. and J.S.Y. conceived, designed and directed the project. L.K.B. and H.P. synthesized and characterized the phosphor powders. S.V.G. directed the fabrication process and calculations. L.K.B., A.S. and J.S.Y. co-wrote the manuscript. All authors have read and agreed to the published version of the manuscript.

Funding: This work was supported by a grant of the Ministry of Science and Higher Education of the Russian Federation for large scientific projects in priority areas of scientific and technological development (grant number 075-15-2020-774).

Institutional Review Board Statement: Not applicable.

Data Availability Statement: Not applicable.

Conflicts of Interest: The authors declare no conflict of interest.

References

1. Dubey, V.; Som, S.; Kumar, V. *Luminescent Materials in Display and Biomedical Applications*; CRC Press: Boca Raton, FL, USA, 2020.
2. Moraitis, P.; Schropp, R.E.I.; van Sark, W.G.J.H.M. Nanoparticles for Luminescent Solar Concentrators—A review. *Opt. Mater.* **2018**, *84*, 636–645. [CrossRef]
3. Yu, M.; Yao, X.; Wang, X.; Li, Y.; Li, G. White-Light-Emitting Decoding Sensing for Eight Frequently-Used Antibiotics Based on a Lanthanide Metal-Organic Framework. *Polymers* **2019**, *11*, 99. [CrossRef] [PubMed]
4. Zhuo, Y.; Brgoch, J. Opportunities for Next-Generation Luminescent Materials through Artificial Intelligence. *J. Phys. Chem. Lett.* **2021**, *12*, 764–772. [CrossRef]
5. Timmermans, G.H.; Hemming, S.; Baeza, E.; van Thoor, E.A.J.; Schenning, A.P.H.J.; Debije, M.G. Advanced Optical Materials for Sunlight Control in Greenhouses. *Adv. Opt. Mater.* **2020**, *8*, 2000738. [CrossRef]
6. Runkle, E.S.; Padhye, S.R.; Oh, W.; Getter, K. Replacing incandescent lamps with compact fluorescent lamps may delay flowering. *Sci. Hortic.* **2012**, *143*, 56–61. [CrossRef]
7. Olle, M.; Viršile, A. The effects of light-emitting diode lighting on greenhouse plant growth and quality. *Agric. Food Sci.* **2013**, *22*, 223–234. [CrossRef]
8. Liu, L.; Keoleian, G.A.; Saitou, K. Replacement policy of residential lighting optimized for cost, energy, and greenhouse gas emissions. *Environ. Res. Lett.* **2017**, *12*, 114034. [CrossRef]
9. Singh, D.; Basu, C.; Meinhardt-Wollweber, M.; Roth, B. LEDs for energy efficient greenhouse lighting. *Renew. Sustain. Energy Rev.* **2015**, *49*, 139–147. [CrossRef]
10. Bayat, L.; Arab, M.; Aliniaeifard, S.; Seif, M.; Lastochkina, O.; Li, T. Effects of growth under different light spectra on the subsequent high light tolerance in rose plants. *AOB Plants* **2018**, *10*, ply052. [CrossRef]
11. Monostori, I.; Heilmann, M.; Kocsy, G.; Rakszegi, M.; Ahres, M.; Altenbach, S.B.; Szalai, G.; Pál, M.; Toldi, D.; Simon-Sarkadi, L.; et al. LED Lighting–Modification of Growth, Metabolism, Yield and Flour Composition in Wheat by Spectral Quality and Intensity. *Front. Plant Sci.* **2018**, *9*, 605. [CrossRef]
12. Rajapakse, N.C.; Shahak, Y. Light-Quality Manipulation by Horticulture Industry. In *Annual Plant Reviews Volume 30: Light and Plant Development*; Blackwell Publishing: Oxford, UK, 2007; pp. 290–312.
13. Kalaitzoglou, P.; van Ieperen, W.; Harbinson, J.; van der Meer, M.; Martinakos, S.; Weerheim, K.; Nicole, C.C.S.; Marcelis, L.F.M. Effects of Continuous or End-of-Day Far-Red Light on Tomato Plant Growth, Morphology, Light Absorption, and Fruit Production. *Front. Plant Sci.* **2019**, *10*, 322. [CrossRef]
14. Zheng, L.; Van Labeke, M.-C. Long-Term Effects of Red- and Blue-Light Emitting Diodes on Leaf Anatomy and Photosynthetic Efficiency of Three Ornamental Pot Plants. *Front. Plant Sci.* **2017**, *8*, 917. [CrossRef]
15. Pavitra, E.; Raju, G.S.R.; Varaprasad, G.L.; Chodankar, N.R.; Rao, M.V.B.; Rao, N.M.; Park, J.Y.; Han, Y.-K.; Huh, Y.S. Desired warm white light emission from a highly photostable and single-component $Gd_2TiO_5:Dy^{3+}/Eu^{3+}$ nanophosphors for indoor illuminations. *J. Alloys Compd.* **2021**, *875*, 160019. [CrossRef]
16. Rama Raju, G.S.; Pavitra, E.; Hwang, S.K.; Song, Y.H.; Park, J.Y.; Chodankar, N.R.; Ranjith, K.S.; Huh, Y.S.; Han, Y.-K. Development of dumbbell-shaped $La_2Si_2O_7:Eu^{3+}$ nanocrystalline phosphors for solid state lighting applications. *Ceram. Int.* **2021**, *47*, 5812–5821. [CrossRef]
17. Piao, X.; Horikawa, T.; Hanzawa, H.; Machida, K.-i. Photoluminescence Properties of $Ca_2Si_5N_8:Eu^{2+}$ Nitride Phosphor Prepared by Carbothermal Reduction and Nitridation Method. *Chem. Lett.* **2006**, *35*, 334–335. [CrossRef]
18. Choi, K.S.; Jee, S.D.; Lee, J.P.; Kim, C.H. A Novel Synthetic Method of $Sr_2Si_5N_8:Eu^{2+}$ from $SrSi_2O_2N_2:Eu^{2+}$ by Carbo-Thermal Reduction and Nitridation. *J. Nanosci. Nanotechnol.* **2013**, *13*, 1867–1870. [CrossRef]
19. Adachi, S. Review—Photoluminescence Properties of Cr^{3+}-Activated Oxide Phosphors. *ECS J. Solid State Sci. Technol.* **2021**, *10*, 026001. [CrossRef]
20. Van Quang, N.; Thi Huyen, N.; Tu, N.; Quang Trung, D.; Duc Anh, D.; Tran, M.T.; Hung, N.D.; Viet, D.X.; Huy, P.T. A high quantum efficiency plant growth LED by using a deep-red-emitting α-$Al_2O_3:Cr^{3+}$ phosphor. *Dalton Trans.* **2021**, *50*, 12570–12582. [CrossRef] [PubMed]
21. Huyen, N.T.; Tu, N.; Quang, N.V.; Quang Trung, D.; Tran, M.T.; Du, N.V.; Hung, N.D.; Viet, D.X.; Trung Kien, N.D.; Huy, P.T. Excellent Quantum Efficiency and Superior Color Purity Red-Emitting $CaAl_{12}O_{19}$–$CaAl_4O_7$–$MgAl_2O_4:Mn^{4+}$ Phosphors for Plant Growth and High Color Rendering Index White Light-Emitting Diode Applications. *ACS Appl. Electron. Mater.* **2022**, *4*, 4322–4331. [CrossRef]
22. Senden, T.; van Dijk-Moes, R.J.A.; Meijerink, A. Quenching of the red Mn^{4+} luminescence in Mn^{4+}-doped fluoride LED phosphors. *Light Sci. Appl.* **2018**, *7*, 8. [CrossRef]
23. Pavitra, E.; Raju, G.S.R.; Park, J.Y.; Hussain, S.K.; Chodankar, N.R.; Rao, G.M.; Han, Y.-K.; Huh, Y.S. An efficient far-red emitting $Ba_2LaNbO_6:Mn^{4+}$ nanophosphor for forensic latent fingerprint detection and horticulture lighting applications. *Ceram. Int.* **2020**, *46*, 9802–9809. [CrossRef]

24. Cao, R.; Fu, T.; Cao, Y.; Jiang, S.; Gou, Q.; Chen, Z.; Liu, P. Tunable emission, energy transfer, and charge compensation in the $CaSb_2O_6:Eu^{3+}$, Bi^{3+} phosphor. *J. Mater. Sci. Mater. Electron.* **2016**, *27*, 3514–3519. [CrossRef]
25. Li, P.; Peng, M.; Yin, X.; Ma, Z.; Dong, G.; Zhang, Q.; Qiu, J. Temperature dependent red luminescence from a distorted Mn^{4+} site in $CaAl_4O_7:Mn^{4+}$. *Opt. Express* **2013**, *21*, 18943–18948. [CrossRef] [PubMed]
26. Amarasinghe, D.K.; Rabuffetti, F.A. Bandshift Luminescence Thermometry Using $Mn^{4+}:Na_4Mg(WO_4)_3$ Phosphors. *Chem. Mater.* **2019**, *31*, 10197–10204. [CrossRef]
27. Brik, M.G.; Pan, Y.X.; Liu, G.K. Spectroscopic and crystal field analysis of absorption and photoluminescence properties of red phosphor $CaAl_{12}O_{19}:Mn^{4+}$ modified by MgO. *J. Alloys Compd.* **2011**, *509*, 1452–1456. [CrossRef]
28. Medić, M.M.; Brik, M.G.; Dražić, G.; Antić, Ž.M.; Lojpur, V.M.; Dramićanin, M.D. Deep-Red Emitting Mn^{4+} Doped Mg_2TiO_4 Nanoparticles. *J. Phys. Chem. C* **2015**, *119*, 724–730. [CrossRef]
29. Hu, J.; Huang, T.; Zhang, Y.; Lu, B.; Ye, H.; Chen, B.; Xia, H.; Ji, C. Enhanced deep-red emission from Mn^{4+}/Mg^{2+} co-doped $CaGdAlO_4$ phosphors for plant cultivation. *Dalton Trans.* **2019**, *48*, 2455–2466. [CrossRef]
30. Sun, Z.; Lu, B.; Ren, G.; Chen, H. Synthesis of Green-Emitting $Gd_2O_2S:Pr^{3+}$ Phosphor Nanoparticles and Fabrication of Translucent $Gd_2O_2S:Pr^{3+}$ Scintillation Ceramics. *Nanomaterials* **2020**, *10*, 1639. [CrossRef]
31. Zhao, J.; Gao, H.; Xu, H.; Zhao, Z.; Bu, H.; Cao, X.; He, L.; Yang, Z.; Sun, J. Structure and photoluminescence of Eu^{3+} doped Sr_2InTaO_6 red phosphor with high color purity. *RSC Adv.* **2021**, *11*, 8282–8289. [CrossRef]
32. Zhou, Q.; Dolgov, L.; Srivastava, A.M.; Zhou, L.; Wang, Z.; Shi, J.; Dramićanin, M.D.; Brik, M.G.; Wu, M. Mn^{2+} and Mn^{4+} red phosphors: Synthesis, luminescence and applications in WLEDs. A review. *J. Mater. Chem. C* **2018**, *6*, 2652–2671. [CrossRef]
33. Haiyang, W.; Xing Wang, D. Phytochrome Signaling Mechanism. *Arab. Book* **2002**, 2004. [CrossRef]
34. Shi, Y.; Wang, Y.; Wen, Y.; Zhao, Z.; Liu, B.; Yang, Z. Tunable luminescence $Y_3Al_5O_{12}:0.06Ce^{3+}$, xMn^{2+} phosphors with different charge compensators for warm white light emitting diodes. *Opt. Express* **2012**, *20*, 21656–21664. [CrossRef] [PubMed]
35. Zhang, H.; Fan, G.; Ruan, F. Charge compensating effect of alkali metal ions R+ (R = Li, Na, K) on the luminescence enhancement of $CaAl_{11}.9P_{0.1}O_{19.1}:Mn^{4+}$ red-emitting phosphor. *Inorg. Chem. Commun.* **2021**, *132*, 108860. [CrossRef]
36. Gupta, S.K.; Sudarshan, K.; Yadav, A.K.; Gupta, R.; Bhattacharyya, D.; Jha, S.N.; Kadam, R.M. Deciphering the Role of Charge Compensator in Optical Properties of $SrWO_4:Eu^{3+}:A$ (A = Li^+, Na^+, K^+): Spectroscopic Insight Using Photoluminescence, Positron Annihilation, and X-ray Absorption. *Inorg. Chem.* **2018**, *57*, 821–832. [CrossRef]

agriculture

MDPI

Article

A Counting Method of Red Jujube Based on Improved YOLOv5s

Yichen Qiao [1], Yaohua Hu [2,*], Zhouzhou Zheng [1], Huanbo Yang [1], Kaili Zhang [1], Juncai Hou [1,*] and Jiapan Guo [3,4]

1 College of Mechanical and Electronic Engineering, Northwest A&F University, Yangling 712100, China
2 College of Optical, Mechanical, and Electrical Engineering, Zhejiang A&F University, Hangzhou 311300, China
3 Bernoulli Institute for Mathematics, Computer Science and Artificial Intelligence, University of Groningen, 9747 AG Groningen, The Netherlands
4 Data Science Center in Health (DASH), University Medical Center Groningen, University of Groningen, 9713 GZ Groningen, The Netherlands
* Correspondence: huyaohua@zafu.edu.cn (Y.H.); houjuncai@nwsuaf.edu.cn (J.H.); Tel.: +86-15291680166 (Y.H.); +86-18792954818 (J.H.)

Abstract: Due to complex environmental factors such as illumination, shading between leaves and fruits, shading between fruits, and so on, it is a challenging task to quickly identify red jujubes and count red jujubes in orchards. A counting method of red jujube based on improved YOLOv5s was proposed, which realized the fast and accurate detection of red jujubes and reduced the model scale and estimation error. ShuffleNet V2 was used as the backbone of the model to improve model detection ability and light the weight. In addition, the Stem, a novel data loading module, was proposed to prevent the loss of information due to the change in feature map size. PANet was replaced by BiFPN to enhance the model feature fusion capability and improve the model accuracy. Finally, the improved YOLOv5s detection model was used to count red jujubes. The experimental results showed that the overall performance of the improved model was better than that of YOLOv5s. Compared with the YOLOv5s, the improved model was 6.25% and 8.33% of the original network in terms of the number of model parameters and model size, and the Precision, Recall, F1-score, AP, and Fps were improved by 4.3%, 2.0%, 3.1%, 0.6%, and 3.6%, respectively. In addition, RMSE and MAPE decreased by 20.87% and 5.18%, respectively. Therefore, the improved model has advantages in memory occupation and recognition accuracy, and the method provides a basis for the estimation of red jujube yield by vision.

Keywords: count red jujubes; red jujube; improved YOLOv5s; ShuffleNet V2 Unit; Stem; BiFPN

Citation: Qiao, Y.; Hu, Y.; Zheng, Z.; Yang, H.; Zhang, K.; Hou, J.; Guo, J. A Counting Method of Red Jujube Based on Improved YOLOv5s. *Agriculture* 2022, *12*, 2071. https://doi.org/10.3390/agriculture12122071

Academic Editors: Vadim Bolshev, Vladimir Panchenko and Alexey Sibirev

Received: 10 October 2022
Accepted: 30 November 2022
Published: 2 December 2022

1. Introduction

Chinese red jujube is a kind of characteristic fruit which is famous for its various nutritional ingredients [1]. With the increasing demand for red jujubes, it is more and more important to count red jujubes so as to provide a basis for the estimation of jujube yield through vision. Due to the increasing supply of red jujubes, the count of red jujubes will play an important role in the planting and production management. Therefore, it is of great significance to realize the count of red jujubes, and it will help improve the utilization rate of red jujubes. However, the development of artificial intelligence, it provides a new way to solve the problem of low fruit production efficiency [2].

It is an important task of orchard management to estimate the fruit yield by counting the number of fruits. Deep learning has become a potential tool for counting the number of fruits, and It enables automatic feature extraction from data sets. At the same time, by extracting the basic parameters of crop growth, intelligent agricultural technology enables farmers to estimate crop yield, thus reasonably arranging the production and processing of red jujubes [3]. Machine learning methods, such as the Watershed algorithm [4] and the kalman filter algorithm [5], are widely used to count fruit. However, because the supervised learning method in machine learning can't capture the nonlinear relationship between input and output variables and the uncertainty of the crop environment, it is difficult for traditional machine learning methods to develop a reliable crop counting model. However, in recent years, the progress of technology has made it possible to develop advanced crop counting models by using deep learning. Shileiliu et al. [6] proposed a light target detection YOLOv5-CS model, which could realize the object detection and accurate counting of green citrus in the natural environment. The map of the model was 98.23%. ZhangYanchao et al. [7] used the YOLOX target detection network to detect and count the holly fruits, and the map was 95%.

Owing to the improvement of computer hardware and the development of computer vision technology, deep learning has been widely used in various industries [8–10]. Object detection algorithm based on deep learning mainly includes One-Stage and Two-Stage. The first type is the detection algorithm based on candidate region, such as R-CNN (Region-Convolutional Neural Networks) [11], Fast R-CNN (Fast Region-Convolutional Neural Networks) [12], Faster R-CNN (Faster Region-Convolutional Neural Networks) [13]. The second kind regards the detection of target position as a regression problem and directly uses CNN (Convolutional Neural Network) for images, such as SSD (Single Shot Multi-Box Detector) [14,15], YOLO (You Only Look Once) [16–19].

Computer vision technology has also been widely used in various fields [20–23]. The image processing technology is one of the key technologies in precision agriculture, and it is mainly used in classification, localization, and yield prediction [24]. Mulyono et al. [25] proposed a texture extraction method based on a gray-level co-occurrence matrix that is followed by a K-nearest neighbor for the classification of litchii. Sutarno et al. [26] adopted similar ideas to extract texture information and then used the learning vector quantization (LVQ) algorithm as the classifier to classify durian based on their color, shape, and texture. The method was difficult to detect the subtle feature changes among different fruits, and the accuracy of fruit classification was 89%. Zhao et al. [27] proposed a matching algorithm that used the sum of absolute transformed differences (SATD) for fruit detection, followed by the support vector machine (SVM) classifier. The accuracy of recognition reached more than 83%. Dorj et al. [4] proposed forecasting the yield of citrus yields. The method preprocessed images by color space conversion and denoising then recognized and detected citrus and counted citrus by the watershed segmentation algorithm. Other researchers have also studied the fruit classification, identification, and count of fruits based on shape invariant moments [28], decision trees [29], and Hough [30] combined with the texture and color of fruits. The above methods use single features or multi-feature combinations with texture features, shape size, and color differences of fruits to recognize fruits. The recognition result is about 93% when the environment is complex, such as light changes, fruit overlap, leaf occlusion, etc. In addition, the traditional machine learning algorithm is limited by the result of the classifier of the algorithm itself, and it is difficult for the algorithm to complete the object detection of fruit in a complex environment [31].

Due to the occlusion of fruit and leaves, the image transformation, and the background switching in complex orchard environments, the deep learning-based object detection algorithm can solve these problems quickly and effectively with its powerful learning ability and feature representation capability. Fu et al. [32] proposed a deep convolutional neural network detection model in which the improved Faster R-CNN was trained end-to-end by using back-propagation, random gradient descent algorithm, and ZFNet (Zeiler and Fergus networks) for kiwifruit detection. The experiment showed that the model could improve the accuracy of fruit recognition to 96%. Liu et al. [33] fused RGB and NIR images to identify kiwifruit by VGG16. The average detection precision of an image was 90.7%, and the detection time was 0.134 s on one image. Wang et al. [34] proposed an improved model of a lightweight detection network of SSD. The model used a modified DenseNet network as the backbone to replace the first three additional layers in SSD and incorporate a multi-level fusion structure. Compared with the original model, the number of parameters of the improved model was reduced by

11.14×10^6, and the average precision was increased by 2.02%. The classical deep learning networks have been successful in fruit identification and detection. There are advantages of high accuracy and efficiency in the identification and detection of fruits. However, the networks are relatively large, which is not conducive to the application of mobile equipment in the agricultural field. Many researchers have already studied the lightweight model. For instance, Li et al. [35] applied the adaptive spatial pyramid to detect the green peppers and the accuracy reached 96.11% in YOLOv4_tiny. Zhang et al. [31] used MobileNet-v3 as the feature extraction network of YOLOv4-LITE. The improved model reduced the model size and improved the detection speed. Therefore, it is feasible to reduce the weight of the model while ensuring the precision of model detection.

The lightweight model will be beneficial to the application of agricultural mobile equipment and realize the intelligence of agricultural equipment. In order to ensure the detection accuracy of the model in complex unstructured orchards and counting fruit, a counting method of red jujube based on improved YOLOv5s was proposed. The main goal of this research was to reduce the size of the model while ensuring its detection accuracy and speed in an embedded device. The effectiveness of counting red jujubes in a complex environment was comprehensively considered from four aspects in this research

(1) ShuffleNet V2 was used as the backbone of the network to extract the feature map and make the model lightweight.
(2) The Stem, a novel data loading module, was proposed to reduce data information loss and improve model detection accuracy.
(3) The original PANet (Path Aggregation Network) structure was improved to BiFPN (Bidirectional Feature Pyramid Network) for multi-scale feature fusion to enhance the model feature fusion capability and improve the model accuracy.
(4) The improved YOLOv5s detection model was used to count red jujubes.

The second section introduced the method of making the dataset, the improved red jujubes detection algorithm, the counting method of red jujubes, and the training of the network. The third section introduced the test results of the model and the analysis compared with other algorithms. In the last section, the counting methods of red jujubes were summarized and prospect.

2. Materials and Methods

In this section, the acquisition and production of the dataset were mainly introduced. Then, a detection algorithm based on the improved yolov5s of red jujube was proposed, and a counting method for red jujubes was presented. Finally, the training method of the network was introduced, as shown in Figure 1.

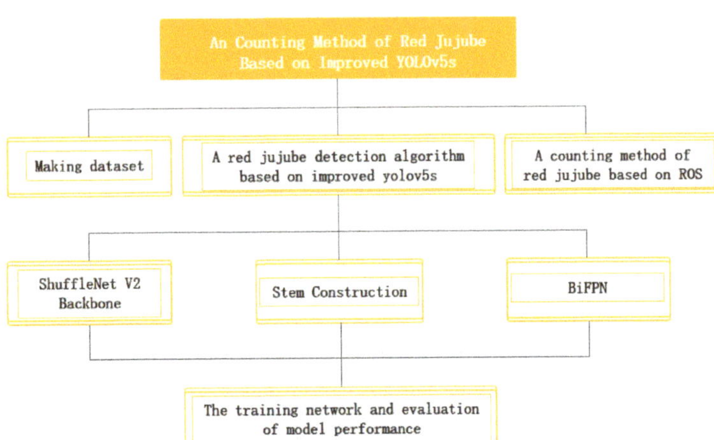

Figure 1. A counting method of red jujube based on improved YOLOv5s.

2.1. Image Data Acquisition

The dataset of red jujube, including Jun jujube and Gray jujube, in this study, was collected from a red jujube orchard from 5 October to 9 October in Alar City, Xinjiang, China.

Images of Jun jujube and Gray jujube were taken in a jujube orchard of the 13th company of a group in Alar City, Xinjiang Uygur Autonomous Region. In order to ensure the reliability of the experimental results, the jujube image dataset was collected, which was under different illumination at 9:00 a.m., 3:00 p.m., and 9:00 p.m. for red jujubes. The resolution of the images was 1080 × 1920 pixels, with a total of 1026 original images, which included illumination changes, leaf shading, and fruit overlap. In order to improve the robustness of this model, each image contained one or more different scenarios. The distribution of the dataset is shown in Table 1.

Table 1. Distribution of dataset of red jujubes.

Dataset	Grey Jujube	Jun Jujube	Total Number
illumination change images	136	190	326
leaf shading images	132	225	357
fruit overlap images	139	204	343
Total Number	407	619	1026

2.2. Data Preprocessing and Augmentation

The collection of data sets would affect the recognition effect of the target detection model. The more sufficient and comprehensive the data set is, the better the generalization ability and robustness of the model. Therefore, the number of samples could be expanded by data amplification. In order to truly simulate the shooting of red jujube in a complex environment and apply it to the detection network, this research used Opencv in python to compress and cut the images into 640 × 640. Then, the images were randomly enhanced by different image processing methods [36], such as rotating 180, mirroring, adding salt and pepper noise which set the threshold to 0.5, and changing the image brightness by setting the threshold to 1.3 and 0.7, as shown in Figure 2. Repeated random image processing on an image many times. After enhancement, a total of 10,000 images were obtained as the data set of the model.

(a) (b) (c) (d) (e) (f)

Figure 2. Image sample after data preprocessing and augmentation. (**a**) original image, (**b**) rotating by 180°, (**c**) Increasing brightness, (**d**) mirroring image, (**e**) adding noise, (**f**) reducing brightness.

2.3. Images Annotation and Dataset Division

In this research, LabelImg was used to label red jujube in the data set with artificial rectangular boxes, as shown in Figure 3. The dataset was divided into 80% training datasets, 10% validation datasets, and 10% test datasets. The final image samples of the training set, verification set, and test set are 8000, 1000, and 1000 respectively.

Figure 3. LabelImg data set annotation.

2.4. Methodologies

The Yolo series are effective in single-stage object detection, and their miniature detection models guarantee higher accuracy, taking into account faster speed and fewer parameters. Therefore, the lightweight detection models of the Yolo series are more suitable to be applied to embedded devices to develop mobile agricultural equipment. However, due to the complexity of the agricultural production environment and the harsh working environment, it is difficult to meet the agricultural production for the simple detection algorithm. Based on YOLOv5s, the original backbone network was replaced by the ShuffleNet V2 backbone network in this research, which significantly reduced the number of parameters of the network. The Foucs were replaced by the Stem to resist partial information missing from the feature map. PANet was replaced by BiFPN to enhance the model feature fusion capability and improve the model accuracy. Finally, the improved YOLOv5s detection network was used to identify the image and count red jujubes.

2.4.1. Yolov5s Network

YOLOv5 is improved by adding some new ideas on the basis of YOLOv4, and its detection accuracy and speed have been greatly improved. The YOLOv5 can be divided into four types according to the size of the model: YOLOv5s, YOLOv5m, YOLOv5l, and YOLOv5x, among which the YOLOv5s model is the smallest. YOLOv5s mainly consists of four parts: Input, Backbone, Neck, and Prediction.

In order to improve the speed and accuracy of the network, the Mosaic data augmentation is used in the YOLOv5 to stitch images by random cropping, scaling, and lining up. YOLOv5s uses adaptive anchor box calculation to set the initial anchor boxes for different datasets and calculates the difference between the bounding boxes and the ground truth. YOLOv5s updates the anchor boxes in the reverse iteration to adaptively calculate the best anchor box for different training sets. To adapt different sizes of images in the dataset, YOLOv5 uses adaptive image scaling to fill the scaled image with the least amount of black edges, which reduces the computation and improves the speed. Backbone will perform information extraction on the feature maps. It mainly includes Focus, CBS, and C3. The input image is sliced by the Focus and convolved by one convolution with 32 kernels, as shown in Figure 4. CBS consists of a convolution, a batch normalization, and the SiLU. The SiLU is defined as follows:

$$SiLU(x) = \frac{x}{1 + exp(-x)} \quad (1)$$

where, x represents the feature map.

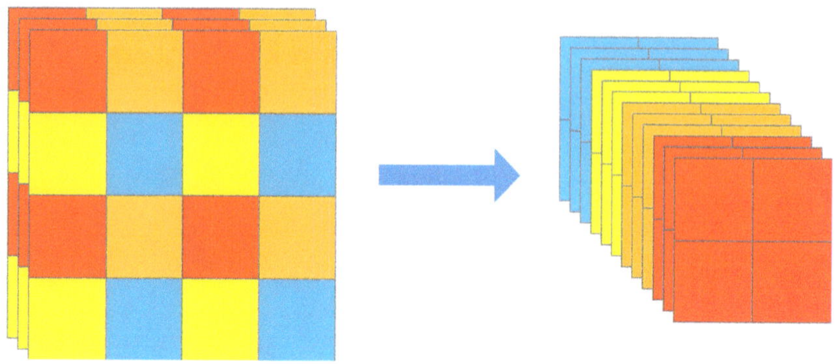

Figure 4. Foucs structrue.

As a new structure of BottlenneckCSP, C3 contains 3 CBS modules and several Bottlenecks. The C3 is used repeatedly in YOLOv5s to extract more information. As shown in Figure 5, the SPP (spatial pyramid pooling) introduces three different pooling kernels of 5×5, 9×9, and 13×13, and it connects different feature maps to expand the respective field, which effectively separates the most important features and improves the accuracy of the model.

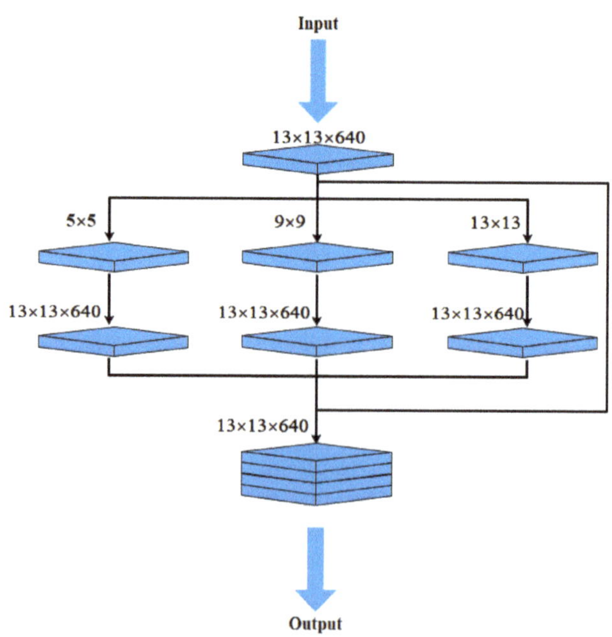

Figure 5. SPP structure.

To utilize most of the backbone information, the Neck of YOLOv5 uses the FPN + PAN. Feature Pyramid Network (FPN) solves the problem of different input feature map sizes by constructing an image pyramid on the feature map. PAN, as the innovative point of path aggregation network (PANet) [37], downsamples the image from FPN and then performs concat on the image. To improve the ability of image recognition and localization, FPN acquires the semantic features of the image from the top, while PAN gets the localization features of the image from the bottom.

There are some regression loss functions used in object detection tasks, such as the Smooth Loss function [16], IOU Loss function [38], GIOU Loss function [39], DIOU Loss function [40], and CIOU_Loss function [41]. In the Prediction, YOLOv5 uses CIOU_Loss as the loss function of the Bounding box. The CIOU_Loss function is defined as follows:

$$L_{CIOU} = 1 - IOU + \frac{\rho^2(b, b^{gt})}{c^2} + \alpha v \tag{2}$$

where, IOU represents the intersection ratio of the prediction box to the object box. b represents the center point of the prediction box. b^{gt} represents the center point of the object box. $\rho^2(b, b^{gt})$ represents Euclidean distance squared between the center point of the prediction box and the center point of the object box. c represents the diagonal length of the two closed boxes. α represents a positive trade-off parameter. v represents the consistency of the aspect ratio.

2.4.2. ShuffleNet V2 Backbone

YOLOv5s reduces the parameters of the model by C3 and improves the speed of the model, but the C3 is very complicated, with a large amount of calculation and still needs a lot of memory. The YOLOv5 lightweight model based on ShuffleNet V2 was designed, which greatly reduced the model parameters. The ShuffleNet V2 backbone was designed by using ShuffleNet V2 Units [42], and the backbone of the original model was replaced by the ShuffleNet V2 backbone.

As a lightweight convolutional neural network that is suitable for application to mobile devices, ShuffleNet V2 was first proposed in 2018. Compared with ShuffleNet V1, ShuffleNet V2 adopts the way of channel Shuffle, which divides the characteristic channels into two parts, ensuring that the input and output channels are the same, One part enters the bottleneck, and the other part does not run. Excessive point convolution will increase computational complexity. ShuffleNet V2 replaces the grouped point convolution with the standard point convolution. ShuffleNet V2 puts the channel shuffle after the dimensional stacking to prevent fragmentation of the model. ShuffleNet V2 replaces element-wise operators with concat to reduce the time of model detection. The basic model units of ShuffleNet V2 are divided into two types. The ShuffleNet V2 Units are shown in Figure 6. ShuffleNet V2 introduces channel shuffle. First, the channels of the input feature map are divided into two branches. The two branches directly connect to the concat. There are two 1×1 point convolution layers and a 3×3 group convolution layer with a stride size of 2 in the other branch. The convolution layers contain a batch normalization layer and ReLu. The other basic model unit of ShuffleNet V2 differs from the previous model, where two convolution layers: a 3×3 group convolution layer with a stride of 2 and a 1×1 point convolution layer. Finally, two images of branches of the same size were spliced together. In order to extract information on different-size feature maps, the ShuffleNet V2 backbone was designed to replace the backbone by using 16 ShuffleNet V2 Units in YOLOv5s.

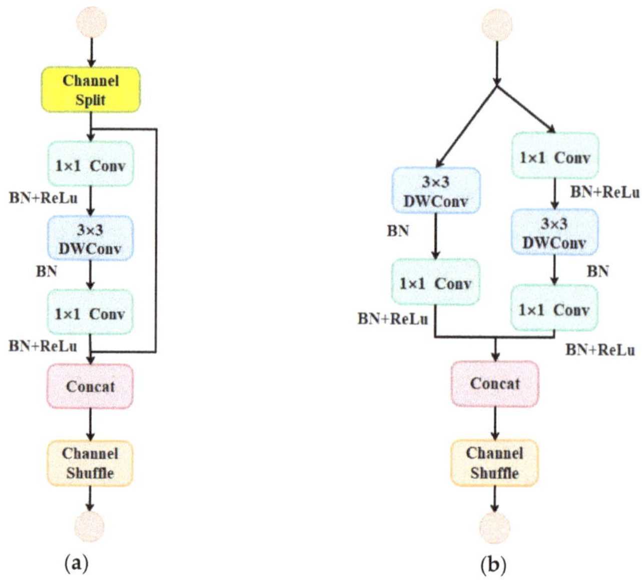

Figure 6. The structure of ShuffleNet-v2 Units. (**a**) the structure of ShuffleNet-v2 Unit1. (**b**) the structure of ShuffleNet-v2 Unit2.

2.4.3. Stem Construction

Inception-v4 [43] was proposed in 2017, which confirmed that residual connectivity largely accelerated the training speed of Inception networks. With reference to the design idea of Inception-v4, the Stem was proposed to rapidly reduce the resolution of the input feature maps, ultimately achieving a top-5 error rate of 3.08% on ILSVRC. The feature map is continuously reduced from 299 × 299 to 35 × 35 by Stem in the InceptionV4 network, and it has many convolution layers, which is better for complex task feature extraction. However, the task is simpler to detect a single target of red jujube, which will cause excessive calculation. The Stem is shown in Figure 7. In order to reduce the parameters of the model, the model could be pruned. Inspired by the idea of fast feature map resolution reduction, four CBS were adopted to make the size of the feature map to be suitable for the network, where 3 × 3 convolutions with the stride of 2 were used in the first and third CBS and 1 × 1 convolution was used in the second and fourth CBS. In contrast to the Foucs, which sliced the feature map into 32 small feature maps before image concat, the Stem used two 3 × 3 convolutions with the stride of 2 to reduce the feature map sizes and concated it with the feature map of the maximum pooling layer, so that the number of parameters was reduced while improving the feature extraction ability of the network and improving the accuracy.

2.4.4. BiFPN

With the deepening of the network level, the semantic information of image features gradually changes from a low dimension to a high dimension. As shown in Figure 8, the PANet structure was used to fuse the multi-scale features of images in the original YOLOv5s detection network. In order to improve the detection accuracy of red jujubes, the BIFPN network, a weighted bidirectional feature pyramid network, was applied to the detection of red jujubes. Compared with the traditional feature fusion network, BiFPN introduced weight to make it more sensitive to important features and makes better use of feature information of different sizes.

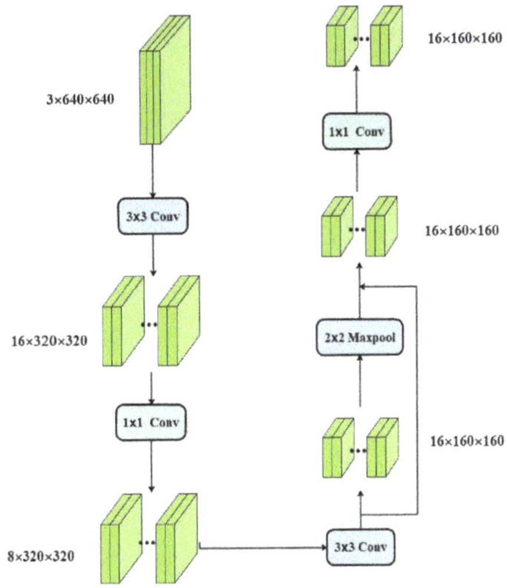

Figure 7. The structure of the Stem.

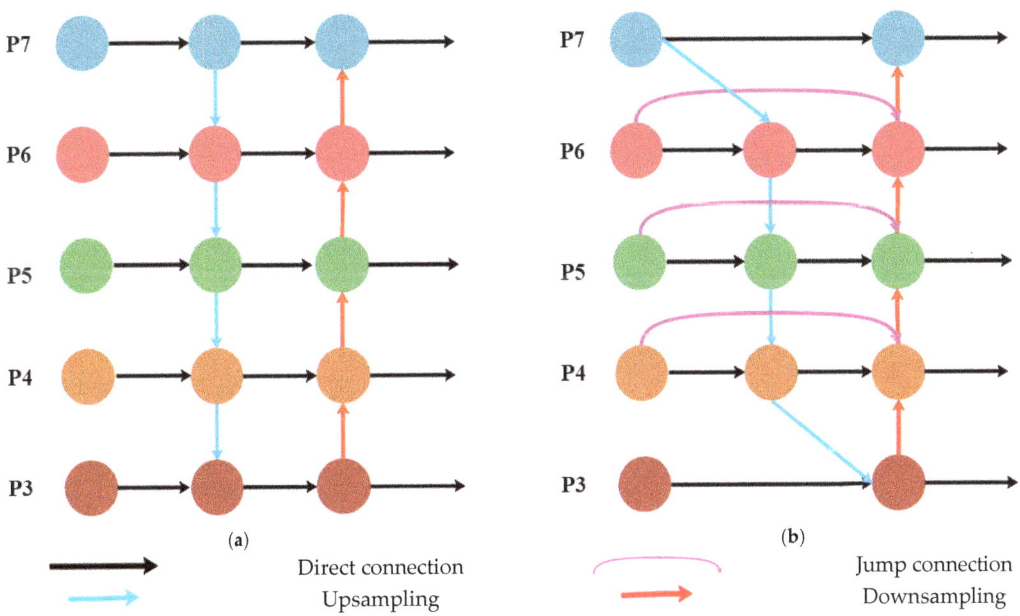

Figure 8. Bi-directional feature fusion network. (**a**) PANet with bi-directional feature fusion network, (**b**) BiFPN with bi-directional feature fusion network.

In this research, BiFPN was introduced in the neck of YOLOv5s, as shown in Figure 9. Because the node, which had only one input edge and no ability of feature fusion, made little contribution to the feature fusion of the network. Therefore, deleting this node had little effect on network feature fusion. When the original input node and the output node were in the same layer, an extra edge was added between the output node and the input node, and feature fusion was realized without increasing too much computational overhead. Different from the PANet structure of YOLOv5s, when performing feature fusion, each bidirectional path was used as a feature network layer, and the feature network layer was reused at the same layer, thus realizing a higher level of feature fusion.

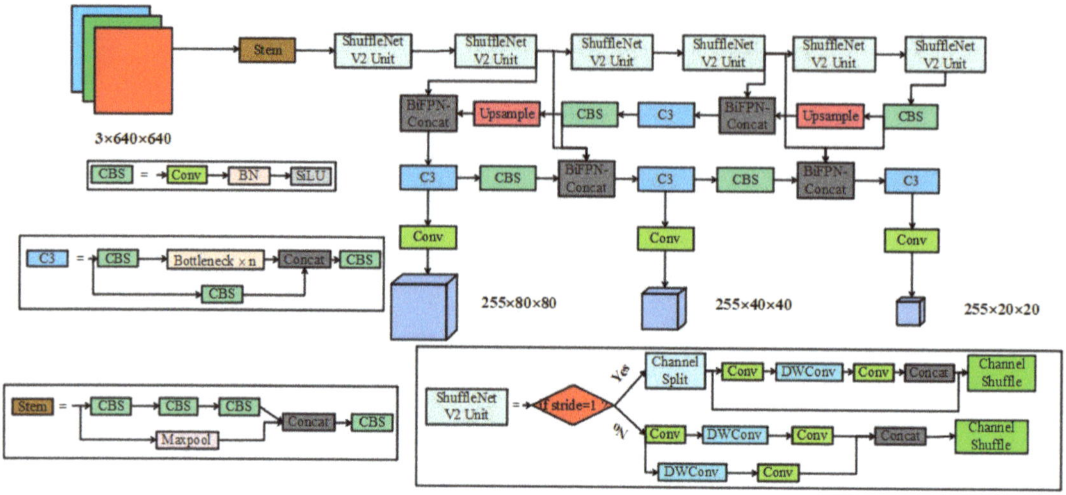

Figure 9. The structure of the improved YOLOv5s model.

2.4.5. Counting Method of Red Jujube

The counting method of red jujubes was based on the improved jujube target detection algorithm. This research used ROS to count red jujubes. The detection steps were as follows: (1) Starting ROS core and publishing topics; (2) the improved YOLOv5s were used to detect the target of jujube fruit, and the target detection frame and corresponding features were obtained; (3) counting the number of target detection frames, as shown in Figure 10a. The detection results are shown in Figure 10b.

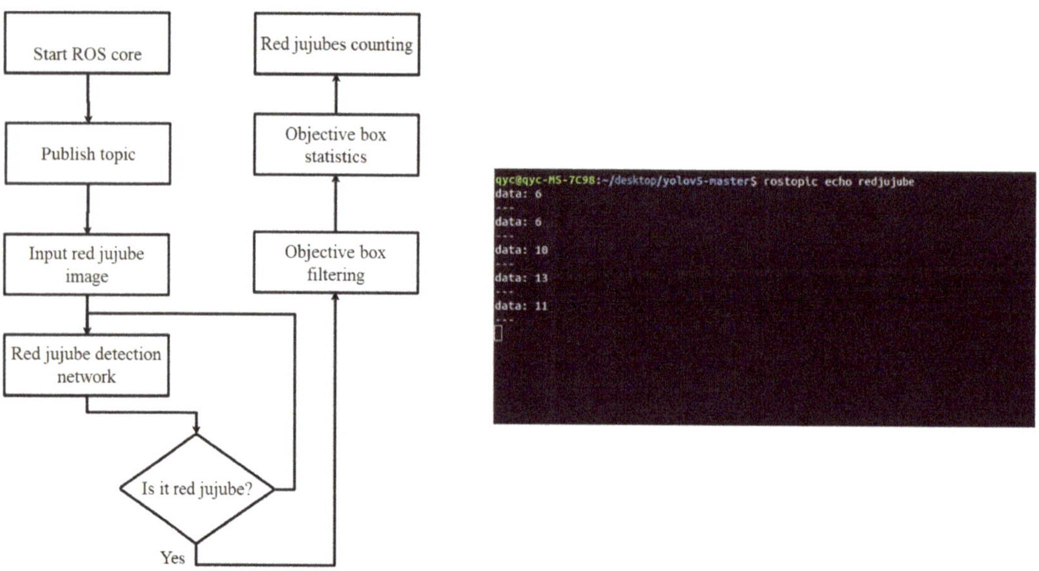

(a) (b)

Figure 10. Counting method of red jujube. (**a**) the process of the red jujube counting method; (**b**) the results of the jujube counting method.

2.5. Test Platform

The experiment was conducted on an improved YOLOv5s architecture with Pytorch based on Python 3.8. The details of the experimental setup are shown in Table 2.

Table 2. Experimental environment.

Configuration	Parameter
CPU	Intel(R) Core(TM) i7-10700K
GPU	NVIDIA GeForce RTX 3070
Accelerated environment	CUDA11.1 CUDNN8.2.1
Development environment	Pycharm2021.3.2
Operating system	Windows 10

The batch size was 4, and the epochs were 400. The adaptive matrix estimation algorithm (Adam) was used to optimize the model. The initial learning rate was 0.001, and the momentum was 0.9. The weight of the model was saved once every training session, and the best weight was also saved.

2.6. Evaluation of Model Performance

In order to evaluate the performance of our model of red jujube, Precision (P), Recall (R), Average Precision (AP), Parameters, Model Size, and detection speed (Fps) were chosen in the article, root mean square error (RMSE) and average absolute percentage error rate (MAPE) were used as evaluation indexes of jujube quality where Recall, Precision, F1-score, RMSE, and MAPE were defined as follows:

$$\text{Precision} = \frac{\text{TP}}{\text{TP} + \text{FP}} \times 100\% \tag{3}$$

$$\text{Recall} = \frac{\text{TP}}{\text{TP} + \text{FN}} \times 100\% \tag{4}$$

$$\text{F1} - \text{score} = \frac{2 \times \text{Precision} \times \text{Recall}}{\text{Precision} + \text{Recall}} \tag{5}$$

$$\text{RMSE} = \sqrt{\frac{1}{m}\sum_{i=1}^{m}(y_i - \hat{y}_i)^2} \tag{6}$$

$$\text{MAPE} = \sum_{i=1}^{m}\frac{|(y_i - \hat{y}_i)/y_i|}{m} \times 100\% \tag{7}$$

where, TP represents the number of true positive samples, FP represents the number of false positive samples, and FN represents the number of false negative samples. The variable y_i represents the actual number of red jujubes in each image, $\hat{y}_i$ represents the number of red jujubes predicted by each image model, and m represents the number of image samples.

3. Results and Discussion

3.1. Performance Comparison Using the Different Improve Method

As shown in Table 3, Recall and Precision were based on a 0.5 threshold. As one of the important indicators for evaluating the model, the area of the Precision-Recall curve was larger, and the AP of the model was higher.

Table 3. The model performance with a different module.

Model	Precision (%)	Recall (%)	F1-Score (%)	AP (%)	Parameters	Model Size (KB)	Fps
YOLOv5s	89.10	90.30	89.70	95.60	7,063,542	14,052	35.10
YOLOv5s + Stem	87.60	93.90	90.60	96.00	7,281,341	14,026	38.40
YOLOv5s + BiFPN	88.60	90.90	89.70	95.30	7,063,542	14,052	39.40
YOLOv5s + ShuffleNet V2	83.80	91.60	87.50	94.00	490,205	1322	35.50
YOLOv5s + Stem + BiFPN	89.70	94.50	92.00	96.20	7,281,341	14,026	39.40
YOLOv5s + Stem + ShuffleNet V2	93.70	89.20	91.40	95.90	441,606	1149	36.30
YOLOv5s + BiFPN + ShuffleNet V2	83.40	92.10	87.50	94.10	490,205	1322	35.50
Our model	93.40	92.30	92.80	96.20	441,606	1149	36.50

ShuffleNet V2 was used as the backbone network of the network, resulting in a reduction in model parameters by 14.41 times and an increase in Fps from 35.10 to 35.47. The improved network could reduce model parameters and increase detection speed. BiFPN was applied to the red jujube detection network. The experimental result showed that BiFPN improved the average accuracy of the network without increasing the parameters of the network. At the same time, it improved the detection speed of the model, with the average accuracy increased by 0.20% and the Fps increased to 39.40. Therefore, BiFPN could enhance the feature fusion ability of YOLOv5s and speed up the detection speed of the model. The Focus was replaced by the Stem, and the improved network has been improved in Recall, F1-score, AP, model size, and Fps, among which the Recall has increased by 3.600%. So, Stem is more effective than Focus in jujube detection. Compared with YOLOv5s, the AP increased by 0.6%, but the parameters increased, which increased the calculation pressure of testing equipment when Stem and BiFPN were used at the same time. When Stem and ShuffleNet V2 were applied at the same time, compared with YOLOv5s, the parameters were greatly reduced, but the detection accuracy was also lower. Our method not only reduced the model parameters but also improved the detection accuracy. The parameters and model size of the improved model was 6.25% and 8.33% of the original network, respectively. The Precision, Recall, F1-score, AP, and Fps were increased by 4.30%, 2.00%, 3.10%, 0.60%, and 3.99%, respectively.

As a lightweight network model, YOLOv5s has high accuracy and can meet the detection of small targets in complex environments, but it is difficult to be satisfied with the identification and localization of red jujubes under limited computation. When locating and recognizing overlapping fruits, the original YOLOv5s tended to easily identify two red jujubes that were mutually obscured as the one red jujube, as shown in Figure 11b. The main reason was that the differences were small between mutually obscured fruits, and the original YOLOv5s did not extract enough feature information about them, causing false detection. In recognition of small red jujube targets, the original YOLOv5s easily missed the red jujubes that were obscured by a large area of leaves or caused by the camera being too far away, as shown in Figure 11e. The main reason was that the environment of outdoor was complex, and the discrimination of the red jujubes was large. The improved model could accurately detect red jujubes and could also accurately identify the blocked jujubes, as shown in Figure 11c, and the number of missed jujubes was obviously less than the original YOLOv5s, as shown in Figure 11f.

Figure 11. The results of different algorithms for the recognition of red jujube. (**a**) the original image of a dense jujube sample. (**b**) the original model to dense jujube detection image. (**c**) the improved model to dense jujube detection image. (**d**) the original image of leaf-obscured jujube. (**e**) the original model to leaf-obscured jujube detection image. (**f**) the improved model to leaf-obscured jujube detection image. Where the red boxes are the label boxes marked manually, and the blue boxes are the test results of model test.

3.2. Performance Comparison Using the Different Lightweight Backbone Networks

In order to embed mobile devices, the ShuffleNet V2 backbone network was used in YOLOv5s in this research. MoblieNet V3, as the improved version of MoblieNet V1 and MoblieNet V2, has a large improvement in detection efficiency. In order to verify the detection performance of the improved model, the MoblieNet V3 network was used as the backbone of YOLOv5s to compare the improved YOLOv5s, which used the ShuffleNet V2 backbone network and YOLOv5s. The results show that after adding MoblieNet V3 as the backbone, the network has a large improvement in Precision, but a large decrease in Recall, resulting in the improved YOLOv5s, which is used the MoblieNet V3 backbone network and the original YOLOv5s in the same AP, as shown in Table 4. In addition, there is a phenomenon of missing the detection of jujube fruit, as shown in Figure 12. The improved YOLOv5s, which is used in the MoblieNet V3 backbone network, has a significant reduction in parameters and Model Size with YOLOv5s. Therefore, using a lightweight network as the backbone reduces the size of the model while maintaining accuracy.

Table 4. The comparison of different backbone networks.

Model	Precision (%)	Recall (%)	AP (%)	Parameters	Model Size (KB)	Fps
YOLOv5s	89.1	90.3	95.6	7,063,542	14,052	35.1
MoblieNet V2-YOLOv5s	81.2	90.3	93.6	2,917,046	5423	23.4
MoblieNet V3-YOLOv5s	94.2	85.8	95.6	3,538,532	7189	22.2
Ghost-YOLOv5s	85.4	92.3	93.4	3,897,605	8492	23.2
ShuffleNet V2-YOLOv5s	83.8	91.6	94.0	490,205	1149	35.5

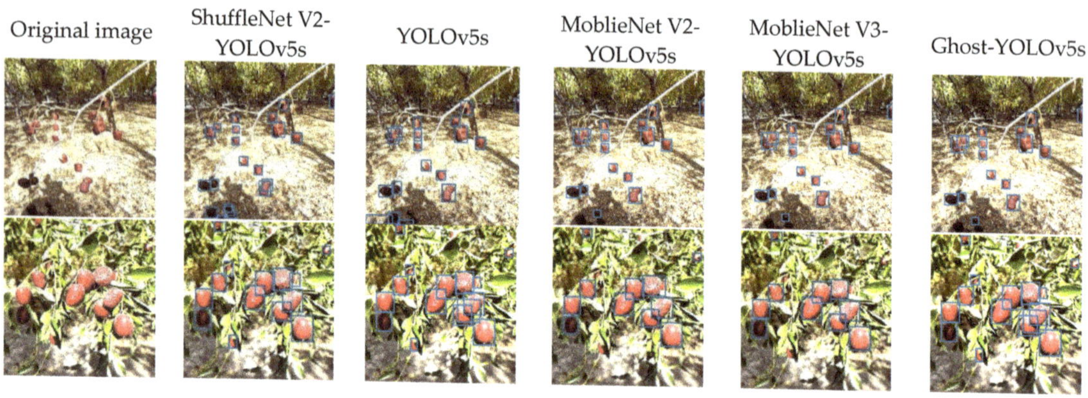

Figure 12. Test results of different lightweight backbone networks. Where the blue boxes are the test results of model test.

The Precision and AP of using ShuffleNet V2 as the backbone network were slightly lower than that of the original YOLOv5s and the improved YOLOv5s using MoblieNet V3 as the backbone network. However, using ShuffleNet V2 as the backbone network could provide a more comprehensive red jujube detection. When MoblieNet V2 and GhostNet were used as a backbone, some red jujubes were missed, as shown in Figure 12. Compared with the other four detection models, the number of parameters using ShuffleNet V2 as the backbone network was only 7.14% of YOLOv5s, and the number of parameters was obviously smaller than other networks. The detection speed using the ShuffleNet V2 backbone network model was also faster than other detection networks, as shown in Table 4. using ShuffleNet V2 as a backbone not only greatly reduced the number of model parameters but also improved the detection speed, which was more suitable for red jujubes counting and related embedded mobile devices.

3.3. *Performance Comparison in Counting Jujubes Using the Different Algorithms*

To verify the effectiveness of improved YOLOv5s for target detection, YOLOv3-tiny, YOLOv4-tiny, Faster R-CNN, SSD, YOLOvx-tiny, and YOLOv7-tiny were selected to compare with improved YOLOv5s. This research experimented with the selected comparison models using datasets of the same size and the same training and test sets. In order to ensure the reliability of the test, the epoch was set to 400, and the batch size was set to 4. In this research, three orchard jujube images were selected to test the yield estimation method. The comparison results are shown in Table 5. The P-R curve of the models is shown in Figure 13.

Table 5. Detection results of red jujubes with different target detection algorithms.

Model	The Number of Actual Jujube						The Number of Predicted Jujube						Precision (%)	Recall (%)	AP (%)	RMSE	MAPE (%)	Model Size (KB)
	1	2	3	4	5	6	1	2	3	4	5	6						
YOLOv5s	10	15	15	9	10	6	9	16	14	8	8	6	89.10	90.30	95.60	1.15	9.07	14,052
YOLOv4-tiny							10	15	11	9	8	6	91.60	89.40	95.90	1.83	7.78	103,012
YOLOv3-tiny							10	14	11	7	8	6	92.30	88.70	95.50	2.04	12.59	481,391
YOLOvx-tiny							8	10	11	7	7	6	86.60	91.30	95.70	3.11	22.04	19,901
YOLOv7-tiny							10	11	12	7	8	6	89.20	90.50	95.10	2.35	14.81	23,674
SSD							8	11	14	7	8	6	88.30	87.10	90.50	2.19	15.93	92,782
Faster R-CNN							9	12	13	7	7	6	64.00	89.30	87.90	2.12	15.93	110,773
Our Model							10	15	13	9	9	6	93.40	92.30	96.20	0.91	3.89	1149

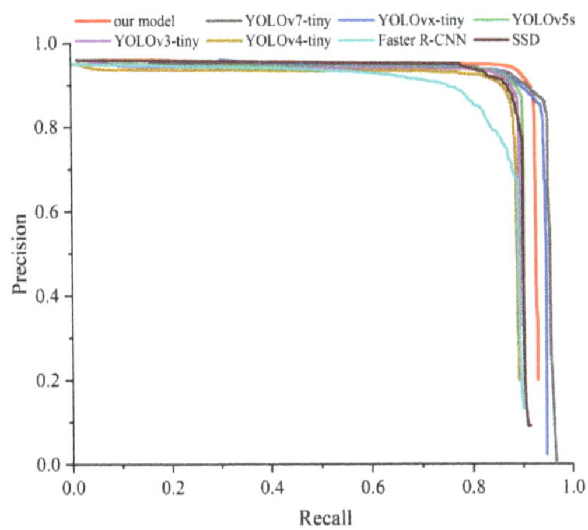

Figure 13. The PR curve of red jujubes with different target detection algorithms.

The P-R curve is a curve with recall as the horizontal coordinate and precision as the vertical coordinate out of the curve, whose area can show the comprehensive performance of the target detection model for red jujubes. Figure 13. shows that the curve areas of YOLOv3-tiny, YOLOv4-tiny, YOLOv5s, YOLOvx-tiny and YOLOv7-tiny are larger than those of SSD and Faster R-CNN. It illustrates that the Yolo series detection networks have higher accuracy and better recognition of red jujubes. YOLOv5s is used as an improved detection network for YOLOv3-tiny and YOLOv4-tiny, but the best detection result is not obtained for red jujubes, as shown in Table 5. The YOLOv4-tiny has better detection results, but the YOLOv5s are smaller in model size and more suitable for being used in agricultural mobile devices. Compared with the classical networks, the improved network not only maintains a better detection performance but also greatly reduces the model size.

Different detection algorithms were used to count red jujubes. YOLOvx-tiny, YOLOv5s, SSD, and Faster R-CNN all showed that the counting results of red jujubes were less than the actual number, as shown in Figure 14 image1. YOLOv7-tiny, YOLOv5s, and Faster R-CNN caused repeated recognition in the process of counting red jujubes, which led to the counting results being higher than the actual number, as shown in Figure 14 image2 and image3. Error counting occurred when SSD counted red jujubes, as shown in Figure 14 image3. When counting image4, only YOLOv4-tiny and Our Model counted accurately. However, Our Model also missed the detection of red jujubes, but compared with other algorithms, the number of missed detection was less, as shown in Figure 14 image5. When counting the Shaded red jujubes, all algorithms could count effectively, as shown in Figure 14 image6.

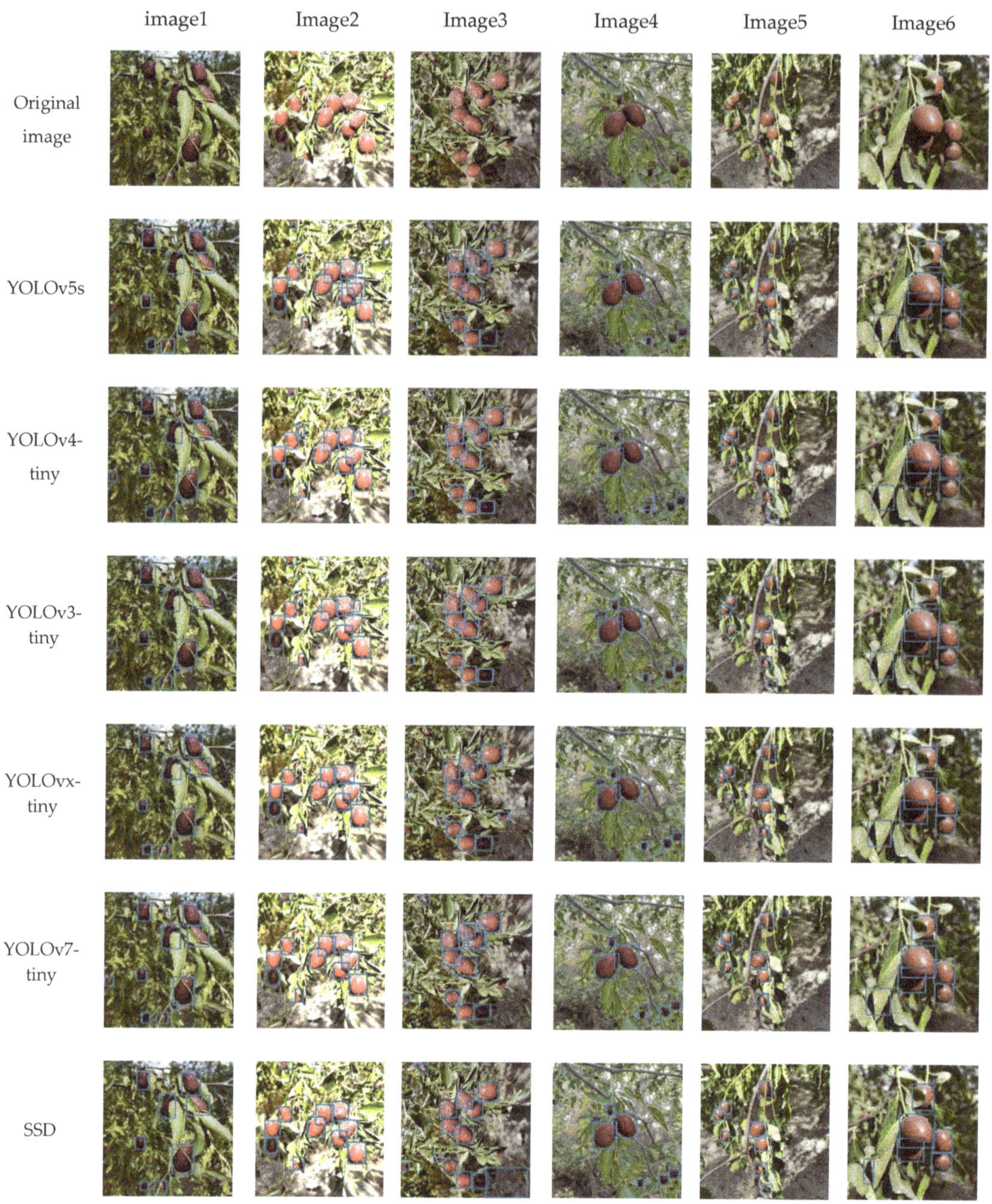

Figure 14. *Cont.*

Figure 14. Test results of different algorithms. Where the blue boxes are the test results of model test.

According to the experimental results, In the detection of red jujube, YOLOv5s, YOLOv4-tiny, and Faster R-CNN all miss the detection, which leads to a decrease in the number of red jujubes. YOLOv3-tiny, SSD, and Faster R-CNN all have error recognition, which leads to the increase in the estimation error of jujube yield by the model, as shown in Figure 14. Faster R-CNN, as one of the representative networks of the two-stage detection model, has good overall detection performance for red jujubes, but the AP is lower compared with other detection networks, And RMSE and MAPE are the maximum values, as shown in Table 5. This difference is mainly manifested in the recognition difficulty of fruits with large leaves shading and poor recognition of overlapping fruits. The reason for the difference is that Faster R-CNN does not build an image feature pyramid and cannot sufficiently extract features for small targets, resulting in insensitivity to small target recognition. For both the single stage detection model Yolo series and SSD, the overall performance is better than Faster R-CNN. Comparing SSD and YOLOv5s, the Precision is reduced by 0.80%. The recall is reduced by 3.20%, and AP is reduced by 5.10%, RMSE is increased by 45.75%, MAPE is increased by 6.86%. The main reasons are: (1) Since YOLOv5s introduces the FPN + PAN, while the detection layer is fused by three levels of feature layers, while all six feature pyramid layers of SSD come from the last layer of FCN, YOLOv5s is better than SSD in detecting red jujubes. (2) Due to the limited number of red jujube and the severe occlusion between red jujubes, it is difficult for the model to learn the various states. Compared with YOLOvx-tiny and YOLOv7-tiny, the AP of the improved network increased by 0.50% and 1.10%, respectively, RMSE decreased by 2.2 and 1.44 respectively, and MAPE decreased by 18.15% and 10.92% respectively. Comparing YOLOv5s, we introduce the ShuffleNet V2 backbone to reduce the size of the model, but the feature extraction ability of the model is limited. The idea of resizing images by convolution layer was adopted, and the Stem was added to enhance the feature extraction ability of the network. The improved model overall outperforms YOLOv5s, with Precision, Recall, and AP improving by 4.3%, 2.0%, and 0.6%. In addition, the model size, RMSE, and MAPE decreased by 91.82%, 20.87%, and 5.18%, respectively. The improved model has the highest Precision, Recall, F1-Score, and AP, and the smallest in model size, RMSE, and MAPE among the comparison networks.

4. Conclusions

In this research, a counting method of red jujube based on improved YOLOv5s was proposed for achieving accurate detection and counting red jujubes while reducing the model size in a complex environment. In order to reduce the number of parameters, ShuffleNet V2 was used as the backbone to make the model lightweight. In addition, the Stem module was designed as an intermediate module between the input and backbone to prevent the information loss caused by the change in feature map size. PANet was replaced by BiFPN for multi-scale feature fusion to enhance the model feature fusion capability and improve the model accuracy. Finally, the improved YOLOv5s detection model was used to count red jujubes. In order to verify the efficiency of the proposed model, YOLOv5s, YOLOv3-tiny, YOLOv4-tiny, SSD, Faster R-CNN, YOLOvx-tiny, and YOLOv7-tiny were used to compare with the improved model. The results showed that the improved model not only greatly reduced the model size but also had better performance in detection results than the comparison networks. Compared with yolov5s, Precision, Recall, and AP are improved by 4.3%, 2%, and 0.6%, respectively. In addition, the model size, RMSE, and MAPE decreased by 91.82%, 42.21%, and 11.47%, respectively. Therefore, the improved YOLOv5s model can not only effectively improve the detection performance of red jujubes but also finish the task of counting red jujubes in agricultural production. The method can provide a basis for estimating the yield of jujube by vision.

In summary, a counting method of red jujube based on improved YOLOv5s was proposed in this research, and the counting effectiveness of the method was verified by experiments. The future work of the red jujube counting method is as follows:

(1) Expand the types of data sets and increase the robustness of the model. There are only two kinds of jujube in the data set used in this research, so it is necessary to add more kinds of jujube fruit data to enhance the robustness of the model.
(2) Construct the model of jujube fruit size and quality. Further, the counting method of red jujubes was used to accurately estimate the yield of red jujubes.

Author Contributions: Data curation, methodology, project administration, writing—original draft, writing—review and editing, Y.Q.; review & editing, supervision, funding acquisition, and project administration, Y.H.; data curation, Z.Z.; formal analysis, H.Y.; formal analysis, K.Z.; review & editing, supervision, funding acquisition, and project administration, J.H. review & editing, supervision, J.G. All authors have read and agreed to the published version of the manuscript.

Funding: Please add: This research was supported by the Talent start-up Project of Zhejiang A&F University Scientific Research Development Foundation (2021LFR066) and the National Natural Science Foundation of China (C0043619, C0043628).

Institutional Review Board Statement: Not applicable.

Informed Consent Statement: Not applicable.

Data Availability Statement: Not applicable.

Conflicts of Interest: The authors declare no conflict of interest.

References

1. Dicianu, D.E.; Butcaru, A.C.; Constantin, C.G.; Dobrin, A.; Stanica, F. Evaluation of some nutritional properties of Chinese jujube (*Zizyphus jujuba* Mill.) fruit organicaly produced in bucharest area. *Sci. Pap. Ser. B Hortic.* **2020**, *64*, 79–84.
2. Kamilaris, A.; Prenafeta-Boldú, F.X. Deep learning in agriculture: A survey. *Comput. Electron. Agric.* **2018**, *147*, 70–90. [CrossRef]
3. Muruganantham, P.; Wibowo, S.; Grandhi, S.; Samrat, N.H.; Islam, N. A Systematic Literature Review on Crop Yield Prediction with Deep Learning and Remote Sensing. *Remote Sens.* **2022**, *14*, 1990. [CrossRef]
4. Dorj, U.-O.; Malrey, L.; Sang-seok, Y. An yield estimation in citrus orchards via fruit detection and counting using image processing. *Comput. Electron. Agric.* **2017**, *140*, 103–112. [CrossRef]
5. Wang, Z.; Kerry, W.; Anand, K. Mango fruit load estimation using a video based MangoYOLO—Kalman filter—Hungarian algorithm method. *Sensors* **2019**, *19*, 2742. [CrossRef]
6. Lyu, S.; Li, R.; Zhao, Y.; Li, Z.; Fan, R.; Liu, S. Green Citrus Detection and Counting in Orchards Based on YOLOv5-CS and AI Edge System. *Sensors* **2022**, *22*, 576. [CrossRef]
7. Zhang, Y.; Zhang, W.; Yu, J.; He, L.; Chen, J.; He, Y. Complete and accurate holly fruits counting using YOLOX object detection. *Comput. Electron. Agric.* **2022**, *198*, 107062. [CrossRef]
8. Li, X.; Du, Y.; Yao, L.; Wu, J.; Liu, L. Design and Experiment of a Broken Corn Kernel Detection Device Based on the YOLOv4-Tiny Algorithm. *Agriculture* **2021**, *11*, 1238. [CrossRef]
9. Gu, Y.; Wang, S.; Yan, Y.; Tang, S.; Zhao, S. Identification and Analysis of Emergency Behavior of Cage-Reared Laying Ducks Based on YOLOv5. *Agriculture* **2022**, *12*, 485. [CrossRef]
10. Zheng, Z.; Yang, H.; Zhou, L.; Yu, B.; Zhang, Y. HLU 2-Net: A Residual U-Structure Embedded U-Net With Hybrid Loss for Tire Defect Inspection. *IEEE Trans. Instrum. Meas.* **2021**, *70*, 1–11.

11. Girshick, R.; Donahue, J.; Darrell, T.; Malik, J. Rich feature hierarchies for accurate object detection and semantic segmentation. In Proceedings of the IEEE Conference on Computer Vision and Pattern Recognition, Columbus, OH, USA, 23–28 June 2014; pp. 580–587.
12. Girshick, R. Fast r-cnn. In Proceedings of the IEEE International Conference on Computer Vision 2015, Santiago, Chile, 7–13 December 2015; pp. 1440–1448.
13. Ren, S.; He, K.; Girshick, R.; Sun, J. Faster R-CNN: Towards Real-Time Object Detection with Region Proposal Networks. *IEEE Trans. Pattern Anal. Mach. Intell.* **2017**, *39*, 1137–1149. [CrossRef] [PubMed]
14. Liu, W.; Anguelov, D.; Erhan, D.; Szegedy, C.; Reed, S.; Fu, C.-Y.; Berg, A.C. Ssd: Single shot multibox detector. In Proceedings of the European Conference on Computer Vision 2016, Amsterdam, The Netherlands, 11–14 October 2016; pp. 21–37.
15. Shen, Z.; Liu, Z.; Li, J.; Jiang, Y.-G.; Chen, Y.; Xue, X. Dsod: Learning deeply supervised object detectors from scratch. In Proceedings of the IEEE International Conference on Computer Vision 2017, Venice, Italy, 22–29 October 2017; pp. 1919–1927.
16. Bochkovskiy, A.; Wang, C.-Y.; Liao, H.-Y.M. YOLOv4: Optimal speed and accuracy of object detection. *arXiv* **2020**, arXiv:2004.10934.
17. Redmon, J.; Divvala, S.; Girshick, R.; Farhadi, A. You only look once: Unified, real-time object detection. In Proceedings of the IEEE Conference on Computer Vision and Pattern Recognition 2016, Las Vegas, NV, USA, 27–30 June 2016; pp. 779–788.
18. Redmon, J.; Farhadi, A. YOLO9000: Better, faster, stronger. In Proceedings of the IEEE Conference on Computer Vision and Pattern Recognition 2017, Honolulu, HI, USA, 21–26 July 2017; pp. 7263–7271.
19. Redmon, J.; Farhadi, A. YOLOv3: An incremental improvement. *arXiv* **2018**, arXiv:1804.02767.
20. Tang, Y.; Chen, M.; Wang, C.; Luo, L.; Li, J.; Lian, G.; Zou, X. Recognition and localization methods for vision-based fruit picking robots: A review. *Front. Plant Sci.* **2020**, *11*, 510. [CrossRef] [PubMed]
21. You, L.; Jiang, H.; Hu, J.; Chang, C.; Chen, L.; Cui, X.; Zhao, M. GPU-accelerated Faster Mean Shift with euclidean distance metrics. *arXiv* **2021**, arXiv:2112.13891.
22. Zhao, M.; Jha, A.; Liu, Q.; Millis, B.A.; Mahadevan-Jansen, A.; Lu, L.; Landman, B.A.; Tyskac, M.J.; Huo, Y. Faster mean-shift: Gpu-accelerated embedding-clustering for cell segmentation and tracking. *arXiv* **2020**, arXiv:2007.14283. [CrossRef]
23. Zhao, M.; Liu, Q.; Jha, A.; Deng, R.; Yao, T.; Mahadevan-Jansen, A.; Tyska, M.J.; Millis, B.A.; Huo, Y. VoxelEmbed: 3D instance segmentation and tracking with voxel embedding based deep learning. In Proceedings of the International Workshop on Machine Learning in Medical Imaging, Strasbourg, France, 27 September 2021; pp. 437–446.
24. Lu, Y.; Young, S. A survey of public datasets for computer vision tasks in precision agriculture. *Comput. Electron. Agric.* **2020**, *178*, 105760. [CrossRef]
25. Mulyono, I.; Lukita, T.; Sari, C.; Setiadi, D.; Rachmawanto, E.; Susanto, A.; Putra, M.; Santoso, D. Parijoto Fruits Classification using K-Nearest Neighbor Based on Gray Level Co-Occurrence Matrix Texture Extraction. *J. Phys. Conf. Ser.* **2020**, *1051*, 012017. [CrossRef]
26. Fauliah, S.P. Implementation of learning vector quantization (lvq) algorithm for durian fruit classification using gray level co-occurrence matrix (glcm) parameters. *J. Phys. Conf. Ser.* **2019**, *1196*, 012040.
27. Zhao, C.; Lee, W.S.; He, D. Immature green citrus detection based on colour feature and sum of absolute transformed difference (SATD) using colour images in the citrus grove. *Comput. Electron. Agric.* **2016**, *124*, 243–253. [CrossRef]
28. Peng, H.; Shao, Y.; Chen, K.; Deng, Y.; Xue, C. Research on multi-class fruits recognition based on machine vision and SVM. *IFAC-PapersOnLine* **2018**, *51*, 817–821. [CrossRef]
29. Wajid, A.; Singh, N.K.; Junjun, P.; Mughal, M.A. Recognition of ripe, unripe and scaled condition of orange citrus based on decision tree classification. In Proceedings of the 2018 International Conference on Computing, Mathematics and Engineering Technologies (iCoMET) 2018, Sukkur, Pakistan, 3–4 March 2018; pp. 1–4.
30. Hussin, R.; Juhari, M.R.; Kang, N.W.; Ismail, R.; Kamarudin, A. Digital image processing techniques for object detection from complex background image. *Procedia Eng.* **2012**, *41*, 340–344. [CrossRef]
31. Zhang, F.; Chen, Z.; Bao, R.; Zhang, C.; Wang, Z. Recognition of dense cherry tomatoes based on improved YOLOv4-LITE lightweight neural network. *Trans. Chin. Soc. Agric. Eng.* **2021**, *37*, 270–278.
32. Fu, L.; Feng, Y.; Majeed, Y.; Zhang, X.; Zhang, J.; Karkee, M.; Zhang, Q. Kiwifruit detection in field images using Faster R-CNN with ZFNet. *IFAC-PapersOnLine* **2018**, *51*, 45–50. [CrossRef]
33. Liu, Z.; Wu, J.; Fu, L.; Majeed, Y.; Feng, Y.; Li, R.; Cui, Y. Improved kiwifruit detection using pre-trained VGG16 with RGB and NIR information fusion. *IEEE Access* **2019**, *8*, 2327–2336. [CrossRef]
34. Wang, Y.; Xue, J. Lightweight object detection method for Lingwu long jujube images based on improved SSD. *Trans. Chin. Soc. Agric. Eng.* **2021**, *37*, 173–182.
35. Li, X.; Pan, J.; Xie, F.; Zeng, J.; Li, Q.; Huang, X.; Liu, D.; Wang, X. Fast and accurate green pepper detection in complex backgrounds via an improved YOLOv4-tiny model. *Comput. Electron. Agric.* **2021**, *191*, 106503. [CrossRef]
36. Novtahaning, D.; Shah, H.A.; Kang, J.-M. Deep Learning Ensemble-Based Automated and High-Performing Recognition of Coffee Leaf Disease. *Agriculture* **2022**, *12*, 1909. [CrossRef]
37. Liu, S.; Qi, L.; Qin, H.; Shi, J.; Jia, J. Path aggregation network for instance segmentation. In Proceedings of the IEEE Conference on Computer Vision and Pattern Recognition 2018, Salt Lake City, UT, USA, 18–23 June 2018; pp. 8759–8768.
38. Wu, S.; Yang, J.; Wang, X.; Li, X. Iou-balanced loss functions for single-stage object detection. *Pattern Recognit. Lett.* **2022**, *156*, 96–103. [CrossRef]

39. Rezatofighi, H.; Tsoi, N.; Gwak, J.; Sadeghian, A.; Reid, I.; Savarese, S. Generalized intersection over union: A metric and a loss for bounding box regression. In Proceedings of the IEEE/CVF Conference on Computer Vision and Pattern Recognition 2019, Long Beach, CA, USA, 15–20 June 2019; pp. 658–666.
40. Zheng, Z.; Wang, P.; Liu, W.; Li, J.; Ye, R.; Ren, D. Distance-IoU loss: Faster and better learning for bounding box regression. In Proceedings of the AAAI Conference on Artificial Intelligence 2020, New York, NY, USA, 7–12 February 2020; pp. 12993–13000.
41. Zheng, Z.; Wang, P.; Ren, D.; Liu, W.; Ye, R.; Hu, Q.; Zuo, W. Enhancing geometric factors in model learning and inference for object detection and instance segmentation. *IEEE Trans. Cybern.* **2022**, *52*, 8574–8586. [CrossRef]
42. Zhang, X.; Zhou, X.; Lin, M.; Sun, J. Shufflenet: An extremely efficient convolutional neural network for mobile devices. In Proceedings of the IEEE Conference on Computer Vision and Pattern Recognition 2018, Salt Lake City, UT, USA, 18–23 June 2018; pp. 6848–6856.
43. Szegedy, C.; Ioffe, S.; Vanhoucke, V.; Alemi, A.A. Inception-v4, inception-resnet and the impact of residual connections on learning. In Proceedings of the Thirty-First AAAI Conference on Artificial Intelligence, San Francisco, CA, USA, 4–9 February 2017.

agriculture

MDPI

Article

Experimental Study on Water Distribution and Droplet Kinetic Energy Intensity from Non-Circular Nozzles with Different Aspect Ratios

Zixin Wang [1], Yue Jiang [1,*], Jialing Liu [1], Hong Li [1] and Hao Li [2]

1 Research Centre of Fluid Machinery Engineering and Technology, Jiangsu University, Zhenjiang 212013, China
2 Institute of Farmland Irrigation, Chinese Academy of Agricultural Sciences, Xinxiang 453002, China
* Correspondence: jyfluid@ujs.edu.cn

Abstract: (1) Background: In sprinkler irrigation systems, the water distribution and droplet kinetic energy are affected by the shape of the nozzle. In this paper, the effects of working pressure and aspect ratio (*L/D*) of circular and non-circular nozzles (diamond and ellipse) on water distribution and droplet kinetic energy intensity were investigated; (2) Methods: The hydraulic performance of a PY15 impact sprinkler with circular and non-circular nozzles was assessed under different working pressures, and the droplet diameter, velocity, and kinetic energy intensity were measured by a 2D video disdrometer. Moreover, the coefficient of variation (*CV*) and form factor (β) were introduced to represent the water distribution and droplet characteristics; (3) Results: The results revealed that, under the same working pressure, the *CV* of the diamond nozzle was the smallest compared with that of the circular and elliptical nozzles, reflecting a more uniform water distribution. The uniformity of water distribution was the best when the *L/D* of the elliptical nozzle was the smallest. In general, the larger the outlet diameter, the larger the wetted radius and water application rate. In addition, the smaller the *L/D*, the smaller the peak water distribution value and the radial increase of the kinetic energy intensity of a single nozzle. The maximum droplet kinetic energy per unit volume of the elliptical nozzle was the smallest compared with that of the circular and diamond nozzles. The circular nozzle at 200 kPa and the diamond and elliptical nozzles at 100 kPa obtained the highest uniformity coefficients of combined kinetic energy intensity distribution, which were 55.93% (circular), 67.59% (diamond), and 57.78% (elliptical) when the combination spacings were 1.0 R, 1.1 R and 1.2 R, and 1.0 R, respectively. Finally, the fitting function of unit volume droplet kinetic energy, distance from the nozzle, *L/D*, and working pressure of non-circular nozzles was established, and a fitting coefficient of 0.92 was obtained, indicating that the fitting equation was accurate; (4) Conclusions: At low working pressures, the elliptic and diamond nozzles showed better water distributions than the circular nozzle. The distal average droplet diameters of the sprinkler with non-circular nozzles were found to be smaller than those produced by the circular nozzle.

Keywords: impact sprinkler; non-circular nozzle; water distribution; aspect ratio; 2D video disdrometer; droplet kinetic energy distribution

Citation: Wang, Z.; Jiang, Y.; Liu, J.; Li, H.; Li, H. Experimental Study on Water Distribution and Droplet Kinetic Energy Intensity from Non-Circular Nozzles with Different Aspect Ratios. *Agriculture* **2022**, *12*, 2133. https://doi.org/10.3390/agriculture12122133

Academic Editors: Vadim Bolshev, Vladimir Panchenko and Alexey Sibirev

Received: 6 October 2022
Accepted: 9 December 2022
Published: 12 December 2022

1. Introduction

Sprinkler irrigation is an effective irrigation method used for reducing water use in agriculture. Sprinkler irrigation systems are generally composed of a water source, a water pump and power equipment, a water delivery pipeline system, and sprinklers. The sprinklers are key pieces of equipment for the implementation of sprinkler irrigation. Their performance not only directly affects the spraying quality, but is also related to the economy of the entire sprinkler irrigation system. Among them, sprinklers with non-circular nozzles can achieve a low-pressure uniform spray, and have the advantage of improving the atomization quality. Therefore, non-circular nozzles have been widely used in agricultural sprinkler irrigation fields [1,2].

Sprinkler hydraulic performance mainly depends on the wetted radius, flow rate, water application rate, uniformity, and droplet diameter [3–6]. Among them, the water application rate and uniformity of the radial water application profile of a sprinkler are important indicators for assessing the uniformity of irrigation property and water distribution, as well as important parameters for evaluating the advantages and disadvantages of sprinkler irrigation systems [7–9]. Bubenzer and Jones [10] reported that the damage degree of water droplets sprayed onto silty soil is a power function relationship between the kinetic energy of water droplets and the sprinkler water application rate. Mohammed and Kohl [11] investigated the distribution trend of the kinetic energy per unit volume of a full-jet nozzle under multiple pressure levels, and revealed that the kinetic energy per unit volume of water droplets along the radial direction increases with decreasing pressure. Christiansen [12] proposed an equation for calculating the uniformity coefficient. El-Wahed et al. [13] surveyed the effects of sprinkler spacing, sprinkler height, and working pressure of a central pivot irrigation system on parameters such as the uniformity coefficient, low value distribution uniformity, and variation coefficient, and determined the best operation condition scheme to reduce the operation costs.

In addition, the droplet diameter, droplet velocity, and droplet kinetic energy of sprinkler irrigation are also important indicators for evaluating the hydraulic performance of a sprinkler, and have an important impact on sprinkler irrigation system quality [14–16]. Xu et al. [17] investigated droplet diameter distribution and developed a normal distribution model, square-root-normal distribution model, logarithmic normal distribution model, and upper limit logarithmic normal distribution model of droplet diameter. Gong et al. [18] used a 2D video disdrometer (2DVD) to study the distribution range of a NelsonD3000 nozzle under different pressure levels, the changing trend of drop diameter, as well as the relationship between drop velocity, angle, and drop diameter. Their results demonstrated that the water droplet diameter and range conformed to the exponential function relationship, and the velocity of water droplets increased logarithmically with the increase of water droplet diameter. Lorenzini [19] investigated the trends of droplet velocity and evaporation under different working pressure levels during sprinkler irrigation. Their results indicated that the air temperature has a significant effect on droplet evaporation, while the air friction should not be mistakenly ignored when calculating droplet evaporation. Ouazaa et al. [20] analyzed the velocity and kinetic energy of droplets using a ballistic model. The results revealed that, under the working pressures of 138 kPa and 69 kPa, the kinetic energy dissipation decreased with increasing nozzle diameter. Yan et al. [4] also analyzed droplet velocity and kinetic energy by using a ballistic model, and reported that the runoff rate, bulk density of soil surface crust, and sediment yield were generally directly proportional to the droplet kinetic energy flux density (DE f) values, while the initiation of runoff, infiltration rate, and infiltration depth prior to runoff were inversely proportional to DE f. Zhu et al. [21] researched single droplet kinetic energy, droplet kinetic energy per unit volume, kinetic energy intensity distribution trend, and the kinetic energy intensity uniformity coefficient under different combination spacings of a full-jet nozzle. The results showed that the relationship between the kinetic energy distribution of a single water droplet and the water droplet diameter in the full-jet nozzle fit well with the developed model and exhibited a power function relationship. Li and Ma [22] investigated the droplet kinetic energy distribution at different measurement points of square and circular nozzles using the flour method and the droplet equation of motion, and analyzed the relationship between nozzle shape and droplet kinetic energy. Moreover, a regression equation able to estimate the total kinetic energy of water droplets from medium-pressure nozzles based on the median diameter was obtained.

The shape of conventional sprinkler nozzles is circular. The newly introduced non-circular nozzles have the advantages of improved water distribution and spraying uniformity, as well as better hydraulic performance under low pressure compared with circular nozzles [23,24]. Wei et al. [25] conducted experiments on the hydraulic performance of non-circular and circular nozzles, produced contour maps of water distribution for a single sprinkler, and obtained the corresponding range values based on water distribution. Chen et al. [2] investigated the effect of the shape coefficient of four different-shaped non-circular nozzles, i.e., diamond, semicircle + triangle, semicircle + rectangle, and star, on droplet diameter, and concluded that the droplet diameter in the end decreases with the increasing shape coefficient. Li et al. [26] experimentally analyzed and discussed the effects of non-circular nozzle shape and pressure on the shape change of a low-pressure jet, as well as the effects of nozzle outlet shape, working pressure, and inlet angle on the shape change of the jet. Zhou et al. [27] designed a variety of non-circular nozzles with the same area, experimentally investigated the uniformity of radial water application profile of the sprinkler, and found that the combined uniformity coefficient of the non-circular nozzles was significantly higher than that of circular nozzles. Jiang et al. [28] used high-speed photography to study the fracture and flow characteristics of non-circular nozzles, and revealed that, under the same working pressure, the triangular jet had the shortest fracture length and jet diffusion angle. Due to the different jet patterns produced by non-circular nozzles, their hydraulic performance is improved under low pressure. While the wetted radius of non-circular nozzles is reduced compared with that of circular nozzles, uniformity of the radial water application profile of the sprinkler is higher and the water droplet distribution is more uniform under low working pressure.

According to the above literature review, it can be understood that, due to their special geometric structure, non-circular nozzles have various jet shapes, which can reduce working pressure and improve their hydraulic performance. A quantitative study on the droplet distribution characteristics of non-circular nozzles under different working pressure levels will provide theoretical support for the application of non-circular nozzles. Nevertheless, there are only a few existing studies concerning the radio water distribution and water droplet kinetic energy distribution characteristics of non-circular nozzles with different aspect ratios (*L/D*: long axis/minor axis). Furthermore, no study has focused on the direct effect relationship between *L/D*, radio water distribution, and water drop kinetic energy. Consequently, in this paper, taking the non-circular nozzle as the research object, three types of nozzles with different shapes are designed, and the radial water distribution and kinetic energy intensity of non-circular nozzles with different aspect ratios under different working conditions are calculated. A 2DVD is used to test the water diameter and velocity of the non-circular nozzles. The distribution trend of the water drop kinetic energy is calculated, and the relationship between *L/D*, water distribution, and water drop kinetic energy is explored. In addition, the influence rule of water distribution and water drop distribution characteristics on nozzle hydraulic performance is determined, and a theoretical basis for further improving the spraying performance and studying the hydraulic characteristics of the nozzle outflow field is provided.

2. Materials and Methods

2.1. Non-Circular Nozzle Design

As shown in Figure 1, a common circular nozzle structure was taken as reference. d0 is its outlet diameter, which was set as 4 mm, 5 mm, and 6 mm. Since the outlet section shape and outlet diameter of non-circular nozzles have an impact on the hydraulic performance of the sprinkler, the non-circular nozzles with elliptical and diamond shapes were designed to have the same flow rate. The non-circular nozzles with different aspect ratios are displayed in Figure 2, and the detailed structural parameters are given in Table 1.

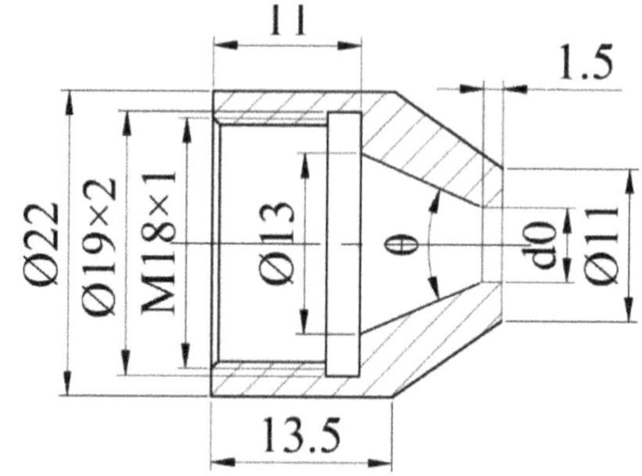

Figure 1. Schematic diagram (cross-section) of a circular nozzle structure with dimensions (mm).

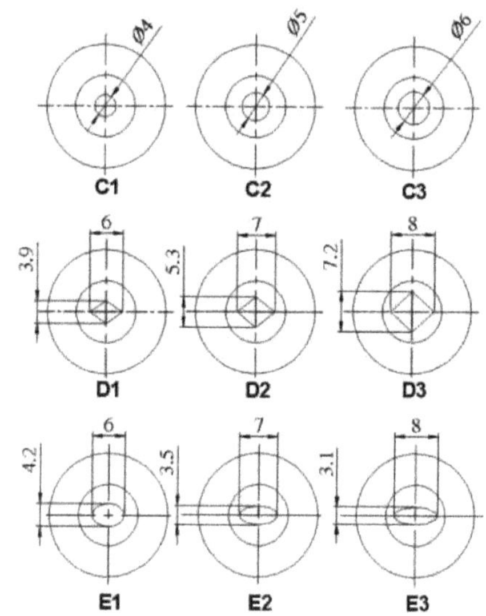

Figure 2. Schematic diagram (right view) of the investigated non-circular nozzles with dimensions (mm).

Table 1. Geometric parameters of the circular and non-circular nozzles (mm).

Inlet Cone Angle	Shape	Number	Outlet Diameter	Long Axis	Minor Axis	Aspect Ratio (*L*/*D*)
45°	Circle	C1	4	/	/	/
		C2	5	/	/	/
		C3	6	/	/	/
	Diamond	D1	4	6	3.9	1.54
		D2	5	7	5.3	1.32
		D3	6	8	7.2	1.11
	Ellipse	E1	5	6	4.2	1.43
		E2	5	7	3.5	2
		E3	5	8	3.1	2.58

In this paper, a PY15 impact sprinkler produced by Jinlong Spray Irrigation Co., China, was used for the experiments. The PY15 nozzle is a rotary impact nozzle with an inlet diameter of 15 mm, working pressure of 200~400 kpa, nozzle diameter of 4~6 mm, and a jet elevation angle of 23°. Each nozzle was manufactured by a wire-cut electric discharge machining (EDM) process and was made of aluminum (Figure 3). Considering the error problem, flow error tests were performed after processing. It was found that the flow error of the nozzles with the same inlet cone angle and outlet diameter was less than 4% under the same pressure level (Table 2). Thus, it was considered that all nozzles followed the same flow principle.

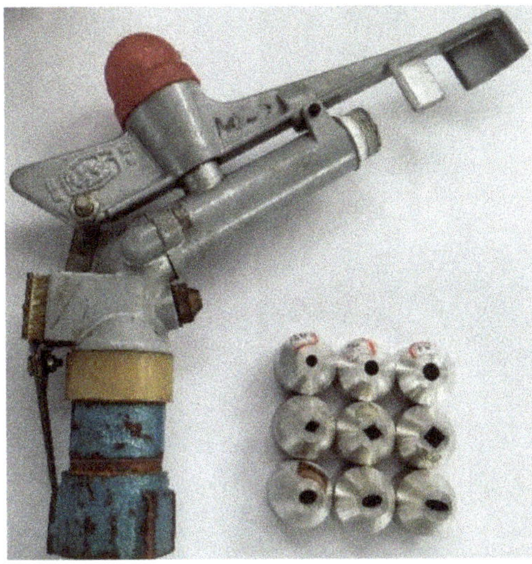

Figure 3. Photograph of the sprinkler and nozzles used in experiments.

Table 2. Flow error analysis of equivalent-flow nozzles (the units of flow: m^3/h).

Outlet Diameter	Number	150 kPa	200 kPa	250 kPa	300 kPa	350 kPa	400 kPa
5 mm	C2	1.067	1.234	1.385	1.516	1.636	1.747
	D2	1.036	1.198	1.347	1.476	1.601	1.709
	Difference	0.031	0.036	0.038	0.04	0.035	0.038
	Error	2.91%	2.92%	2.74%	2.64%	2.14%	2.18%
	E1	1.035	1.207	1.353	1.485	1.604	1.715
	Difference	0.032	0.027	0.032	0.031	0.032	0.032
	Error	3.00%	2.19%	2.31%	2.04%	1.96%	1.83%
	E2	1.075	1.261	1.411	1.554	1.673	1.79
	Difference	0.008	0.027	0.026	0.038	0.037	0.043
	Error	0.75%	2.19%	1.88%	2.51%	2.26%	2.46%
	E3	1.064	1.237	1.386	1.523	1.642	1.755
	Difference	0.003	0.003	0.001	0.007	0.006	0.008
	Error	0.28%	0.24%	0.07%	0.46%	0.37%	0.46%
4 mm	C1	0.693	0.81	0.898	0.981	1.067	1.142
	D1	0.675	0.78	0.88	0.975	1.052	1.131
	Difference	0.018	0.03	0.018	0.006	0.015	0.011
	Error	2.60%	3.70%	2.00%	0.61%	1.41%	0.96%
6 mm	C3	1.461	1.694	1.895	2.079	2.247	2.401
	D3	1.487	1.726	1.926	2.118	2.288	2.453
	Difference	0.026	0.032	0.031	0.039	0.041	0.052
	Error	1.78%	1.89%	1.64%	1.88%	1.82%	2.17%

2.2. Experimental Equipment

A 2DVD system manufactured by Joanneum Research Co., Graz, Styria, Austria, was employed to measure the size, shape, direction, aggregation state, and falling velocity of single droplets. The 2DVD system comprised two subsystems. The imaging system is composed of two background illumination sources, two line-scan cameras, and other components (Figure 4). The number of pixels of the camera was 512. The test area was 10,000 mm^2 (100 × 100 mm). The particle diameter measurement range was 0.2–8.0 mm and the vertical particle velocity range was 0–10.0 m/s. The data analysis and display system consisted of the View_HYD software (v8.010) designed by Joanneum Research Co., Graz, Styria, Austria, which was used to display the data generated by the 2DVD, record the measured droplet volumes, and calculate the rainfall velocity.

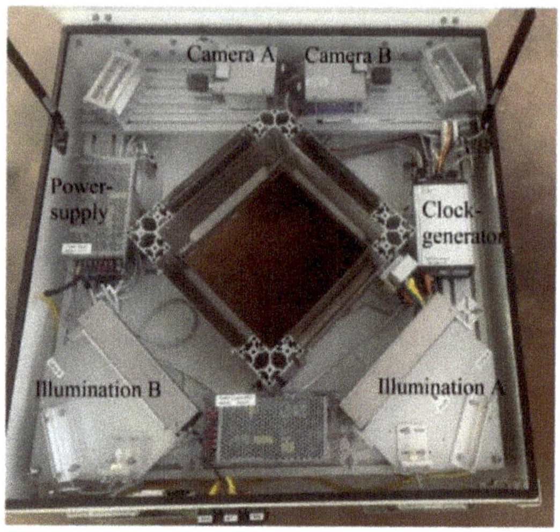

Figure 4. 2DVD system used to measure the characteristics of rainfall drops.

2.3. Test Method

The water flow distribution and droplet kinetic energy distribution experiments were performed in an indoor no-wind sprinkler irrigation hall with a diameter of 44 m. The PY15 impact sprinkler installation height was 1.4 m. Full-round spraying took place under the working pressure levels of 100 kPa, 150 kPa, 200 kPa, 250 kPa, and 300 kPa. Since there was no wind in the sprinkler hall, the instruments were placed along a linear path (ray). The data collected on this ray could be used to represent the data over the entire rotation. As it can be seen in Figures 5 and 6, rain gauges (opening diameter 85 mm) were placed in the radial direction of the jet at 1.0 m intervals. In order to ensure that the collected water would not overflow, the rain gauge was designed as a cylinder with a wedge-shaped opening edge without any defects. The test started after 10 min to ensure stable operation, and lasted for 1 h to determine the water application rate at each point. In addition, after the nozzle operated stably for 10 min, the 2DVD was moved by a distance of 2.0 m, and the diameter and velocity at each point were measured successively. The test duration at each point was at least 5 min. This is to ensure that the number of water droplets collected at each test location is higher than 5000, which is enough to provide appropriate water droplet statistics. In order to reduce the random and edge effect measurement errors caused by droplet splashing, the 3σ criterion of the normal distribution law of random errors is adopted to statistically test the particle number corresponding to the velocity value in each diameter range [29]. According to the relationship between droplet diameter and droplet falling speed, the diameter range was determined, so as to identify and eliminate the gross error, and then, the spray water drop strike kinetic energy was calculated.

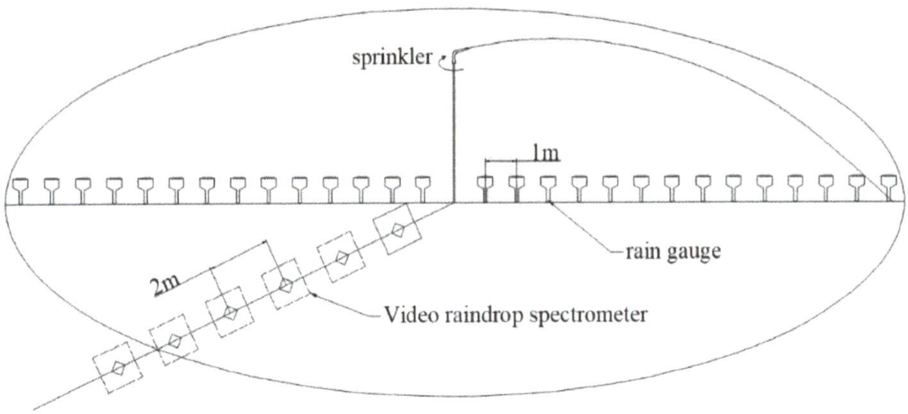

Figure 5. Schematic layout of the experimental system.

Figure 6. The test site.

2.4. Data Processing

2.4.1. Coefficient of Variation

The coefficient of variation (CV) is the ratio of the standard deviation of the water depth in each rainfall gauge to the arithmetic mean deviation [30]. The worse the uniformity of the radial water application profile of the sprinkler, the greater the measured CV. The relationship of the coefficient of variation CV is given as follows:

$$CV = \frac{SD}{MN} \times 100\%, \tag{1}$$

where CV is the coefficient of variation (%), SD is the standard deviation of the water depth received by all rain gauges (mm), and MN is the arithmetic mean of the water in all rain drums.

2.4.2. Kinetic Energy of Droplets per Unit Volume

The kinetic energy of droplets per unit volume refers to the ratio of the sum of the kinetic energy of individual droplets at different measuring points to the total volume [31]; the relationship is as follows:

$$E_{ks} = \frac{\sum\limits_{j=1}^{m} E_{sd_j}}{1000 \sum\limits_{j=1}^{m} \frac{1}{6}\pi \bullet m_j \bullet d_j^3}, \tag{2}$$

$$E_{sd} = \frac{\sum\limits_{i=1}^{n} \frac{1}{12}\pi \bullet \rho_\omega \bullet d_i^3 \bullet m_{di} \bullet V_{di}^2}{\sum\limits_{i=1}^{n} m_{di}}, \tag{3}$$

where d is the droplet diameter (mm), E_{ks} is the kinetic energy of water droplets per unit volume (J/L), E_{sd_j} is the kinetic energy of a single water droplet with diameter d (J/L), m_j is the number of particles corresponding to the velocity of water droplets with diameter d, and j is the water droplet diameter class.

2.4.3. Kinetic Energy Intensity

The kinetic energy intensity represents the magnitude of the kinetic energy at the measured point per unit time. It can be determined by the water droplet diameter, droplet velocity, and water application rate [31], and the relationship is as follows:

$$K = \frac{\sum\limits_{j=1}^{m} E_{sd_j}}{1000 \sum\limits_{j=1}^{m} \frac{1}{6}\pi \bullet d_j^3} \times \frac{h_j}{3600}, \tag{4}$$

where K is the spray kinetic energy intensity at the distance j from the nozzle (W/m^2), and h_j is the water application rate at different distances from the nozzle (mm/h).

For an overlapping sprinkler irrigation system, the uniformity coefficient of kinetic energy intensity distribution can be calculated by the Christensen average, which can reflect the distribution of rainfall energy in a sprinkler irrigation system. Therefore, the uniform coefficient of kinetic energy intensity distribution can comprehensively evaluate the advantages and disadvantages of different systems in terms of potential runoff. The equation is:

$$CU_K = \left(1 - \frac{\sum\limits_{k=1}^{N} (K_k - \overline{K})}{\sum\limits_{k=1}^{N} K_k}\right) \times 100\%, \tag{5}$$

where CU_k is the uniformity coefficient of the kinetic energy intensity (%), N is the total number of measuring points, and $\overline{K}$ is the average kinetic energy intensity (W/m^2).

3. Results and Analysis

3.1. Water Discrete Degree Analysis

In order to study the difference in hydraulic performance between the non-circular nozzle and circular nozzle under low pressure and the performance of the non-circular nozzle under medium pressure, the test pressure of the circular nozzle was set to 100~200 kpa and that of the non-circular nozzle was set to 100~300 kpa. Table 3 shows the *CV* for nozzles with different shapes under different working pressure levels. According to Table 3, under the same working pressure, the *CV* of the circular nozzle was the largest, while that of the diamond nozzle was the smallest. It indicated that the water distribution of the diamond nozzle was the most uniform, while that of the round nozzle was the least uniform. The *CV* of the C2 nozzle decreased with pressure; the *CV* of the D2 and E1 nozzles increased at first and then decreased, reaching a maximum at 150 kPa, decreasing at 250 kPa, and increasing at 300 kPa. Within the pressure range of 100–200 kPa, the *CV*s of both the D2 and E1 nozzles were far smaller than that of the C2 nozzle, indicating that the water distribution of the non-circular nozzles at low pressure was more uniform than that of the circular nozzles.

Table 3. Coefficient of variation for nozzles with different shapes under different working pressure levels.

Nozzle Shape (Outlet Diameter 5 mm, Inlet Cone Angle 45°)	Working Pressure (kPa)				
	100	150	200	250	300
Circle C2	1.11	0.79	0.59	/	/
Diamond D2	0.30	0.37	0.36	0.34	0.35
Ellipse E1	0.40	0.47	0.44	0.26	0.37

Table 4 displays the *CV* for diamond nozzles with different *L/Ds*. According to Table 4, the *CV* of the D2 nozzle (*L/D* = 1.32) was the lowest, indicating that the water distribution under this *L/D* was the most uniform. The *CV* of the D1 and D3 nozzles exhibited a decreasing trend. Their maximum value appeared at 100 kPa, far exceeding that of the D2 nozzle, and the water distribution was uneven. Therefore, the diamond nozzle should be designed with an *L/D* of 1.32. Table 5 presents the *CV* for elliptical nozzles with different *L/Ds*. According to Table 5, the *CV* followed the E2 > E3 > E1 sequence, and the E1 nozzle with the smallest *L/D* had the best water distribution uniformity.

Table 4. Coefficient of variation for diamond nozzles with different *L/Ds*.

Nozzle Number	Working Pressure (kPa)				
	100	150	200	250	300
D1 (*L/D* = 1.54)	0.82	0.54	0.46	0.37	0.36
D2 (*L/D* = 1.32)	0.30	0.37	0.36	0.34	0.35
D3 (*L/D* = 1.11)	0.68	0.49	0.43	/	/

Table 5. Coefficient of variation for elliptical nozzles with different *L/Ds*.

Nozzle Number	Working Pressure (kPa)				
	100	150	200	250	300
E1 (*L/D* = 1.43)	0.40	0.47	0.44	0.26	0.37
E2 (*L/D* = 2.00)	0.76	0.50	0.58	0.48	0.43
E3 (*L/D* = 2.58)	0.55	0.47	0.48	0.43	0.42

According to Tables 3–5, the maximum *CV* appeared at the low pressures of 100 kPa and 150 kPa, indicating that, at low pressure, the water distribution is more uneven. When the pressure reached 300 kPa, the *CV*s of all nozzles had little differences. Under low pressure, the water distribution of the non-circular nozzles (i.e., diamond and elliptical) was more uniform than that of the circular nozzle. In order to further verify the uniformity of the low-pressure water distribution of the non-circular nozzles, the radial water distribution of a single nozzle under different working pressures was investigated.

3.2. Radial Water Application Profiles

Figure 7 shows the radial water distribution curves of three nozzles (C2, D2, and E1) with an outlet diameter of 5 mm. It can be observed that the water distribution of the different nozzles was concentrated at the middle and distal end of the wetted radius, and the water application rate dropped rapidly to 0 after the peak application rate was reached. Moreover, with the increase of working pressure, the wetted radius increased, the peak application rate value decreased, and the distance from the nozzle corresponding to the peak application rate value increased. In general, the lower the working pressure, the more uneven the water distribution. Under the same working pressure, the wetted radius of the elliptical nozzle was the shortest, while that of the circular nozzle was the longest. The peak application rate was obtained in the following order: diamond < elliptical < circular. According to Figure 7 and Table 3 above, under the same working pressure, the water distribution of the diamond nozzle was the most uniform, while that of the circular nozzle was the least uniform.

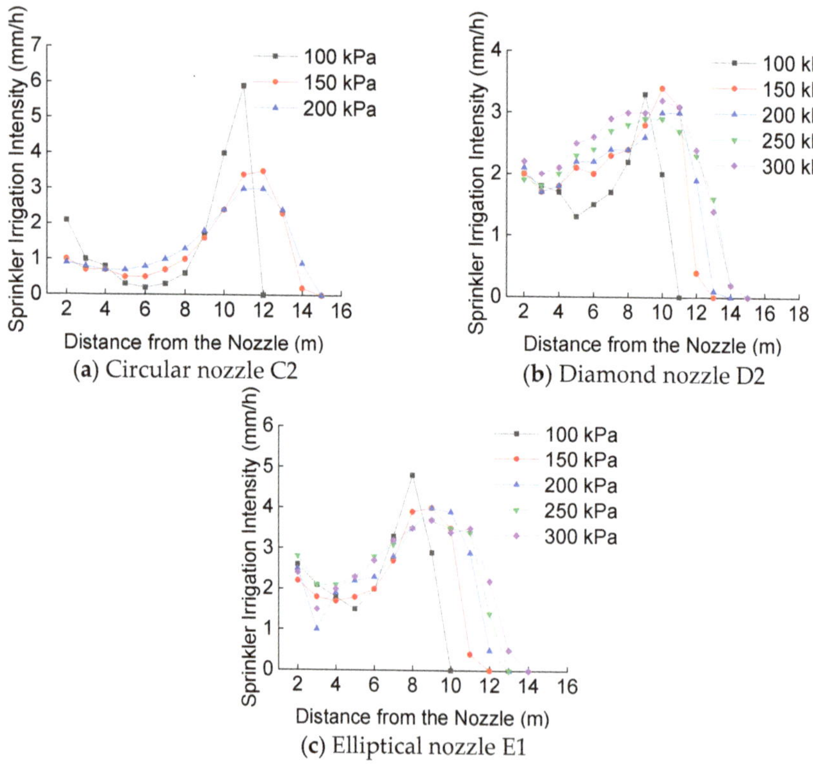

Figure 7. Radial water distribution of different nozzles under different working pressures. (C2 refers to the circular nozzle with a diameter of 5 mm; D2 refers to the diamond nozzle with an aspect ratio of 1.32; E2 refers to the elliptic nozzle with an aspect ratio of 2).

Figure 8 shows the radial water distribution of diamond nozzles D1, D2, and D3 with different outlet diameters and *L/Ds* in the order of D1 > D2 > D3 under different working pressure levels. It can be observed that the wetted radius increased with decreasing *L/D*. Furthermore, the water application rate at the measuring point 2 m away from the sprinkler increased gradually with decreasing *L/D*. The amount of water was mainly concentrated at 8~12 m. The radial distribution of the water application rate exhibited an overall trend of increasing first and then decreasing. After the peak value was reached, it quickly decayed to

0. The higher the working pressure, the stronger the phenomenon that the water application rate decreased with increasing *L/D*.

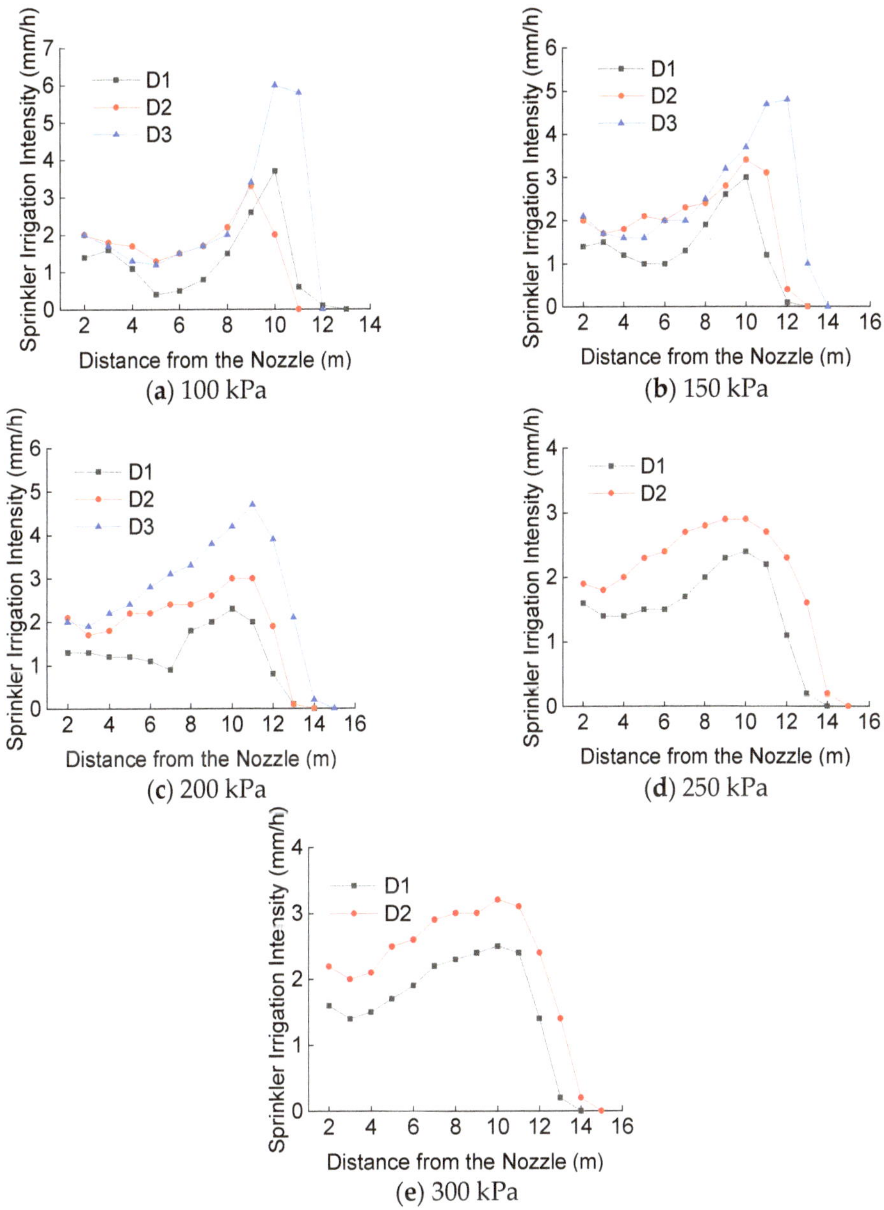

Figure 8. Radial water distribution of diamond nozzles with different outlet diameters under different working pressure levels. (D1, D2, and D3 refer to the diamond nozzles with aspect ratios of 1.54, 1.32, and 1.11, respectively).

Figure 9 shows the radial water distribution of equal flow elliptical nozzles (E1, E2, and E3) with different aspect ratios under different pressure levels. Combined with Table 5, it can be found that the E1 nozzle with the smallest *L/D* had the most uniform water distribution, while the E3 nozzle with the largest *L/D* and a slit-like shape had the largest peak water application rate and the smallest wetted radius. In general, the smaller the aspect ratio, the closer its shape to a circle, the smaller the peak application rate value, and the more uniform the water distribution. The sprinkler water application rate of the E3 nozzle with the largest *L/D* was much higher than that of the other two nozzles before reaching the peak application rate, and it declined faster, soon after reaching the peak application rate. The wetted radius decreased with increasing *L/D* and decreasing working pressure. At 150~300 kPa, the distances from the initial position of the E1, E2, and E3 nozzles when they reached the peak application rate were 9 m, 10 m, and 8 m, respectively. This is because when the *L/D* was too large, the nozzle shape tended to be a "slit" and the wetted radius decreased.

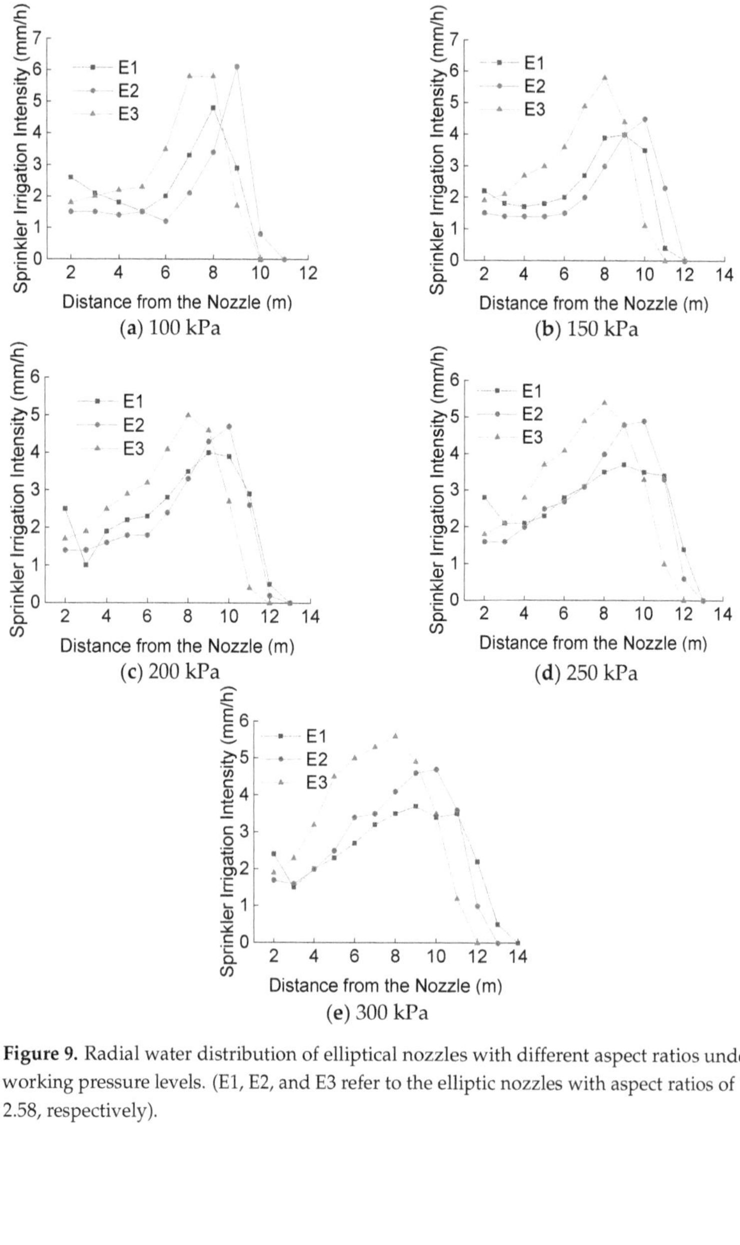

Figure 9. Radial water distribution of elliptical nozzles with different aspect ratios under different working pressure levels. (E1, E2, and E3 refer to the elliptic nozzles with aspect ratios of 1.43, 2, and 2.58, respectively).

3.3. Water Distribution Uniformity Coefficient for Combined Sprinkler Irrigation

After having determined the water distribution of the different single sprinkler heads, the main reasons affecting the uniformity of combined sprinklers are the combination mode of sprinkler heads and the combination spacing. The MATLAB software was employed to simulate and calculate the combination uniformity (*CU*) coefficients of each non-circular nozzle under different spacing in a rectangular combination arrangement at low pressure (100~200 kPa), and verify the effect of non-circular nozzles on *CU*. To avoid missing areas of water application, the combination spacing was selected as 1.0 R, 1.1 R, 1.2 R, and 1.3 R, where R is the effective spraying radius of the wetted radius of the sprinkler. The *CU* coefficients of the different nozzles under different working pressures and combination spacing are listed in Table 6.

Table 6. Combination uniformity coefficients of each nozzle under different pressures and combination spacing.

Shape	Number	Pressure (kPa)	1.0 R	1.1 R	1.2 R	1.3 R	1.4 R
Circle	C2	100	54.55	37.87	25.76	29.37	19.61
		150	58.59	58.25	50.11	37.32	26.49
		200	65.26	61.75	56.92	46.71	36.15
Diamond	D1	100	47.34	45.92	27.04	20.42	27.61
		150	63.17	59.24	47.61	45.23	46.36
		200	67.43	64.32	56.31	51.70	53.51
	D2	100	68.47	60.32	65.95	54.83	47.44
		150	69.56	70.84	63.44	55.59	54.77
		200	71.89	72.28	66.68	63.15	62.21
	D3	100	66.64	52.13	46.88	46.12	33.98
		150	66.19	62.70	60.19	48.71	41.50
		200	69.26	68.37	62.37	59.29	59.84
Ellipse	E1	100	67.92	60.02	61.47	43.52	38.02
		150	66.62	64.65	63.00	53.30	45.88
		200	68.43	65.36	58.76	53.86	54.38
	E2	100	56.98	44.87	49.03	37.15	26.24
		150	66.05	57.09	62.21	50.34	38.82
		200	64.27	59.30	49.05	46.60	45.33
	E3	100	62.46	62.12	52.83	38.58	43.59
		150	66.04	67.15	60.13	51.79	52.57
		200	68.51	68.72	60.87	59.05	56.46

It can be seen in Table 6 that, for the same nozzle, the *CU* coefficient increased with increasing working pressure and decreased with increasing combination spacing. Among the three different-shaped nozzles, the best combination spacing for the C2 nozzle was 1.0 R, and the *CU* coefficient was the highest (65.26%) when the working pressure was 200 kPa and the combination spacing was 1.0 R. The highest uniformity coefficient (72.28%) was presented by the D2 nozzle, under a working pressure of 200 kPa and combination spacing of 1.1 R. The highest uniformity coefficient (68.72%) of the elliptical nozzle combination was presented by the E3 nozzle, under a working pressure of 200 kPa and a combination spacing of 1.1 R. The C2 nozzle could achieve 65% uniformity at 200 kPa and 1.0 R, while the non-circular nozzles could meet the requirements at 100 kPa, indicating that the combined sprinkler irrigation with non-circular nozzles is more uniform under low pressure. The three-dimensional (3D) distribution diagrams when the *CU* coefficient of each nozzle combination was the best are presented in Figure 10.

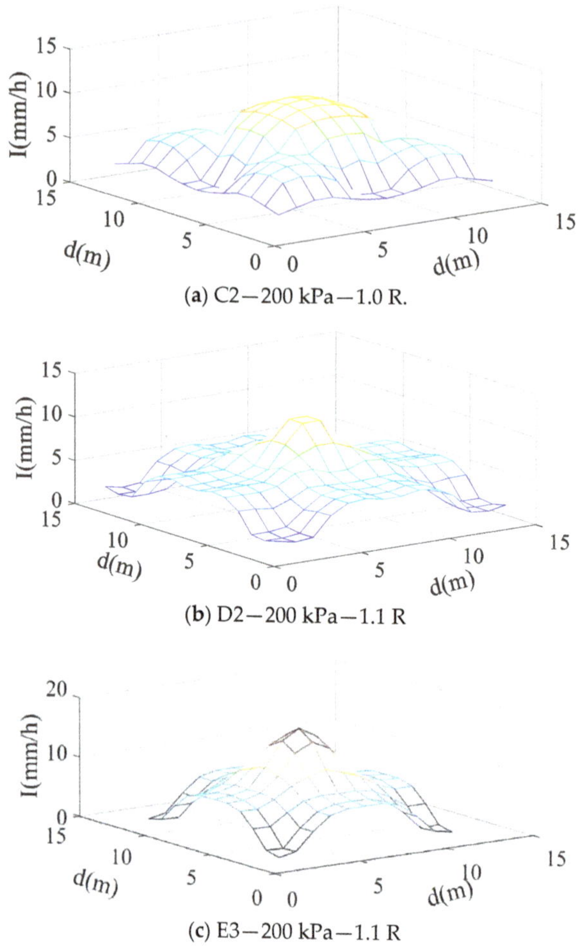

Figure 10. 3D water distribution under combined sprinkler irrigation. (C2 refers to the circular nozzle with a diameter of 5 mm; D2 refers to the diamond nozzle with an aspect ratio of 1.32; E3 refers to the elliptic nozzle with an aspect ratio of 2.58).

3.4. Kinetic Energy Intensity Distribution of Single Nozzles

Figure 11 displays the kinetic energy intensity distribution of each nozzle under different working pressure levels using B-spline curves. According to Figure 11a, the maximum kinetic energy intensity of the C2 nozzle at 100 kPa, 150 kPa, and 200 kPa was at a distance of 11 m, 12 m, and 12 m, respectively. According to Figure 11b, the maximum kinetic energy intensity of the D2 nozzle at 100 kPa, 150 kPa, 200 kPa, 250 kPa, and 300 kPa was in all cases found at a distance of 10 m. According to Figure 11c, the maximum kinetic energy intensity of the E1 nozzle at 100 kPa, 150 kPa, 200 kPa, 250 kPa, and 300 kPa was at a distance of 8 m, 10 m, 10 m, 10 m, and 10 m, respectively.

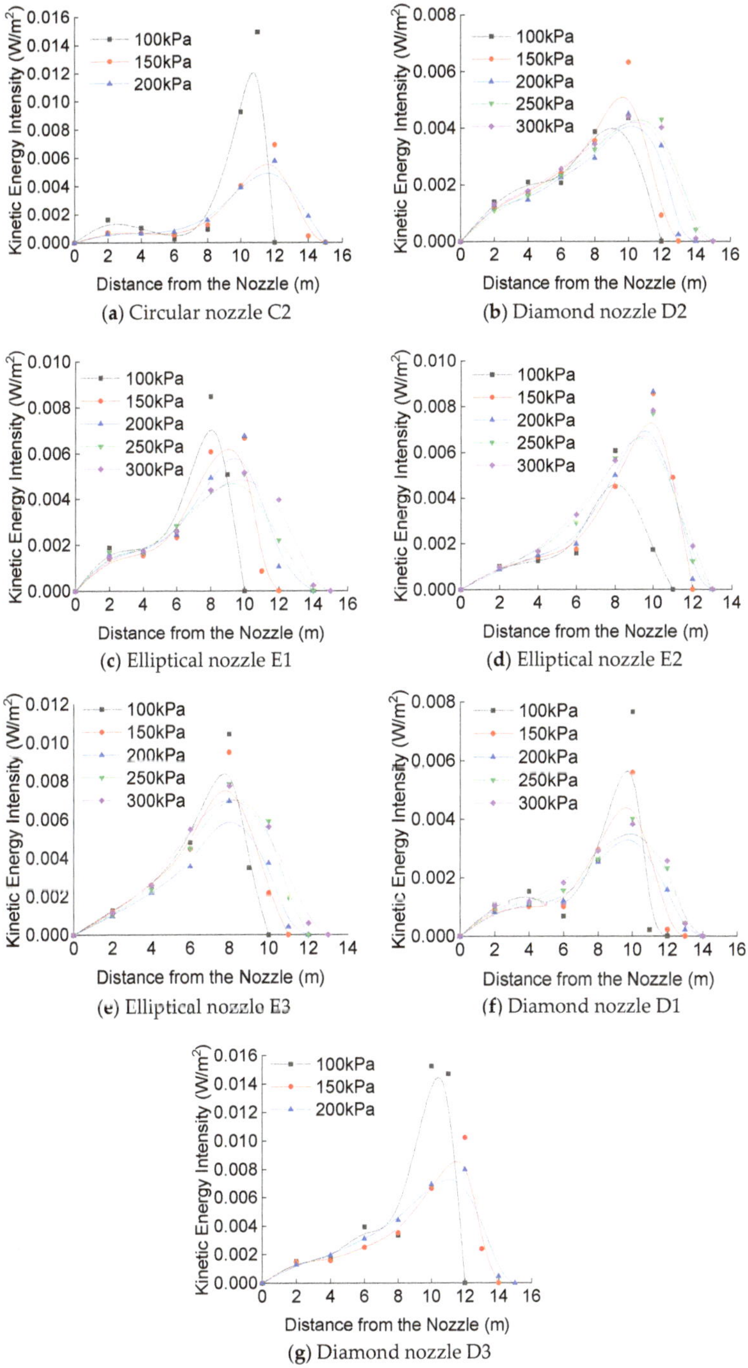

Figure 11. Kinetic energy intensity of the different single nozzles as a function of the distance from the nozzle under different working pressure levels. (C2 refers to the circular nozzle with a diameter of 5 mm; D1, D2, and D3 refer to the diamond nozzles with aspect ratios of 1.54, 1.32, and 1.11, respectively; E1, E2, and E3 refer to the elliptic nozzles with aspect ratios of 1.43, 2, and 2.58, respectively).

Based on Figure 11a–c, it can be seen that, within 0–8 m, the slope of the radial profile of the diamond nozzle was the largest, and that of the circular nozzle was the smallest. That is, the further the nozzle shape from a circle, the larger the radial increase of the kinetic energy intensity. In Figure 11c–e, it can be observed that, with the increase of the *L/D*, the peak value of kinetic energy intensity increased, and the distance from the nozzle position to the peak value position under each pressure was shorter. Within 0–6 m, the E3 nozzle (largest *L/D*) was closest to a straight line, and the slope of the radial profile was the largest, while the E1 nozzle (smallest *L/D*) had the smallest slope of the radial profile. This indicated that the larger the L/D, the greater the radial increase of the kinetic energy intensity. According to Figure 11b,f,g, the larger the outlet diameter, the higher the peak value of the kinetic energy intensity. Under each working pressure, the slopes of the kinetic energy intensity curves of the nozzles with an outlet diameter of 5 mm were the closest before and after reaching the peak value.

In the kinetic energy intensity curves of each nozzle plotted by B-splines, it can be observed that the peak value of the kinetic energy intensity corresponded to a distance from the nozzle position that increased with increasing working pressure, which ranged between 100 and 200 kPa. The distribution trend of the kinetic energy intensity of each nozzle along the radial direction under each working pressure was the same. It increased first and then decreased with increasing distance from the nozzle position, and then decreased rapidly to 0 after reaching the peak value of kinetic energy intensity. At 100 kPa and 150 kPa, the kinetic energy near the end of the wetted radius increased sharply, resulting in a high impact intensity on the soil surface structure, which can cause soil compaction and surface runoff.

3.5. Droplet Diameter Distribution

3.5.1. Radial Distribution of Droplet Diameter under Different Pressures

The commonly used methods for calculating the droplet diameter at home and abroad include the number-weighted method, the water-weighted average method, and the median diameter method [31,32]. In this study, the water-weighted average method was used to calculate the average droplet diameter at each measuring point. The Exp2PMod1 exponential fitting model was used, and its equation is as follows:

$$\mathrm{d} = \mathrm{a}e^{bl}, \tag{6}$$

where a and b are fitting coefficients, d is the droplet diameter, and l is the distance from the nozzle position.

After fitting the measured data, it was found that the fitting correlation coefficient R2 of each nozzle shape under each working pressure was between [0.92, 0.99], which was larger than 0.9, indicating that the fitting accuracy of this exponential function was high. The results for each nozzle are presented in Figure 12. As it can be observed in Figure 12, in all cases, the higher the pressure, the smaller the slope of the exponential function curve and the smaller the average droplet diameter at the same measuring point. This indicated that the increasing trend of the droplet diameter along the distance from the nozzle decreased with increasing pressure and the jet break-up was more severe. Among them, the slope of the radial profile of the C2 nozzle decreased the most. As shown in Table 5, among these three nozzles under the same pressure, the slope of the radial profile of the diamond nozzle was the largest and that of the circular nozzle was the smallest. It indicated that the droplet diameter of the diamond nozzle increased the most along the distance from the nozzle position.

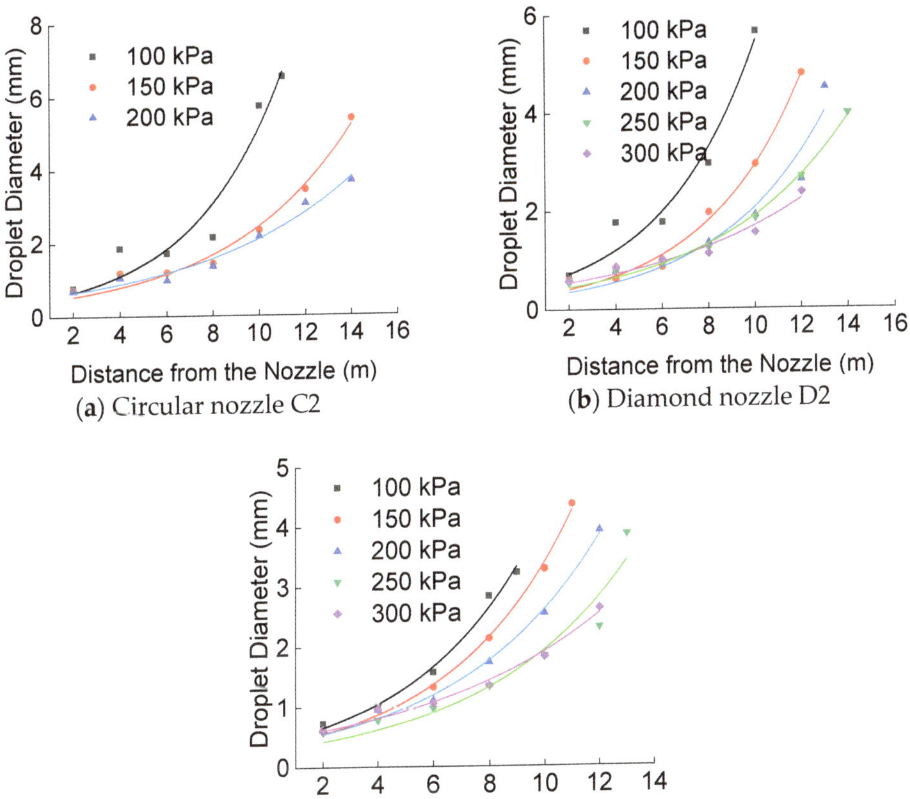

(**a**) Circular nozzle C2

(**b**) Diamond nozzle D2

(**c**) Elliptical nozzle E1

Figure 12. Relationship between average droplet diameter and distance from the nozzle. (C2 refers to the circular nozzle with a diameter of 5 mm; D2 refers to the diamond nozzle with an aspect ratio of 1.32; E1 refers to the elliptic nozzle with an aspect ratio of 1.43).

In addition, when the droplet diameter was 3 mm, the distance from the C2 nozzle at 100 kPa, 150 kPa, and 200 kPa was 7.9 m, 11 m, and 12.4 m, respectively; the distance from the D2 nozzle at 100 kPa, 150 kPa, 200 kPa, and 250 kPa was 7.6 m, 10 m, 11.7 m, and 12.5 m, respectively; and the distance from the E1 nozzle at 100 kPa, 150 kPa, 200 kPa, and 250 kPa was 8.5 m, 9.5 m, 10.7 m, and 12.3 m, respectively. At 300 kPa, the droplet diameters of the diamond and elliptical nozzles were less than 3 mm. In general, droplets with a smaller diameter tend to drift and have evaporation losses, while larger-diameter droplets can cause greater damage to the soil surface, which is not conducive to water and soil conservation and crop growth. Thus, the droplet diameter range suitable for spraying is within 1~3 mm [15]. Consequently, the E1 nozzle had the optimal wetted radius except for the 100 kPa case, where the C2 nozzle had the optimal wetted radius. When the distance was larger than 8 m, the pressure had a significant effect on droplet diameter.

3.5.2. Radial Distribution of Droplet Diameter under Different Aspect Ratios

Figure 13 depicts the relationship between mean droplet diameter and distance from the nozzle for different *L/D*s at different working pressure levels. For the nozzles with the same *L/D*, the slope of the exponential curve of the droplet diameter decreased with increasing pressure, i.e., the radial increase trend of the droplet diameter decreases with increasing pressure. Under the same pressure, shape, inlet cone angle, outlet diameter, and different aspect ratios, the droplet diameter at each measuring point was almost the same within 6 m from the nozzle position. The relationship between average droplet diameter distance from the nozzle generally followed the following order of E3 > E2 > E1, i.e., the larger the *L/D*, the larger the average droplet diameter. The E3 nozzle (largest *L/D*) had the shortest wetted radius and the largest overall droplet diameter. The E1 nozzle (smallest *L/D*) had the smallest droplet diameter along the radial direction and a relatively large wetted radius. Therefore, nozzles with a small *L/D* should be selected in sprinkler spraying.

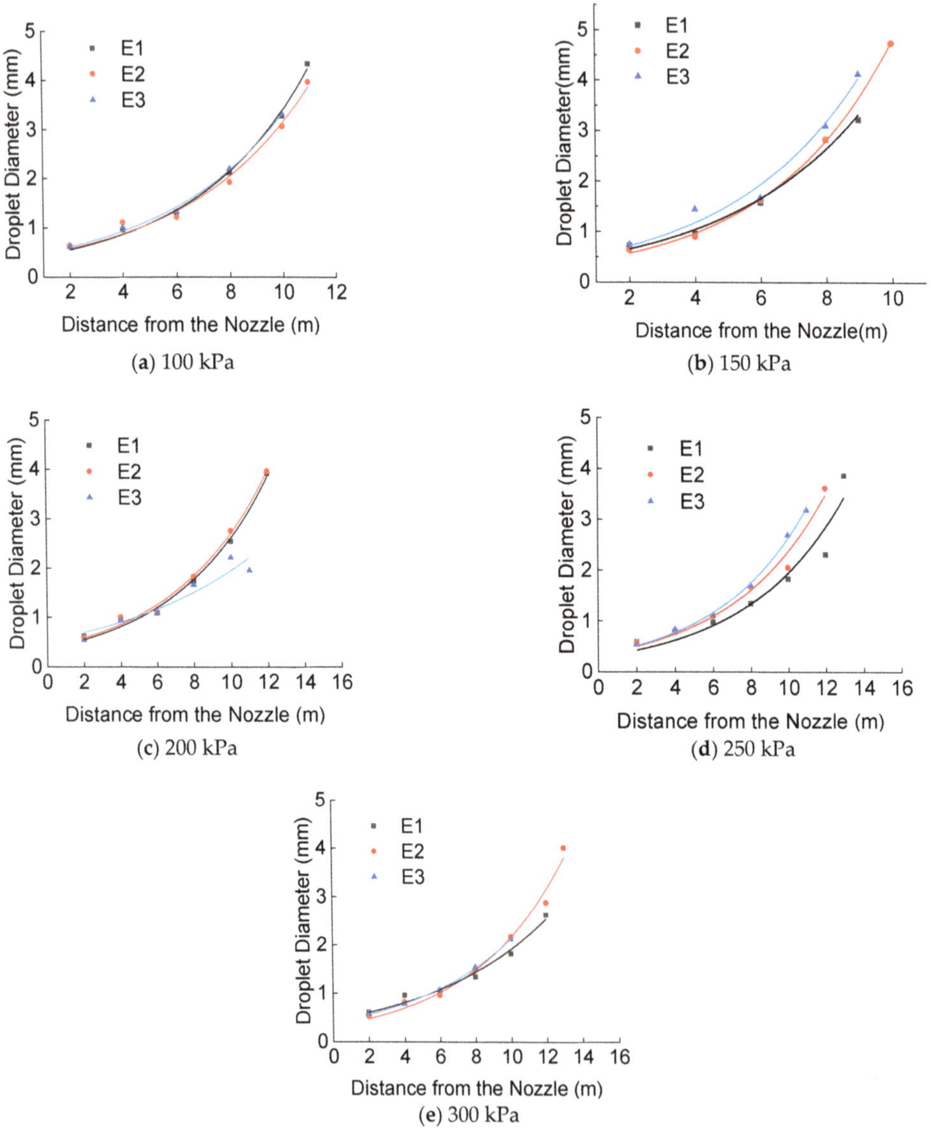

Figure 13. Relationship between mean droplet diameter and distance from the nozzle for different aspect ratios at different working pressure levels. (E1, E2, and E3 refer to the elliptic nozzles with aspect ratios of 1.43, 2, and 2.58, respectively).

3.6. Droplet Velocity Distribution

3.6.1. Droplet Velocity Distribution of Nozzles with Different Shapes

The velocity of spraying droplets is an important factor to determine the kinetic energy of striking droplets. 2DVD was used to measure the velocity of droplets with different diameters sprayed from different nozzles.

There is a logarithmic function that can express the relationship between droplet diameter and droplet velocity. The relationship is as follows [18]:

$$v = a - b\ln(d + c)a, \tag{7}$$

where a, b, and c are fitting coefficients, v is the droplet velocity, and d is the droplet diameter.

Figure 14 shows the results where the fitting correlation coefficient R2 was between [0.93, 0.97]. As it can be observed, for the nozzles with the same inlet cone angle and outlet diameter, the velocity curve slope of the circular nozzle was the largest and that of the elliptical nozzle was the smallest, i.e., with the increase of the droplet diameter, the increase rate of the velocity of the elliptical nozzle droplets was the smallest. The droplet velocity increased with increasing droplet diameter, and the increasing trend gradually decreased, indicating that droplet diameter is an important factor affecting droplet velocity. In general, the fitted curves of the three nozzles with equal flow rate were almost identical. It shows that, when the flow rate, inlet cone angle, and outlet diameter are the same, the nozzle outlet shape has little effect on the relationship between average droplet diameter and droplet velocity.

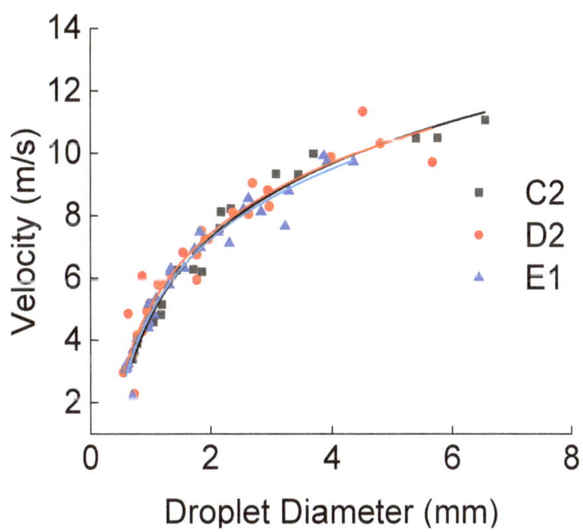

Figure 14. Relationship between average droplet diameter and velocity sprayed from nozzles with different shapes. (C2 refers to the circular nozzle with a diameter of 5 mm; D2 refers to the diamond nozzle with an aspect ratio of 1.32; E1 refers to the elliptic nozzle with an aspect ratio of 1.43).

3.6.2. Droplet Velocity Distribution under Different Aspect Ratios

The relationship between the diameter and velocity of droplets sprayed from elliptical nozzles with different aspect ratios is shown in Figure 15. The fitting correlation coefficient R2 was between [0.86, 0.98]. As it can be observed in Figure 15, for elliptical nozzles with the same inlet cone angle and outlet diameter, and different aspect ratios, the velocity curve slope followed the following order: E3 > E2 > E1, i.e., the velocity curve slope increased with increasing *L/D*. In addition, the magnitude of droplet velocity increased with increasing droplet diameter. At the same droplet diameter, the droplet velocity of the E2 nozzle was always the maximum. When the droplet diameter was less than 2.5 mm, the droplet velocity of the elliptical nozzle followed the E1 > E3 order. When the droplet diameter was greater than 2.5 mm, the droplet velocity followed the opposite order (E3 > E1). This indicated that there was an aspect ratio between the maximum and minimum aspect ratios. At this aspect ratio, the velocity of droplets with the same diameter was the highest, and therefore, the kinetic energy of the striking droplets was the highest.

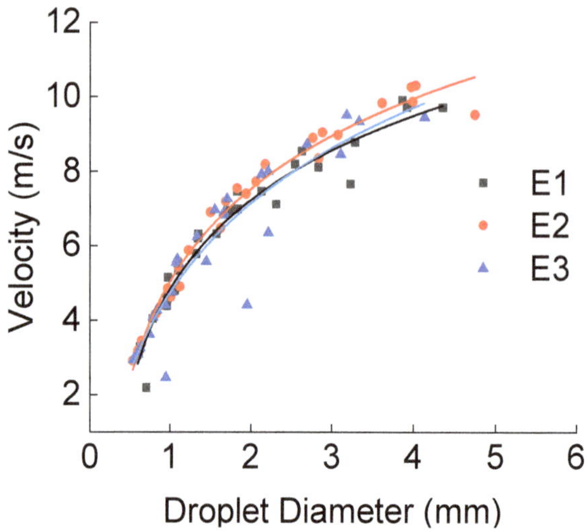

Figure 15. Relationship between average droplet diameter and velocity sprayed from elliptical nozzles with different aspect ratios. (E1, E2, and E3 refer to the elliptic nozzles with aspect ratios of 1.43, 2, and 2.58, respectively).

The relationship between the diameter and velocity of droplets sprayed from diamond nozzles with different aspect ratios is demonstrated in Figure 16. The fitting correlation coefficient R2 was between [0.93, 0.96]. As it can be seen in Figure 16, for nozzles with the same shape and inlet cone angle, the velocity curve slope followed the following order: D3 > D2 > D1, i.e., the velocity curve slope decreased with increasing *L/D*. In general, the smaller the *L/D*, the higher the droplet velocity, which increased with increasing droplet diameter. When the droplet diameter was less than 5 mm, the D1 > D2 > D3 order was followed, indicating that, when the droplet diameter is the same, the smaller the *L/D*, the lower the droplet velocity.

3.7. Kinetic Energy per Unit Volume Radial Profiles

Based on the kinetic energy distribution trend of droplets per unit volume, the kinetic energy per unit volume at different measuring points was calculated. Figure 17 shows the radial distribution of kinetic energy per unit volume of droplets sprayed from different nozzles under different pressure levels. As it can be seen in Figure 17, the higher the working pressure, the lower the kinetic energy per unit volume of the droplets at the same position, the smaller the increase of kinetic energy, and the less the damage to crops and soil surface. At 2 m, the kinetic energy per unit volume of droplets from each nozzle was not significantly different under all working pressures. According to Figure 17a–c, the maximum kinetic energy per unit volume of the elliptical nozzle was the lowest among the nozzles with equal flow and the same inlet cone angle and outlet diameter under each working pressure. At 100 kPa, the maximum kinetic energy per unit volume of the circular nozzle was the highest. According to Figure 17c–e, the larger the *L/D*, the smaller the wetted radius. The maximum kinetic energy per unit volume of the E1 nozzle (smallest

L/D) at 100 kPa was the lowest. At 150–300 kPa, the E3 nozzle (largest *L/D*) had the lowest kinetic energy per unit volume. In Figure 17b,f,g, it can be observed that, the smaller the outlet diameter, the lower the kinetic energy per unit volume.

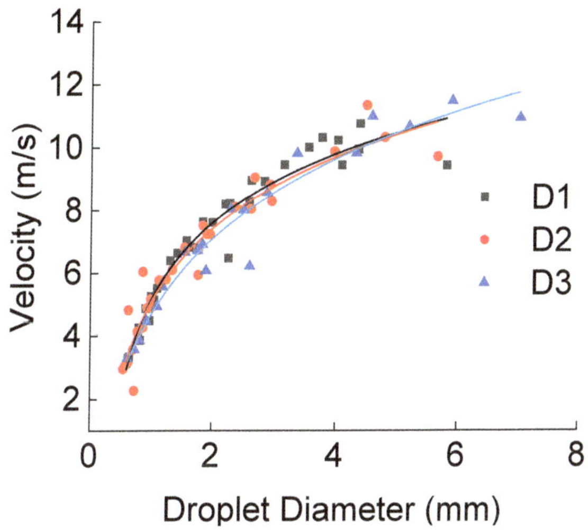

Figure 16. Relationship between droplet velocity and diameter sprayed from diamond nozzles with different aspect ratios. (D1, D2, and D3 refer to the diamond nozzles with aspect ratios of 1.54, 1.32, and 1.11, respectively).

In order to further investigate the radial distribution characteristics of the droplet kinetic energy per unit volume of the non-circular nozzles, in this paper, regression analysis on the kinetic energy per unit volume of each nozzle under various pressures was performed, and a distribution model of the kinetic energy based on the distance from the nozzle was established [30]. The relationship is given in the following equation:

$$E_{ks} = \mathrm{al} + \mathrm{b}, \tag{8}$$

where a and b are fitting coefficients, E_{ks} is the droplet kinetic energy per unit volume, and l is the distance from the nozzle position.

After fitting the measured data, it was found that the fitting correlation coefficient R2 of the kinetic energy per unit volume for the non-circular nozzles was between [0.89, 0.99], indicating that the fitting accuracy was high.

The data of the droplet kinetic energy per unit volume of the three elliptical nozzles with different aspect ratios under different working pressures were uniformly regressed, and the kinetic energy per unit volume was further analyzed. A mathematical model of the relationship between droplet kinetic energy per unit volume E_{ks}, distance from the nozzle l, aspect ratio β, and working pressure P for non-circular nozzles was established. The functional relationship is as follows:

$$E_{ks} = 2.775P^{-0.318}l + 0.0926\beta + 0.932\ (R^2 = 0.924), \tag{9}$$

The fitting coefficient of the droplet kinetic energy per unit volume for the elliptical nozzles with different aspect ratios was 0.92, indicating that the fitting accuracy was high. According to this regression equation, we can obtain the influence of the change of aspect ratio on the kinetic energy of water droplets under different working pressures.

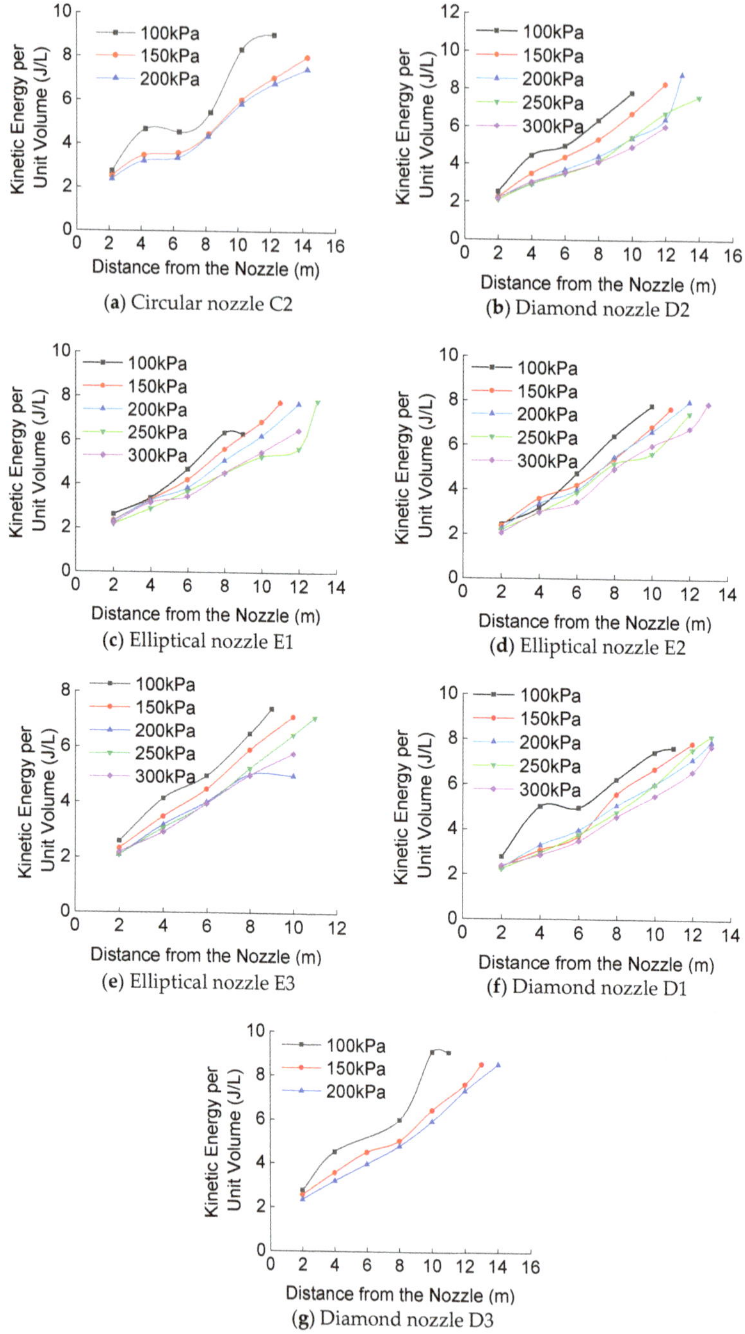

(**a**) Circular nozzle C2
(**b**) Diamond nozzle D2
(**c**) Elliptical nozzle E1
(**d**) Elliptical nozzle E2
(**e**) Elliptical nozzle E3
(**f**) Diamond nozzle D1
(**g**) Diamond nozzle D3

Figure 17. Kinetic energy per unit volume as a function of the distance from each nozzle under different working pressure levels. (C2 refers to the circular nozzle with a diameter of 5 mm; D1, D2, and D3 refer to the diamond nozzles with aspect ratios of 1.54, 1.32, and 1.11, respectively; E1, E2, and E3 refer to the elliptic nozzles with aspect ratios of 1.43, 2, and 2.58, respectively).

3.8. Uniformity Coefficient of Kinetic Energy Intensity Distribution of Combined Sprinkler

In general, the sprinklers in sprinkler irrigation systems are arranged in square and triangular shapes. In this study, the working pressure of the sprinkler was 100 kPa, 150 kPa, and 200 kPa, and a square arrangement was simulated. The uniformity coefficient of the kinetic energy intensity of combined sprinkler irrigation was simulated and calculated by using the MATLAB software. The combination spacing was selected as 1.0 R, 1.1 R, 1.2 R, and 1.3 R, where R is the wetted radius. The calculation results are listed in Table 7.

Table 7. Uniformity coefficient of the kinetic energy intensity distribution for each nozzle under different combination spacing.

Shape	Number	Pressure (kPa)	1.0 R	1.1 R	1.2 R	1.3 R
		100	40.07	29.60	18.53	18.53
Circle	C2	150	49.85	45.61	22.77	10.37
		200	55.93	51.15	30.90	22.51
		100	49.32	42.63	11.27	11.27
	D1	150	51.78	46.68	19.08	19.08
		200	63.52	45.24	45.48	49.98
		100	63.62	67.59	67.59	40.31
Diamond	D2	150	56.34	52.41	34.58	34.58
		200	65.36	59.13	49.19	54.15
		100	52.32	53.37	53.37	18.18
	D3	150	57.12	53.72	36.09	28.76
		200	59.39	59.51	39.88	39.60
		100	55.65	48.67	48.67	29.94
	E1	150	57.02	49.22	32.61	32.61
		200	57.02	51.99	33.90	33.90
		100	49.20	38.37	38.37	20.99
Ellipse	E2	150	55.02	53.66	28.17	28.17
		200	51.84	45.69	21.36	21.36
		100	52.81	41.12	41.12	35.87
	E3	150	51.82	39.31	39.31	35.47
		200	57.78	49.95	49.95	39.98

It can be observed that the uniformity coefficients of the combined kinetic energy intensity of the nozzles under the three pressure levels were different under different spacing. The results indicated that the best combination spacing for the C2 nozzle was 1.0 R, and the best kinetic energy intensity distribution uniformity coefficients at 100 kPa, 150 kPa, and 200 kPa were 40.07%, 49.85%, and 55.93%, respectively. The best kinetic energy intensity distribution coefficient among the diamond nozzles was exhibited by the D2 nozzle. When the outlet diameter was 5 mm, the working pressure was 100 kPa, and the best combination spacing was 1.1 R and 1.2 R, the optimal kinetic energy intensity distribution uniformity coefficient was 67.59%. The optimal combination spacing for the elliptical nozzles was 1.0 R. The optimal uniformity coefficient of the kinetic energy intensity distribution was exhibited by the E3 nozzle (largest *L/D*) at the working pressure of 100 kPa (57.78%). In addition, it was found that the uniformity coefficient of the kinetic energy intensity distribution of the combined sprinkler increased gradually with increasing pressure. The 3D distributions of the combined sprinkler irrigation kinetic energy intensity of the nozzles with the best uniformity coefficient per shape type are exhibited in Figure 18.

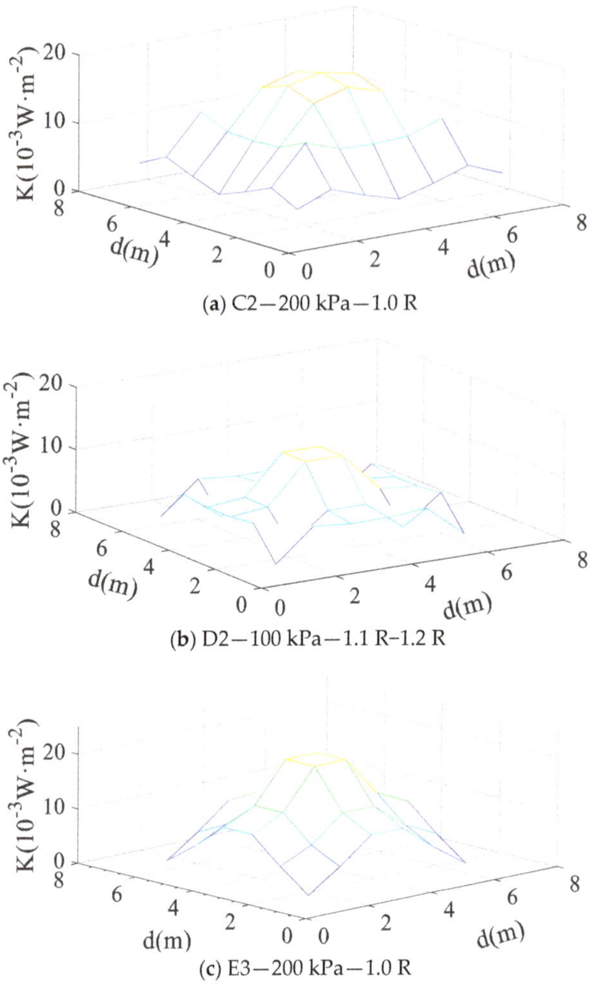

(**a**) C2—200 kPa—1.0 R

(**b**) D2—100 kPa—1.1 R-1.2 R

(**c**) E3—200 kPa—1.0 R

Figure 18. Distribution of kinetic energy intensity under combined sprinkler irrigation for each nozzle shape with the best uniformity coefficient. (C2 refers to the circular nozzle with a diameter of 5 mm; D2 refers to the diamond nozzle with an aspect ratio of 1.32; E3 refers to the elliptic nozzle with an aspect ratio of 2.58).

4. Discussion

Most of the research on the distribution characteristics of water droplets in non-circular nozzles focuses on the influence of the shape and pressure on the jet shape. In this study, a video raindrop spectrometer is used to supplement the research on the influence of the shape and pressure of non-circular nozzles on the distribution characteristics of water droplets, such as diameter, velocity, and kinetic energy. The diameter of water droplets increases in the radial direction and decreases with the increase of pressure. The larger the diameter of the outlet is, the greater the speed of water droplets increases with the diameter of water droplets. When the diameter of water droplets is the same, the larger the diameter of outlet, the smaller the velocity of water droplets. The impact kinetic energy per unit volume of droplets at the same position and its growth range decrease with the increase of pressure.

In addition, the influence of the aspect ratio of a non-circular nozzle on spraying hydraulic performance and water droplet distribution characteristics is studied in this paper. The shape coefficient of the non-circular nozzle increases with the increase of aspect ratio, and the range decreases with the increase of aspect ratio. Under five working pressures, the diameter of water droplets in the diamond nozzle increases the most along the radial direction. The larger the aspect ratio is, the greater the speed of water droplets increases with the diameter of water droplets. With the increase of droplet diameter, the growth rate in droplet velocity of the elliptical nozzle is the smallest, while that of circular nozzle is the largest. The nozzle with the outlet diameter of 5 mm has the smallest average droplet diameter, and the droplet diameter decreases the most along the longitudinal direction.

5. Conclusions

A prediction model which can accurately reflect the droplet diameter, velocity, and kinetic energy distribution per unit volume of three types of nozzles was developed, and a fitting function of the relationship between *L/D* and kinetic energy per unit volume for non-circular nozzles was established. The relationship between droplet diameter and kinetic energy and the distance from the nozzle is an exponential and a linear function, respectively, and the fitting correlation coefficients were between [0.92, 0.99] and [0.89, 0.99], respectively. The distribution of droplet velocity and diameter was found to be logarithmic, and the fitting coefficient was between [0.86 and 0.98]. The relationship between *L/D* and kinetic energy was linear, and the fitting coefficient was 0.924. In all cases, the fitting accuracy was high.

Among the tested nozzles, the sprinkler with the circular nozzle generated the largest *CV*, while the elliptical and diamond nozzles generated similar *CV*s. When operating at the same working pressure, the diamond nozzle exhibited the smallest peak application rate value and the largest radially increasing trend in droplet diameter, while the elliptical nozzle had the lowest maximum droplet kinetic energy per unit volume. The more the nozzle shape deviated from a circle, the larger the radial increase of the kinetic energy intensity. For elliptical nozzles with equal flow and area, the smaller the *L/D*, the smaller the average droplet diameter and velocity, the lower the peak sprinkler water application rate, the more uniform the water distribution, the larger the wetted radius, and the smaller the radial increase of the kinetic energy intensity. The water distribution uniformity was the best when the *L/D* was the smallest.

Under a working pressure of 200 kPa and a combination spacing of 1.0 R, the uniformity coefficients of combined sprinkler irrigation and combined kinetic energy intensity distribution with circular nozzles were the highest, i.e., 65.26% and 55.93%, respectively. The uniformity coefficient of combined sprinkler irrigation with diamond nozzles was the highest (72.28%) when the *L/D* was 1.32, the working pressure was 200 kPa, and the combination spacing was 1.1 R. The combined kinetic energy intensity distribution of the diamond nozzle was the most uniform (67.59%) when the outlet diameter was 5 mm, the working pressure was 100 kPa, and the combination spacing was 1.1 R and 1.2 R. Among the three elliptical nozzles with the same flow rate and different *L/D*, the one with the largest *L/D* had the highest uniformity coefficient (68.72%) when the working pressure was 200 kPa and the combination spacing was 1.1 R. When the combination spacing was 1.0 R, the distribution uniformity coefficient of combination kinetic energy intensity was also the highest (57.78%). In general, the larger the *L/D* of the nozzle, the lower the maximum droplet kinetic energy per unit volume.

Author Contributions: Conceptualization, Y.J. and H.L. (Hong Li); methodology, Y.J., Z.W. and J.L.; validation, Z.W.; formal analysis, Y.J., Z.W. and J.L.; investigation, Y.J., H.L. (Hong Li) and H.L. (Hao Li); data curation, J.L.; writing—original draft preparation, J.L.; writing—review and editing, Y.J. and Z.W.; visualization, J.L.; supervision, Y.J. and H.L. (Hao Li); project administration, Y.J. and H.L. (Hong Li). All authors have read and agreed to the published version of the manuscript.

Funding: This research was funded by the Postgraduate Research & Practice Innovation Program of Jiangsu Province, grant number SJCX22_1870, the Jiangsu Province and Education Ministry Co-sponsored Synergistic Innovation Center of Modern Agricultural Equipment, grant number XTCX2018, the Changzhou Key Research and Development Program, grant number CE20222024, Zhenjiang Key Research and Development Program, grant number CN2022003, and the Youth Talent Development Program of Jiangsu University.

Institutional Review Board Statement: Not applicable.

Data Availability Statement: The data presented in this study are available on request from the corresponding author.

Acknowledgments: Thanks for grateful to the Postgraduate Research & Practice Innovation Program of Jiangsu Province (SJCX22_1870), the Jiangsu Province and Education Ministry Co-sponsored Synergistic Innovation Center of Modern Agricultural Equipment (XTCX2018), the Changzhou Key Research and Development Program (No. CE20222024), Zhenjiang Key Research and Development Program (No. CN2022003) and the Youth Talent Development Program of Jiangsu University.

Conflicts of Interest: The authors declare no conflict of interest.

References

1. Li, J. Study on the atomizing condition of non-circular nozzles. *J. Hydraul. Eng.* **1991**, *3*, 28–32+64.
2. Chen, C.; Yuan, S.; Li, H.; Wang, C. Effect of Non-circle Nozzle on Hydraulic Performance of Impact Variable-rate Sprinkler. *Trans. Chin. Soc. Agric. Mach.* **2011**, *42*, 111–115.
3. Bao, Y.; Liu, J.; Liu, X.; Tian, K.; Zhang, Q. Experimental study on the effects of pressure on water distribution model of low-pressure sprinkler. *J. Drain. Irrig. Mach. Eng.* **2016**, *34*, 81–85.
4. Yan, H.J.; Bai, G.; He, J.Q.; Lin, G. Influence of Droplet Kinetic Energy Flux Density from Fixed Spray-Plate Sprinklers on Soil Infiltration, Runoff and Sediment Yield. *Biosyst. Eng.* **2011**, *110*, 213–221. [CrossRef]
5. Shi, Y.; Zhu, X.; Hu, G.; Zhang, A.; Li, J. Influence of different working conditions on water distribution in sprinkler irrigation. *J. Drain. Irrig. Mach. Eng.* **2021**, *39*, 318–324.
6. Liu, H.; Kang, Y. Effects of Droplets Kinetic Energy on Soil Infiltration Rate and Surface Runoff under Sprinkler Irrigation. *J. Irrig. Drain.* **2002**, *21*, 71–74+79. [CrossRef]
7. Jiang, Y.; Li, H.; Xiang, Q.; Chen, C. Experimental study on breakup length and range of free jet of non-circle jet nozzle. *J. Irrig. Drain.* **2014**, *33*, 149–153. [CrossRef]
8. William, E. Hart Sprinkler Distribution Analysis With a Digital Computer. *Trans. ASAE* **1963**, *6*, 0206–0208. [CrossRef]
9. Yuan, S.; Li, H.; Wang, X. Status, problems, trends and suggestions for water-saving irrigation equipment in China. *J. Drain. Irrig. Mach. Eng.* **2015**, *33*, 78–92.
10. Bubenzer, G.D.; Jones, B.A., Jr. Drop Size and Impact Velocity Effects on the Detachment of Soils Under Simulated Rainfall. *Trans. ASAE* **1971**, *14*, 0625–0628. [CrossRef]
11. Mohammed, D.; Kohl, R.A. Infiltration Response to Kinetic Energy. *Trans. ASAE* **1987**, *30*, 0108–0111. [CrossRef]
12. Christiansen, J.E. *Irrigation by Sprinkling*; University of California: Berkeley, CA, USA, 1942; Volume 4.
13. Abd El-Wahed, M.H.; Medici, M.; Lorenzini, G. Harvesting Water in a Center Pivot Irrigation System: Evaluation of Distribution Uniformity with Varying Operating Parameters. *J. Eng. Thermophys.* **2015**, *24*, 143–151. [CrossRef]
14. Li, Y.; Liu, J. Prospects for development of water-saving irrigation equipment and technology in China. *J. Drain. Irrig. Mach. Eng.* **2020**, *38*, 738–742.
15. Wu, P.; Zhu, D.; Lv, H.; Zhang, L. Hydraulics problems in farmland irrigation. *J. Drain. Irrig. Mach. Eng.* **2012**, *30*, 726–732.
16. Tu, Q.; Li, H.; Wang, X.; Li, Y.; Jiang, Y. Comparison and selection of small-scale irrigation machines with multiple sprinklers based on grey relational analysis. *J. Jiangsu Univ.* **2014**, *35*, 656–662.
17. Xu, H.; Gong, S.; Jia, R.; Liu, X. Study on droplet size distribution of ZY sprinkler head. *J. Hydraul. Eng.* **2010**, *41*, 1416–1422. [CrossRef]
18. Gong, X.; Zhu, D.; Zhang, L.; Zhang, Y.; Ge, M.; Yang, W. Drop Size Distribution of Fixed Spray-plate Sprinklers with Two-dimensional Video Disdrometer. *Trans. Chin. Soc. Agric. Mach.* **2014**, *45*, 128–133+148.
19. Lorenzini, G. Water droplet dynamics and evaporation in an irrigation spray. *Trans. ASABE* **2006**, *49*, 545–549. [CrossRef]
20. Ouazaa, S.; Burguete, J.; Paniagua, M.P.; Salvador, R.; Zapata, N. Simulating Water Distribution Patterns for Fixed Spray Plate Sprinkler Using the Ballistic Theory. *Span. J. Agric. Res.* **2014**, *12*, 850. [CrossRef]
21. Zhu, X.; Liu, X.; Liu, J.; Yuan, S.; Bao, Y. Droplet kinetic energy distribution regulation of complete fluidic sprinkler. *Trans. Chin. Soc. Agric. Eng.* **2015**, *31*, 26–31.
22. Li, J.; Ma, F. Effect of nozzle shape on the spraydrop kinetic energy from sprinklers. *J. Irrig. Drain.* **1997**, *16*, 3–8.
23. Chen, D. Shape and structure of special nozzle. *Water Sav. Irrig.* **1982**, *3*, 37.
24. Li, Y.; Liu, J.; Guo, Z.; Liu, X.; Lou, Y. Research and development of hydraulic performance of energy-saving special-shaped nozzle. *J. Irrig. Drain.* **1990**, *2*, 43–50. [CrossRef]

25. Wei, Y.; Yuan, S.; Li, H.; Xiang, Q.; Chen, C. Hydraulic Performance Experiment of the Variable-rate Sprinkler with Non-circle Nozzle. *Trans. Chin. Soc. Agric. Mach.* **2011**, *42*, 70–74.
26. Li, D.; Lu, X.; Zhao, X. Experimental Study on Low Pressure Jet Characteristic of the Non-circle Jet Nozzle. *Light Ind. Mach.* **2006**, *24*, 18–20.
27. Zhou, X.; Li, H.; Jiang, Y. Study on water distribution uniformity of non-circular nozzles at low pressure. *J. Drain. Irrig. Mach. Eng.* **2017**, *35*, 448–453.
28. Jiang, Y.; Li, H.; Hua, L.; Zhang, D.; Issaka, Z. Experimental Study on Jet Breakup Morphologies and Jet Characteristic Parameters of Non-Circular Nozzles under Low-Intermediate Pressures. *Appl. Eng. Agric.* **2019**, *35*, 617–632. [CrossRef]
29. Liu, X. Droplets Distribution Characteristic Study on Complete Fluidic Sprinkler. Master's Thesis, Jiangsu University, Zhenjiang, China, 2016.
30. Zhu, X.; Shi, Y.; Hu, G.; Liu, J. Dynamic Simulation and Test of Water Distribution of Fluidic Sprinkler. *J. Irrig. Drain.* **2020**, *39*, 74–83. [CrossRef]
31. Zhu, X.; Liu, G.; Liu, J.; Jiang, J.; Tian, K. Droplets distribution research of impact sprinkler based on Laser Precipitation Monitor. *J. Drain. Irrig. Mach. Eng.* **2015**, *33*, 908–914.
32. Liu, J.; Yuan, S.; Li, H.; Zhu, X. Analysis and Experiment on Influencing Factors of Range and Spraying Uniformity for Complete Fluidic Sprinkler. *Trans. Chin. Soc. Agric. Mach.* **2008**, *39*, 51–54.

agriculture

MDPI

Article

Yolov5s-CA: An Improved Yolov5 Based on the Attention Mechanism for Mummy Berry Disease Detection

Efrem Yohannes Obsie [1], Hongchun Qu [2,*], Yong-Jiang Zhang [3,4], Seanna Annis [3] and Francis Drummond [3,5]

1 College of Computer Science, Chongqing University of Posts and Telecommunications, Chongqing 400065, China
2 College of Automation, Chongqing University of Posts and Telecommunications, Chongqing 400065, China
3 School of Biology and Ecology, University of Maine, Orono, ME 04469, USA
4 Climate Change Institute, University of Maine, Orono, ME 04469, USA
5 Cooperative Extension, University of Maine, Orono, ME 04469, USA
* Correspondence: hcchyu@gmail.com

Abstract: Early detection and accurately rating the level of plant diseases plays an important role in protecting crop quality and yield. The traditional method of mummy berry disease (causal agent: *Monilinia vaccinii-corymbosi*) identification is mainly based on field surveys by crop protection experts and experienced blueberry growers. Deep learning models could be a more effective approach, but their performance is highly dependent on the volume and quality of labeled data used for training so that the variance in visual symptoms can be incorporated into a model. However, the available dataset for mummy berry disease detection does not contain enough images collected and labeled from a real-field environment essential for making highly accurate models. Complex visual characteristics of lesions due to overlapping and occlusion of plant parts also pose a big challenge to the accurate estimation of disease severity. This may become a bigger issue when spatial variation is introduced by using sampling images derived from different angles and distances. In this paper, we first present the "cut-and-paste" method for synthetically augmenting the available dataset by generating additional annotated training images. Then, a deep learning-based object recognition model Yolov5s-CA was used, which integrates the Coordinated Attention (CA) module on the Yolov5s backbone to effectively discriminate useful features by capturing channel and location information. Finally, the loss function GIoU_loss was replaced by CIoU_loss to improve the bounding box regression and localization performance of the network model. The original Yolov5s and the improved Yolov5s-CA network models were trained on real, synthetic, and combined mixed datasets. The experimental results not only showed that the performance of Yolov5s-CA network model trained on a mixed dataset outperforms the baseline model trained with only real field images, but also demonstrated that the improved model can solve the practical problem of diseased plant part detection in various spatial scales with possible overlapping and occlusion by an overall precision of 96.30%. Therefore, our model is a useful tool for the estimation of mummy berry disease severity in a real field environment.

Keywords: wild blueberry; *Vaccinium angustifolium*; *Monilinia vaccinii-corymbosi*; deep learning; coordinated attention; synthetic data; prediction accuracy

Citation: Obsie, E.Y.; Qu, H.; Zhang, Y.-J.; Annis, S.; Drummond, F. Yolov5s-CA: An Improved Yolov5 Based on the Attention Mechanism for Mummy Berry Disease Detection. *Agriculture* **2023**, *13*, 78. https://doi.org/10.3390/agriculture13010078

Academic Editors: Vadim Bolshev, Vladimir Panchenko and Alexey Sibirev

Received: 4 November 2022
Revised: 22 December 2022
Accepted: 23 December 2022
Published: 27 December 2022

1. Introduction

In agriculture, plant diseases cause an estimated 10–15% annual loss of the world's major crops [1]; 70–80% of these diseases are caused by pathogenic fungi that have an adverse effect on crop growth, quality and yield. Therefore, disease management is important to agricultural systems including the wild lowbush blueberry production system. Wild blueberry (mainly *Vaccinium angustifolium* Aiton) is a perennial shrub that spreads by underground rhizomes, with aerial shoots occurring every 2–30 cm. Wild blueberry plants are not planted [2,3] but grow naturally in rocky hills and sandy fields, and are managed to form a carpet for berry production [4]. Wild blueberry is one of the most important crops in Maine, USA, and the Canadian provinces of Quebec and the Maritimes, and the crop is a major source of income for growers in the regions [5,6]. The state of Maine is one of the largest producers of wild blueberries of the world, accounting for 97% of the total production in the US [7–9]. The yield and quality of blueberries are impacted by several factors, but one of the most important is mummy berry disease caused by the fungus *Monilinia vaccinii-corymbosi* [10]. *Monilinia vaccinii-corymbosi* ascospores attack opening flower clusters and axillary buds in the spring and kills infected tissues [11]. These tissues then produce secondary asexual spores that infect healthy flowers and the fungus colonizes the developing fruit. High levels of infection can kill up to 90% of the leaves and flower buds during the early part of the growing season [10,12]. The infection of the developing fruit directly affects yield and the loss of flowers and leaves can indirectly reduce yield [8]. The loss of yield (berry weight harvested) can be substantial and poses an economic challenge to growers.

The current method of early warning monitoring for mummy berry disease is based on the prediction of potential infection periods determined by weather conditions and development stages of the plants and fungus [13]. If a high likelihood of infection is predicted, based on the duration of leaf wetness and suitable air temperature, growers are advised to protect their crops from infection with the application of fungicides [14]. Follow-up field scouting by crop protection experts and experienced blueberry growers is often implemented to determine the effectiveness of forecasting infection and fungicide applications. However, monitoring for the presence and rating the level of disease is extremely time-consuming and labor-intensive since infected plants can be scattered in patches around the field and so typically multiple transects are used to observe many individual stems across a field. It can also be prone to error due to confusion between mummy berry disease symptoms and those from frost damage or other diseases such as *Botrytis* blight. These are some of the main reasons why researchers are looking for alternative methods to identify diseases in the field [15–17]. Previous studies involving other crops and diseases using traditional machine learning algorithms have mainly relied on manual extraction of features from image texture, color and shape [18] to locate disease. However, the symptoms of the same disease may have different visual characteristics, such as during different stages of infection, when infecting flowers, leaves or fruit, and possible occlusions and high spatial variations among individual plants. Therefore, when there is variation in environmental conditions and symptom traits, the generalization ability of these algorithms decreases significantly.

In recent years, with the rapid advancement of computer vision techniques and deep learning, various methods of plant disease detection and classification techniques have been developed in agriculture resulting in highly accurate results [16,19,20]. Despite its success in achieving superior performance in plant disease detection, deep neural network architectures depend heavily on the availability of large quantities of training data that are characterized by variation to accurately "learn the breadth of behavior" for proper training of a model. However, the available dataset for wild blueberry plant disease detection does not contain the abundance of images collected and labeled from a real-field environment which is essential for making highly accurate models [19]. Levels of mummy berry symptoms vary by field characteristics, weather, and inoculum level and symptoms of the first stage of infection of leaves and flower buds only last for one to three weeks depending upon the field and weather. Clean and background-free images of diseased and healthy plant parts also are difficult to obtain in blueberry fields. Accurately labeling images for model training is also very labor-intensive. To address the problem of data scarcity in training deep learning models, researchers have proposed various techniques to generate synthetic images based on the available dataset to obtain diverse and inexpensive training data [21,22] rather than field collecting and annotating training images which is an expensive and time-consuming task.

Although computer vision techniques have greatly improved for plant disease detection, practical problems such as the small size of lesions, occlusion of shoots, interference of complex background, uncontrollable light conditions in fields, etc., remain unsolved for mummy berry disease identification. For instance, masses of conidia (a sign on leaves and flowers of primary mummy berry infection) on blueberry shoots are tiny (<33 μm long; in [11] and only account for a very small portion of a field taken image, which makes it unlikely to be automatically identified by computer vision techniques. Moreover, the much branched and dense structure of blueberry bushes often occlude small diseased plant parts such as those exhibiting conidia. Multiple shoots or branches also complicate the background of field-taken images, which also poses a challenge to disease detection. An example of field-taken images of mummy berry disease and conidia on shoots are shown in Figure 1. In addition, disease traits (such as size, color, and portion) in the field obtained sample images for disease detection and severity rating may vary considerably due to the changes in camera shooting angle and distance. These highly spatial variations could inevitably degrade the performance of identification, despite the most advanced object detection algorithms having been employed [23].

Figure 1. An example of field-taken images of mummy berry disease with the complex background of blueberry bushes. The red square marks the area in the left panel. The same image is zoomed in and depicted in the right panel, where yellow squares identify the presence of conidia.

In deep learning, as in human vision, the attention mechanism tends to focus on key regions of the input objects by ignoring irrelevant information. Recent studies have demonstrated the remarkable effectiveness of attention-based methods for boosting deep learning networks and have proven their usefulness in a variety of computer vision tasks, such as object detection [20,24,25]. CBAM [26] is a widely used attention mechanism that combines channel and spatial attention. SE [27], on the other hand, focuses on the relationship between channels to learn each image feature based on the loss function, increases the weight of relevant image features, and decreases the weight of irrelevant image features to achieve the best results. In plant disease detection, the lightness of the model determines whether it can be deployed to embedded devices, which is of great importance for growers to monitor the growth and disease status of blueberries in real-time in the field [15]. Considering the limited computational power and storage capacity of mobile or embedded devices, SE and CBAM attention mechanisms are still the most popular attention methods. However, SE neglects the importance of location information and CBAM only captures local relationships and cannot model long-range dependencies essential for capturing object structure in visual tasks [28]. In contrast, coordinate attention (CA) considers both inter-channel relationships and position information.

Therefore, in order to overcome the problems in the current plant disease detection methods, and solve the limitation of data scarcity for mummy berry disease detection of the wild blueberry plant in a real-field environment, we have implemented the cut and paste method [29] for synthetically augmenting the available dataset to generate annotated training images for object detection tasks, which reduces the effort required to collect and manually annotate huge datasets. Thereafter, we improved the backbone of the original Yolov5s network model by integrating the lightweight coordinate attention (CA) module to effectively highlight the important features by capturing the channel and location information to improve a mummy berry disease detection network model in a real natural field environment with little extra computational cost. The main contributions of this study are summarized as follows:

- The coordinate attention (CA) module is integrated into a Yolov5s backbone. This allows the network to increase the weight of key features and pay more attention to visual features related to disease to improve the performance of disease detection in various spatial scales.
- The loss function, General Intersection over Union (GIoU), is replaced by the loss function, Complete Intersection over Union (CIoU) to enhance bounding box regression and localization performance in identifying diseased plant parts with a complex background.
- A synthetic dataset generation method is presented that can reduce the effort of collecting and annotating large datasets and boost the performance of identification by artificially increasing available features in deep model training.

2. Related Work

The scope of the research presented in this article is related to the use of data augmentation processes to create synthetic image datasets and object detection models to identify mummy berry disease affecting wild blueberry productivity. Thus, the literature review presented in this section is divided into two subsections. The first subsection lists the techniques reported in the literature on the use of data augmentation to create synthetic image datasets, while the second subsection details the various machine learning and deep learning algorithms reported in the literature for plant disease identification.

2.1. Data Augmentation

To build robust deep-learning models, it is important to ensure that validation error during training is minimized with the training error. The approach that has been reported in the literature to be successful is the data augmentation technique [30].

Recently, a method of data augmentation crop-and-paste has become popular in object detection [31] and instance segmentation. Khoreva et al. [32] used the cut-and-paste method to generate pairs of synthetic images for video object segmentation. However, object positions are sampled uniformly and changes between image pairs need only be guaranteed to be kept small, which does not work for image-level instance segmentation. A copy–paste method was proposed by Ghiasi et al. [33], which randomly selects a segment object and pastes it at a random location onto the background image data without considering its visual state. The high performance and efficiency of this method was experimentally verified. The authors of [34] presented a simple yet effective approach that took an object detection VOC2007 dataset and cut out objects according to their ground truth labels and pasted them onto images with different backgrounds. With this naive approach, the authors showed a significant improvement in object detection models such as YOLO [35] and SSD [36]. Khalil et al. [37] proposed a new method for augmenting annotated training datasets used for object detection tasks, which aims at relocating objects based on their segmentation masks to a new background that comprise changes in the property of the object such as: image spatial location, surrounding context, and scale. In [31], the authors proposed a context model to place segmented objects at backgrounds with proper context. They demonstrated that this technique can improve object detection on a Pascal VOC dataset. However, the method requires extra model training and off-line data preprocessing. A method of annotated instance masks with a location probability map is explored in [38] to augment the training dataset that can effectively improve the generalization ability of the dataset. Abayomi-Alli et al. [39] proposed a novel histogram transformation approach that improved the accuracy of deep learning models by generating synthetic images from low-quality test images to enhance the number of images in a cassava leaf disease dataset by applying Gaussian blurring, motion blurring, resolution down-sampling, and overexposure with a Modified MobileNetV2 neural network model. Nair and Hinton [40] expanded and enriched their training data by random crop and horizontal reflection. They also applied PCA (principal component analysis) on the color space to change the intensity of the RGB channel (red, green, blue color model). Furthermore, geometric and color transformation were also performed on the dataset. However, the method is based upon simple transformations and cannot simulate higher levels of complexity inherent in the field environment.

Other recent works on image analysis [41,42] built and trained models on purely synthetic rendered 2D and 3D scenes. However, it is difficult to guarantee that models trained on synthetic images will generalize well to real field-collected data, as the process introduces significant changes in image statistics. To solve this problem, Gupta et al. [43] adopted a different approach by embedding real segmented objects into natural images. This reduces the presence of artifacts. The authors in [44] estimated the scene geometry and spatial orientation before synthetically placing objects to generate realistic training examples for the task of object instance detection.

2.2. Deep Learning for Plant Disease Detection

With the aim of developing effective plant disease detection systems, there has been an increasing number of research studies focused on plant disease classification and detection in recent years. Qu and Sun [15] proposed a lightweight deep learning model that can be deployed on embedded devices to detect mummy berry disease in a real environment. The model uses MobileNetV1 as the main network and adopts multi-scale feature extraction which combines dilated and depth-wise convolution in a parallel manner. In addition, at the end of the model, a feature filtering module-based channel attention mechanism is employed to improve classification performance. Fuentes et al. [45] presented a method of detection and identification of diseases and pests of tomatoes captured by camera equipment with different levels of resolution. To find a suitable deep learning architecture, the Fuentes et al. study combined three main families of detectors: fast region-based convolutional neural network (FAST R-CNN), region-based fully convolutional network (R-FCN), and single shot multibox detector (SSD) with VGG net and residual net to effectively identify nine different types of diseases and pests. Roy and Bhaduri [46] developed a deep learning based multi-class apple plant disease detection method and achieved 91.2% mean average precision and 95.9% a F1-score. The model was modified to optimize accuracy and validated by detecting diseases under complex orchard scenarios. Qi et al. [20] proposed a method for the recognition of tomato virus disease based on an improved SE-Yolov5 network model. A squeeze-and-excitation (SE) module was added to a Yolov5 model to focus the network on the effective features of tomato virus visual features. This approach improved the performance of the network.

3. Materials and Methods

In this section, we briefly present a field-collected image dataset that was used for model training and evaluation, as well as for generating synthetic images. We then introduced the system of synthetic dataset generation methods for object detection tasks. This section concludes with the description of an improved Yolov5 model based on attention mechanism and evaluation metrics.

3.1. Data Source

The first step in developing a deep learning model is to prepare a dataset. As the primary source of data in this study, images of healthy and diseased flowers, fruits, and leaves of the blueberry crop in a field environment with complex backgrounds were obtained from the University of Maine wild blueberry experimental fields at Blueberry Hill Farm, Jonesboro, ME, USA [47]. However, the total number of field images collected for training a deep learning network was not adequate. Therefore, to achieve high performance and reduce the risk of overfitting a predictive model for mummy berry disease detection, we first produced annotated synthetic images with a complex background that mimicked real field situations. Then we collected blueberry images with mummy berry disease from online sources such as the National Ecological Observatory Network (www.bugwood.org, accessed on 23 April 2022), and Google Images (www.google.com, accessed on 2 May 2022) to incorporate variety in training images, as deep learning models show enhanced results and higher generalization ability on the availability of a large dataset. A total of 459 field images of blueberries with mummy berry disease were obtained from the University of Maine wild blueberry experimental fields and online sources. Based on field images, a total of 1661 annotated images were produced by the synthetic data generation method (Table A1 in Appendix A).

3.2. Synthetic Data Generation

In this study, we applied the cut and paste technique [29] to create synthetic images and related annotations by random scaling, rotation, and adding segmented images of interest to the background. Unlike Mixup and CutMix, our method only copied the exact pixels that corresponded to an object, as opposed to all the pixels in the object's bounding box. To generate a synthetic dataset with our cut and paste method, we randomly selected images of 55 flowers, 48 fruits, and 58 leaves of diseased blueberry plant tissue from the field dataset (discussed in Section 3.1) and created masks. Then a total of 83 "healthy" background photographic images with only healthy uninfected flowers and leaves were collected in a lowbush blueberry field at the University of Maine, Blueberry Hill Farm (Jonesboro, ME, USA). In order to make the background more complex, seven distractor object images of healthy fruits were obtained from online sources, and then masks were created. Objects of interest masks were created using Adobe Photoshop software, unlike a previous study [29] that automated this process by training a machine learning model to segment and extract the objects.

Once the image data was ready, we randomly selected the background image and resized it to 1080 × 1320 pixels and 1320 × 1080 pixels, vertically and horizontally, respectively. Then, to make the background of the synthetic dataset diverse, we randomly selected at most 10 segmented distractor images and randomly resized, rotated and added them to the background iteratively. Under field conditions in agriculture production systems, occlusion problems are common challenges that need to be considered. Hence, in generating a synthetic dataset, a newly added image can partially or fully overlap a previously added image. Therefore, to control the degree of overlap and include cases of occlusion in the synthetic dataset, the threshold value for the degree of overlap was set at 25%. Finally, in an iterative process, we randomly chose a maximum of 15 segmented images of diseased leaves, flowers, and fruits and randomly resized and rotated them, and then added the new background images on top of the background distractor images (see Figure 2).

3.3. Coordinate Attention Module

When detecting mummy berry disease, the infection can be randomly distributed on the plant stem, resulting in a mixture of overlapping occlusions, and the infected region may occupy a relatively small proportion of the image area, leading to missed or incorrect detection. In our study, we introduce the coordinate attention (CA) module to help the deep learning model focus on the most significant information related to infection and ignore minor features. The CA mechanism is an efficient and lightweight module that embeds position information into the attention map. The model can obtain information about a large area without introducing additional computational costs. The coordinate attention block can be considered a computational unit that increases the expressive power of the learned features. It takes an intermediate feature tensor: $\mathbf{X} = [x_1, x_2, \ldots, x_C] \in \mathbb{R}^{C \times H \times W}$ as input; and outputs a transformed tensor with enhanced representations: $\mathbf{Y} = [y_1, y, \ldots, y_C]$ of the same size to $\mathbf{X}$.

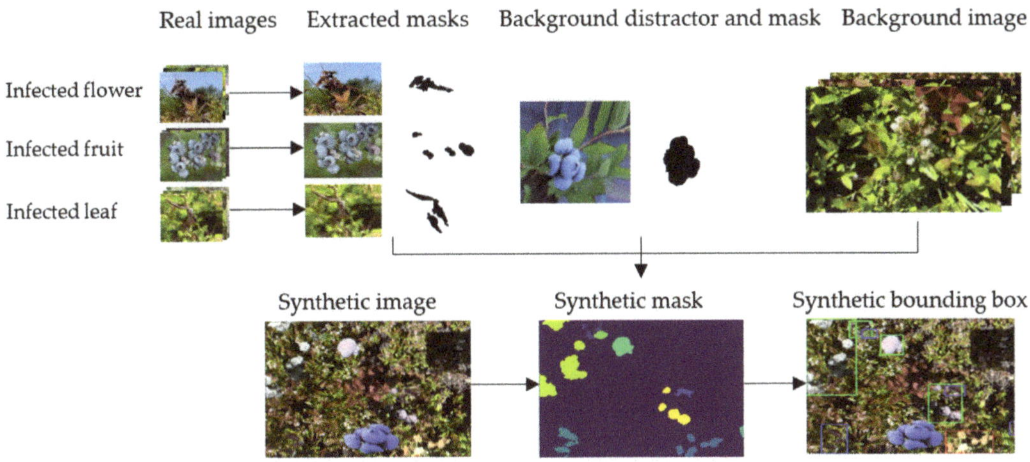

Figure 2. The procedure of synthetic image dataset generation.

In the structure of the coordinate attention module, the operation is divided into two steps: (1) coordinate information embedding; and (2) coordinate attention generation (Figure 3). The first step factors global pooling as given in Equation (1) into two 1D feature encoding operations that encode each channel along the horizontal and vertical directions, respectively.

$$z_c^h(h) = \frac{1}{W}\sum_{0 \leq i < W} x_c(h,i) \; z_c^w(w) = \frac{1}{H}\sum_{0 \leq i < H} x_c(j,w) \quad (1)$$

where X denotes the input, $z_c^h(h)$ and $z_c^w(w)$ indicate the outputs of the $c-$ th channel at height h and width w, respectively. The second step concatenates the feature maps produced and sends them to the shared 1×1 convolutional transformation F_1 to obtain the intermediate feature map, f, as formulated in Equation (2),

$$f = \delta\left(F_1\left(\left[Z^h, Z^w\right]\right)\right) \quad (2)$$

where [. , .] denotes the concatenation operation along the spatial dimension, and δ is a non-linear activation function. The feature map f is then split along the spatial dimension into two separate tensors f^h and f^w, followed by another two 1×1 convolutional functions F_h and F_w, which are determined by Equation (3),

$$g^h = \sigma\left(F_h\left(f^h\right)\right), g^w = \sigma(F_w(f^w)) \quad (3)$$

where σ denotes the sigmoid activation function. The final attention weight Y is generated according to Equation (4),

$$y_c(i,j) = x_c(i,j) \times g_c^h(i) \times g_c^w(j) \quad (4)$$

Therefore, in this study, we integrated the coordinate attention (CA) module on the Yolov5 backbone. This offers three obvious advantages: (1) it captures cross-channel and position-sensitive information that helps models accurately locate and recognize objects of interest; (2) having a lightweight property that is less lightweight than other attention mechanisms [26,27]; and (3) flexibility to be plugged into object detection models such as Yolov5 with little additional computational overhead.

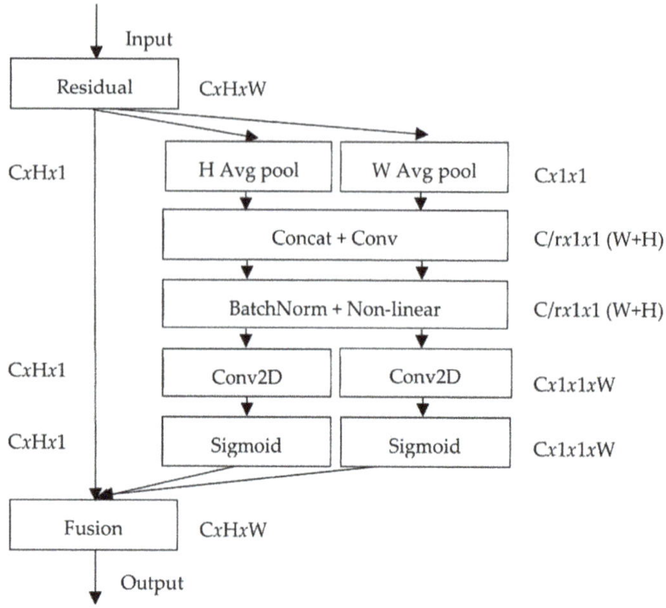

Figure 3. Structure of the coordinate attention (CA) module.

3.4. Yolov5 Method

Object detection is a computer vision technique for locating instances of a certain class of objects in an image. Recent object detection methods can be categorized into two main types: one-stage and two-stage. One-stage methods prioritize inference speed and include a series of YOLO detection methods [15,35,48–50], SSD [36,51], and RetinaNet [52]. Typical two-stage methods prioritize detection accuracy and include R-CNN [53], Fast R-CNN [54], Mask R-CNN [55], Cascade R-CNN [56], and others.

Yolov5 is the latest generation of one-stage object detection network models of the YOLO series proposed by Ultralytics in May 2020 see [57]. Based on the network depth and width of feature maps, Yolov5 can be divided into four models, namely Yolov5s, Yolov5m, Yolov5l, and Yolov5x [23]. Compared with two-stage detection network models, Yolov5 greatly improves the running speed of the model while maintaining detection accuracy. This not only meets the needs of real-time detection, but also has the advantage of a small structure size. The Yolov5 network model is an improved model based on Yolov3 with improvements such as multi-scale prediction, which can simultaneously detect images of different sizes [20]. Therefore, we proposed a lightweight mummy berry disease detection network model based on Yolov5s by improving the network backbone with an attention mechanism. The architecture of the improved Yolov5s-CA network model is shown in Figure 3.

3.5. Improvement of Yolov5s-CA Network Model

Figure 3 shows the structure of the improved Yolov5s-CA network model to detect mummy berry disease. It can be seen that a lightweight module CA [28] was introduced into the backbone of Yolov5s to strengthen the feature representation ability of the network and select useful information, which enhances detection performance. The network structure of Yolov5s-CA consists of four parts: input, backbone, neck, and head.

The backbone of the Yolov5s-CA network model contains Conv, C3, CA, and Spatial Pyramid Pooling Fusion (SPPF). The Conv is the basic convolution unit, which performs two-dimensional convolution, regularization, and activation operations on the input. The C3 module is located in both the backbone and neck. The C3 module with a shortcut structure is implemented in the backbone of the network. It divides the input tensor equally into two branches and performs convolution operations. One branch passes through a Conv module and then passes through multiple residual structures to avoid degradation problems in the deep computational process. The other branch directly combines the two branches to form a Conv module. As shown in Figure 4, the CA modules are integrated into the backbone following the C3 module to highlight and select the most important disease-related visual features and improve the representation ability of the object detection model to detect mummy berry disease in a field environment. The last layer of the backbone, Spatial Pyramid Pooling Fast (SPPF), shown in Figure 5, comprises three MaxPool layers of 5×5 kernel sizes in series and passes the input through the MaxPool layers in turn and performs a concatenation operation on the output before performing a Conv operation. The SPPF structure can achieve similar feature ex-traction results as SPP, but SPPF runs faster. The image can learn features at multiple scales with the help of MaxPool layers and jump connections, and then increase the representativeness of the feature map by combining global and local features.

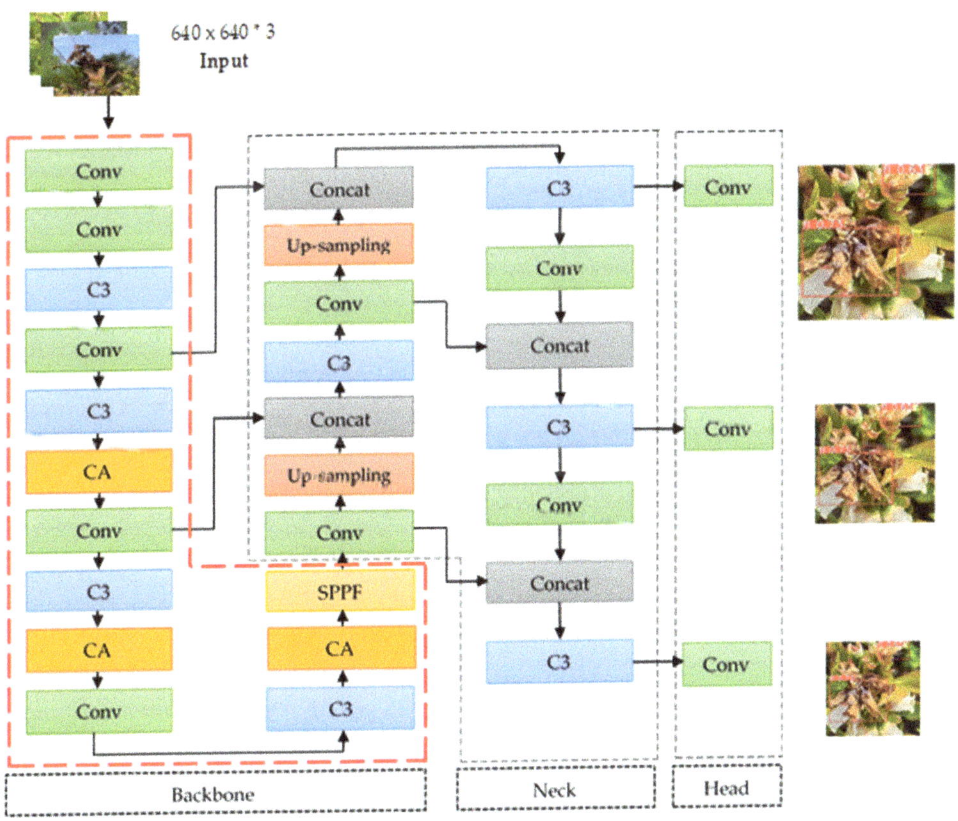

Figure 4. The improved Yolov5s-CA network model structure.

The neck module is a feature aggregation layer between the head and the backbone. It collects as much information as possible from the backbone before feeding it to the head. It consists of two parts: the Feature Pyramid Network (FPN) and the Path Aggregation Network (PAN). The FPN structure transmits semantically robust features from the top-down, while the PAN transmits information in a bottom-up pyramid to strengthen the feature representation capabilities of the network model. In addition, C3 modules were added to enhance the network's feature extraction capability, and the C3 at the neck replaces the residual structure with multiple Conv modules.

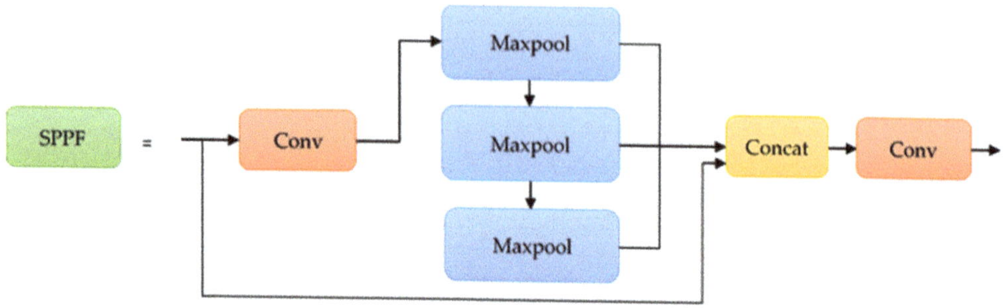

Figure 5. The SPPF module.

The head outputs a vector containing the object category probability, the object scores, and the position of the bounding box. The loss function of Yolov5s consists of three parts: the confidence loss, the classification loss and the position loss of the target and prediction box. The original Yolov5s uses GIoU_loss as a bounding box regression loss function to evaluate the distance between the predicted box and the ground truth box. It can be expressed in the following formulae represented in Equations (5)–(7).

$$IoU = \frac{A \cap B}{A \cup B} \tag{5}$$

$$GIoU = IoU - \frac{A^c - u}{A^c} \tag{6}$$

$$L_{GIoU} = 1 - GIoU \tag{7}$$

where A is the predicted box, B is the ground truth box, IoU represents the intersection ratio of the predicted box and the ground truth box, A^c represents the intersection of the predicted box and the ground truth box, u represents the smallest circumscribed rectangle of the predicted box and the ground truth box, and L_{GIoU} is the $GIoU$ Loss.

Compared with the IoU_loss function, the GIoU loss function can solve the problem of non-overlapping bounding boxes. However, GIoU loss cannot solve the problem that the prediction frame is inside the target frame and the size of the prediction frame is the same. On the other hand, CIoU loss considers the scale information of the aspect ratio of the bounding box and measures it from the three viewpoints: (1) overlapping area, (2) center point distance, and (3) aspect ratio, which makes the prediction box regression more efficient. Therefore, in this study, we use CIoU loss as the regression loss function represented in Equations (8)–(10).

$$L_{loc} = 1 - IoUI(B, B_{gt}) + \frac{d^2}{c^2} + av \tag{8}$$

$$a = \frac{v}{1 - IoU + v} \tag{9}$$

$$v = \frac{4}{\pi^2}\left(arctan, \frac{w^{gt}}{h^{gt}} - arctan\frac{w}{h}\right)^2 \tag{10}$$

where w is the width and h is the height of the prediction box and w^{gt} and h^{gt} are the width and height of the ground truth box, respectively.

3.6. Model Evaluation

Three metrics were used to evaluate the performance of the models. First, we used precision (P), defined as the proportion of true positives to the total number of positive detections (Equation (11)). Second, we used recall (R), defined as the proportion of true positives to the total number of actual objects (Equation (12)). Third, mean average precision ($mAP_{@0.5}$) was used, which represents the mean value of AP for different categories with a threshold of 0.5% when *mAP* (Equation (13)) is converted to percent.

$$Precision = \frac{TP}{TP + FP} \tag{11}$$

$$Recall = \frac{TP}{TP + FN} \tag{12}$$

$$mAP = \frac{\sum AP}{n} \tag{13}$$

In Equations (11)–(13), TP is the number of correctly detected disease regions, FP is the number of healthy regions of plants that have not been detected as having disease, FN is the number of incorrectly detected disease regions, AP is the area under the precision-recall curve and n is the number of classes.

The experiments were carried out following the improved Yolov5s-CA model (Figure 3). To implement the mummy berry disease detection model, we used Pytorch version 1.11.0. The code was written, edited, and run using Google Colab Pro's notebook, a subscription-based service provided by Google Research that allows users to write and run Python code in web browsers. The hardware configuration that we used was: NVIDIA Tesla P100 GPU, 16 GB RAM, 127 GB hard disk, and CUDA version 11.2. The hyperparameters of the two models were set uniformly. The initial learning rate of the model was set to 0.01, and the momentum of the learning rate to 0.9. The batch size was set to process 16 images per iteration. The resolution of the input image was set to 640 × 640 pixels, and the number of epochs was set to 300. The training, validation, and test set images were in the ratio of 8:1:1 with no overlap between the three sets. To demonstrate the effectiveness of improving the original Yolov5s, we conducted experiments with and without modifying the backbone of Yolov5s with an attention mechanism. Each experiment was validated on the field-collected test dataset.

4. Results

We designed and conducted five experiments. The first experiment was designed to compare the two Yolov5s models (Yolov5s vs. Yolov5s-CA) on disease detection when they were only trained on the field-collected data. The second experiment compared the two models when they were trained only on the synthetic data. The third experiment compared the two models when they were trained on a combined dataset of synthetic and field-collected data. The fourth experiment compared the detection speeds of the two models. The fifth experiment compared the detection of the two models at different spatial scales (camera shooting distances).

4.1. Comparison of Disease Detection Models Trained Only on the Field-Collected Dataset

This experiment developed a baseline model by evaluating the effect of varying the amount of training data on the model's performance. To this end, the improved Yolov5s-CA and Yolov5s models were evaluated only on field-collected images. The precision of the Yolov5s-CA model is 70.2%, the recall is 61.3% and $mAP_{@0.5}$ is 65.8%, which shows an increase of 2.7%, 0.5% and 1.1% in precision, recall and $mAP_{@0.5}$, respectively, compared to the Yolov5s model (Table 1). Increasing the amount of the field-collected training data from 10–100% in all cases leads to an increase in the performance of the model (Figure 6).

Table 1. Performance comparisons of models trained only on real field dataset.

Models	Precision (%)	Recall (%)	mAP @0.5 (%)
Yolov5s	67.5	60.8	64.7
Yolov5s-CA	**70.2** [1]	**61.3** [1]	**65.8** [1]

[1] Bold type reflects the best precision, recall and $mAP_{@0.5}$ values.

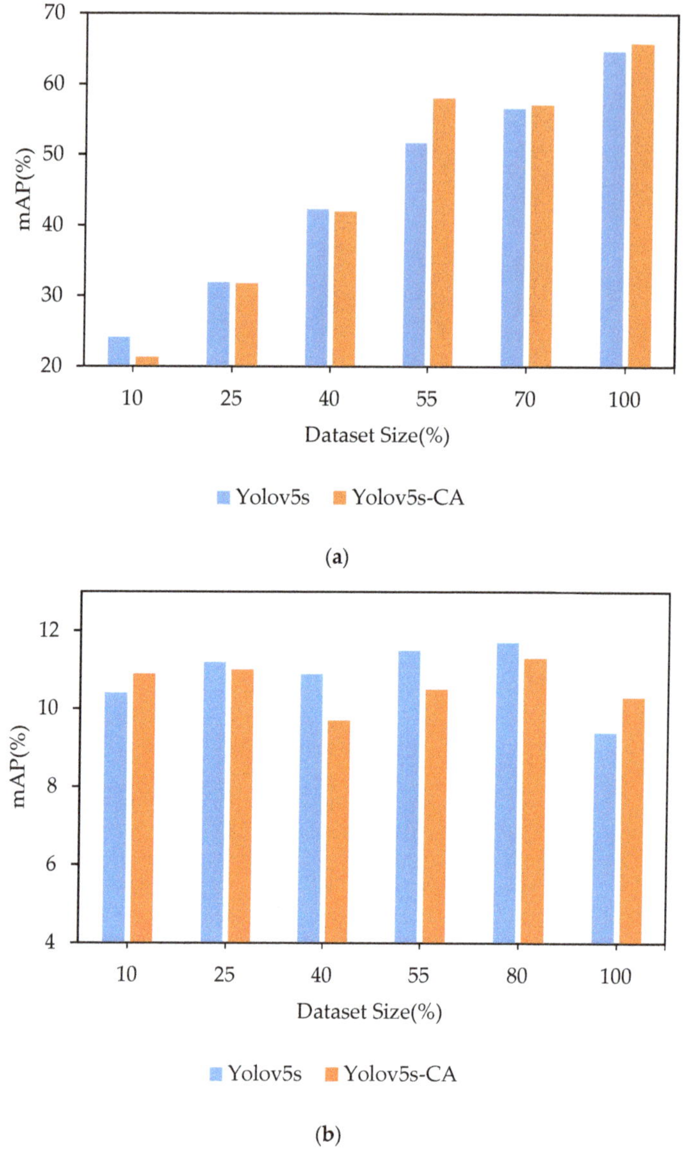

Figure 6. Effects of training data size. (**a**) Field-collected dataset. (**b**) Synthetic dataset.

The comparison of experimental indicators shows that the performance of the improved Yolov5s-CA model is higher than Yolov5s, which confirms the effectiveness of integrating the attention mechanism on the backbone of the Yolov5s model. This approach suppresses less important features and improves the rate of correct detection.

4.2. Comparison of Disease Detection Models Trained Only on the Synthetic Dataset

This experiment evaluated models trained exclusively with synthetically generated images, in contrast to the results in Section 4.1 illustrating the performance of the models trained only on a limited number of field-collected images. We created a synthetic dataset containing 1661 images (method in Section 3.2). Similar to Section 4.1, we varied the amount of synthetic training images to investigate their effect on model performance (Figure 6).

Increasing the training data size to more than 80% of the total available images has no contribution in terms of improving the performance of the model. Compared with Yolov5s, the recall of the improved Yolov5s-CA model increased by 4.9%; however, precision and $mAP_{@0.5}$ values decreased by 5.9% and 0.4%, respectively (Table 2).

Table 2. Performance comparisons of the model trained only on the synthetic dataset.

Models	Precision (%)	Recall (%)	mAP @0.5 (%)
Yolov5s	**30** [1]	14.9	**11.7** [1]
Yolov5s-CA	24.1	**19.8** [1]	11.3

[1] Bold type reflects the best precision, recall and $mAP_{@0.5}$ values.

Moreover, as illustrated in Tables 1 and 2, when the precision, recall and $mAP_{@0.5}$ values of models trained on field-collected and synthetic datasets are compared, a model trained only on a synthetic dataset generalized poorly compared to the field-collected dataset. This suggests that, although synthetic images are fast to generate, the domain gap between the synthetic and the field-collected data prevents the disease detection model trained only on the synthetic dataset from achieving the same performance as the field-trained models.

4.3. Comparison of Disease Detection Models Trained on a Combination of Synthetic and Field-Collected Datasets

In this experiment, we explored the effects of varying the amount of field-collected and synthetic data in mixed model training datasets. We evaluated the models on 10%, 25%, 40%, 55%, and 70% field-collected images with 80% synthetic images. The aim is to achieve baseline detection performance with less field-collected data and more synthetic data. The evaluation results are shown in Table 3.

Table 3. Performance comparison of the model trained on a combination of synthetic and real field datasets.

Dataset Size	Models	Precision (%)	Recall (%)	mAP $_{@0.5}$ (%)
Synthetic + Real field 10%	Yolov5s	37.2	33	27.9
	Yolov5s-CA	55.8	33	35
Synthetic + Real field 25%	Yolov5s	40.4	40.7	35.2
	Yolov5s-CA	45.9	43.8	41.2
Synthetic + Real field 40%	Yolov5s	47.6	43.5	42.4
	Yolov5s-CA	62	49.2	48.8
Synthetic + Real field 55%	Yolov5s	62.6	47.3	52.4
	Yolov5s-CA	69.6	48.9	54.2
Synthetic + Real field 70%	Yolov5s	62.6	55.9	61.1
	Yolov5s-CA	71.4	59.2	66.3
Synthetic + Real field 100%	Yolov5s	71.6	54	62.3
	Yolov5s-CA	**75.2** [1]	**61.2** [1]	**68.2** [1]

[1] Bold type reflects the best precision, recall and $mAP_{@0.5}$ values.

The improved Yolov5s-CA and Yolov5s models were trained on the mixed datasets, and precision, recall and $mAP_{@0.5}$ values were calculated for the two models. The precision of the Yolov5s-CA model is 71.4%, the recall is 59.2% and $mAP_{@0.5}$ is 66.3% (Table 3), which indicates an increase in precision, recall and $mAP_{@0.5}$ by 8.8%, 3.3%, and 5.2%, respectively, compared to Yolov5s model. Figures 7–9 show comparative results of the model prediction.

Figure 7. Comparison of detection results focused on the plant part. Yolov5s (**a**–**c**) and Yolov5s-CA (**d**–**f**).

Figure 8. Comparison of detection results focused on the plant stem.Yolov5s (**a**–**c**) and Yolov5s-CA (**d**–**f**).

Figure 9. Comparison of detection results focused on the clone. Yolov5s (**a–c**) and Yolov5s-CA (**d–f**).

In addition, when the number of field-collected images in the training datasets increased, a slight increase in the experimental indicators of the models was observed. In particular, the mixed model trained on 70% of field-collected images had a better detection performance and outperforms the baseline model trained on only field-collected images (see Section 4.1 with 1.2% precision and 0.5% $mAP_{@0.5}$).

4.4. Comparison of Detection Speed of the Models

We compared the detection speed of the improved Yolov5s-CA and the original Yolov5s model (Table 4). The predicted inference speeds of the Yolov5s-CA model are 11.4 ms, 12.3 ms, and 11.9 ms, which are 2.3 ms, 1.6 ms, and 1.4 ms longer than the Yolov5s for the field-collected, synthetic, and mixed datasets, respectively. In addition, despite the increase in the parameters of the model, the size of Yolov5s-CA model is only 0.1 MB larger than the Yolov5s model. The detection speed or frames per second (FPS) of the Yolov5s-CA model was slightly lower, but had an improved performance compared to the Yolov5s model (Table 4). Therefore, the improved Yolov5s-CA model can ensure real-time performance with relatively little additional detection time and nearly no computational overhead.

Table 4. Performance comparison of model detection speed.

Models	Datasets	Frame Per Second (FPS)	Inference Time (ms)	Parameters	Model Size (MB)
Yolov5s	Real field	109.89	9.1	7,027,720	13.7
	Synthetic	93.46	10.7		
	Mixed	95.24	10.5		
Yolov5s-CA	Real field	87.72	11.4	7,063,400	13.8
	Synthetic	81.30	12.3		
	Mixed	84.03	11.9		

4.5. Comparison of Detection at Different Spatial Scales

To compare the detection results of the models, nine images similar to those shown in Figures 7–9 were selected from the test set as these images represented detection scenarios at different spatial scales (camera shooting distances) in the dataset. In the figures, the labels fl, fr, and lf represent infected flower, infected fruit, and infected leaf, respectively. The improved Yolov5s-CA network model proposed in this study was the superior model to detect diseased plant parts at different camera shooting distances (Figures 7–9). There is almost no difference between the improved network model and Yolov5s in detecting large target plant parts taken at close distances (Figure 7a–f). The detection results focused on the plant stem (Figure 8a–f) show that there is a difference between the two network models in detecting small plant parts from the image. As shown in Figure 8d–f, the improved network model can accurately detect small target plant parts with occlusion which could not be detected by the original Yolov5s model. For the clone-level detection results shown in Figure 9a–f between the two models, the Yolov5s model has more wrong and missed detections than the improved network model. In Figure 9a, Yolov5s predicted two wrong detection and nine correct detections, while the improved network model in Figure 9d predicted thirteen correct detections with one wrong detection. Both models predicted three correct detections in Figure 9b,e, but Yolov5s predicted two wrong detections while the improved network model had only one wrong detection. Additionally, in Figure 9f, the improved network model predicted twelve correct detections, while, Yolov5s predicted eight correct detections (Figure 9c). However, although the improved model still has satisfactory detection ability with some degree of occlusion, and overlap of leaves, as shown in Figure 9c,f, both models failed to detect small diseased leaves in the image at long distances.

To further verify the effectiveness of the improved Yolov5s-CA model proposed in the present study, nine test sets representing different spatial scale detection scenarios were analyzed (Table 5). There were 78 mummy berry disease objects in nine test sets. The number of objects detected by these methods was 47 and 54 for Yolov5s and Yolov5s-CA, respectively, of which mummy berry disease was 41 for Yolov5s and 52 for Yolov5s-CA. The recall rate, accuracy, and misdetection rate of the methods were 52.56%, 87.23%, and 12.77% for Yolov5s and 66.67%, 96.30%, and 3.70% for the improved Yolov5s-CA.

From Table 5 and Figure 8d–f, it can be seen that the detection is the best in the plant stem scenario with a recall and precision rate of 80.95% and 100.00%, respectively. The plant parts taken at close distances are also accurately detected. Both methods can correctly detect plant parts in the image and their recall rate is 77.78%. In addition, the proposed method can effectively detect mummy berry disease objects at a long distance in the clone with, a recall rate of 58.33%,and a precision of 93.33%.

The loss and mAP curves for the two network models tested in the present study are shown in Appendix A. The loss curves of both models had a downward trend and the values of the loss function decreased rapidly when tested against the real field and mixed datasets (Figures A1b and A3b). However, when the network iterations reach approximately 150, the loss curves gradually exhibited a slowed rate of change and stabilized. In contrast, the loss curves in Figure A2b using the synthetic dataset had a downward trend, but only after approximately 25 iterations, the loss curves showed an upward trend indicating noisy movements and no improvement in the values of the loss function. Analysis of the loss function from Figures A1 and A3 shows that the integrated attention module on the Yolov5s backbone can effectively accelerate the network convergence speed and improve the model performance.

Table 5. Detail detection results of mummy berry disease at different spatial scales.

Models		Spatial Plant Scales			Total
		Plant Part	Plant Stem	Clone [1]	
Yolov5s	Number of objects detected correctly	7	14	20	41
	Number of annotations	9	21	48	78
	Recall rate (%)	77.78	66.67	41.67	52.56
	Precision rate(%)	87.50	93.33	83.33	87.23
Yolov5s-CA	Number of objects detected correctly	7	17	28	52
	Number of annotations	9	21	48	78
	Recall rate (%)	77.78	80.95	58.33	66.67
	Precision rate(%)	100.00	100.00	93.33	96.30

[1] Clone is a term that refers to a genetically distinct plant (range: <1–>25 m diameter).

5. Discussion

In the present study, a deep learning model based on the improved Yolov5s for automatic detection of mummy berry disease in a real wild blueberry field environment is proposed. In order to highlight important information that is relevant to the current task and improve the effectiveness of the network model, the coordinate attention (CA) module was introduced on the backbone structure of the original Yolov5s. In addition, to overcome the problem of data scarcity, we present a method for generating synthetic training images for object detection models, which greatly reduces the effort required to collect and annotate large datasets.

The overall performance of the improved network model was better than the original Yolov5s. A one-way ANOVA test on precision found a significant difference between the means of the two network models ($F_{(1299)} = 18.069$, $p < 0.001$). The precision of the improved network model reached 71.4%, which is 1.2% higher than Yolov5s precision. This result is consistent with previous studies conducted to recognize plant diseases. Yan et al. [58] compared the original Yolov5s network model with the improved Yolov5s for real-time apple disease target detection, and the improved Yolov5s model $mAP_{@0.5}$ increased by 5.1%. Similar results and comparisons with Yolov5 models were shown in a study [59], where the authors found that with the joint efforts of the coordinate attention module and Softpool pooling, the multi-scale feature fusion (MFF) convolutional neural network (CNN) obtained the optimal detection accuracy with a 1.6% improvement compared to Yolov5s. Another study [60] developed an accurate apple fruitlet detection method with a small model size and the channel pruned Yolov5s model provided an effective method to detect apple fruitlets under different conditions. For tomato disease detection, the study in [20] used a mobile phone to collect images of tomato disease in a greenhouse and the improved SE-Yolov5 mAP@0.5 was 1.78% higher than the Yolov5 model.

The performance of our improved network model was evaluated on the field-collected, synthetic and mixed datasets. Compared to training the object detection model only on synthetic images, we found a detection model with satisfactory performance on field-collected images, but a significant increase in performance was achieved when trained on a mixed dataset of field-collected and synthetic images. Our proposed Yolov5s-CA network model trained on a mixed dataset of 70% real field images and 80% of synthetic images outperformed, by 1.2% precision and 0.5% $mAP_{@0.5}$ values, the baseline model trained using only field-collected images. The results indicated that labeled real-world field-collected datasets are key to improving performance by overcoming domain gaps when training a plant disease detection model with synthetic datasets.

The improved Yolov5s network model has improved disease prediction performance under a certain degree of occlusion, leaf overlap, and different spatial scale scenarios (Table 5, Figures 7–9 and Figure A4). This is because the integrated coordinate attention (CA) mechanism at the backbone of the Yolov5s network model suppresses less relevant information and highlights key disease-related visual features to help identify mummy berry disease in a field environment. The lightweight coordinate attention (CA) module captures long-term dependencies in one space, retains accurate disease location information in the other, and forms a pair of direction-aware and position-aware feature maps, which can help the model locate and identify potential targets more precisely and enhance the representation capability of effective information. In addition, the CIoU loss used in this study takes into account the overlap area, the center point distance and the aspect ratio similarity between the actual box and the prediction box, which improves the network's regression accuracy and sensitivity to small disease organs [61]. The advantages of our method become even more obvious when dealing with scenarios of large spatial scale where a huge number of interacting and overlapping plant parts are present in a clone level image. Therefore, the effectiveness of the improved network model for mummy berry disease detection makes it clearly better than the l Yolov5s family and meets the needs of real-time detection of mummy berry disease under field conditions.

In general, promising results were obtained for training object detection models by combining a small number of field-collected images with synthetic datasets. The presented synthetic image generation method is essential when the collection and annotation of a large dataset are expensive and/or prohibitive. In addition, the coordinate attention (CA) module integrated into the Yolov5s backbone has contributed to the detection of mummy berry disease in a commercial lowbush blueberry field environment by efficiently discriminating important features.

6. Conclusions

This study focused on detecting mummy berry disease in a real natural environment based on the deep learning method and proposed an improved Yolov5s network model. By integrating the coordinate attention module into the backbone of Yolov5s, the visual features associated with mummy berry disease are well focused and extracted, which boosts the performance of the model in identifying disease symptoms. In addition, we presented the cut-and-paste method for synthetically augmenting the available dataset to generate annotated training images which greatly reduces the effort required to collect and annotate large datasets. To test the generalization ability of the improved network model and prove the usefulness of the synthetic dataset to enhance the performance of deep learning-based object detection models, quantitative performance comparisons of the improved network model and Yolov5s trained on field-collected, synthetic and mixed datasets were conducted (Tables 1–3). Compared to the baseline model with a 100% real field dataset, the synthetic dataset combined with 70% of real field outperformed the baseline model (Table 3). In all three datasets tested, the overall performance of the improved Yolov5s-CA network model is superior to that of the Yolov5s model with only slightly higher computational costs. Moreover, the improved Yolov5s network model has improved the disease prediction performance in occlusion, leaf overlaps, and different spatial scales. In general, the effectiveness of the improved network model for mummy berry disease detection is better than the original Yolov5s and meets the needs of real-time detection of mummy berry disease under field conditions. However, as the synthetic data generation process and the network model were trained on small numbers of field-collected images with limited variability in disease symptoms and camera shooting distances, some missed or incorrect detection cases were observed. In addition, the presented cut–paste synthetic data generation method is highly influenced by the quality of segmentation of the object from the image.

In the future, taking images using high-resolution cameras at different shooting distances will contribute to creating a more robust model, as well as solving the limitations of missed or incorrect detection over different occlusions and spatial scales. Furthermore, we will automate the segmentation process to extract the object from the image. Finally, we will work on implementing the models to run on a cloud server so that web and mobile applications can access it to make predictions.

Author Contributions: E.Y.O.: Conceptualization, methodology, software, formal analysis, writing—original draft preparation. H.Q.: conceptualization, methodology, formal analysis, writing—original draft preparation and review and editing, supervision, project administration, funding acquisition. Y.-J.Z.: writing—review and editing, data curation. S.A.: field collection of images, writing—review and editing. F.D.: writing—review and editing. All authors have read and agreed to the published version of the manuscript.

Funding: We acknowledge funding by the National Natural Science Foundation of China (61871061). This also a publication of the Project of Advanced Scientific Research Institute of CQUPT under Grant E011A2022329. YJZ is supported by the USDA National Institute of Food and Agriculture (Hatch Project number ME0-22021) through the Maine Agricultural & Forest Experiment Station.

Institutional Review Board Statement: Not applicable.

Informed Consent Statement: Not applicable.

Data Availability Statement: The dataset is available upon request from the corresponding author at hcchyu@gmail.com.

Conflicts of Interest: The authors declare no conflict of interest.

Appendix A

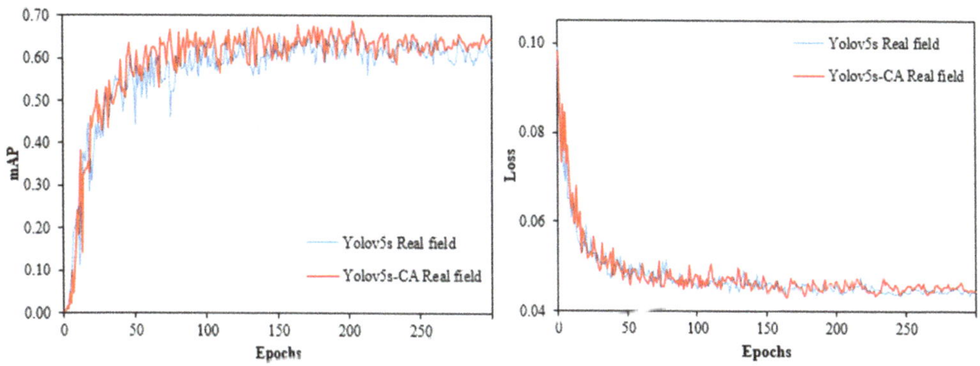

(a) (b)

Figure A1. Experimental results of each model on the field-collected dataset. (**a**) $mAP_{@0.5}$ curves for the real field dataset. (**b**) Loss curves for the real field dataset.

Table A1. Number of images in each data subset.

	Train	Validation	Test
Real field	367	46	46
Synthetic	1661	-	-

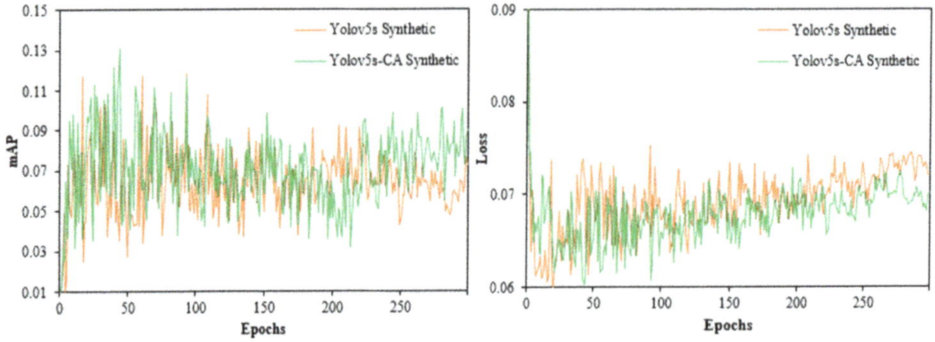

(**a**) (**b**)

Figure A2. Experimental results of each model on the synthetic dataset. (**a**) mAP$_{@0.5}$ curves for the synthetic dataset. (**b**) Loss curves for the synthetic dataset.

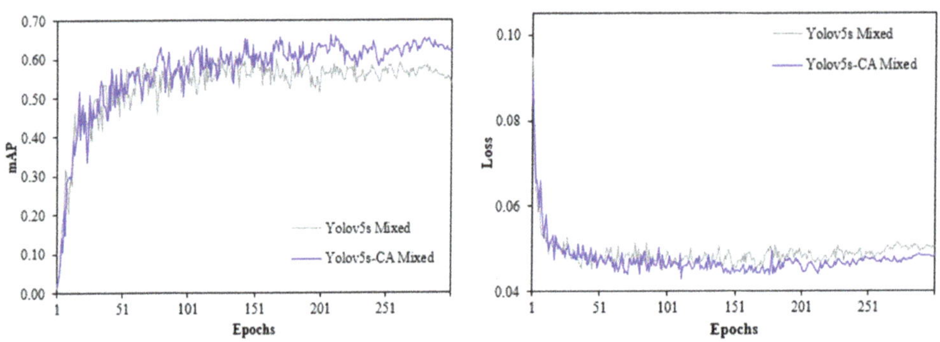

(**a**) (**b**)

Figure A3. Experimental results of each model on the mixed dataset. (**a**) mAP$_{@0.5}$ curves for the mixed dataset. (**b**) Loss curves for the mixed dataset.

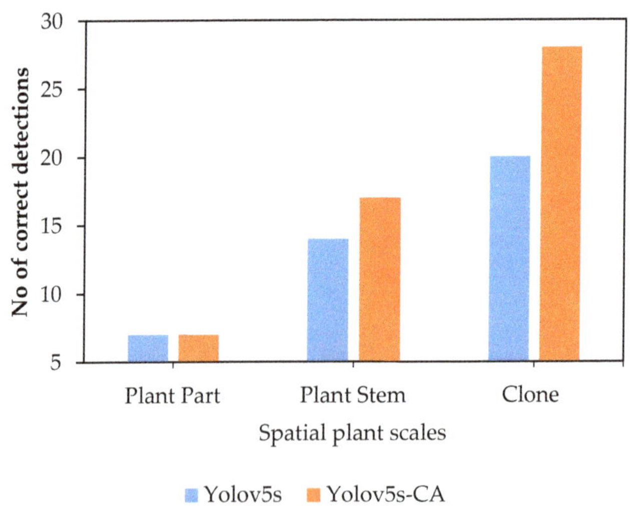

Figure A4. Comparison of correct detection between the Yolov5s and Yolov5s-CA models.

References

1. Chatterjee, S.; Kuang, Y.; Splivallo, R.; Chatterjee, P.; Karlovsky, P. Interactions among Filamentous Fungi Aspergillus Niger, Fusarium Verticillioides and Clonostachys Rosea: Fungal Biomass, Diversity of Secreted Metabolites and Fumonisin Production. *BMC Microbiol.* **2016**, *16*, 83. [CrossRef] [PubMed]
2. Asare, E.; Hoshide, A.K.; Drummond, F.A.; Criner, G.K.; Chen, X. Economic Risk of Bee Pollination in Maine Wild Blueberry, Vaccinium Angustifolium. *J. Econ. Entomol.* **2017**, *110*, 1980–1992. [CrossRef] [PubMed]
3. Tasnim, R.; Calderwood, L.; Tooley, B.; Wang, L.; Zhang, Y.-J. Are Foliar Fertilizers Beneficial to Growth and Yield of Wild Lowbush Blueberries? *Agronomy* **2022**, *12*, 470. [CrossRef]
4. Seireg, H.R.; Omar, Y.M.K.; Abd El-Samie, F.E.; El-Fishawy, A.S.; Elmahalawy, A. Ensemble Machine Learning Techniques Using Computer Simulation Data for Wild Blueberry Yield Prediction. *IEEE Access* **2022**, *10*, 64671–64687. [CrossRef]
5. Hanes, S.P.; Collum, K.K.; Hoshide, A.K.; Asare, E. Grower Perceptions of Native Pollinators and Pollination Strategies in the Lowbush Blueberry Industry. *Renew. Agric. Food Syst.* **2015**, *30*, 124–131. [CrossRef]
6. Drummond, F. Reproductive Biology of Wild Blueberry (Vaccinium Angustifolium Aiton). *Agriculture* **2019**, *9*, 69–80. [CrossRef]
7. Strik, B.C.; Yarborough, D. Blueberry Production Trends in North America, 1992 to 2003, and Predictions for Growth. *Horttechnology* **2005**, *15*, 391–398. [CrossRef]
8. Jones, M.S.; Vanhanen, H.; Peltola, R.; Drummond, F. A Global Review of Arthropod-Mediated Ecosystem-Services in Vaccinium Berry Agroecosystems. *Terr. Arthropod Rev.* **2014**, *7*, 41–78. [CrossRef]
9. Obsie, E.Y.; Qu, H.; Drummond, F. Wild Blueberry Yield Prediction Using a Combination of Computer Simulation and Machine Learning Algorithms. *Comput. Electron. Agric.* **2020**, *178*, 105778. [CrossRef]
10. Penman, L.N.; Annis, S.L. Leaf and Flower Blight Caused by Monilinia Vaccinii Corymbosi on Lowbush Blueberry: Effects on Yield and Relationship to Bud Phenology. *Phytopathology* **2005**, *95*, 1174–1182. [CrossRef]
11. Batra, L.R. Monilinia Vaccinii-Corymbosi (Sclerotiniaceae): Its Biology on Blueberry and Comparison with Related Species. *Mycologia* **1983**, *75*, 131–152. [CrossRef]
12. Mcgovern, K.B.; Annis, S.L.; Yarborough, D.E. Efficacy of Organically Acceptable Materials for Control of Mummy Berry Disease on Lowbush Blueberries in Maine. *Int. J. Fruit Sci.* **2012**, *12*, 188–204. [CrossRef]
13. Annis, S.L.; Slemmons, C.R.; Hildebrand, P.D.; Delbridge, R.W. An Internet-Served Forecast System for Mummy Berry Disease in Maine Lowbush Blueberry Fields Using Weather Stations with Cellular Telemetry. In Proceedings of the Phytopathology; The American Phytopathological Society: Saint Paul, MN, USA, 2013; Volume 103, p. 8.
14. Annis, S.; Schwab, J.; Tooley, B.; Calderwood, L. 2022 Pest Management Guide: Disease. 2022. Available online: https://extension.umaine.edu/blueberries/wp-content/uploads/sites/41/2022/02/2022-fungicide-chart.pdf (accessed on 25 September 2022).
15. Wang, X.; Liu, J.; Zhu, X. Early Real-Time Detection Algorithm of Tomato Diseases and Pests in the Natural Environment. *Plant Methods* **2021**, *17*, 43. [CrossRef]
16. Singh, V.; Misra, A.K. Detection of Plant Leaf Diseases Using Image Segmentation and Soft Computing Techniques. *Inf. Process. Agric.* **2017**, *4*, 41–49. [CrossRef]
17. Qu, H.; Sun, M. A Lightweight Network for Mummy Berry Disease Recognition. *Smart Agric. Technol.* **2022**, *2*, 100044. [CrossRef]
18. Sullca, C.; Molina, C.; Rodríguez, C.; Fernández, T. Diseases Detection in Blueberry Leaves Using Computer Vision and Machine Learning Techniques. *Int. J. Mach. Learn. Comput.* **2019**, *9*, 656–661. [CrossRef]
19. Arsenovic, M.; Karanovic, M.; Sladojevic, S.; Anderla, A.; Stefanovic, D. Solving Current Limitations of Deep Learning Based Approaches for Plant Disease Detection. *Symmetry* **2019**, *11*, 939. [CrossRef]
20. Qi, J.; Liu, X.; Liu, K.; Xu, F.; Guo, H.; Tian, X.; Li, M.; Bao, Z.; Li, Y. An Improved YOLOv5 Model Based on Visual Attention Mechanism: Application to Recognition of Tomato Virus Disease. *Comput. Electron. Agric.* **2022**, *194*, 106780. [CrossRef]
21. Dewi, C.; Chen, R.-C.; Liu, Y.-T.; Jiang, X.; Hartomo, K.D. Yolo V4 for Advanced Traffic Sign Recognition with Synthetic Training Data Generated by Various GAN. *IEEE Access* **2021**, *9*, 97228–97242. [CrossRef]
22. Abbas, A.; Jain, S.; Gour, M.; Vankudothu, S. Tomato Plant Disease Detection Using Transfer Learning with C-GAN Synthetic Images. *Comput. Electron. Agric.* **2021**, *187*, 106279. [CrossRef]
23. Ultralytics. Yolov5. Available online: https://github.com/ultralytics/yolov5 (accessed on 18 July 2022).
24. Shi, C.; Lin, L.; Sun, J.; Su, W.; Yang, H.; Wang, Y. A Lightweight YOLOv5 Transmission Line Defect Detection Method Based on Coordinate Attention. In Proceedings of the 2022 IEEE 6th Information Technology and Mechatronics Engineering Conference (ITOEC), Chongqing, China, 4–6 March 2022; Volume 6, pp. 1779–1785.
25. Guo, R.; Zuo, Z.; Su, S.; Sun, B. A Surface Target Recognition Algorithm Based on Coordinate Attention and Double-Layer Cascade. *Wirel. Commun. Mob. Comput.* **2022**, *2022*, 6317691. [CrossRef]
26. Woo, S.; Park, J.; Lee, J.-Y.; Kweon, I.S. Cbam: Convolutional Block Attention Module. In Proceedings of the European Conference on Computer Vision (ECCV), Munich, Germany, 8–14 September 2018; pp. 3–19.
27. Hu, J.; Shen, L.; Sun, G. Squeeze-and-Excitation Networks. In Proceedings of the IEEE Conference on Computer Vision and Pattern Recognition, Salt Lake City, UT, USA, 18–22 June 2018; pp. 7132–7141.
28. Hou, Q.; Zhou, D.; Feng, J. Coordinate Attention for Efficient Mobile Network Design. In Proceedings of the IEEE/CVF Conference on Computer Vision and Pattern Recognition, Nashville, TN, USA, 20–25 June 2021; pp. 13713–13722.
29. Dwibedi, D.; Misra, I.; Hebert, M. Cut, Paste and Learn: Surprisingly Easy Synthesis for Instance Detection. In Proceedings of the IEEE International Conference on Computer Vision, Venice, Italy, 22–29 October 2017; pp. 1301–1310.

30. Shorten, C.; Khoshgoftaar, T.M. A Survey on Image Data Augmentation for Deep Learning. *J. Big Data* **2019**, *6*, 1–48. [CrossRef]
31. Dvornik, N.; Mairal, J.; Schmid, C. Modeling Visual Context Is Key to Augmenting Object Detection Datasets. In Proceedings of the European Conference on Computer Vision (ECCV), Munich, Germany, 8–14 September 2018; pp. 364–380.
32. Khoreva, A.; Benenson, R.; Ilg, E.; Brox, T.; Schiele, B. Lucid Data Dreaming for Video Object Segmentation. *Int. J. Comput. Vis.* **2019**, *127*, 1175–1197. [CrossRef]
33. Ghiasi, G.; Cui, Y.; Srinivas, A.; Qian, R.; Lin, T.-Y.; Cubuk, E.D.; Le, Q.V.; Zoph, B. Simple Copy-Paste Is a Strong Data Augmentation Method for Instance Segmentation. In Proceedings of the IEEE/CVF Conference on Computer Vision and Pattern Recognition, Nashville, TN, USA, 20–25 June 2021; pp. 2918–2928.
34. Rao, J.; Zhang, J. Cut and Paste: Generate Artificial Labels for Object Detection. In Proceedings of the International Conference on Video and Image Processing, Singapore, 27–29 December 2017; pp. 29–33.
35. Redmon, J.; Farhadi, A. YOLO9000: Better, Faster, Stronger. In Proceedings of the IEEE Conference on Computer Vision and Pattern Recognition, Honolulu, HI, USA, 21–26 July 2017; pp. 7263–7271.
36. Liu, W.; Anguelov, D.; Erhan, D.; Szegedy, C.; Reed, S.; Fu, C.-Y.; Berg, A.C. Ssd: Single Shot Multibox Detector. In Proceedings of the European Conference on Computer Vision, Amsterdam, The Netherlands, 11–14 October 2016; pp. 21–37.
37. Khalil, O.; Fathy, M.E.; El Kholy, D.K.; El Saban, M.; Kohli, P.; Shotton, J.; Badr, Y. Synthetic Training in Object Detection. In Proceedings of the 2013 IEEE International Conference on Image Processing, Melbourne, VIC, Australia, 15–18 September 2013; pp. 3113–3117.
38. Fang, H.-S.; Sun, J.; Wang, R.; Gou, M.; Li, Y.-L.; Lu, C. Instaboost: Boosting Instance Segmentation via Probability Map Guided Copy-Pasting. In Proceedings of the IEEE/CVF International Conference on Computer Vision, Seoul, Republic of Korea, 27–28 October 2019; pp. 682–691.
39. Abayomi-Alli, O.O.; Damaševičius, R.; Misra, S.; Maskeliūnas, R. Cassava Disease Recognition from Low-quality Images Using Enhanced Data Augmentation Model and Deep Learning. *Expert Syst.* **2021**, *38*, e12746. [CrossRef]
40. Nair, V.; Hinton, G.E. Rectified Linear Units Improve Restricted Boltzmann Machines. In Proceedings of the 27th International Conference on Machine Learning, Haifa, Israel, 21–24 June 2010; pp. 807–814.
41. Su, H.; Qi, C.R.; Li, Y.; Guibas, L.J. Render for Cnn: Viewpoint Estimation in Images Using Cnns Trained with Rendered 3d Model Views. In Proceedings of the IEEE International Conference on Computer Vision, Santiago, Chile, 7–13 December 2015; pp. 2686–2694.
42. Movshovitz-Attias, Y.; Kanade, T.; Sheikh, Y. How Useful Is Photo-Realistic Rendering for Visual Learning? In Proceedings of the European Conference on Computer Vision, Amsterdam, The Netherlands, 11–14 October 2016; pp. 202–217.
43. Gupta, A.; Vedaldi, A.; Zisserman, A. Synthetic Data for Text Localisation in Natural Images. In Proceedings of the IEEE Conference on Computer vision and Pattern Recognition, Las Vegas, NV, USA, 27–30 June 30 2016; pp. 2315–2324.
44. Georgakis, G.; Mousavian, A.; Berg, A.C.; Kosecka, J. Synthesizing Training Data for Object Detection in Indoor Scenes. *arXiv* **2017**, arXiv:1702.07836.
45. Fuentes, A.; Yoon, S.; Kim, S.C.; Park, D.S. A Robust Deep-Learning-Based Detector for Real-Time Tomato Plant Diseases and Pests Recognition. *Sensors* **2017**, *17*, 2022. [CrossRef] [PubMed]
46. Roy, A.M.; Bhaduri, J. A Deep Learning Enabled Multi-Class Plant Disease Detection Model Based on Computer Vision. *AI* **2021**, *2*, 26. [CrossRef]
47. Chen, Y.-Y.; Pahadi, P.; Calderwood, L.; Annis, S.; Drummond, F.; Zhang, Y.-J. Will Climate Warming Alter Biotic Stresses in Wild Lowbush Blueberries? *Agronomy* **2022**, *12*, 371. [CrossRef]
48. Redmon, J.; Divvala, S.; Girshick, R.; Farhadi, A. You Only Look Once: Unified, Real-Time Object Detection. In Proceedings of the IEEE Conference on Computer Vision and Pattern Recognition, Las Vegas, NV, USA, 27–30 June 30 2016; pp. 779–788.
49. Redmon, J.; Farhadi, A. Yolov3: An Incremental Improvement. *arXiv* **2018**, arXiv:1804.02767.
50. Bochkovskiy, A.; Wang, C.-Y.; Liao, H.-Y.M. Yolov4: Optimal Speed and Accuracy of Object Detection. *arXiv* **2020**, arXiv:2004.10934.
51. Fu, C.Y.; Liu, W.; Ranga, A.; Tyagi, A.; Berg, A.C. Dssd: Deconvolutional Single Shot Detector. *arXiv* **2017**, arXiv:1701.06659.
52. Wang, Y.; Wang, C.; Zhang, H.; Dong, Y.; Wei, S. Automatic Ship Detection Based on RetinaNet Using Multi-Resolution Gaofen-3 Imagery. *Remote Sens.* **2019**, *11*, 531. [CrossRef]
53. Girshick, R.; Donahue, J.; Darrell, T.; Malik, J. Rich Feature Hierarchies for Accurate Object Detection and Semantic Segmentation. In Proceedings of the IEEE Conference on Computer Vision and Pattern Recognition, Columbus, OH, USA, 23–28 June 2014; pp. 580–587.
54. Girshick, R. Fast R-Cnn. In Proceedings of the Proceedings of the IEEE International Conference on Computer Vision, Santiago, Chile, 7–13 December 2015; pp. 1440–1448.
55. He, K.; Gkioxari, G.; Dollár, P.; Girshick, R. Mask R-Cnn. In Proceedings of the Proceedings of the IEEE international conference on Computer Vision, Venice, Italy, 22–29 October 2017; pp. 2961–2969.
56. Cai, Z.; Vasconcelos, N. Cascade R-Cnn: Delving into High Quality Object Detection. In Proceedings of the Proceedings of the IEEE Conference on Computer Vision and Pattern Recognition, Salt Lake City, UT, USA, 18–22 June 2018; pp. 6154–6162.
57. Liu, K.; Tang, H.; He, S.; Yu, Q.; Xiong, Y.; Wang, N. Performance Validation of YOLO Variants for Object Detection. In Proceedings of the 2021 International Conference on Bioinformatics and Intelligent Computing, Harbin, China, 22–24 January 2021; pp. 239–243.

58. Yan, B.; Fan, P.; Lei, X.; Liu, Z.; Yang, F. A Real-Time Apple Targets Detection Method for Picking Robot Based on Improved YOLOv5. *Remote Sens.* **2021**, *13*, 1619. [CrossRef]
59. Li, Y.; Sun, S.; Zhang, C.; Yang, G.; Ye, Q. One-Stage Disease Detection Method for Maize Leaf Based on Multi-Scale Feature Fusion. *Appl. Sci.* **2022**, *12*, 7960. [CrossRef]
60. Wang, D.; He, D. Channel Pruned YOLO V5s-Based Deep Learning Approach for Rapid and Accurate Apple Fruitlet Detection before Fruit Thinning. *Biosyst. Eng.* **2021**, *210*, 271–281. [CrossRef]
61. Dong, X.; Yan, S.; Duan, C. A Lightweight Vehicles Detection Network Model Based on YOLOv5. *Eng. Appl. Artif. Intell.* **2022**, *113*, 104914. [CrossRef]

agriculture

MDPI

Article

Optimal Design of Agricultural Mobile Robot Suspension System Based on NSGA-III and TOPSIS

Zhanghao Qu [1], Peng Zhang [1], Yaohua Hu [2], Huanbo Yang [1], Taifeng Guo [1], Kaili Zhang [1] and Junchang Zhang [1,*]

1 College of Mechanical and Electronic Engineering, Northwest A&F University, Xianyang 712100, China
2 College of Optical, Mechanical, and Electrical Engineering, Zhejiang A&F University, Hangzhou 311300, China
* Correspondence: zhangjc@nwafu.edu.cn; Tel.: +86-137-5989-8701

Abstract: The stability of vehicles is influenced by the suspension system. At present, there are many studies on the suspension of traditional passenger vehicles, but few are related to agricultural mobile robots. There are structural differences between the suspension system of agricultural mobile robots and passenger vehicles, which requires structural simplification and modelling concerning suspension of agricultural mobile robots. This study investigates the optimal design for an agricultural mobile robot's suspension system designed based on a double wishbone suspension structure. The dynamics of the quarter suspension system were modelled based on Lagrange's equation. In our work, the non-dominated sorting genetic algorithm III (NSGA-III) was selected for conducting multi-objective optimization of the suspension design, combined with the Technique for Order Preference by Similarity to the Ideal Solution (TOPSIS) to choose the optimal combination of parameters in the non-dominated solution set obtained by NSGA-III. We compared the performance of NSGA-III with that of other multi-objective evolutionary algorithms (MOEAs). Compared with the second-scoring solution, the score of the optimal solution obtained by NSGA-III increased by 4.92%, indicating that NSGA-III has a significant advantage in terms of the solution quality and robustness for the optimal design of the suspension system. This was verified by simulation in Adams that our method, which utilizes multibody dynamics, NSGA-III and TOPSIS, is feasible to determine the optimal design of a suspension system for an agricultural mobile robot.

Keywords: multi-objective evolutionary algorithms; double wishbone mechanisms; multibody dynamics; pareto solution set

Citation: Qu, Z.; Zhang, P.; Hu, Y.; Yang, H.; Guo, T.; Zhang, K.; Zhang, J. Optimal Design of Agricultural Mobile Robot Suspension System Based on NSGA-III and TOPSIS. *Agriculture* 2023, 13, 207. https://doi.org/10.3390/agriculture13010207

Academic Editors: Vadim Bolshev, Vladimir Panchenko and Alexey Sibirev

Received: 21 December 2022
Revised: 11 January 2023
Accepted: 12 January 2023
Published: 14 January 2023

1. Introduction

The smoothness of a vehicle during driving is one of its important evaluation indicators. When a vehicle is driven in irregular road conditions, uncomfortable vibrations may be transmitted to the driver. As unpaved roads make up the majority of an agricultural mobile robot's working environment, considerable of vibration can be transmitted to the operating equipment carried by an agricultural mobile robot, reducing its working precision and shortening its lifespan. As an important system in a vehicle that has a vibration mitigation function, a properly designed suspension system can absorb some of the vibrations and reduce the impact on the driver or working equipment, and has therefore long been the subject of research and optimization by academics. There have been many studies on conventional vehicle suspensions. The optimal design of an agricultural robot suspension system is essentially a constrained optimization problem. Lagrange multiplier methods, evolutionary algorithms and machine learning are commonly applied to solve constrained optimization problems. These methods are utilized in many areas, such as path planning [1], lesion diagnosis [2–4], defect detection [5,6], structural design [7] and resource allocation [8,9]. There are very few studies on the optimal design of agricultural robot suspensions. Therefore, it is helpful and instructive to carry out research on the optimal design of the suspension of agricultural robots by investigating the existing research on the suspension optimization design of passenger cars.

Ciro Moreno Ramírez et al. [10] investigated the effect of two different structures of suspension systems on vehicle dynamics, and investigated the stability of the systems using root trajectory analysis methods. Li et al. [11] used the response surface method to establish an approximate model of the MacPherson suspension systems and verified the reliability of it, and then used the NSGA-II algorithm to optimize the key hard-point coordinates of the suspension. To optimize the design of the pertinent parameters, Qian and Jin [12] examined the performance of an SUV double wishbone suspension and applied the D-optimal method based on the response surface model. Guang Li et al. [13] used the NSGA-II algorithm to optimize the suspension parameters of high-speed locomotives. The locomotive dynamics model established by SIMPACK software was used to optimize the design of six parameters by the NSGA-II algorithm, and the three evaluation indexes designed were minimum indicators. None of the above studies established an accurate non-linear mathematical model, which may lead to problems such as inaccurate solution results or reduced solution efficiency in the solution process.

Chen et al. [14] established a mathematical model of the double wishbone suspension matching an in-wheel motor with a cosine matrix method. Zhang and Li [15] established a nonlinear mathematical model of a double wishbone independent suspension using a kinematic analysis method, and then used simulation analysis in Adams software and bench testing to demonstrate that the developed mathematical model accurately expressed the dynamic properties of the suspension. The nonlinear motion equation for flexible double wishbone suspension was developed by Abdelrahman et al. [16] using the concept of imaginary displacement, but the main focus of this study was on how flexible structures respond to various road unevenness, vehicle speed, and material damping coefficients. By utilising laser scanning to create 3D models of the suspension parts, and Adams to simulate their dynamics, Prastiyo and Fiebig [17] compared the benefits of linear and progressive double wishbone suspensions and found no discernible differences between them, which showed that it is more practical to use a traditional linear double wishbone suspension. A mathematical model of the suspension system was created by the aforementioned analysis, but further optimization of the current suspension system has not been done.

A mathematical model of an independent steering-suspension guidance mechanism was developed by Chen et al. [18] based on the theory of spatial mechanics, and the structural hard-point coordinates of the suspension were optimized using sensitivity analysis. Sancibrian et al. [19] developed a model containing a large number of structural parameters, mainly the lengths of the individual links, for the structure of a double wishbone suspension. Seven functional parameters were applied to evaluate the design parameters. This study was based on the gradient descent method to find the optimal solution. Shi et al. [20] investigated the issue of optimizing the MacPherson suspension's hard-point coordinates, established the relationship between the hard-point coordinates of the suspension and the evaluation index based on a dynamics model and support vector regression (SVR) established in the Adams/Car software, and designed a double-loop multi-objective particle swarm algorithm to optimize the hard-point coordinates of the suspension. The results showed that the improved algorithm outperformed the traditional multi-objective particle swarm algorithm and the genetic algorithm. Totu and Alexandru [21] presented a comprehensive design method based on the least squares approach for the optimal design of an innovative racing car suspension system. This study used Adams/View and Adams/Insight to build a regression model and carry out the optimization design. Zhu et al. [22] provided a sliding mode control method for the equivalent two-degrees-of-freedom model and calibrate the unknown parameters in the equivalent two-degrees-of-freedom model by parameter identification for a quarter double wishbone suspension model. In this study, the equivalent two-degrees-of-freedom suspension model was constructed by parameter identification, allowing for the analysis of the suspension response without the need to create a dynamic model. The above studies did not consider the influence of suspension performance parameters (such as damping and stiffness) on suspension performance during the optimization process.

Issa and Samn [23] used the Harris Hawk Optimization (HHO) algorithm to optimize the damping and stiffness coefficients of passive suspension, which they simplified to a two-degrees-of-freedom model. The result showed that the optimized passive suspension performance was improved. Huang et al. [24] simplified the double wishbone suspension to a single-wheel two-degrees-of-freedom model and employed evolutionary algorithms to find its parameters. Based on the suspension equivalent two-degrees-of-freedom model and the seat-passenger equivalent eight-degrees-of-freedom model, Papaioannou and Koulocheris [25] improved the solution speed and quality of evolutionary algorithms by dividing the optimization targets into primary optimization objectives and auxiliary ones. Drehmer et al. [26] modeled the motion of the whole vehicle by introducing the influence on the driver's seat to establish an 8-degrees-of-freedom motion model. Particle swarm optimization algorithms and sequential quadratic programming algorithms were used to optimize the damping and stiffness of the four independent suspensions of the complete vehicle under different road conditions. Gobbi et al. [27] used a 2-degrees-of-freedom linear model to analytically describe the dynamic behaviour of a vehicle travelling on a randomly contoured road, and respectively optimized the damping and stiffness of the passive suspension and the damping, stiffness and controller gain of the active suspension based on multi-objective planning theory and robust design theory. Kwon et al. [28] developed a mathematical model of a hydro-pneumatic suspension system and a whole vehicle model for the design of hydro-pneumatic suspension in heavy vehicles. This study developed an agent model to reduce the computational effort in the optimal design process, and used it for multi-objective optimization solutions. Zheng et al. [29] designed an active, tuned inertial damper (TID) suspension systems based on a combination of active actuator and inertializer, and proposed a parameter optimization method based on an analytical solution. Stability algebraic analysis was carried out using Hurwitz's criterion and it was verified that the parameters obtained with this method could guarantee the stability of the suspension. Yang et al. [30] used a decomposition-based multidisciplinary optimization approach for the optimal design of passive suspension systems based on an equivalent two-degrees-of-freedom model. In terms of the selection of design parameters, this study focused on the refinement of the parameters affecting damping and stiffness in the suspension system, such as the number of coils of the spring, the spring diameter and the piston diameter, among other structural parameters. Li et al. [31] proposed a dimensionless hybrid index based on safety probability to evaluate suspension performance for the optimal design of energy-harvesting suspension systems, transformed the multi-objective optimization problem into a single-objective optimization problem, and then used a genetic algorithm to solve the multi-parameter optimization problem. Truong and Dao [32] proposed a hybrid HNSGA-III & MOPSO algorithm based on the MOPSO algorithm and NSGA-III for the optimal design of the stiffness and damping of a powertrain suspension model. HNSGA-III & MOPSO algorithms showed better efficiency and solution quality than MOPSO and NSGA-III in this study. Grotti et al. [33] proposed a multi-objective archive-based Quantum Particle Swarm Optimizer (MOQPSO) algorithm for the optimization of suspension systems. The results were compared with NSGA-II algorithm and COGA-II algorithm. It was shown that the MOQPSO algorithm proposed in this study could obtain better non-dominated set solutions than the NSGA-II and COGA-II algorithms. Bingul and Yildiz [34] carried out a multi-objective optimized design of a non-linear suspension system for an electric vehicle based on the NSGA-II algorithm. Three suspension systems, including passive suspension, active suspension based on PD control, and active suspension based on FL control, were optimized. This study designed seven minimum optimization objectives to evaluate the degree of influence of the active suspension on the driver under road vibration conditions. Prasad et al. [35] optimized and improved the stiffness and PID control parameters of a semi-active suspension system in a quarter suspension model. They proposed a fast convergence optimization algorithm for saving computational costs.

Compared with NSGA-II algorithm, the geometry-inspired genetic algorithm proposed in this study converges quickly under small population conditions. The above studies did not consider the influence of suspension structural parameters on suspension performance during the optimization process.

Gialleonardo et al. [36] used the NSGA-II algorithm for optimal design of control strategy parameters for active suspensions. They designed six minimum targets to evaluate the parameters of three different active control strategies and to compare the performance of them. Xu et al. [37] used a multi-objective particle swarm algorithm to optimize the design of key parameters of a straw back-throwing device. This study showed that the multi-objective evolutionary algorithm is effective and feasible for the optimal design of mechanical structures. Jiang et al. [38] used the Kriging model and NSGA-II algorithm for the multi-objective lightweight design of two parts for a control arm and torsion beam, which are widely used in passenger car suspensions. Dang et al. [39] used NSGA-III to optimize the parameters of the rotational speed and opening length of a fertiliser spreader with the objectives of accuracy, uniformity, adjustment time, and crushing rate. They chose the single-objective evolutionary algorithm GA and the multi-objective evolutionary algorithm MOEA-D-DE for comparison. The evaluation metrics showed that NSGA-III has significant advantages in solving this kind of problem. Li et al. [40] proposed a multi-objective optimal control method for an active suspension system aimed at solving the negative vibration problem generated by the in-wheel motor of an electric vehicle. An integrated model considering the electromechanical coupling between the electromagnetic excitation of the motor and the transient dynamics of the vehicle was established and developed. The Pareto solution set for the optimal parameters of the active suspension system was solved using the particle swarm optimization algorithm. Chen et al. [41] conducted a study on the multi-objective optimization problem of high-speed train suspension systems using the NSGA-II algorithm to optimize four suspension parameters with three minimum objectives. The research above did not offer a clear strategy on how to select the optimal solution from the Pareto solution set. Some studies [13,31–34,36,37,41] lacked comparisons with other multi-objective optimization algorithms.

To address the above problems, our study proposes an optimal design method for the suspension of the agricultural mobile robot based on NSGA-III and TOPSIS. First, a dynamics model of the suspension system of the agricultural mobile robot was established based on the first-class Lagrange equation, which includes structural and performance parameters of the suspension system. Four minimum objectives were established for evaluating the performance of the suspension system. Second, the dynamics model was multi-objective optimized using eight multi-objective optimization algorithms including NSGA-III. The optimal solution in the Pareto solution set obtained by the multi-objective optimization algorithm was selected using the TOPSIS method. Finally, the optimal solutions set obtained by the multiple evolutionary algorithms was scored and ranked using the TOPSIS method. The results show that the NSGA-III algorithm obtained the optimal parameter combinations in this study. The top five parameter combinations were simulated by Adams software to verify the feasibility and effectiveness of the method.

This work aimed to find the optimum design of an agricultural robot passive suspension system using the NSGA-III algorithm and TOPSIS method. The main contributions and innovations of this study are as follows:

1. A suspension system dynamics model based on the first type of Lagrangian equations containing structural and performance parameters was developed to facilitate the optimization process for the overall optimal design of both types of parameters.
2. NSGA-III was used to compute the multi-objective optimization problem, resulting in a more significant diversity and convergence of feasible solutions. A comparison with various multi-objective evolutionary algorithms was also made, showing the advantages of NSGA-III in solving this problem.
3. The optimal solution in the Pareto solution set was selected by scoring it using the TOPSIS method. In the comparison to multiple algorithms, the optimal solution set

composed was again scored and ranked using the TOPSIS method, and the scoring was more reasonable.

2. Materials and Methods

2.1. Suspension Dynamics Model

2.1.1. Mechanical Model Simplification

The study was carried out concerning the suspension system of the agricultural mobile robot. The quarter model of the suspension system is shown in Figure 1, where the slider is fixed to the ABC bar, indicating that the robot's frame is able to move freely in the vertical direction. The BD and CF bars indicate the wishbone in the double wishbone suspension, the AE bar indicates the spring damped shock absorber, the DEF bar indicates the steering system, and G indicates the wheels.

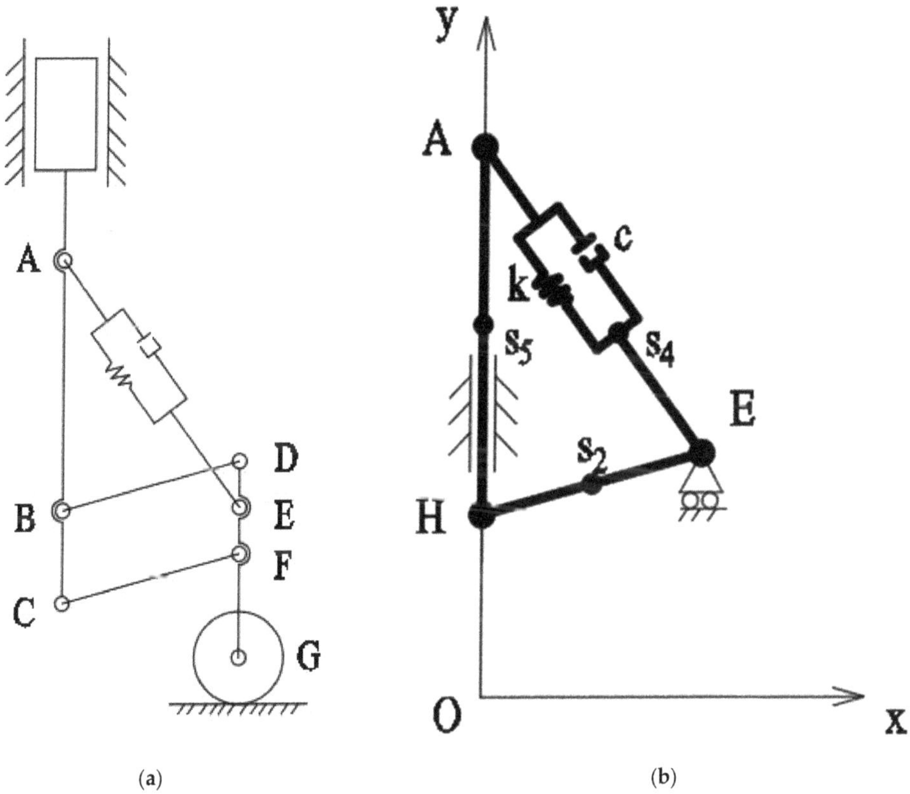

Figure 1. The quarter model of a suspension system: (**a**) model diagram; (**b**) simplified model diagram. (A–H indicate the hinge connection points between the parts).

For analysis of the quarter suspension system model shown in Figure 1a, the BD bar and CF bar motion forms and motion parameters are the same, so a simplified model of the quarter suspension system can be obtained, as shown in Figure 1b.

2.1.2. Dynamic Model

In Figure 1b, H is the BC midpoint. For the motion analysis of the simplified model shown in Figure 1b, the generalised coordinates can be taken as $q = x, x_E, y_E, y_H, y_A$ (m). The active member is the frame AH bar, the wheel bracket E is constrained by the ground, the centre of mass of each member is $s_i (i = 2,4,5)$, the mass of each member is $m_i (i = 2,4,5)$ (kg), the velocity of motion at the centre of mass of each member is $v_i (i = 2,4,5)$ (ms^{-1}), the angular velocity at the centre of mass of each member is $\omega_i (i = 2,4)$ (rads^{-1}), the spring deformation is x, the wheel bracket E is subjected to the support reaction force F(N) from the ground, and its motion analysis is shown in Figure 2.

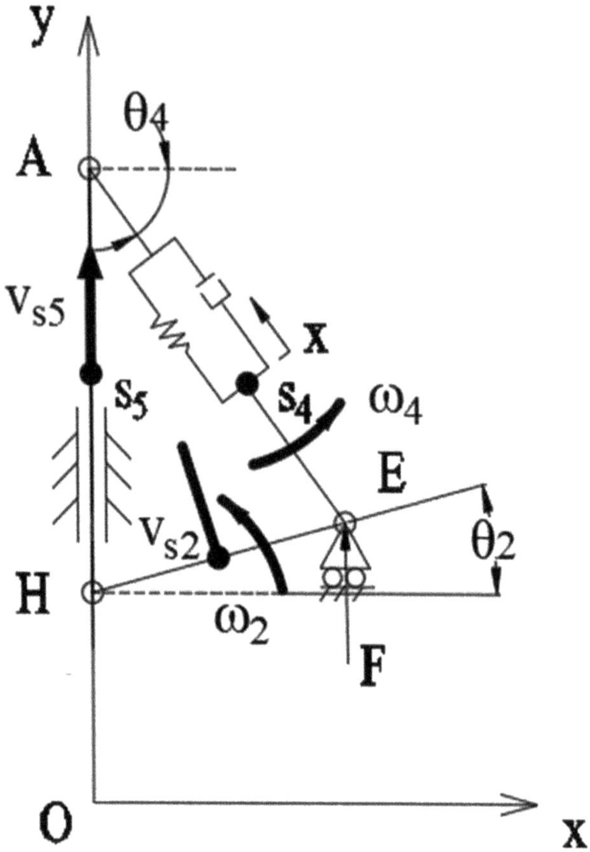

Figure 2. Motion analysis diagram of the quarter simplified model of suspension system. (A, E and H indicate the hinge connection points between the parts).

The dynamics of the suspension system can be modelled on the basis of the first type of Lagrange equation as follows:

$$\frac{d}{dt}\left(\frac{\partial L}{\partial \dot{q}_i}\right) - \frac{\partial L}{\partial q_i} + \sum_{a=1}^{4} \lambda_a \frac{\partial C_a}{\partial q_i} = Q_i \tag{1}$$

where L is the Lagrangian function of the system, $L = E_k - E_p$.

E_k is the kinetic energy of the system.

E_p is the potential energy of the system.

q_i are generalized coordinates.

Q_i for broad forces.

The structural constraints are

$$C_1 : (x_E - x_H)^2 + (y_E - y_H)^2 = l_2{}^2 \tag{2}$$

$$C_2 : y_A - y_H = l_{ah} \tag{3}$$

$$C3 : (x_E - x_A)^2 + (y_E - y_A)^2 = (l - x)^2 \tag{4}$$

$$C_4 : y_E = Y_E \tag{5}$$

The Lagrangian function of the system is

$$\begin{aligned} L = & -\frac{g}{2}(m_2(y_E + y_H) + m_4(y_A + y_E) + m_5(y_A + y_H)) - \frac{1}{2}kx^2 \\ & + \frac{m_4(l-x)^2\left(\frac{\dot{y}_A - \dot{y}_E}{x_E} - \frac{\dot{x}_E(y_A - y_E)}{x_E^2}\right)^2}{6\left(\frac{(y_A - y_E)^2}{x_E^2} + 1\right)^2} + \frac{l_2^2 m_2\left(\frac{\dot{y}_E - \dot{y}_H}{x_E} - \frac{\dot{x}_E(y_E - y_H)}{x_E^2}\right)^2}{6\left(\frac{(y_E - y_H)^2}{x_E^2} + 1\right)^2} \\ & + \frac{1}{2}m_2\left(\frac{1}{4}\dot{x}_E^2 + \frac{1}{4}(\dot{y}_E + \dot{y}_H)^2\right) + \frac{1}{2}m_4\left(\frac{1}{4}\dot{x}_E^2 + \frac{1}{4}(\dot{y}_A + \dot{y}_E)^2\right) + \frac{1}{8}m_5(\dot{y}_A + \dot{y}_H)^2 \end{aligned} \tag{6}$$

Generalized forces Q_i

$$\left.\begin{array}{l} Q_1 = -c\dot{x} \\ Q_2 = 0 \\ Q_3 = (m_2 + m_4 + m_5 + m_b)g \\ Q_4 = 0 \\ Q_5 = 0 \end{array}\right\} \tag{7}$$

According to Equations (1)–(7), we can obtain

$$M_{11}\dot{x}_E\dot{y}_A + M_{12}\dot{y}_E\dot{y}_A + M_{13}\dot{x}_E\dot{y}_E + M_{14}\dot{x}_E^2 + M_{15}\dot{y}_A^2 + M_{16}\dot{y}_E^2 + kx + 2\lambda_3(l - x) = -c\dot{x} \tag{8}$$

$$M_{21} + M_{22}\dot{x} + M_{23}\dot{x}_E + M_{24}(\dot{y}_A - \dot{y}_E) + \frac{(m_2+m_4)}{4}\ddot{x}_E + 2(\lambda_1 + \lambda_3)x_E + M_{25}(\dot{y}_E - \dot{y}_H) = 0 \tag{9}$$

$$\left.\begin{array}{l} M_{31} + M_{32}\dot{x}_E^2 + M_{33}\dot{x}_E^2 + M_{34}\dot{y}_A^2 + M_{35}\dot{x}_E\dot{y}_E + M_{36}\dot{y}_E^2 + M_{37}\dot{x}_E\dot{x} + M_{38}\dot{y}_A\dot{x} \\ +M_{39}\dot{y}_E\dot{x} + M_{310}\dot{x}_E\dot{y}_A + M_{311}\dot{y}_E\dot{y}_A + M_{312}\dot{x}_E\dot{y}_H + M_{313}\dot{y}_E\dot{y}_H + M_{314}\ddot{y}_A \\ +M_{315}\ddot{y}_E + M_{316}\dot{y}_H^2 + M_{317}\ddot{x}_E + M_{318}\ddot{y}_H = g(m_2 + m_4 + m_5 + m_b) \end{array}\right\} \tag{10}$$

$$M_{41} + M_{42}\dot{x}_E^2 + M_{43}(\dot{y}_E^2 + \dot{y}_H^2) + M_{44}\ddot{y}_H + M_{45}\ddot{x}_E + M_{46}\ddot{y}_E + \frac{m_5}{4}\ddot{y}_A + M_{47}\dot{x}_E\dot{y}_H + M_{48}\dot{y}_E\dot{y}_H + M_{49}\dot{x}_E\dot{y}_E = 0 \tag{11}$$

$$\left.\begin{array}{l} M_{51} + M_{52}\dot{x}_E^2 + M_{53}\dot{y}_A^2 + M_{54}\dot{x}_E\dot{y}_E + M_{55}\dot{y}_E^2 + M_{56}\dot{x}_E\dot{x} + M_{57}\dot{y}_A\dot{x} + M_{58}\dot{y}_E\dot{x} \\ +M_{59}\dot{x}_E\dot{y}_A + M_{510}\dot{y}_E\dot{y}_A + M_{511}\dot{x}_E\dot{y}_A + M_{512}\ddot{x}_E + M_{513}\ddot{y}_A + M_{514}\ddot{y}_E + \frac{m_5}{4}\ddot{y}_H = 0 \end{array}\right\} \tag{12}$$

where

$$M_{13} = \frac{2m_4(l-x)(y_A - y_E)}{3C_2^2 x_E^3}, M_{14} = \frac{m_4(l-x)(y_A - y_E)^2}{3C_2^2 x_E^4}$$

$$M_{15} = \frac{m_4(l-x)}{3C_2^2 x_E^2}, M_{16} = \frac{m_4(l-x)}{3C_2^2 x_E^2}$$

$$\begin{aligned} M_{21} = & -\frac{2C_1^2 m_4(l-x)^2(y_A - y_E)^2}{3C_2^3 x_E^3} + \frac{2C_1 C_4 m_4(l-x)^2(y_A - y_E)}{3C_2^3 x_E^2} \\ & - \frac{C_1 C_3 m_4(l-x)^2}{3C_2^2} - \frac{C_{10} l_2^2 m_2(y_E - y_H)}{3C_6^2 x_E^2} \\ & - \frac{C_9 m_4(l-x)^2(y_A - y_E)}{3C_2^2 x_E^2} - \frac{2C_5^2 l_2^2 m_2(y_E - y_H)^2}{3C_6^3 x_E^3} \\ & - \frac{C_5 C_7 l_2^2 m_2}{3C_6^2} + \frac{2C_5 C_8 l_2^2 m_2(y_E - y_H)}{3C_6^3 x_E^2} \end{aligned}$$

$$M_{22} = \frac{2C_1 m_4 (l-x)(y_A - y_E)}{3{C_2}^2 {x_E}^2}$$
$$M_{23} = \frac{2C_5 {l_2}^2 m_2 (y_E - y_H)}{3{C_6}^2 {x_E}^3} + \frac{2C_1 m_4 (l-x)^2 (y_A - y_E)}{3{C_2}^2 {x_E}^3}$$
$$M_{24} = -\frac{C_1 m_4 (l-x)^2}{3{C_2}^2 {x_E}^2}, M_{25} = -\frac{C_5 {l_2}^2 m_2}{3{C_6}^2 {x_E}^2}$$
$$M_{31} = \frac{g}{2}(m_2 + m_4) + 2\lambda_1 (y_E - y_H) + 2\lambda_3 (y_E - y_A) + \lambda_4$$
$$M_{32} = (\frac{(y_A - y_E)^2}{C_2 {x_E}^2} - 1)\frac{2m_4 (l-x)^2 (y_A - y_E)}{3{C_2}^2 {x_E}^4}$$
$$M_{33} = (1 - \frac{(y_E - y_H)^2}{C_6 {x_E}^2})\frac{2{l_2}^2 m_2 (y_E - y_H)}{3{C_6}^2 {x_E}^4}$$
$$M_{34} = \frac{2m_4 (l-x)^2 (y_A - y_E)}{3{C_2}^3 {x_E}^4}$$
$$M_{35} = (\frac{2(y_A - y_E)^2}{C_2 {x_E}^2} - 1)\frac{2m_4 (l-x)^2}{3{C_2}^2 {x_E}^3} + (\frac{2(y_E - y_H)^2}{C_6 {x_E}^2} - 1)\frac{2{l_2}^2 m_2}{3{C_6}^2 {x_E}^3}$$
$$M_{36} = \frac{2m_4 (l-x)^2 (y_A - y_E)}{3{C_2}^3 {x_E}^4} - \frac{2{l_2}^2 m_2 (y_E - y_H)}{3{C_6}^3 {x_E}^4}$$
$$M_{37} = -\frac{2m_4 (l-x)(y_A - y_E)}{3{C_2}^2 {x_E}^3}, M_{38} = \frac{2m_4 (l-x)}{3{C_2}^2 {x_E}^2}, M_{39} = -\frac{2m_4 (l-x)}{3{C_2}^2 {x_E}^2}$$
$$M_{310} = (1 - \frac{2(y_A - y_E)^2}{C_2 {x_E}^2})\frac{2m_4 (l-x)^2}{3{C_2}^2 {x_E}^3}, M_{311} = -\frac{4m_4 (l-x)^2 (y_A - y_E)}{3{C_2}^3 {x_E}^4}$$
$$M_{312} = (1 - \frac{2(y_E - y_H)^2}{C_6 {x_E}^2})\frac{2{l_2}^2 m_2}{3{C_6}^2 {x_E}^3}, M_{313} = \frac{4{l_2}^2 m_2 (y_E - y_H)}{3{C_6}^3 {x_E}^4}$$
$$M_{314} = \frac{m_4}{4} - \frac{m_4 (l-x)^2}{3{C_2}^2 {x_E}^2}, M_{315} = \frac{m_4 (l-x)^2}{3{C_2}^2 {x_E}^2} + \frac{{l_2}^2 m_2}{3{C_6}^2 {x_E}^2} + \frac{m_2}{4} + \frac{m_4}{4}$$
$$M_{316} = -\frac{2{l_2}^2 m_2 (y_E - y_H)}{3{C_6}^3 {x_E}^4}$$
$$M_{317} = \frac{m_4 (l-x)^2 (y_A - y_E)}{3{C_2}^2 {x_E}^3} - \frac{{l_2}^2 m_2 (y_E - y_H)}{3{C_6}^2 {x_E}^3}$$
$$M_{318} = \frac{m_2}{4} - \frac{{l_2}^2 m_2}{3{C_6}^2 {x_E}^2}$$
$$M_{41} = \frac{g}{2}(m_2 + m_5) - \lambda_2 - 2\lambda_1 (y_E - y_H)$$
$$M_{42} = (\frac{(y_E - y_H)^2}{C_6 {x_E}^2} - 1)\frac{2{l_2}^2 m_2 (y_E - y_H)}{3{C_6}^2 {x_E}^4}$$
$$M_{43} = \frac{2{l_2}^2 m_2 (y_E - y_H)}{3{C_6}^3 {x_E}^4}, M_{44} = \frac{{l_2}^2 m_2}{3{C_6}^2 {x_E}^2} + \frac{m_2}{4} + \frac{m_5}{4}$$
$$M_{45} = \frac{{l_2}^2 m_2 (y_E - y_H)}{3{C_6}^2 {x_E}^3}, M_{46} = \frac{m_2}{4} - \frac{{l_2}^2 m_2}{3{C_6}^2 {x_E}^2}$$
$$M_{47} = (\frac{2(y_E - y_H)^2}{C_6 {x_E}^2} - 1)\frac{2{l_2}^2 m_2}{3{C_6}^2 {x_E}^3}, M_{48} = -\frac{4{l_2}^2 m_2 (y_E - y_H)}{3{C_6}^3 {x_E}^4}$$
$$M_{49} = (1 - \frac{2(y_E - y_H)^2}{C_6 {x_E}^2})\frac{2{l_2}^2 m_2}{3{C_6}^2 {x_E}^3}$$
$$M_{51} = \frac{g}{2}(m_4 + m_5) + \lambda_2 - 2\lambda_3 (y_E - y_A)$$

$$M_{52} = (1 - \frac{(y_A-y_E)^2}{C_2 x_E^2})\frac{2m_4(l-x)^2(y_A-y_E)}{3C_2^2 x_E^4}$$
$$M_{53} = -\frac{2m_4(l-x)^2(y_A-y_E)}{3C_2^3 x_E^4}$$

$$M_{54} = (1 - \frac{2(y_A-y_E)^2}{C_2 x_E^2})\frac{2m_4(l-x)^2}{3C_2^2 x_E^3}, M_{55} = -\frac{2m_4(l-x)^2(y_A-y_E)}{3C_2^3 x_E^4}$$

$$M_{56} = \frac{2m_4(l-x)(y_A-y_E)}{3C_2^2 x_E^3}, M_{57} = -\frac{2m_4(l-x)}{3C_2^2 x_E^2}, M_{58} = \frac{2m_4(l-x)}{3C_2^2 x_E^2}$$

$$M_{59} = \frac{4m_4(l-x)^2(y_A-y_E)^2}{3C_2^3 x_E^5}, M_{510} = \frac{4m_4(l-x)^2(y_A-y_E)}{3C_2^3 x_E^4}$$

$$M_{511} = -\frac{2m_4(l-x)^2}{3C_2^2 x_E^3}, M_{512} = -\frac{m_4(l-x)^2(y_A-y_E)}{3C_2^2 x_E^3}$$

$$M_{513} = \frac{m_4(l-x)^2}{3C_2^2 x_E^2} + \frac{m_4}{4} + \frac{m_5}{4}, M_{514} = \frac{m_4}{4} - \frac{m_4(l-x)^2}{3C_2^2 x_E^2}$$

$$C_1 = \frac{\dot{y}_A - \dot{y}_E}{x_E} - \frac{(y_A-y_E)\dot{x}_E}{x_E^2} C_2 = 1 + \frac{(y_A-y_E)^2}{x_E^2}$$

$$C_3 = \frac{2(y_A-y_E)\dot{x}_E}{x_E^3} - \frac{\dot{y}_A - \dot{y}_E}{x_E^2}$$

$$C_4 = \frac{2(y_A-y_E)(\dot{y}_A-\dot{y}_E)}{x_E^2} - \frac{2(y_A-y_E)^2\dot{x}_E}{x_E^3}$$

$$C_5 = \frac{\dot{y}_E - \dot{y}_H}{x_E} - \frac{(y_E-y_H)\dot{x}_E}{x_E^2}$$

$$C_6 = 1 + \frac{(y_E-y_H)^2}{x_E^2}$$

$$C_7 = \frac{2(y_E-y_H)\dot{x}_E}{x_E^3} - \frac{\dot{y}_E - \dot{y}_H}{x_E^2}$$

$$C_8 = \frac{2(y_E-y_H)(\dot{y}_E-\dot{y}_H)}{x_E^2} - \frac{2(y_E-y_H)^2\dot{x}_E}{x_E^3}$$

$$C_9 = \frac{2(y_A-y_E)\dot{x}_E^2}{x_E^3} - \frac{2\dot{x}_E(\dot{y}_A-\dot{y}_E)}{x_E^2} - \frac{(y_A-y_E)\ddot{x}_E}{x_E^2} + \frac{\ddot{y}_A - \ddot{y}_E}{x_E}$$

$$C_{10} = \frac{2(y_E-y_H)\dot{x}_E^2}{x_E^3} - \frac{2\dot{x}_E(\dot{y}_E-\dot{y}_H)}{x_E^2} - \frac{(y_E-y_H)\ddot{x}_E}{x_E^2} + \frac{\ddot{y}_E - \ddot{y}_H}{x_E}$$

$$m_2 = \rho_a l_2$$

$$m_4 = \rho_d l$$

$$m_5 = \rho_f l_{ah}$$

$$l_{ah} = \sqrt{l^2 - l_2^2}$$

The relevant symbols in the quarter suspension dynamics model are defined as shown in Table 1.

Table 1. Parameter setting of the quarter suspension model.

Parameter	Symbol	Value	Unit
Body mass	m_b	50.000	kg
Length of damper	l	- *	m
Length of horizontal arm	l_2	- *	m
Damping coefficient	c	- *	$N \cdot s \cdot m^{-1}$
Stiffness coefficient	k	- *	$N \cdot m^{-1}$
Mass coefficient of damper	ρ_d	3.566	$kg \cdot m^{-1}$
Mass coefficient of horizontal arm	ρ_a	1.614	$kg \cdot m^{-1}$
Mass coefficient of frame	ρ_f	2.400	$kg \cdot m^{-1}$

* These values are given by the optimization procedure.

Suspension system parameters include damper length (l), wishbone length (l_2), damping factor (c) and stiffness factor (k). The selection of the suspension system parameters affects the objectives of the mobile robot's own weight, stability, the degree of fluctuation and the deformation of the shock absorber. These four objectives are described as follows.

1. The total mass of the quarter suspension system is used as a measure of the degree of lightness, as shown in Equation (13).

$$J_1 = Min(M_{sum}) \tag{13}$$

where $M_{sum} = \sum m_i$.

2. The height of the centre of mass at steady state is used as a measure of suspension stability, as shown in Equation (14).

$$J_2 = Min(Y_{cm}) \tag{14}$$

where $Y_{cm} = \frac{\sum m_i y_i}{\sum m_i}$.

3. The standard deviation of the suspension centre-of-mass height time response curve is used as a measure of the degree of system fluctuation, as shown in Equation (15).

$$J_3 = Min(\sigma) \tag{15}$$

where $\sigma = \sqrt{\frac{\sum_{t=1}^{T} \left[(Y_{cm})_t - \overline{Y}_{cm}\right]^2}{T-1}}$.

4. The ratio of the deformation of the shock absorber x to the wishbone l_2 is chosen as the deformation factor to measure the deformation of the shock absorber as shown in Equation (16).

$$J_4 = Min(DF) \tag{16}$$

where $DF = \frac{x}{l_2}$.

The focus of the optimal design problem for the suspension system of an agricultural mobile robot is the selection of the optimal combination of parameters for the quarter suspension model damper length (l), wishbone length (l_2), damping factor (c) and stiffness factor (k) to achieve the optimal objective function, i.e., the optimal suspension own weight, stability, degree of system fluctuation and damper deformation.

2.2. Suspension Structure Optimization Based on Non-Dominated Sorting Genetic Algorithm III (NSGA-III) and Technique for Order Preference by Similarity to Ideal Solution (TOPSIS)

In single-objective optimization problems, since only one objective function needs to be considered optimal, the optimal solution can be found by some common mathematical methods. In multi-objective optimization problems, however, there may be certain constraints between the various objective functions. When one of the objectives reaches optimality, the performance of the other objectives may be unacceptable. Therefore, solving the multi-objective problem for the optimal solution is very difficult. Further, in the solution of multi-objective optimization problems, the non-dominated solutions set, the Pareto set, is usually used to represent the acceptable better solution. The solutions of the Pareto set are non-dominated by any other solution, and are considered as equally optimal solutions. With the TOPSIS method, it is possible to select the optimal solution needed from the non-dominated solution set.

2.2.1. Non-Dominated Sorting Genetic Algorithm III (NSGA-III)

The non-dominated genetic sorting algorithm is a multi-objective evolutionary algorithm based on a genetic algorithm that simulates the evolutionary behaviour of natural populations in terms of "survival of the fittest" and the reproduction of offspring.

In natural selection, individuals in a population that are well adapted to nature are more likely to survive and thus reproduce their offspring, passing on the genes for good traits from generation to generation. The process of reproduction produces offspring with different traits from their parents due to crossover and variation, ensuring a diversity of traits in the population. Through multiple generations of evolution, the individuals that eventually survive in the population are those that are well adapted to nature.

Compared to NSGA, NSGA-II reduces the computational complexity of nondominated sorting, introduces and elitism strategy which could help prevent the loss of good solutions, and doesnot require specifying the sharing parameter. With the above improvements, the iterative convergence speed of NSGA-II is improved, and the computational complexity is reduced from $O(MN^3)$ to $O(MN^2)$. NSGA-II, when faced with high-dimensional multi-objective optimization problems with more than three objectives, suffers from the shortcomings of convergence. Compared to NSGA-II, NSGA-III, based on the reference point selection mechanism, is effective in reducing computational complexity and improving convergence for high-dimensional multi-objective optimization problems with a large number of objectives. NSGA-III improves on the population update selection mechanism of NSGA-II by providing and adaptively updating a number of well-distributed reference points to help maintain diversity among population members. With these improvements, the computational complexity of NSGA-III is the greater of $O(MN^2)$ or $O(N^2 \log^{M-2} N)$.

2.2.2. Technique for Order Preference by Similarity to Ideal Solution (TOPSIS)

The TOPSIS method is a ranking method that approximates an ideal solution and makes full use of the information from raw data to accurately reflect the gaps between individual evaluation objects. It is a method of ranking a limited number of evaluation objects according to their proximity to an idealised target, and is an evaluation of the relative merits of the available objects. Therefore, TOPSIS is a common and effective method for multi-objective decision analysis.

In this study, since there are a finite number of evaluation objects in the non-dominated solution set, and each object has four known evaluation indicators, the TOPSIS method of scoring is as follows.

Assume that there are n solutions in the non-dominated solution set, i.e., there are n evaluation objects and four evaluation indicators, consisting of a standardised matrix.

$$Z = \begin{bmatrix} z_{11} & z_{12} & z_{13} & z_{14} \\ z_{21} & z_{22} & z_{23} & z_{24} \\ \vdots & \vdots & \vdots & \vdots \\ z_{n1} & z_{n2} & z_{n3} & z_{n4} \end{bmatrix} \quad (17)$$

Define the minimum value

$$Z^- = (Z_1{}^-, Z_2{}^-, Z_3{}^-, Z_4{}^-) = (\min\{z_{11}, z_{21}, \cdots, z_{n1}\}, \min\{z_{12}, z_{22}, \cdots, z_{n2}\}, \min\{z_{13}, z_{23}, \cdots, z_{n3}\}, \min\{z_{14}, z_{24}, \cdots, z_{n4}\})$$

Define the distance of the $i(i = 1, 2, \cdots, n)$ evaluation object from the minimum value $D_i^- = \sqrt{\sum_{j=1}^{4} (Z_j^- - z_{ij})^2}$

The unnormalised score of the $i(i = 1, 2, \cdots, n)$ evaluation object can be calculated $S_i = \frac{1}{D_i^-}$

Normalised scores for the $i(i = 1, 2, \cdots, n)$ evaluation objects $\overline{S}_i = \frac{S_i}{\sum_{i=1}^{n} S_i}$.

The individual with the highest score based on the normalised score is selected as the optimal solution.

2.2.3. Parameters Optimization of the Suspension System Based on NSGA-III and TOPSIS

This section describes the use of NSGA-III to optimize the structural parameters of a robotic suspension system and to select the optimal solution in a non-dominated solutions set using the TOPSIS method. The objective of the optimized robotic suspension system problem is to find optimal values for the system parameters (damper length, cross arm length, stiffness and damping coefficient), aiming to minimise the four evaluation metrics developed (total suspension mass, suspension centre of mass height, suspension fluctuation coefficient and suspension compression coefficient) to obtain the highest score. Algorithm 1 describes the process of optimizing the optimum parameter values for the quarter passive robot suspension system.

Algorithm 1: Parameters Optimization of the Suspension System Based on NSGA-III and TOPSIS

1: Input the ranges of (l), (l_2), (k) and (c)
2: Initialize (N) search agents of NSGA-III with four values of $l^{(1:N)}$, $l_2{}^{(1:N)}$, $k^{(1:N)}$ and $c^{(1:N)}$
3: Calculate the indicators $(M_{sum}{}^{(1:N)}, Y_{CM}{}^{(1:N)}, \sigma^{(1:N)}, DF^{(1:N)})$ of search agents using the multibody dynamic modal.
4: Determine the nondominated solution set ($l^{F_1}, l_2{}^{F_1}, k^{F_1}$ and c^{F_1}) that have the lowest indicators $(M_{sum}{}^{(1:N)}, Y_{CM}{}^{(1:N)}, \sigma^{(1:N)}, DF^{(1:N)})$.
5: Update the search agents values ($l^{(1:N)}, l_2{}^{(1:N)}, k^{(1:N)}$ and $c^{(1:N)}$) based on NSGA-III genetic operations.
6: Repeat from step 3 for T iterations.
7: Output the nondominated solution set ($l^{F_1}, l_2{}^{F_1}, k^{F_1}$ and c^{F_1}).
8: Calculate the scores of the nondominated solution set ($l^{F_1}, l_2{}^{F_1}, k^{F_1}$ and c^{F_1}) by the indicators $(M_{sum}{}^{F_1}, Y_{CM}{}^{F_1}, \sigma^{F_1}, DF^{F_1})$ based on TOPSIS.
9: Select the highest score solution as the best global solution ($l^{gbest}, l_2{}^{gbest}, k^{gbest}$ and c^{gbest}).

3. Results and Discussion

To investigate the advantages of multi-objective evolutionary algorithms in multi-objective optimization problems, we compared the performance of NSGA-III with some multi-objective evolutionary algorithms. At the same time, we built a quarter suspension model in Adams software to perform a dynamics analysis of the parameter optimization results and compare the time response of the parameter combinations in the optimal solution set for the suspension dynamics analysis.

3.1. *Comparison of Multi-Objective Evolutionary Algorithms' (MOEA) Results*

The non-dominated ranking genetic algorithm II (NSGA-II), the covariance matrix adaptive evolution strategy (CMA-ES), the third generation generalised difference algorithm (GDE3), the decomposition-based multi-objective evolution algorithm (MOEAD), the multi-objective particle swarm optimization algorithm (OMOPSO), the speed-constrained multi-objective particle swarm optimization algorithm (SMOPSO) and the strength Pareto Evolutionary Algorithm (SPEA2) as representatives of multi-objective evolutionary algorithms were compared with NSGA-III. The parameter settings for the experimental tests are shown in Table 2.

Table 2. Parameter setting of the optimization test.

Parameter	Lower Bound	Upper Bound	Unit
l	0.150	0.450	m
l_2	0.035	0.300	m
c	1.000	1000.000	$N \cdot s \cdot m^{-1}$
k	0.100	30,000.000	$N \cdot m^{-1}$

The differential equation model based on Equations (8)–(12) was optimally solved within the parameter intervals shown in Table 2 using the Wolfram Language NDSolve function and the multi-objective evolutionary algorithm described above. NDSolve function is a general-purpose numerical solver for differential equations. It can solve many ordinary differential equations (ODEs) and some partial differential equations (PDEs). The set of Pareto solutions obtained from each evolutionary algorithm solution was selected using the TOPSIS method to find the optimal solution. The optimal solutions obtained by the eight evolutionary algorithms were integrated and scored again using the TOPSIS method.

The experimental results of the optimization of the multi-objective dynamics model using NSGA-III on a quarter suspension model compared to other multi-objective evolutionary algorithms are shown in Tables 3–5. Each algorithm was run five times and standard deviations analysed to determine the stability of the algorithm. The experimental results show that the NSGA-III achieved optimal evaluation metrics for best, worst and average results with the lowest standard deviation. Meanwhile, among the 40 optimal solution sets obtained by the eight multi-objective evolutionary algorithms, the optimal solution obtained by the NSGA-III achieved the highest score in the TOPSIS composite score. This indicates that the NSGA-III outperforms other comparative algorithms in terms of solution quality and robustness in solving multi-objective suspension optimization design problems. A comparison of the computational time of NSGA-III with other multi-objective evolutionary algorithms is presented in Table 6. All optimizations were conducted on a laptop with Intel i7-6700HQ (2.60 GHz) and 16 GB RAM. Both NSGA-III and MOEAD took longer than the others in terms of computation time. However, there was no real-time requirement for the task of suspension design. Therefore, the average computation time of 183.31 s for NSGA-III is acceptable, which does not affect the excellent performance of NSGA-III in accomplishing the design problem targeted in our study.

Table 3. Optimum score of the NSGA-III in comparison with other evolutionary algorithms.

Algorithm	Best	Worst	Average	St. Dev
CMA-ES	0.02879	0.01780	0.02394	0.00424
GDE3	0.03361	0.01547	0.02198	0.00648
MOEAD	0.03835	0.01957	0.02782	0.00655
OMOPSO	0.02954	0.01375	0.02094	0.00537
SMPSO	0.03858	0.01252	0.02681	0.00896
SPEA2	0.03278	0.01483	0.02250	0.00753
NSGA-II	0.02954	0.01387	0.02141	0.00554
NSGA-III	0.04048	0.03017	0.03460	0.00381

Table 4. Based on TOPSIS method, the optimal solution sets integrated by eight evolutionary algorithms were comprehensively scored, and the top five solutions are listed.

Rank	Algorithm	Score	l (m)	l_2 (m)	k (Nm^{-1})	c (Nsm^{-1})
1	NSGA-III	0.04048	0.21790	0.08181	19,201.10593	385.76418
2	SMPSO	0.03858	0.21371	0.08497	16,474.23182	188.98510
3	MOEAD	0.03835	0.22376	0.07908	21,444.41076	300.06606
4	NSGA-III	0.03749	0.21944	0.09227	14,775.16239	237.24490
5	GDE3	0.03361	0.23349	0.06862	24,567.76130	404.21663

Table 5. The top five scoring solutions correspond to the four design objective values based on Equations (13)–(16), respectively.

Rank	Algorithm	M_{sum} (kg)	Y_{cm} (m)	σ ($\times 10^{-5}$ m)	DF
1	NSGA-III	1.39373	0.33824	4.33187	0.31139
2	SMPSO	1.36980	0.33203	6.27938	0.34897
3	MOEAD	1.42793	0.34342	3.33967	0.29398
4	NSGA-III	1.40929	0.33234	6.18176	0.32603
5	GDE3	1.47897	0.35391	2.31888	0.28621

Table 6. Computational time of the NSGA-III in comparison with other evolutionary algorithms.

Algorithm	Shortest (s)	Longest (s)	Average (s)	St. Dev
CMA-ES	37.86	43.88	40.76	2.04
GDE3	37.58	54.15	47.44	5.54
MOEAD	174.59	216.50	193.76	14.05
OMOPSO	40.62	52.04	44.05	4.35
SMPSO	45.54	62.82	51.32	6.49
SPEA2	41.24	54.02	46.28	4.22
NSGA-II	42.95	50.79	46.85	2.51
NSGA-III	173.21	195.65	183.31	8.78

3.2. Simulation Experiments

Further comparative validation of the above optimization results was carried out. A parametric quarter suspension model was built in Adams software and the dynamics of the optimized parameter combinations were analysed to compare results.

The Adams model used in the dynamic simulation is shown in Figure 3, and its relevant parameters are shown in Table 7. In the Adams model, we used RigidBody: Link to represent frame and horizontal arm in the suspension system; RigidBody: Cylinder to represent damper, and RigidBody: Box to represent the body, under-spring mass (e.g., wheel) and the ground. Artificial masses of all entities were used to ensure that the model parameters in the Adams simulation were the same as the PDE model parameters, involving the parameterized frame, horizontal arm and damper corresponding RigidBody: Link density calculated by linear density shown in Table 1. Table 8 compares the parameters of the globally optimal solution (obtained from NSGA-III) in the PDE model with those in the Adams model. Since the parameters of the Adams model were calculated based on the parameters of the PDE model, the quality parameters of the two models were exactly the same. In Adams View, add fixed, revolute and translational joints to the model at the corresponding position.

Table 7. Main parameters of the Adams parametric model for dynamics simulation.

Parameter	Symbol	Value	Unit
Mass of body	${m_b}^{Sim}$	50.000	kg
Length of damper	l^{Sim}	- *	m
Length of horizontal arm	${l_2}^{Sim}$	- *	m
Damping coefficient	c^{Sim}	- *	$N \cdot s/m$
Stiffness coefficient	k^{Sim}	- *	N/m
Density of damper	${\rho_d}^{Sim}$	11,350.931	kg/m^3
Density of horizontal arm	${\rho_a}^{Sim}$	29,494.779	kg/m^3
Density of frame	${\rho_f}^{Sim}$	5328.001	kg/m^3

* These values relate to the optimization results.

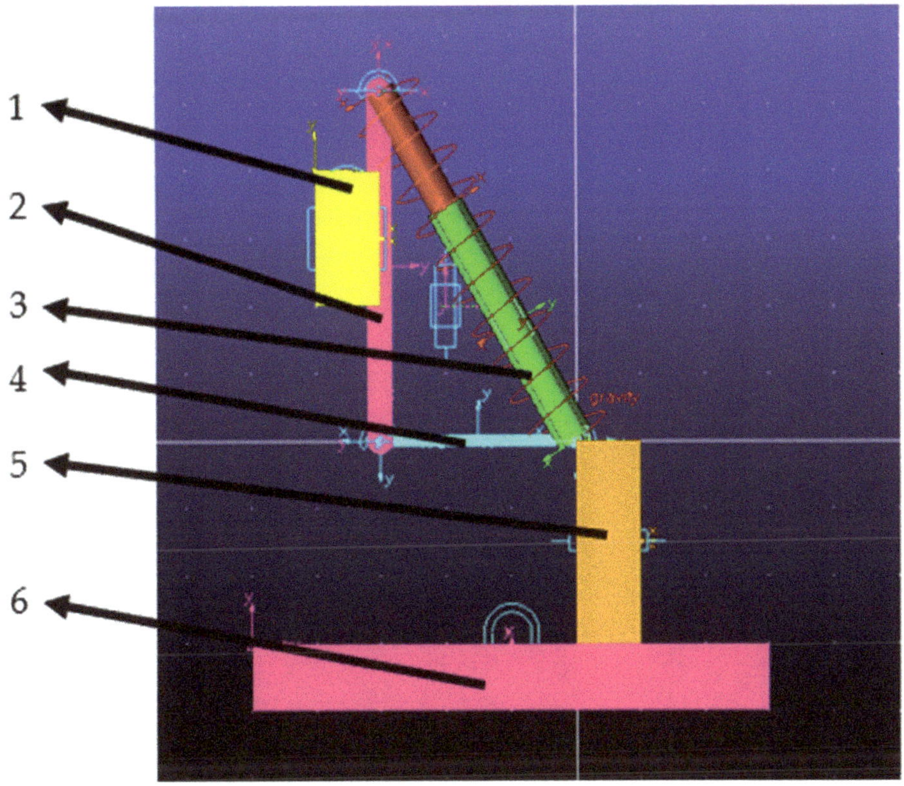

Figure 3. Adams parametric model for dynamics simulation (1. Body, 2. Frame, 3. Damper, 4. Horizontal arm, 5. Under-spring mass, 6. Ground).

Table 8. Parameters of the globally optimal solution (obtained by NSGA-III) in the partial differential equation model compared with those in the Adams model.

Parameter	PDE Value (*kg*)	Adams Value (*kg*)
Mass of body	50.00000	50.00000
Mass of damper	0.77703	0.77703
Mass of horizontal arm	0.13204	0.13204
Mass of frame	0.55860	0.55860

A step signal was applied at the wheel-ground contact position, the amplitude of which was the sum of the suspension gravity and the design load of the individual wheels. Three parameters: suspension centre of mass acceleration, suspension centre of mass height and damper spring deflection coefficient, were evaluated for comparison over a period of 1.5 s. The parameter response curves over time are shown in Figures 4–6.

Analysis of Figure 4 shows that the optimal combination of suspension parameters derived from NSGA-III can quickly reduce the body acceleration while keeping the acceleration within a small fluctuation range, which has a positive impact on maintaining smoothness in the work of agricultural mobile robots. By analysing Figure 5, it can be seen that the NSGA-III optimal combination of suspension parameters allows the suspension to reach a steady state in the shortest possible stabilisation time, while keeping the suspension's centre of mass variation range to a minimum, which is beneficial to the mobile robot's ability to effectively maintain and quickly recover a steady state when encountering bumps. Analysis of Figure 6 shows that the optimal combination of suspension parameters obtained by NSGA-III enables the spring deflection coefficient of the suspension to be kept at a low value. The deflection coefficient was chosen as a measure of spring deflection rather than a direct measure of deflection because an increase in the length of the cross-arm causes an increase in spring deflection when other parameters are held constant. The use of a deflection factor avoids this problem affecting the optimization process. The results of the combined dynamics simulations show that the NSGA-III combined with the TOPSIS method is indeed the best solution in the Pareto solution set of the multi-objective evolutionary algorithm for the multi-objective suspension optimization problem.

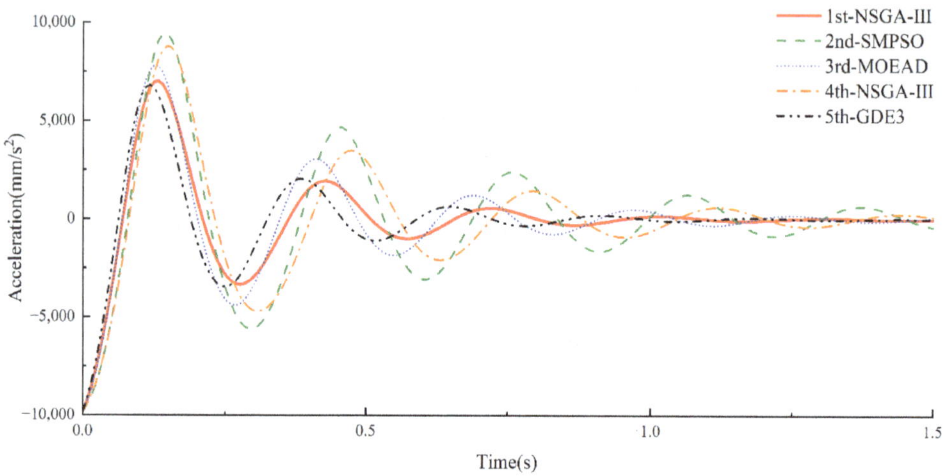

Figure 4. The quarter suspension modal barycenter's acceleration time response curve.

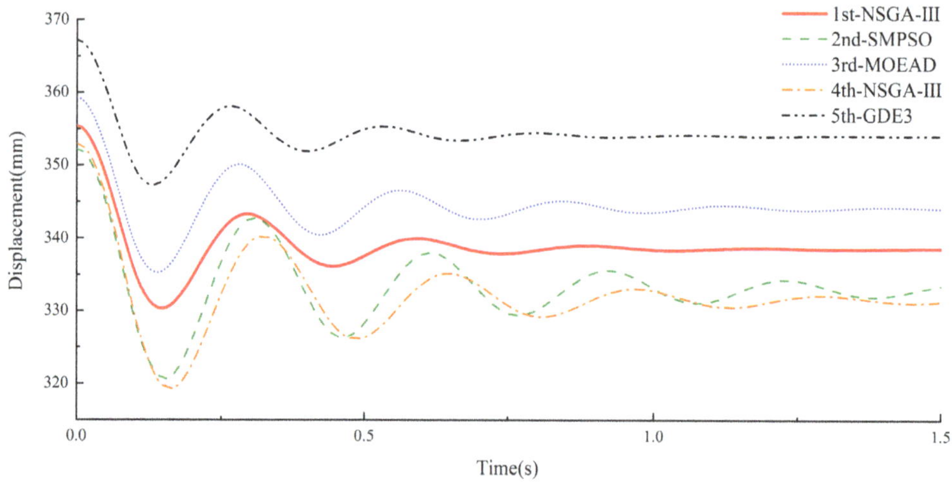

Figure 5. The quarter suspension modal barycenter's height displacement time response curve.

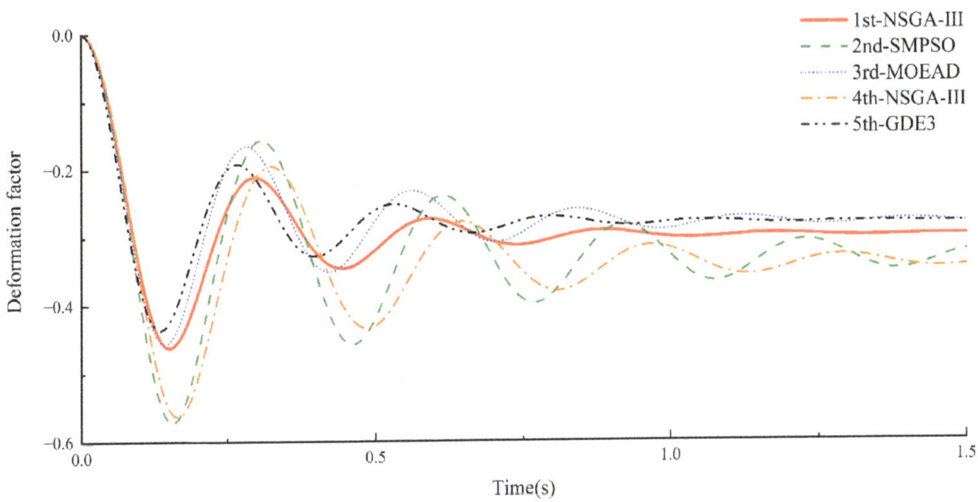

Figure 6. The quarter suspension modal spring deformation factor time response curve.

Combining Tables 4 and 5 and Figures 4–6, it is obvious that no solution is optimal for all evaluation criteria. For example, when we wish to reduce the deformation factor (DF) of the damper, we can do so by increasing the length of the horizontal arm (l_2). However, at the same time, increasing the horizontal length (l_2) leads to an increase in the total mass of the suspension (M_{sum}). The conflict between the objectives makes it difficult to select an optimal solution from the non-dominated solution set, which is why we introduced the TOPSIS method for optimal solution selection. The TOPSIS method measured the advantages and disadvantages of different non-dominated solutions in a scientific and rational way and scored them. Table 4 shows that one of the solutions obtained by NSGA-III after analysis using the TOPSIS method received the highest score. The four objective values corresponding to the top five scoring solutions are shown in Table 5. We identified that both of the four objectives of the highest scoring NSGA-III solution were not the smallest, i.e., optimal. The smallest total mass (M_{sum}) and the lowest centre-of-mass height (Y_{cm}) appeared in the solution obtained by SMPSO and the smallest degree of centre-of-mass fluctuation(σ) and the smallest degree of damper deformation (DF) appeared in the solution obtained by GDE3. Meanwhile, the time response curves for the NSGA-III derived solution in Figures 4–6, respectively, were not optimal. In the centre of-mass acceleration time response curves and the damper deformation factor time response curves, the GDE3 solution exhibited better attenuation than NSGA-III, while both the peak acceleration and deformation factor were smaller than the NSGA-III solution. In the centre-of-mass height time response curve, another solution derived by NSGA-III (ranked 4th) was able to obtain the lowest centre-of-mass height, which meant that it has the best stability.

The above results do not seem to indicate that the solution found by NSGA-III is optimal, but when the four objectives were considered collectively, the solution found by NSGA-III showed superiority over the other solutions. We calculated the difference between each objective and their minimum of the sample set according to the TOPSIS method. This meant that any solution will only receive a higher score if all four objectives are as close as possible to their minimum of the sample set. The SMPSO derived solutions had the largest degree of centre-of-mass fluctuation (σ) and damper deformation factor (DF), the GDE3 derived solutions had the second largest total mass (M_{sum}) and the largest height of mass (Y_{cm}); therefore, their scores were weakened. Under the comprehensive scoring of the TOPSIS method, the solution obtained by NSGA-III had the highest score, i.e., the optimal solution.

4. Conclusions

We proposed a multi-objective optimization design method for the suspension of agricultural robots that can balance different performance indicators and obtain the optimal parameter combination of the suspension. We utilized the Lagrangian equation to establish the partial differential equation(PDE) model of the agricultural robot double wishbone suspension, including structural parameters and performance parameters, and utilized the multi-objective evolutionary algorithm NSGA-III and TOPSIS method to carry out the agricultural robot double wishbone optimum design of the suspension. We established four evaluation indicators to evaluate the performance of the suspension, including the total mass of the suspension system, the barycenter's height of the suspension in a stable state, the fluctuation degree of the suspension under the step response, and the deformation degree of the damper. We selected eight typical multi-objective evolutionary algorithms to solve the multi-objective suspension optimization design problem, took advantage of the TOPSIS method to score the non-dominated solution set, and selected the solution with the highest score as the optimal solution. The results show that the optimal solution obtained by NSGA-III and TOPSIS method has advantages in a comprehensive performance. Therefore, we come to the conclusion that NSGA-III combined with the TOPSIS method can effectively obtain a high-quality design of an agricultural mobile robot's suspension system.

Author Contributions: Conceptualization, methodology, writing—original draft preparation, writing—review and editing Z.Q.; methodology and visualization, P.Z.; review & editing, supervision, funding acquisition and project administration, Y.H.; validation and formal analysis, H.Y. and T.G.; software, K.Z.; review & editing, supervision, funding acquisition and project administration, J.Z. All authors have read and agreed to the published version of the manuscript.

Funding: This research was supported by the Talent start-up Project of Zhejiang A&F University Scientific Research Development Foundation (2021LFR066) and the National Natural Science Foundation of China (32171894(C0043619), 31971787(C0043628)).

Institutional Review Board Statement: Not applicable.

Informed Consent Statement: Not applicable.

Data Availability Statement: Not applicable.

Conflicts of Interest: The authors declare no conflict of interest.

References

1. Kok, J.; Gonzalez, L.F.; Kelson, N. FPGA implementation of an evolutionary algorithm for autonomous unmanned aerial vehicle on-board path planning. *IEEE Trans. Evol. Comput.* **2012**, *17*, 272–281. [CrossRef]
2. Zhao, M.; Jha, A.; Liu, Q.; Millis, B.A.; Mahadevan-Jansen, A.; Lu, L.; Landman, B.A.; Tyska, M.J.; Huo, Y. Faster Mean-shift: GPU-accelerated clustering for cosine embedding-based cell segmentation and tracking. *Med. Image Anal.* **2021**, *71*, 102048. [CrossRef] [PubMed]
3. You, L.; Jiang, H.; Hu, J.; Chang, C.H.; Chen, L.; Cui, X.; Zhao, M. GPU-accelerated Faster Mean Shift with euclidean distance metrics. In Proceedings of the 2022 IEEE 46th Annual Computers, Software, and Applications Conference (COMPSAC), Los Alamitos, CA, USA, 27 June–1 July 2022; pp. 211–216.
4. Zhao, M.; Liu, Q.; Jha, A.; Deng, R.; Yao, T.; Mahadevan-Jansen, A.; Tyska, M.J.; Millis, B.A.; Huo, Y. VoxelEmbed: 3D instance segmentation and tracking with voxel embedding based deep learning. In Proceedings of the International Workshop on Machine Learning in Medical Imaging; Springer: Cham, Switzerland; pp. 437–446.
5. Zheng, Z.; Hu, Y.; Yang, H.; Qiao, Y.; He, Y.; Zhang, Y.; Huang, Y. AFFU-Net: Attention feature fusion U-Net with hybrid loss for winter jujube crack detection. *Comput. Electron. Agric.* **2022**, *198*, 107049. [CrossRef]
6. Zheng, Z.; Yang, H.; Zhou, L.; Yu, B.; Zhang, Y. HLU 2-Net: A Residual U-Structure Embedded U-Net With Hybrid Loss for Tire Defect Inspection. *IEEE Trans. Instrum. Meas.* **2021**, *70*, 1–11.
7. Rong, X.; Xu, R.; Feng, S.; Qiu, T. Research on the laterally loaded piles under huge moment based on Lagrange multipliers method. *Rock Soil Mech.* **2017**, *38*, 2605–2612. [CrossRef]
8. Bertsekas, D.P. *Constrained Optimization and Lagrange Multiplier Methods*; Academic Press: Cambridge, MA, USA, 2014.
9. Everett, H., III. Generalized Lagrange multiplier method for solving problems of optimum allocation of resources. *Oper. Res.* **1963**, *11*, 399–417. [CrossRef]

10. Moreno Ramírez, C.; Tomás-Rodríguez, M.; Evangelou, S.A. Dynamic analysis of double wishbone front suspension systems on sport motorcycles. *Nonlinear Dyn.* **2018**, *91*, 2347–2368. [CrossRef]
11. Li, W.; Wang, S.; Zhou, B.; Zhang, L.; Ma, Y. Macpherson Suspension Parameter Optimization Based on Response Surface Method and NSGA-II Algorithm. *J. Hunan Univ. Nat. Sci.* **2011**, *38*, 27–32.
12. Qian, B.; Jin, Z. Performance Analysis and Optimization Design of SUV Double Wishbone Suspension. *Mech. Sci. Technol. Aerosp. Eng.* **2022**, *41*, 505–510. [CrossRef]
13. Li, G.; Wu, R.; Deng, X.; Shen, L.; Yao, Y. Suspension parameters matching of high-speed locomotive based on stability/comfort Pareto optimization. *Veh. Syst. Dyn.* **2022**, *60*, 3848–3867. [CrossRef]
14. Chen, X.; Wan, G.; Yu, Z.; Zhou, P. Design of Double-Wishbone Suspension Mechanism for Independent Suspension/In-Wheel Motor Module. *Automot. Eng.* **2004**, *26*, 513–515,520. [CrossRef]
15. Zhang, B.; Li, Z. Mathematical modeling and nonlinear analysis of stiffness of double wishbone independent suspension. *J. Mech. Sci. Technol.* **2021**, *35*, 5351–5357. [CrossRef]
16. Abdelrahman, A.; Nabawy, A.; Abdelhaleem, A.; Alieldin, S.; Eltaher, M. Nonlinear dynamics of viscoelastic flexible structural systems by finite element method. *Eng. Comput.* **2020**, *38*, 169–190. [CrossRef]
17. Prastiyo, W.; Fiebig, W. Multibody simulation and statistical comparison of the linear and progressive rate double wishbone suspension dynamical behavior. *Simul. Model. Pract. Theory* **2021**, *108*, 102273. [CrossRef]
18. Chen, X.; Liang, D.; Wu, X.; Huang, L. Design and Analysis of Suspension Guiding Mechanism of Independent Steer-by-wire System. *J. Tongji Univ. Nat. Sci.* **2014**, *42*, 1567–1571. [CrossRef]
19. Sancibrian, R.; Garcia, P.; Viadero, F.; Fernandez, A.; De-Juan, A. Kinematic design of double-wishbone suspension systems using a multiobjective optimisation approach. *Veh. Syst. Dyn.* **2010**, *48*, 793–813. [CrossRef]
20. Shi, Q.; Peng, C.; Chen, Y.; He, J.; Li, P.; Chen, J. Robust kinematics design of MacPherson suspension based on a double-loop multi-objective particle swarm optimization algorithm. *Proc. Inst. Mech. Eng. Part D J. Automob. Eng.* **2019**, *233*, 3263–3278. [CrossRef]
21. Toţu, V.; Alexandru, C. Multi-Criteria Optimization of an Innovative Suspension System for Race Cars. *Appl. Sci.* **2021**, *11*, 4167. [CrossRef]
22. Zhu, B.; Qin, W.; Shangguan, W.; Rakheja, S. Modeling and control of a quarter-car model with double wishbones and its equivalent two degrees of freedom model. *J. Vib. Shock* **2019**, *38*, 6–14.
23. Issa, M.; Samn, A. Passive vehicle suspension system optimization using Harris Hawk Optimization algorithm. *Math. Comput. Simul.* **2022**, *191*, 328–345. [CrossRef]
24. Huang, T.; Zhou, J.; Meng, F.; Guo, X. Model Modification of 1/4 Suspension System and Analysis of Suspension Damping Performance. *Mech. Sci. Technol. Aerosp. Eng.* **2022**, *41*, 465–547.
25. Papaioannou, G.; Koulocheris, D. An approach for minimizing the number of objective functions in the optimization of vehicle suspension systems. *J. Sound Vib.* **2018**, *435*, 149–169. [CrossRef]
26. Drehmer, L.R.C.; Paucar Casas, W.J.; Gomes, H.M. Parameters optimisation of a vehicle suspension system using a particle swarm optimisation algorithm. *Veh. Syst. Dyn.* **2015**, *53*, 449–474. [CrossRef]
27. Gobbi, M.; Levi, F.; Mastinu, G. Multi-objective stochastic optimisation of the suspension system of road vehicles. *J. Sound Vib.* **2006**, *298*, 1055–1072. [CrossRef]
28. Kwon, K.; Seo, M.; Kim, H.; Lee, T.H.; Lee, J.; Min, S. Multi-objective optimisation of hydro-pneumatic suspension with gas–oil emulsion for heavy-duty vehicles. *Veh. Syst. Dyn.* **2020**, *58*, 1146–1165. [CrossRef]
29. Zheng, G.; Weirui, W.; Guangping, L.; Daogong, R. Design, Parameter Optimisation, and Performance Analysis of Active Tuned Inerter Damper (TID) Suspension for Vehicle. *J. Sound Vib.* **2022**, *525*, 116750.
30. Yang, L.; Gobbi, M.; Mastinu, G.; Previati, G.; Ballo, F. Multi-Disciplinary Optimisation of Road Vehicle Chassis Subsystems. *Energies* **2022**, *15*, 2172. [CrossRef]
31. Li, S.; Xu, J.; Gao, H.; Tao, T.; Mei, X. Safety probability based multi-objective optimization of energy-harvesting suspension system. *Energy* **2020**, *209*, 118362. [CrossRef]
32. Trường, N.H.; Dao, D.-N. New hybrid between NSGA-III with multi-objective particle swarm optimization to multi-objective robust optimization design for Powertrain mount system of electric vehicles. *Adv. Mech. Eng.* **2020**, *12*, 1687814020904253. [CrossRef]
33. Grotti, E.; Mizushima, D.M.; Backes, A.D.; de Freitas Awruch, M.D.; Gomes, H.M. A novel multi-objective quantum particle swarm algorithm for suspension optimization. *Comput. Appl. Math.* **2020**, *39*, 1–29. [CrossRef]
34. Bingül, Ö.; Yıldız, A. Fuzzy logic and proportional integral derivative based multi-objective optimization of active suspension system of a 4× 4 in-wheel motor driven electrical vehicle. *J. Vib. Control* **2022**, 10775463211062691. [CrossRef]
35. Prasad, V.; Pawaskar, D.N.; Seshu, P. Controller design and multi-objective optimization of heavy goods vehicle suspension system by geometry-inspired GA. *Struct. Multidiscip. Optim.* **2021**, *64*, 89–111. [CrossRef]
36. Di Gialleonardo, E.; Facchinetti, A.; Bruni, S. Control of an integrated lateral and roll suspension for a high-speed railway vehicle. *Veh. Syst. Dyn.* **2022**, 1–27. [CrossRef]
37. Xu, H.; Hu, Z.; Zhang, P.; Gu, F.; Wu, F.; Song, W.; Wang, C. Optimization and Experiment of Straw Back-Throwing Device of No-Tillage Drill Using Multi-Objective QPSO Algorithm. *Agriculture* **2021**, *11*, 986. [CrossRef]
38. Jiang, R.; Jin, Z.; Liu, D.; Wang, D. Multi-objective lightweight optimization of parameterized suspension components based on NSGA-II algorithm coupling with surrogate model. *Machines* **2021**, *9*, 107. [CrossRef]

39. Dang, Y.; Ma, H.; Wang, J.; Zhou, Z.; Xu, Z. An Improved Multi-Objective Optimization Decision Method Using NSGA-III for a Bivariate Precision Fertilizer Applicator. *Agriculture* **2022**, *12*, 1492. [CrossRef]
40. Li, Z.; Zheng, L.; Ren, Y.; Li, Y.; Xiong, Z. Multi-objective optimization of active suspension system in electric vehicle with In-Wheel-Motor against the negative electromechanical coupling effects. *Mech. Syst. Signal Process.* **2019**, *116*, 545–565. [CrossRef]
41. Chen, X.; Yao, Y.; Shen, L.; Zhang, X. Multi-objective optimization of high-speed train suspension parameters for improving hunting stability. *Int. J. Rail Transp.* **2022**, *10*, 159–176. [CrossRef]

agriculture

MDPI

Article

Pyrolysis of Amaranth Inflorescence Wastes: Bioenergy Potential, Biochar and Hydrocarbon Rich Bio-Oil Production

Julia Karaeva [1,*], Svetlana Timofeeva [1], Svetlana Islamova [1], Kseny Bulygina [2], Firdavs Aliev [3], Vladimir Panchenko [4] and Vadim Bolshev [5,*]

1 Institute of Power Engineering and Advanced Technologies, FRC Kazan Scientific Center, Russian Academy of Sciences, 420111 Kazan, Russia
2 A.E. Arbuzov Institute of Organic and Physical Chemistry, FRC Kazan Scientific Center, Russian Academy of Sciences, 420088 Kazan, Russia
3 Institute of Geology and Petroleum Technologies, Kazan Federal University, 420008 Kazan, Russia
4 Department of Theoretical and Applied Mechanics, Russian University of Transport, 127994 Moscow, Russia
5 Laboratory of Power Supply and Heat Supply, Federal Scientific Agroengineering Center VIM, 109428 Moscow, Russia
* Correspondence: julieenergy@list.ru (J.K.); vadimbolshev@gmail.com (V.B.)

Abstract: Many agro-industrial companies grow amaranth for the subsequent production of amaranth oil, flour, cereals, flakes, and bran. After the grain is extracted, waste in the form of inflorescences remains, which can be used to obtain useful new products. This work investigated the use of pyrolysis to recycle amaranth inflorescence wastes (AIW). Thermochemical conversion experiments in an inert medium were carried out in a laboratory setup at 550 °C and a heating rate of 10 °C/min. It was found that the AIW pyrolysis produced 37.1 wt.% bio-oil, 35.8 wt.% pyrogas and 27.1 wt.% biochar. The oil fraction of the obtained bio-oil contains 41.8% of hydrocarbons. Thermogravimetric analysis of AIW was performed in the temperature range from 40 to 1000 °C at heating rates of 10, 15, and 20 °C/min in argon medium (75 mL/min). The kinetic parameters were determined by the model-free Friedman, Ozawa-Flynn-Wall, and Kissinger-Akahira-Sunose methods. The average activation energy values are in the range of 208.44–216.17 kJ/mol, and they were used to calculate the thermodynamic parameters. The results indicate that the pyrolysis application will allow efficient conversion of AIW into value-added products.

Keywords: amaranth inflorescence wastes; pyrolysis; thermogravimetric analysis; hydrocarbon rich bio-oil

Citation: Karaeva, J.; Timofeeva, S.; Islamova, S.; Bulygina, K.; Aliev, F.; Panchenko, V.; Bolshev, V. Pyrolysis of Amaranth Inflorescence Wastes: Bioenergy Potential, Biochar and Hydrocarbon Rich Bio-Oil Production. *Agriculture* **2023**, *13*, 260. https://doi.org/10.3390/agriculture13020260

Academic Editor: Giorgio Testa

Received: 22 December 2022
Revised: 13 January 2023
Accepted: 18 January 2023
Published: 20 January 2023

1. Introduction

Amaranth is of great importance for world food security, especially for developing countries in Africa and Asia [1,2]. These plants are pseudocereals that were very important for ancient civilizations [3]. Currently, interest in this culture is growing for a number of reasons. First, amaranth can grow in a wide range of weather conditions and is drought-tolerant. Secondly, the growing demand for a healthy diet encourages the use of this plant. Amaranth is recognized as a rich and inexpensive source of dietary fiber, minerals, vitamins, proteins, and antioxidants [2,4]. One of the most common cereal species is Amaranthus cruentus [5]. After the grain is extracted, waste in the form of inflorescences remains, which can be used to obtain useful new products.

Pyrolysis is a technology widely used for waste disposal [6–8]. The uniqueness of this process lies in the simultaneous production of gaseous, liquid and solid products. The sphere of use of these products is quite wide, including the chemical industry [9], energy [10,11], and agro-industrial complex [12,13]. The process of biomass pyrolysis can be carried out in a decentralized manner, which is especially important for agriculture [14]. It is necessary to assess the energy potential of the resulting pyrolysis products in order to assess the possibility of creating a local non-volatile enterprise. It should be noted that biooil is of particular interest since its composition is very complex and largely depends on the feedstock [15]. It is important to find such a source of biomass, which initially, without the use of catalysts during thermochemical conversion and joint pyrolysis with polymers, contains a significant amount of hydrocarbons.

Currently, thermogravimetric analysis (TGA) is used for a detailed assessment of the pyrolysis process [16–18]. TGA data are used to study the kinetics of the thermochemical conversion. This allows a comprehensive study of pyrolysis reactions, revealing the characteristic mechanism and predicting the degree of complexity of the reaction, which is necessary for designing pyrolysis apparatuses and evaluating the possibilities of using products [19]. The kinetics of the amaranth inflorescence wastes (AIW) pyrolysis process was studied using the model-free methods of Friedman, OFW (Ozawa-Flynn-Wall) and KAS (Kissinger-Akahira-Sunose), since they have shown themselves to be effective in assessing the kinetics of biomass pyrolysis reactions [6,20–26]. Currently, numerous studies are being carried out on the use of biomass as a raw material for thermal decomposition [27,28]; however, there are few works on the pyrolysis of amaranth [8,29–31], and no study has yet been reported on the thermal decomposition characteristics of the inflorescences of this plant.

This study is aimed at solving the following problems: (a) determining the material balance of the pyrolysis process of a new type of plant waste; (b) study of the composition and quality of the resulting pyrolysis products to assess their subsequent use; (c) analysis of the features of thermal decomposition of waste according to TGA data at heating rates of 10, 15 and 20 °C/min in an inert atmosphere; (d) determination of kinetic triplets for the main stage of pyrolysis—isolation of volatile components using model-free methods; and (e) determination of thermodynamic functions for subsequent design, optimization and scaling of the parameters of the pyrolysis reactor. Thus, the cultivation of amaranth and the subsequent pyrolysis of the remaining waste will improve not only food, but also energy security, which is especially important for countries with adverse climatic conditions.

2. Materials and Methods

2.1. *Amaranth Inflorescence Wastes*

AIW samples were taken from a farm after harvest (Russia). Inflorescences were dried at room temperature. All samples of AIW were manually cut into small pieces with the help of blades and then finely powdered using a mixer-cum-grinder. All powdered samples were kept in airtight containers for use in further experiments.

2.2. *Physicochemical Characterization*

The proximate analysis of all samples was performed to estimate the volatile matter, ash content, moisture, and fixed carbon using appropriate ASTM protocols (E1755-01, 2020; ASTM E1756-08, 2020; E871-82, 2019; E872-82, 2019; D1762-84, 2021). The percentages of carbon, hydrogen, nitrogen, and sulfur were determined using the CHNS analyzer (Euro EA 3000, Eurovector, S.p.A., Milan, Italy) and oxygen was calculated by difference. The higher heating value (HHV) was calculated according to the formula presented in the work [32]. The content of macro- and microelements was studied using the energy-dispersive fluorescence X-ray spectrometer (EDX-800HS2, Shimadzu, Kyoto, Japan) by a semi-quantitative method. Gas chromatography–mass spectrometry of the pyrolysis liquid were carried out on spectrometer (GCMS-QP2010, Shimadzu, Japan) on HP-5MS column (0.25 μm, 30 m). The evolved gas was analyzed by a gas chromatograph Chromatec-Crystal 5000.2 (Chromatec, Yoshkar-Ola, Russia) using GOST 32507−2013 and ASTM D 5134-98, 2008. The gas samples were delivered to the given machine from the autoclave's gas output through special heat-resistant tubing. The gas separation was carried out in capillary column with a length of 30 m and two absorption chambers. Chromatography was run in following temperature mode: 90 degrees for 4 min, from 90 °C to 250 °C with the heating rate of 10 °C/min. The gas carrier was helium and the stream velocity was 2.5 mL/min.

2.3. Pyrolysis Experimental Procedure

Pyrolysis was carried out on a laboratory setup described in [8]. The initial temperature was 25 °C, the heating rate was 10 °C/min, and the temperature of the pyrolysis process itself was 550 °C. The material balance was determined according to the method presented in [8,33]. Each experiment was repeated at least three times.

2.4. Thermogravimetric Analysis (TGA)

A thermogravimetric analyzer (STA 449 F1 Jupiter, Netzsch, Selb, Germany) was applied to record the mass loss as a function of temperature during the pyrolysis of AIW. The sample was placed in a crucible and heated from 40 to 1000 °C at three different heating rates (10, 15, 20 °C/min) with argon flowing at 75 mL/min. To ensure the repeatability of the experiment with an error of 1.5%, the experimental conditions were repeated at least three times. The results showed that the TG and DTG curves were almost identical, which consequently gave very low standard deviations for the calculated kinetic parameters obtained. The results presented here are a set of those experiments that satisfy the above conditions.

2.5. Kinetic Analysis

The mechanism of pyrolysis is characterized by a rather complex set of competitive and parallel reactions, which is also complicated by the variability of the lignocellulosic composition of the biomass [34]. The global pyrolysis reaction is expressed by the following equation [35]:

$$\text{Biomass (solid)} \overset{k(T)}{\rightarrow} \text{Volatiles (condencable + noncondencable)} + \text{Char}, \quad (1)$$

where $k(T)$ is reaction rate constant, which is expressed by the Arrhenius equation:

$$k(T) = Ae^{-E_\alpha/RT}, \quad (2)$$

where E_α is the activation energy (kJ/mol); T is temperature (K); R is the universal gas constant (8.314 J/mol·K); and A is the pre-exponential factor (1/s).

The biomass conversion rate α is defined as the mass fraction of the degraded sample. It can be calculated for each point of TGA according to the equation [36–38]:

$$\alpha = \frac{m_0 - m}{m_0 - m_f}, \quad (3)$$

where m_0 is the initial sample weight of the sample (mg), m is actual weight to each point of analysis (mg); and m_f is the final weight of the sample after pyrolysis (mg).

Generalized fundamental expression for non-isothermal TGA experiments at linear heating rate:

$$\beta = \frac{dT}{dt} = \frac{dT}{d\alpha}\frac{d\alpha}{dt}, \quad (4)$$

Or

$$\frac{d\alpha}{dt} = \frac{A}{\beta} \cdot \exp\left(\frac{-E_\alpha}{RT}\right) f(\alpha), \quad (5)$$

where $f(\alpha)$ is function of conversion.

The analytical form of the function $f(\alpha)$ depends on the thermal decomposition mechanism. Integration of Equation (5) makes it possible to analyze the kinetic data obtained by the TGA method. Integration can be performed using model-free (isoconversion) methods [38,39], which estimate the activation energy ($E\alpha$) when changing the degree of conversion α. These methods are also called "multi-curve" since they require the use of several kinetic curves for analysis [40].

2.6. Model-Free Methods

In this work, the modeless methods of Friedman, OFW, and KAS were used to analyze the kinetic parameters [6,20].

The Friedman's method is a differential method which is expressed by the equation:

$$\ln\left[\beta_i\left(\frac{d\alpha}{dT}\right)_{\alpha,i}\right] = \ln[A_\alpha f(\alpha)] - \frac{E_\alpha}{RT}, \tag{6}$$

where the subscript i is given heating rate value, and subscript α is given conversion degree.

The OFW method is an integral method which is expressed by the equation:

$$\ln(\beta_i) = \ln\left(\frac{A_\alpha E_\alpha}{R_g(\alpha)}\right) - 5.331 - 1.052\frac{E_\alpha}{RT_{\alpha i}}, \tag{7}$$

The KAS method used for kinetic determination is given in equation:

$$\ln\left(\frac{\beta_i}{T_{\alpha i}^2}\right) = \ln\left(\frac{A_\alpha R}{E_\alpha\, g(\alpha)}\right) - \frac{E_\alpha}{RT_{\alpha i}}, \tag{8}$$

where $g(\alpha)$ is constant with given conversion value.

2.7. Reaction Model Determination for AIW Pyrolysis

The master-plot method is used to predict solid state mechanisms in the thermal decomposition of biomass. The master graph is built either in a differential or in a differential-integral form [37]. Various models are fitted to solid-phase kinetic data based on such reaction mechanisms as nucleation, geometric shape, diffusion, and reaction order [19,40]. The theoretical master plots do not depend on the heating rate, but strictly depend on the kinetic model used to model the reaction [41]. To construct a differential graph, a comparison is used at the control point α = 0.5 [42].

$$\frac{\frac{d\alpha}{d\theta}}{\left(\frac{d\alpha}{d\theta}\right)_{0.5}} = \frac{f(\alpha)}{f(0.5)}, \tag{9}$$

where $\frac{f(\alpha)}{f(0.5)}$– theoretically determined from the function, the expressions for which are given in [43]; θ denotes the reaction time taken to attain a particular α at infinite temperature. The left side of expression (9) is the experimental curve calculated using the following equation:

$$\frac{\frac{d\alpha}{d\theta}}{\left(\frac{d\alpha}{d\theta}\right)_{0.5}} = \frac{\frac{d\alpha}{dt}}{\left(\frac{d\alpha}{dt}\right)_{0.5}} \cdot \frac{\exp\left(\frac{E}{RT}\right)}{\exp\left(\frac{E}{RT_{0.5}}\right)}, \tag{10}$$

where $T_{0.5}$ is the reaction temperature at α = 0.5.

2.8. Thermodynamic Parameters

Estimating thermodynamic parameters is a useful tool for understanding biomass pyrolysis, determining the feasibility of a thermal decomposition process, and calculating energy performance. Enthalpy change ΔH (kJ/mol), Gibbs free energy ΔG (kJ/mol), and entropy change ΔS (J/mol·K) were calculated according to the equations derived from the activation complex theory (Eyring Theory) using the following formulas [44–46]:

$$\Delta H = E_\alpha - R \cdot T_{peak}, \tag{11}$$

$$\Delta G = E_\alpha + \left(R \cdot T_{peak} \cdot \ln\frac{K_B \cdot T_{peak}}{h \cdot A}\right), \tag{12}$$

$$\Delta S = \frac{\Delta H - \Delta G}{T_{peak}}, \quad (13)$$

where T_{peak} is the temperature corresponding to the maximum mass loss rate, °C; K_B is Boltzmann constant ($1.38 \cdot 10^{-23}$ J/K); h is Planck's constant ($6.626 \cdot 10^{-34}$ J·s).

3. Results and Discussion

3.1. Results of Proximate and Ultimate Analyses

To assess the possibility of using AIW as a bioenergy raw material, the main physical and chemical characteristics were considered (Table 1).

Table 1. The results of the proximate and ultimate analyses of AIW sample.

Analysis	Values
Proximate (wt.%)—based on air-dried basis:	
Moisture	7.42 ± 0.02
Volatile matter	74.65 ± 0.30
Ash	8.76 ± 0.01
Fixed carbon	9.17 ± 0.06
HHV, MJ/kg	17.87
Ultimate (wt.%)—based on dry basis:	
Carbon	41.83 ± 0.26
Hydrogen	6.81 ± 0.08
Nitrogen	4.71 ± 0.13
Oxygen	37.89 ± 0.17

Humidity and ash content in the AIW sample corresponds to the range of values typical for commercial biomass fuels (humidity up to 25.6 wt.%, ash content up to 9.8 wt.%) [47]. The test sample has a high content of volatile substances; therefore, it is suitable for various thermochemical processes due to its high flammability. The obtained value of volatile substances is comparable with the values obtained for other agricultural wastes suitable for energy use [47,48]. In addition, this means that the AIW sample is more reactive than traditional energy sources such as coal. The HHV of the sample corresponds to the commercial fuel olive stone (17.88 MJ/kg), energy crops–thistle (17.75 MJ/kg) [47], as well as such biomass as: apple tree branches (17.82 MJ/kg), feijoa leaves (17.81 MJ/kg), hazelnut tree leaves (17.87 MJ/kg), kiwi branches (17.81 MJ/kg), and olive stone (17.88 MJ/kg) [49].

3.2. Pyrolysis Products Yields and Their Quality

The pyrolysis products of AIW are shown in Figure 1. The presented values are consistent with the data obtained from the pyrolysis of rice husks [34], switchgrass [50], algal waste [51], and poultry litter [52]. The maximum mass fraction of 37.1 wt.% is characteristic of the pyrolysis liquid. In connection with the subsequent use of pyrolysis liquid for energy purposes, it was separated into oil and aqueous fractions. It is important to use a homogeneous fuel to ensure timely ignition, as well as efficient atomization in the combustion zone and maintaining flame stability in combustion devices [53].

The aqueous fraction of the pyrolysis liquid contains 85.72% water, 10.4% acetic acid, and 3.88% unidentified components. Water is the main component in the liquid, which is explained by the humidity of the AIW sample, dehydration reactions at temperatures below 550 °C, and the occurrence of secondary cracking reactions of oxygen-containing macromolecular compounds at high temperatures [54]. The oil fraction has a diverse and rich composition. Approximately 70.85% of the relative content of the total peak area was identified (GC-MS analysis). The identified compounds were classified into the following main chemical categories: hydrocarbons, phenols, alcohols, ketones, ethers, and N-containing heterocycles. Components with a peak area ≥ 1% are presented in Table 2. Saturated hydrocarbons tetratetracontane and tetracontane are present in large quantities.

It is known that the oil fraction from red amaranth seeds is a rich source of squalene, so the content of 2,6,10,14,18,22-Tetracosahexaene, 2,6,10,15,19,23-hexamethyl-, (all-E) equals 5.44% [55]. All identified compounds (including peak area ≤ 1%) are grouped and shown in Figure 2. The oil fraction contains 41.8% hydrocarbons, which characterizes it as a high-quality fuel.

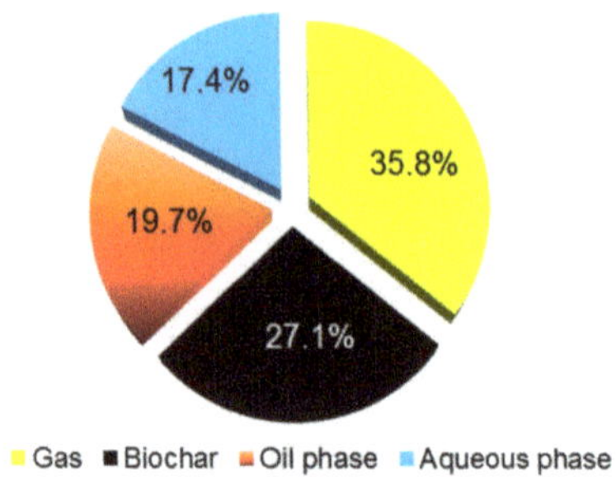

Figure 1. AIW pyrolysis products.

Table 2. The main components of the oil fraction (peak area ≥ 1%).

N°	Area, %	Name	Formula	Mw, g/mol
1	12.36	Tetratetracontane	$C_{44}H_{90}$	619.8
2	9.70	Tetracontane	$C_{40}H_{82}$	563.1
3	8.20	1-Octacosanol	$C_{28}H_{58}O$	410.8
4	5.44	2,6,10,14,18,22-Tetracosahexaene, 2,6,10,15,19,23-hexamethyl-, (all-E)-	$C_{30}H_{50}$	410.7
5	2.19	Octacosanoic acid, methyl ester	$C_{29}H_{58}O_2$	438.8
6	2.00	Phenol	C_6H_5OH	94.11
7	1.80	Pentadecane	$C_{15}H_{32}$	212.41
8	1.50	Triacontanoic acid, methyl ester	$C_{31}H_{62}O_2$	466.82
9	1.43	Tetracosane	$C_{24}H_{50}$	338.7
10	1.40	Phenol, 2-methoxy-	$C_7H_8O_2$	124.12
11	1.24	Pyridine, 3-methyl-	C_6H_7N	93.13
12	1.11	Octadecane	$C_{18}H_{38}$	254.49

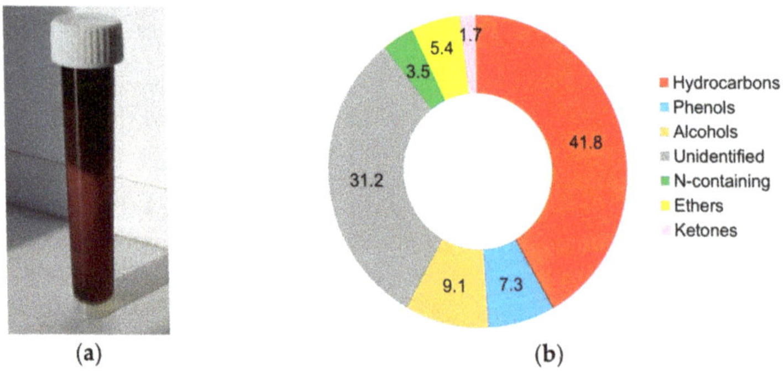

Figure 2. Pyrolysis Liquid: (**a**) photograph and (**b**) composition of the oil fraction.

It should also be noted that there are no organic acids in the oil fraction; they are present only in the composition of esters. Accordingly, the pH value is high and the liquid is characterized by an alkaline reaction, which is also important for the design of power plants.

The concentrations of the detected pyro-gas components, converted to nitrogen-free composition, are shown in Table 3. It was found that the predominant components in

AIW pyrolysis are CO_2 and CO. The total concentration of these gases reaches 94.44%. The combustible part of the pyrolysis gas includes 52.7% of the components, which is consistent with the data of other authors [52,56,57].

Table 3. Pyro—gas composition.

Component	CO_2	CO	CH_4	C_2H_6	C_xH_y	C_2H_4	H_2
Concentration, %	47.30	47.14	3.51	1.17	0.75	0.12	0.01

The main physicochemical characteristics of AIW biochar (Table 4) correspond to biochars obtained by pyrolysis (process temperature 500 °C) of different biomass [58]. The elemental composition of AIW biochar is typical, in which the content of carbon is in the range of 50–87.2%, hydrogen 0.7–4.5%, nitrogen 0.08–6.94%, and oxygen 6–30% [59]. Figure 3 shows the microelement composition of the ash of the solid carbonaceous residue. The predominant components of the ash were K and Ca, and their total content was 81.8% of the total mass.

Table 4. The results of the proximate and ultimate analyses for biomass biochars.

Analysis	Biomass				
	AIW	Maize Stalk [58]	Lantana Camara [58]	Pine Needles [58]	Black Gram [58]
Proximate (wt.%)					
Volatile matter	21.34 ± 0.03	20.67	22.56	27.62	23.56
Ash	20.49 ± 0.01	19.7	15.7	13.5	23.3
Moisture	4.4 ± 0.19	11.5	6.13	8.05	12.41
Fixed carbon	53.77 ± 0.0,9	48.13	55.61	50.83	40.73
HHV *, MJ/kg	20.92	23.7	25.87	22.33	21.06
Ultimate (wt.%)					
carbon	56.56 ± 0.17	61.9	70.5	65.8	56.7
hydrogen	3.09 ± 0.05	3.56	2.69	2.13	3.14
nitrogen	4.12 ± 0.01	1.17	0.86	0.78	1.24
oxygen	15.75 ± 0.14	13.67	10.25	17.79	15.62

* calculated.

Figure 3. Biochar: (**a**) photography; (**b**) elemental composition of mineral part.

Thus, the studied biochar can serve as a direct source of potassium, which is the most important element—a biophile, the removal of which with the harvest of agricultural crops is always greater than that of phosphorus and nitrogen. An analysis of the literature showed that elevated values of K, Mg, and Ca in the solid pyrolysis product make it possible to use it for liming and neutralizing acidic soils [60,61].

3.3. Thermal Degradation Analysis

The results of pyrolysis of AIW samples at heating rates of 10, 15, and 20 °C/min in an argon atmosphere are shown in Figure 4. The TG curves are the change in weight loss with temperature, and the DTG curves are the rate of weight loss with temperature. According to the shape of the curves, it can be judged that the thermal degradation of the studied AIW sample occurs similarly to the general trend of biomass pyrolysis.

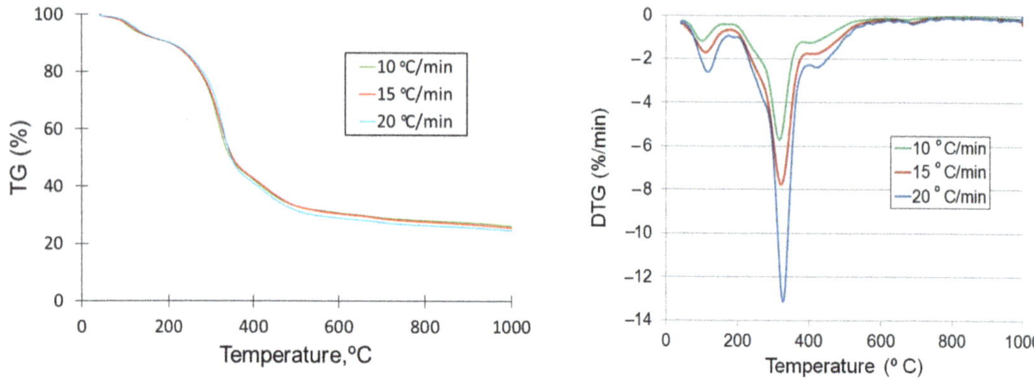

Figure 4. (**a**) TG—curves; (**b**) DTG—curves.

Based on the analysis of the obtained TG data, the AIW pyrolysis process can be divided into 3 main stages (Table 5).

Table 5. Main stages of thermal decomposition.

N°	Pyrolysis Stage	Heating Rate (°C/min)	Starting Temperature (°C)	Ending Temperature (°C)	Temperature Peak (°C)
I	Moisture evaporation	10	40	191.26	103.1
		15	40	190.76	115.4
		20	40	191.77	126.7
II	Devolatilization	10	191.26	529.5	317.7
		15	190.76	544.48	322.6
		20	191.77	558.95	328.5
III	Degradation of char and minerals	10	529.5	1000	668.2
		15	544.48	1000	685.6
		20	558.95	1000	690.6

The first stage in the temperature range from 40 °C to 190 °C corresponded to the process of evaporation of physically bound moisture from samples of AIW. It is also possible to release light volatile components at this stage [46]. The average weight loss at this stage was 9.25 wt.% for three heating rates (Table 6). The first stage has a small peak characterized by an endothermic reaction, which is associated with the absorption of heat in the process of moisture evaporation [44].

The main stage, corresponding to the main pyrolysis, occurred in the temperature range from 190 °C to the temperature range of 530–560 °C for three heating rates and was accompanied by the main loss of organic matter mass. During this stage, there was an active decomposition of the biomass components and the release of volatile substances associated with the thermal destruction of hemicellulose, cellulose, and lignin [62,63]. The average weight loss during the release of volatiles was 59.63 wt.%. As can be seen from Figure 4, rapid weight loss begins above a temperature of 190 °C, which is associated with the rapid breakdown of thermally unstable components of hemicellulose and extractives [37,64].

Hemicellulose consists of short chain heteropolysaccharides and is an amorphous and branched structure [8,19,33,37,39,41,65,66]. Furthermore, with an increase in the pyrolysis temperature, cellulose is involved in the degradation process, which is characterized by a higher decomposition temperature (315–400 °C) due to the presence of a long polymer of glucose units and a large number of hydrogen bonds in its composition [67]. Cellulose, due to its chemical structure, is more resistant to thermal degradation; its decomposition is typical for the temperature range of 270–350 °C [68,69].

Table 6. Mass loss characteristics of AIW obtained from TGA analysis.

Heating Rate (°C/min)	Mass Loss, wt.%			Residual Mass, wt.%
	Moisture Evaporation	Devolatilization	Degradation of Char and Minerals	
10	9.14	58.65	5.79	26.24
15	9.33	59.17	5.89	25.61
20	9.28	61.08	4.88	24.76
Average, %	9.25	59.63	5.52	25.54

On the DTG-curves (Figure 4) at the stage of devolatilization, one can note the maximum temperature peak, which has values of 317.7, 322.6, and 328.5 °C for the three heating rates. This peak is characterized by an endothermic reaction. In addition, a small temperature exothermic peak is found at 402.7–422.1 °C, which can be associated with the beginning of the decomposition of lignin in the test sample. The literature data indicate that the onset of lignin decomposition for various types of biomass occurs in the temperature range of 280–550 °C [16]. The mechanism of lignin pyrolysis is more complex than that of cellulose and hemicellulose; it includes reactions of free radicals [70,71]. Due to the fact that lignin has the highest thermal stability, it decomposes slowly throughout the thermal degradation up to a temperature of 900 °C [16].

The third stage, which is typical for the temperature range of 529.5 °C and up to 1000 °C for three heating rates, is associated with the process of degradation of char and minerals. At this stage, after the completion of the release of volatiles and the main thermal destruction, the process of enrichment with carbon and the formation of the structure of carbonaceous matter continue. Although small, inorganic minerals in biomass can have a significant effect on the pyrolysis process. In this regard, the process of thermal degradation of mineral components is primarily associated with the decomposition of $CaCO_3$ in the temperature range from 780 to 1000 °C [72]. In addition, pyrolysis products can interact with inorganic elements in the residual carbonaceous matter [73]. In this case, the mineral components act as catalysts in the reactions of gas formation from pyrolysis products [72]. The residual fraction as a result of AIW pyrolysis was 25.5 wt.% for the three heating rates. As a result of the experiments, it was revealed that the nature of the TG and DTG curves of the studied samples is similar to the biomass of herbaceous plants, which were reported in [16,17,74].

3.4. Kinetic Analysis

In this work, a kinetic analysis was carried out for the main stage of pyrolysis—devolatilization—since at this stage, the maximum mass loss occurs [75]. AIW kinetic parameters were determined using three model-free methods: Friedman, KAS, and OFW, based on TGA data. Figure 5 shows the results of linear regression in the range of conversions from 0.1 to 0.9 of the kinetic analysis of the total thermal decomposition reactions of the AIW samples. Straight line slope data obtained from each model-free method were used to calculate the $E\alpha$ values presented in Table 7.

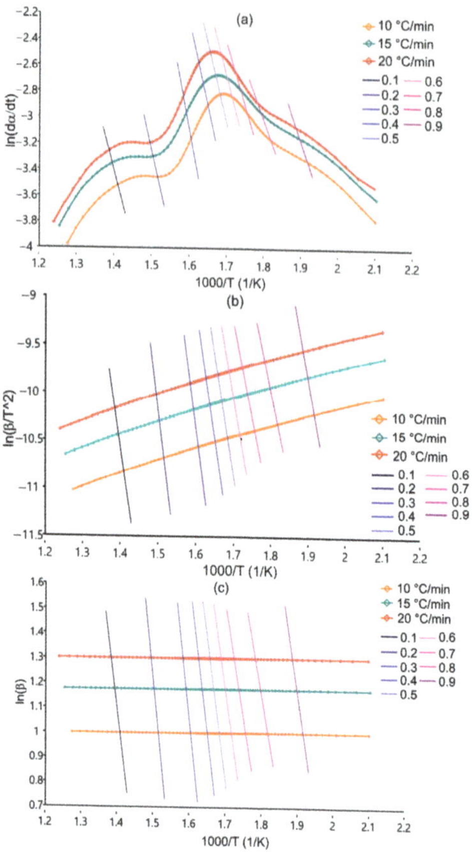

Figure 5. Plots for determination *Eα* of AIW pyrolysis using: (**a**) Friedman; (**b**) KAS; (**c**) OFW.

Table 7. Values of activated energy according to different methods.

α	Friedman		KAS		OFW	
	Eα (kJ/mol)	Log A (1/s)	Eα (kJ/mol)	Log A (1/s)	Eα (kJ/mol)	Log A (1/s)
0.1	164.48	13.22	185.27	15.72	185.42	15.68
0.2	152.52	11.44	156.78	12.26	157.00	12.23
0.3	184.77	14.16	167.31	12.93	167.50	12.90
0.4	212.88	16.47	189.57	14.76	189.74	14.72
0.5	231.03	17.80	209.08	16.31	209.23	16.26
0.6	250.60	19.10	225.76	17.52	225.91	17.48
0.7	291.94	21.85	253.73	19.48	253.88	19.44
0.8	244.29	16.61	265.25	19.30	265.43	19.26
0.9	213.02	13.17	223.21	14.75	223.43	14.70
Average	216.17	15.98	208.44	15.89	208.61	15.85

The dynamics of *Eα* values obtained by the Friedman, KAS, and OFW methods highlight the complexity of the AIW sample kinetics. *Eα* gradually increases until reaching its maximum at a conversion rate of 0.8 for the OFW and KAS methods, and *α* conversion rate of 0.7 for the Friedman method. A similar trend in *Eα* values was found during pyrolysis of such biomass as bark of *Ficus natalensis* [7], water hyacinth [76], elephant grass [77], and mustard stalk [78].

The pre-exponential factor A characterizes the frequency of collisions of reacting molecules. This indicator makes it possible to explain the chemistry of reactions, which is important for optimizing the pyrolysis process [36]. Almost all obtained A values are in the range from 10^4 to 10^9, which indicates a low reactivity of the test sample and the occurrence of a surface reaction, as well as a tight junctional complex (closed complex) [36,39].

The $E\alpha$ values are in the range of 152.52–291.94 kJ/mol (Friedman), 156.78–265.25 kJ/mol (KAS), and 157.00–265.43 kJ/mol (OFW). The value of $E\alpha$ shows a measure of the minimum energy required to start a chemical reaction, as well as a potential measure of reactivity [79,80]. According to the literature data, the KAS and OFW methods are less accurate than the Friedman method [39,79], since it does not contain assumptions and approximations [39,81]. It should be noted that the $E\alpha$ values calculated by the KAS, OFW, and Friedman methods for the AIW sample agree with each other. The average value of $E\alpha$ obtained by the Friedman method is only 3.7% higher than that calculated by the OFW and KAS methods. Comparative analysis of $E\alpha$ values for different types of biomass is presented in Table 8.

Table 8. Comparison of biomasses activation energy.

Fuel	Heating Rate (K/min)	Used Methods	Activation Energy (kJ/mol)	Reference
AIW	10, 15, and 20	Friedman, KAS, OFW	216.17 208.44 208.61	Present Study
Cotton stalk	10–40	KAS, OFW	223–230 213–240	[82]
Sugarcane leaves	5–40	Friedman, KAS, OFW	239.58 226.75 226.97	[21]
Prosopis juliflora fuelwood	2–25	Friedman, KAS, OFW	219.3 204.0 203.2	[22]
Phyllanthus emblica seeds	10–50	Friedman, KAS, OFW	189.95 184.77 195.10	[23]
Camphor branch	2.5, 5, and 10	Ozawa	190	[83]
Microalgae *Chlorella vulgaris*	10–40	Kissinger, Friedman, OFW, KAS, Vyazovkin, DAEM	135.6–337.1	[24]
Digested biomass wastes	10, 15, and 20	Friedman, KAS	202.55 202.21	[75]
Sorghum bicolor	2, 5, and 8	Friedman and KAS	226.6	[84]
Pea waste	10–40	KAS, OFW	212.71 211.55	[85]
Basswood waste	20–40	KAS, OFW	197.2 207.9	[86]

Figure 6 compares the theoretical differential plots of $f(\alpha))/f(0.5)$ versus α with the experimental plot of $(d\alpha/d\theta)/(d\alpha/d\theta)$ 0.5 versus α for a heating rate of 10 °C/min to draw a conclusion about the reaction mechanism of solid phase pyrolysis.

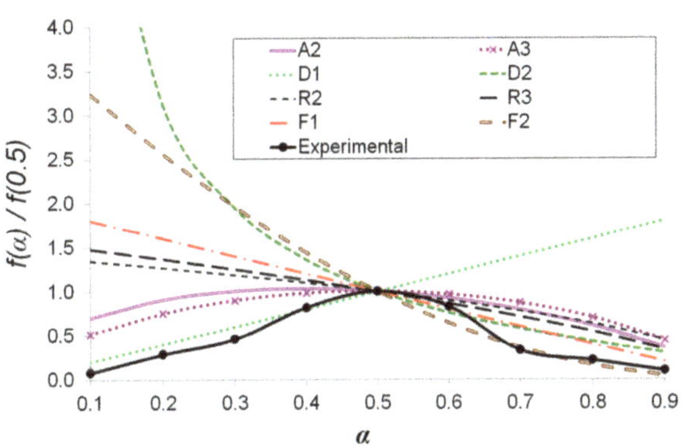

Figure 6. Comparison of experimental and theoretical master plots for samples AIW.

In the range of α from 0.1 to 0.5, the AIW degradation mechanism refers to the one-dimensional diffusion (D1) process, i.e., heat transfer in the sample occurs by diffusion. When α values are greater than 0.5, the AIW degradation mechanism tends to random nucleation with one nucleus in a single particle (F1). In the range of α from 0.7 to 0.9, the mechanism is reduced to random nucleation with two nuclei in the individual particle (F2). The F1 and F2 degradation mechanisms are initiated from random points that act as growth centers for the development of the degradation reaction [87]. Similar results were obtained for other types of biomass [88]. A slight discrepancy between the experimental curves of the master plot for the studied AIW samples can be explained by the deviation of the ideal conditions adopted in the kinetic models from the actual pyrolysis conditions.

3.5. Thermodynamic Analysis

To design, optimize, and scale the parameters of the pyrolysis reactor, it is also necessary to know the thermodynamic properties of the feedstock used. Thermodynamic parameters were determined for a heating rate of 10 °C/min (Figure 7). In biomass pyrolysis, ΔH is the total energy required for biomass decomposition into solid, liquid, and gaseous products [45,89,90]. The ΔH values for the studied AIW sample were in the range of 152.28–287.03 kJ/mol according to the Friedman method, 151.86–260.34 kJ/mol according to the KAS method, and 152.05–260.52 kJ/mol according to the FWO method. Positive values of ΔH indicate the endothermic nature of biomass pyrolysis, which implies the need for energy from an external heat source [6]. The difference between the average values of $E\alpha$ and ΔH is insignificant, approximately 5 kJ/mol (for all methods), which indicates that the studied AIW sample is suitable for pyrolysis [43,87,88,90–94].

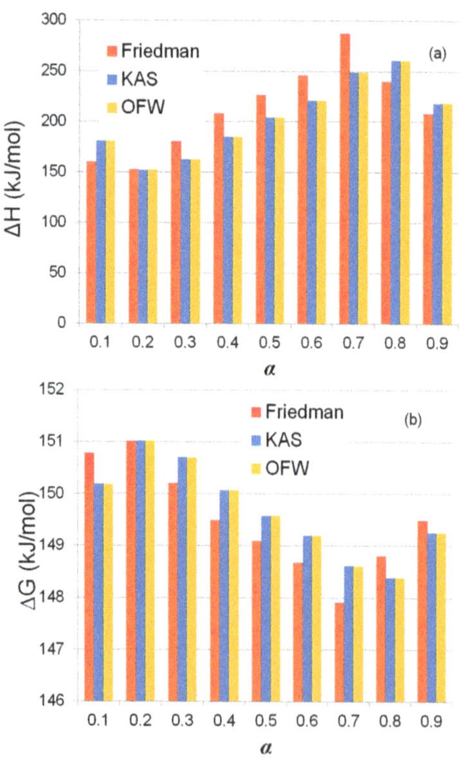

Figure 7. *Cont.*

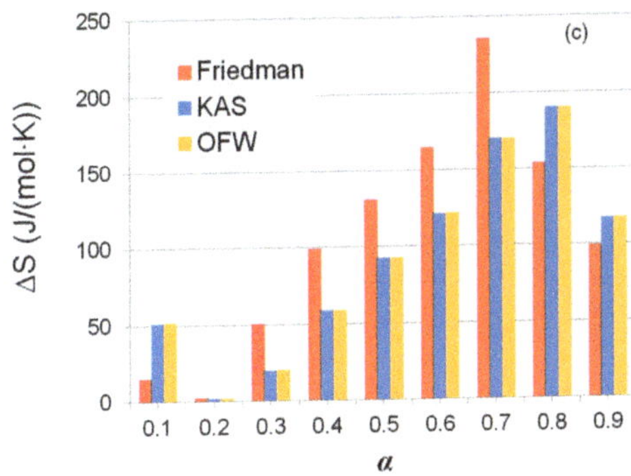

Figure 7. (**a**) ΔH, (**b**) ΔG, and (**c**) ΔS of the pyrolysis of AIW.

The change in ΔG makes it possible to judge the energy available in biomass [6,49,95,96]. The ΔG values for the studied AIW sample ranged from 147.90 to 150.78 kJ/mol by the Friedman method, 148.38–150.19 kJ/mol by the KAS method, and 148.38–150.18 kJ/mol by the OFW method. For most known biomasses and their mixtures, the ΔG values are positive [87,90,95–98]. The resulting average value of ΔG is 149.61 kJ/mol of the AIW sample. It is comparable to the ΔG values for pseudo-hemicelluloses of cocoa shell (143.19 kJ/mol) [99], torrefied biomass of *Acacia nilotica* T-250 (159.97 kJ/mol) [100], and red macroalgae *Gelidium floridanum* for stage 1 (147.25 kJ/mol) [44], but higher than mustard stalk (128 kJ/mol) [78]. The data obtained indicate the high energy potential of AIW.

Entropy is a function of the state of a thermodynamic system, which characterizes the direction of spontaneous processes and is a measure of their irreversibility. The change in ΔS serves as a measure of the change in the order of a thermodynamic system. The entropy of the system is the higher the greater the degree of disorder of the system. Thus, if the process goes in the direction of increasing the disorder of the system, then ΔS is a positive value. To increase the degree of order in the system, it is necessary to expend energy [90,93]. The ΔS values of amaranth samples range from 2.15 to 235.45 J/mol·K by the Friedman method and 1.42-189.78 J/mol·K by the KAS and OFW methods. Throughout the conversion process, the ΔS values were positive for the three model-free methods, indicating a high reactivity of the biomass and a rapid formation of the activated complex. It should be noted that the degree of disorder in the resulting products was quite high, and this is typical of the pyrolysis process [48,100]. The mean ΔS of the AIW sample was 105.41 J/mol·K by the Friedman method, 91.15 J/mol·K by the KAS method, and 91.46 J/mol·K by the OFW method. The ΔS value is comparable with the values obtained for mixtures of sugarcane bagasse, water hyacinth *Eichhornia crassipes* and yellow oleander *Thevetia Peruviana* [101].

4. Conclusions

In this work, a study was made of the pyrolysis of a new type of plant waste using TGA and experiments in a laboratory installation for thermochemical processing. The physicochemical parameters of the studied raw materials correspond to the range of values typical for commercial biomass fuels. The test sample has a high content of volatile substances and high reactivity. The maximum specific gravity in the pyrolysis products of 37.1% corresponds to the pyrolysis liquid. The maximum mass fraction in the pyrolysis products of 37.1 wt.% corresponds to the pyrolysis liquid. At the same time, the oil fraction contains 41.8% hydrocarbons, which characterizes it as a high-quality fuel. Analysis of the features of thermal decomposition of waste was determined at heating rates of 10, 15, and 20 °C/min in an inert atmosphere. The main stage of thermochemical degradation is devolatilization. The kinetic parameters for this stage were determined using the model-free methods of Friedman, OFW, and KAS. The one-dimensional diffusion model (D1), then random nucleation with two nuclei in the individual particle (F1), and random nucleation with two nuclei on the individual particle (F2) were recommended to describe the mechanism of AIW thermal destruction. The average activation energy values are in the range of 208.44–216.17 kJ/mol, and they were used to calculate the thermodynamic parameters. The results indicate that the pyrolysis application will allow the efficient conversion of AIW into value-added products.

Author Contributions: Conceptualization, J.K.; methodology, J.K.; software, S.T. and V.P.; validation, V.B. and V.P.; formal analysis, S.I. and F.A.; investigation, S.T., S.I., K.B. and F.A.; resources, J.K., V.B. and V.P.; data curation, J.K., S.T., V.B. and V.P.; writing—original draft preparation, S.T., S.I., K.B. and F.A.; writing—review and editing, J.K. and V.B.; visualization, S.T.; supervision, V.B. and V.P.; project administration, J.K.; funding acquisition, V.B. and V.P;. All authors have read and agreed to the published version of the manuscript.

Funding: This research received no external funding.

Institutional Review Board Statement: Not applicable.

Data Availability Statement: Not applicable.

Conflicts of Interest: The authors declare no conflict of interest.

References

1. Emmanuel, O.C.; Babalola, O.O. Amaranth production and consumption in South Africa: The challenges of sustainability for food and nutrition security. *Int. J. Agric. Sustain.* **2022**, *20*, 449–460. [CrossRef]
2. Sulaiman, M.I.; Andini, R. Potential of Amaranth in Alleviating Malnutrition in Indonesia. In *Nutritional Value of Amaranth*; Waisundara, Y., Ed.; IntechOpen: London, UK, 2020.
3. Schmidt, D.; Verruma-Bernardi, M.R.; Forti, V.A.; Borges, M.T.M.R. Quinoa and Amaranth as Functional Foods: A Review. *Food Rev. Int.* **2021**, *37*, 1–20. [CrossRef]
4. Adhikary, D.; Khatri-Chhetri, U.; Slaski, J. Amaranth: An Ancient and High-Quality Wholesome Crop. In *Nutritional Value of Amaranth*; Waisundara, Y., Ed.; IntechOpen: London, UK, 2020.
5. Aderibigbe, O.R.; Ezekiel, O.O.; Owolade, S.O.; Korese, J.K.; Sturm, B.; Hensel, O. Exploring the potentials of underutilized grain amaranth (*Amaranthus* spp.) along the value chain for food and nutrition security: A review. *Crit. Rev. Food Sci. Nutr.* **2022**, *62*, 656–669. [CrossRef] [PubMed]
6. Wen, Y.; Shi, Z.; Wang, S.; Mu, W.; Jönsson, P.G.; Yang, W. Pyrolysis of raw and anaerobically digested organic fractions of municipal solid waste: Kinetics, thermodynamics, and product characterization. *Chem. Eng. J.* **2021**, *415*, 129064. [CrossRef]
7. Farooq, A.; Ashraf, M.; Aslam, Z.; Anwar, A.; Jiang, S.; Farooq, A.; Liu, L. Pyrolytic conversion of a novel biomass Ficus na-talensis barkcloth: Physiochemical and thermo-kinetic analysis. *Biomass Conv. Bioref.* **2021**, *11*. [CrossRef]
8. Karaeva, J.V.; Timofeeva, S.S.; Islamova, S.I.; Gerasimov, A.V. Pyrolysis kinetics of new bioenergy feedstock from anaerobic digestate of agro-waste by thermogravimetric analysis. *J. Environ. Chem. Eng.* **2022**, *10*, 107850. [CrossRef]
9. Figueirêdo, M.B.; Hita, I.; Deuss, P.J.; Venderbosch, R.H.; Heeres, H.J. Pyrolytic lignin: A promising biorefinery feedstock for the production of fuels and valuable chemicals. *Green Chem.* **2022**, *24*, 4680–4702. [CrossRef]
10. Yang, Q.; Zhou, H.; Bartocci, P.; Fantozzi, F.; Mašek, O.; Agblevor, F.A.; Wei, Z.; Yang, H.; Chen, H.; Lu, X.; et al. Prospective contributions of biomass pyrolysis to China's 2050 carbon reduction and renewable energy goals. *Nat. Commun.* **2021**, *12*, 1698.
11. Mlonka-Mędrala, A.; Evangelopoulos, P.; Sieradzka, M.; Zajemska, M.; Magdziarz, A. Pyrolysis of agricultural waste biomass towards production of gas fuel and high-quality char: Experimental and numerical investigations. *Fuel* **2021**, *296*, 120611. [CrossRef]
12. Vuppaladadiyam, A.K.; Vuppaladadiyam, S.S.V.; Sahoo, A.; Murugavelh, S.; Anthony, E.; Bhaskar, T.; Zheng, Y.; Zhao, M.; Duan, H.; Zhao, Y.; et al. Bio-oil and biochar from the pyrolytic conversion of biomass: A current and future perspective on the trade-off between economic, environmental, and technical indicators. *Sci. Total. Environ.* **2023**, *857*, 159155. [CrossRef]
13. Lobzenko, I.; Burachevskaya, M.; Zamulina, I.; Barakhov, A.; Bauer, T.; Mandzhieva, S.; Sushkova, S.; Minkina, T.; Tereschenko, A.; Kalinichenko, V.; et al. Development of a Unique Technology for the Pyrolysis of Rice Husk Biochar for Promising Heavy Metal Remediation. *Agriculture* **2022**, *12*, 1689. [CrossRef]
14. Ansari, K.B.; Gaikar, V.G. Investigating production of hydrocarbon rich bio-oil from grassy biomass using vacuum pyrolysis coupled with online deoxygenation of volatile products over metallic iron. *Renew. Energy* **2019**, *130*, 305–318. [CrossRef]
15. Giorcelli, M.; Das, O.; Sas, G.; Försth, M.; Bartoli, M. A Review of Bio-Oil Production through Microwave-Assisted Pyrolysis. *Processes* **2021**, *9*, 561. [CrossRef]
16. Chen, H.; Liu, Z.; Chen, X.; Chen, Y.; Dong, Z.; Wang, X.; Yang, H. Comparative pyrolysis behaviors of stalk, wood and shell biomass: Correlation of cellulose crystallinity and reaction kinetics. *Bioresour. Technol.* **2020**, *310*, 123498. [CrossRef] [PubMed]
17. Nour, M.; Amer, M.; Elwardany, A.; Attia, A.; Li, X.; Nada, S. Pyrolysis, kinetics, and structural analyses of agricultural residues in Egypt: For future assessment of their energy potential. *Clean. Eng. Technol.* **2021**, *2*, 100080. [CrossRef]

18. Pu, X.; Wei, M.; Chen, X.; Wang, L.; Deng, L. Thermal Decomposition Characteristics and Kinetic Analysis of Chicken Manure in Various Atmospheres. *Agriculture* **2022**, *12*, 607. [CrossRef]
19. Tian, B.; Wang, X.; Zhao, W.; Xu, L.; Bai, L. Pyrolysis behaviors, kinetics and gaseous product evolutions of two typical bi-omass wastes. *Catal. Today* **2021**, *374*, 77–85. [CrossRef]
20. Ahmad, M.S.; Mehmood, M.A.; Taqvi, S.T.H.; Elkamel, A.; Liu, C.-G.; Xu, J.; Rahimuddin, S.A.; Gull, M. Pyrolysis, kinetics analysis, thermodynamics parameters and reaction mechanism of *Typha latifolia* to evaluate its bioenergy potential. *Bioresour. Technol.* **2017**, *245*, 491–501. [CrossRef]
21. Kumar, M.; Sabbarwal, S.; Mishra, P.; Upadhyay, S. Thermal degradation kinetics of sugarcane leaves (*Saccharum officinarum* L) using thermo-gravimetric and differential scanning calorimetric studies. *Bioresour. Technol.* **2019**, *279*, 262–270. [CrossRef]
22. Chandrasekaran, A.; Ramachandran, S.; Subbiah, S. Determination of kinetic parameters in the pyrolysis operation and thermal behavior of *Prosopis juliflora* using thermogravimetric analysis. *Bioresour. Technol.* **2017**, *233*, 413–422. [CrossRef]
23. Mishra, R.K.; Mohanty, K. Kinetic analysis and pyrolysis behaviour of waste biomass towards its bioenergy potential. *Bioresour. Technol.* **2020**, *311*, 123480. [CrossRef] [PubMed]
24. Soria-Verdugo, A.; Goos, E.; García-Hernando, N.; Riedel, U. Analyzing the pyrolysis kinetics of several microalgae species by various differential and integral isoconversional kinetic methods and the Distributed Activation Energy Model. *Algal Res.* **2018**, *32*, 11–29. [CrossRef]
25. Ermolaev, D.V.; Timofeeva, S.S.; Islamova, S.I.; Bulygina, K.S.; Gilfanov, M.F. A comprehensive study of thermotechnical and thermogravimetric properties of peat for power generation. *Biomass Conv. Bioref.* **2019**, *9*, 767–774. [CrossRef]
26. Karaeva, J.V.; Timofeeva, S.S.; Bashkirov, V.N.; Bulygina, K.S. Thermochemical processing of digestate from biogas plant for recycling dairy manure and biomass. *Biomass Conv. Bioref.* **2021**, *13*, 685–695. [CrossRef]
27. Wang, G.; Dai, Y.; Yang, H.; Xiong, Q.; Wang, K.; Zhou, J.; Li, Y.; Wang, S. A Review of Recent Advances in Biomass Pyrolysis. *Energy Fuels* **2020**, *34*, 15557–15578. [CrossRef]
28. Lewandowski, W.M.; Ryms, M.; Kosakowski, W. Thermal Biomass Conversion: A Review. *Processes* **2020**, *8*, 516. [CrossRef]
29. Karaeva, J.V.; Timofeeva, S.S.; Kovalev, A.A.; Kovalev, D.A.; Gilfanov, M.F.; Grigoriev, V.S.; Litti, Y.V. Co-pyrolysis of agri-cultural waste and estimation of the applicability of pyrolysis in the integrated technology of biorenewable hydrogen produc-tion. *Int. J. Hydrogen Energy* **2022**, *47*, 11787–11798. [CrossRef]
30. Sun, Y.; Liu, K.; Kun, L.; Hou, C.; Liu, J.; Huang, R.; Cao, C.; Song, W. Nitrogen, Sulfur Co-doped Carbon Materials Derived from the Leaf, Stem and Root of Amaranth as Metal-free Catalysts for Selective Oxidation of Aromatic Hydrocarbons. *ChemCatChem* **2019**, *11*, 1010–1016. [CrossRef]
31. Gao, S.; Geng, K.; Liu, H.; Wei, X.; Zhang, M.; Wang, P.; Wang, J. Transforming organic-rich amaranthus waste into nitro-gen-doped carbon with superior performance of the oxygen reduction reaction. *Energy Environ. Sci.* **2015**, *8*, 221–229. [CrossRef]
32. Tinwala, F.; Mohanty, P.; Parmar, S.; Patel, A.; Pant, K.K. Intermediate pyrolysis of agro-industrial biomasses in bench-scale pyrolyser: Product yields and its characterization. *Bioresour. Technol.* **2015**, *188*, 258–264. [CrossRef]
33. Biswas, B.; Pandey, N.; Bisht, Y.; Singh, R.; Kumar, J.; Bhaskar, T. Pyrolysis of agricultural biomass residues: Comparative study of corn cob, wheat straw, rice straw and rice husk. *Bioresour. Technol.* **2017**, *237*, 57–63. [CrossRef] [PubMed]
34. Rego, F.; Dias, A.P.S.; Casquilho, M.; Rosa, F.C.; Rodrigues, A. Pyrolysis kinetics of short rotation coppice poplar biomass. *Energy* **2020**, *207*, 118191. [CrossRef]
35. Bhardwaj, G.; Kumar, M.; Mishra, P.K.; Upadhyay, S.N. Kinetic analysis of the slow pyrolysis of paper wastes. *Biomass Conv. Bioref.* **2021**, *11*, 1–14. [CrossRef]
36. Shahbeig, H.; Nosrati, M. Pyrolysis of biological wastes for bioenergy production: Thermo-kinetic studies with machine-learning method and Py-GC/MS analysis. *Fuel* **2020**, *269*, 117238. [CrossRef]
37. Singh, R.K.; Pandey, D.; Patil, T.; Sawarkar, A.N. Pyrolysis of banana leaves biomass: Physico-chemical characterization, thermal decomposition behavior, kinetic and thermodynamic analyses. *Bioresour. Technol.* **2020**, *310*, 123464. [CrossRef]
38. Jiang, G.; Wei, L. Analysis of pyrolysis kinetic model for processing of thermogravimetric analysis data. In *Phase Change Materials and Their Applications*; IntechOpen: London, UK, 2018. [CrossRef]
39. Yuan, X.; He, T.; Cao, H.; Yuan, Q. Cattle manure pyrolysis process: Kinetic and thermodynamic analysis with isoconversional methods. *Renew. Energy* **2017**, *107*, 489–496. [CrossRef]
40. Khawam, A. Application of Solid-State Kinetics to Desolvation Reactions. Ph.D. Thesis, University of Iowa, Iowa, IA, USA, 2007.
41. Sánchez-Jiménez, P.E.; Pérez-Maqueda, L.A.; Perejón, A.; Criado, J.M. Generalized master plots as a straightforward approach for determining the kinetic model: The case of cellulose pyrolysis. *Thermochim. Acta* **2013**, *552*, 54–59. [CrossRef]
42. Gotor, F.J.; Criado, J.M.; Malek, J.; Koga, N. Kinetic Analysis of Solid-State Reactions: The Universality of Master Plots for Analyzing Isothermal and Nonisothermal Experiments. *J. Phys. Chem. A* **2000**, *104*, 10777–10782. [CrossRef]
43. Brown, M.E. *Handbook of Thermal Analysis and Calorimetry*; Principles and Practice; Elsevier Science: Amsterdam, The Netherlands, 1998; Volume 1, p. 722.
44. Alves, J.L.F.; Da Silva, J.C.G.; Filho, V.F.S.; Alves, R.F.; Ahmad, M.S.; Ahmad, M.S.; Galdino, W.V.A.; De Sena, R.F. Bioenergy potential of red macroalgaeGelidium floridanumby pyrolysis: Evaluation of kinetic triplet and thermodynamics parameters. *Bioresour. Technol.* **2019**, *291*, 121892. [CrossRef]
45. Varma, A.K.; Lal, N.; Rathore, A.K.; Katiyar, R.; Thakur, L.S.; Shankar, R.; Mondal, P. Thermal, kinetic and thermodynamic study for co-pyrolysis of pine needles and styrofoam using thermogravimetric analysis. *Energy* **2020**, *218*, 119404. [CrossRef]

46. El-Sayed, S.A.; Mostafa, M.E. Kinetics, thermodynamics, and combustion characteristics of Poinciana pods using TG/DTG/DTA techniques. *Biomass Conv. Bioref.* **2021**, *11*, 1–25. [CrossRef]
47. García, R.; Pizarro, C.; Lavín, A.G.; Bueno, J.L. Characterization of Spanish biomass wastes for energy use. *Bioresour. Technol.* **2012**, *103*, 249–258. [CrossRef] [PubMed]
48. Emiola-Sadiq, T.; Zhang, L.; Dalai, A.K. Thermal and Kinetic Studies on Biomass Degradation via Thermogravimetric Anal-ysis: A Combination of Model-Fitting and Model-Free Approach. *ACS Omega* **2021**, *6*, 22233–22247. [CrossRef] [PubMed]
49. Güleç, F.; Pekaslan, D.; Williams, O.; Lester, E. Predictability of higher heating value of biomass feedstocks via proximate and ultimate analyses—A comprehensive study of artificial neural network applications. *Fuel* **2022**, *320*, 123944. [CrossRef]
50. Imam, T.; Capareda, S. Characterization of bio-oil, syn-gas and bio-char from switchgrass pyrolysis at various temperatures. *J. Anal. Appl. Pyrolysis* **2012**, *93*, 170–177. [CrossRef]
51. Aboulkas, A.; Hammani, H.; El Achaby, M.; Bilal, E.; Barakat, A.; El harfi, K. Valorization of algal waste via pyrolysis in a fixed-bed reactor: Production and characterization of bio-oil and bio-char. *Bioresour. Technol.* **2017**, *243*, 400–408. [CrossRef]
52. Pandey, D.S.; Katsaros, G.; Lindfors, C.; Leahy, J.J.; Tassou, S.A. Fast Pyrolysis of Poultry Litter in a Bubbling Fluidised Bed Reactor: Energy and Nutrient Recovery. *Sustainability* **2019**, *11*, 2533. [CrossRef]
53. Panchasara, H.; Ashwath, N. Effects of Pyrolysis Bio-Oils on Fuel Atomisation—A Review. *Energies* **2021**, *14*, 794. [CrossRef]
54. Alvarez, J.; Lopez, G.; Amutio, M.; Artetxe, M.; Barbarias, I.; Arregi, A.; Bilbao, J.; Olazar, M. Characterization of the bio-oil obtained by fast pyrolysis of sewage sludge in a conical spouted bed reactor. *Fuel Process. Technol.* **2016**, *149*, 169–175. [CrossRef]
55. Nasirpour-Tabrizi, P.; Azadmard-Damirchi, S.; Hesari, J.; Piravi-Vanak, Z. *Amaranth Seed Oil Composition*; Waisundara, V.Y., Ed.; IntechOpen: London, UK, 2020.
56. Griessacher, T.; Antrekowitsch, J.; Steinlechner, S. Charcoal from agricultural residues as alternative reducing agent in metal recycling. *Biomass Bioenergy* **2012**, *39*, 139–146. [CrossRef]
57. Peer, V.; Frantík, J.; Kielar, J.; Mašek, D. Substrates for slow pyrolysis. In Proceedings of the XXI International Scientific Conference—The Application of Experimental and Numerical Methods in Fluid Mechanics and Energy, Teplice, Slovakia, 25–27 April 2018.
58. Das, S.K.; Ghosh, G.K.; Avasthe, R.K.; Sinha, K. Compositional heterogeneity of different biochar: Effect of pyrolysis temper-ature and feedstocks. *J. Environ. Manag.* **2021**, *278*, 2. [CrossRef] [PubMed]
59. Tangmankongworakoon, N. An approach to produce biochar from coffee residue for fuel and soil amendment purpose. *Int. J. Recycl. Org. Waste Agric.* **2019**, *8* (Suppl. S1), 37–44. [CrossRef]
60. Baidoo, I.K.; Sarpong, D.B.; Bolwig, S.; Ninson, D. Biochar amended soils and crop productivity: A critical and meta-analysis of literature. *Int. J. Dev. Sustain.* **2016**, *5*, 414–432.
61. Yuan, J.-H.; Xu, R.-K.; Qian, W.; Wang, R.-H. Comparison of the ameliorating effects on an acidic ultisol between four crop straws and their biochars. *J. Soils Sediments* **2011**, *11*, 741–750. [CrossRef]
62. Anca-Couce, A. Reaction mechanisms and multi-scale modelling of lignocellulosic biomass pyrolysis. *Prog. Energy Combust. Sci.* **2016**, *53*, 41–79. [CrossRef]
63. Pielsticker, S.; Gövert, B.; Umeki, K.; Kneer, R. Flash Pyrolysis Kinetics of Extracted Lignocellulosic Biomass Components. *Front. Energy Res.* **2021**, *9*, 497. [CrossRef]
64. Senneca, O.; Cerciello, F.; Russo, C.; Wütscher, A.; Muhler, M.; Apicella, B. Thermal treatment of lignin, cellulose and hemi-cellulose in nitrogen and carbon dioxide. *Fuel* **2020**, *271*, 117656. [CrossRef]
65. Basu, P. *Biomass Gasification, Pyrolysis and Torrefaction*, 3rd ed.; Academic Press: Cambridge, MA, USA, 2018; p. 564.
66. Shen, D.K.; Gu, S.; Bridgwater, A.V. The thermal performance of the polysaccharides extracted from hardwood: Cellulose and hemicellulose. *Carbohydr. Polym.* **2010**, *82*, 39–45. [CrossRef]
67. Usino, D.O.; Ylitervo, P.; Moreno, A.; Sipponen, M.H.; Richards, T. Primary interactions of biomass components during fast pyrolysis. *J. Anal. Appl. Pyrolysis* **2021**, *159*, 105297. [CrossRef]
68. Yu, J.; Paterson, N.; Blamey, J.; Millan, M. Cellulose, xylan and lignin interactions during pyrolysis of lignocellulosic biomass. *Fuel* **2017**, *191*, 140–149. [CrossRef]
69. Shen, D.; Xiao, R.; Gu, S.; Luo, K. The pyrolytic behavior of cellulose in lignocellulosic biomass: A review. *RSC Adv.* **2011**, *1*, 1641–1660. [CrossRef]
70. Qiao, Y.; Wang, B.; Ji, Y.; Xu, F.; Zong, P.; Zhang, J.; Tian, Y. Thermal decomposition of castor oil, corn starch, soy protein, lignin, xylan, and cellulose during fast pyrolysis. *Bioresour. Technol.* **2019**, *278*, 287–295. [CrossRef] [PubMed]
71. Wang, S.; Lin, H.; Ru, B.; Sun, W.; Wang, Y.; Luo, Z. Comparison of the pyrolysis behavior of pyrolytic lignin and milled wood lignin by using TG–FTIR analysis. *J. Anal. Appl. Pyrolysis* **2014**, *108*, 78–85. [CrossRef]
72. Peng, C.; Zhang, G.; Yue, J.; Xu, G. Pyrolysis of black liquor for phenols and impact of its inherent alkali. *Fuel Process. Technol.* **2014**, *127*, 149–156. [CrossRef]
73. Lin, F.; Waters, C.L.; Mallinson, R.G.; Lobban, L.L.; Bartley, L.E. Relationships between Biomass Composition and Liquid Products Formed via Pyrolysis. *Front. Energy Res.* **2015**, *3*, 45. [CrossRef]
74. Laougé, Z.B.; Merdun, H. Pyrolysis and combustion kinetics of *Sida cordifolia* L. using thermogravimetric analysis. *Bioresour. Technol.* **2020**, *299*, 122602. [CrossRef]
75. Vuppaladadiyam, A.K.; Antunes, E.; Sanchez, P.B.; Duan, H.; Zhao, M. Influence of microalgae on synergism during co-pyrolysis with organic waste biomass: A thermogravimetric and kinetic analysis. *Renew. Energy* **2020**, *167*, 42–55. [CrossRef]

76. Mallick, D.; Bora, B.J.; Baruah, D.; Barbhuiya, S.A.; Banik, R.; Garg, J.; Sarma, R. Mechanistic investigation and thermal deg-radation of *Eichhornia crassipes* using Thermogravimetric analysis. *SSRN Electron. J.* **2020**. [CrossRef]
77. Baruah, D.; Mallick, D.; Kalita, P.; Moholkar, V.S. A Detailed Study of Pyrolysis Kinetics of Elephant Grass Using Thermogravimetric Analysis. In Proceedings of the 2nd International Conference on Energy Power and Environment, Shillong, India, 1–2 June 2018.
78. Patidar, K.; Singathia, A.; Vashishtha, M.; Sangal, V.K.; Upadhyaya, S. Investigation of kinetic and thermodynamic parameters approaches to non-isothermal pyrolysis of mustard stalk using model-free and master plots methods. *Mater. Sci. Energy Technol.* **2021**, *5*, 6–14. [CrossRef]
79. Fernandez, A.; Ortiz, L.R.; Asensio, D.; Rodriguez, R.; Mazza, G. Kinetic analysis and thermodynamics properties of air/steam gasification of agricultural waste. *J. Environ. Chem. Eng.* **2020**, *8*, 103829. [CrossRef]
80. Wang, S.; Wen, Y.; Shi, Z.; Niedzwiecki, L.; Baranowski, M.; Czerep, M.; Mu, W.; Kruczek, H.P.; Jönsson, P.G.; Yang, W. Effect of hydrothermal carbonization pretreatment on the pyrolysis behavior of the digestate of agricultural waste: A view on kinetics and thermodynamics. *Chem. Eng. J.* **2022**, *431*, 133881. [CrossRef]
81. Ozawa, T. Estimation of activation energy by isoconversion methods. *Thermochim. Acta* **1992**, *203*, 159–165. [CrossRef]
82. Mandapati, R.N.; Ghodke, P.K. Kinetics of pyrolysis of cotton stalk using model-fitting and model-free methods. *Fuel* **2021**, *303*, 121285. [CrossRef]
83. Hu, S.; Jess, A.; Xu, M. Kinetic study of Chinese biomass slow pyrolysis: Comparison of different kinetic models. *Fuel* **2007**, *86*, 2778–2788. [CrossRef]
84. Hilten, R.; Vandenbrink, J.; Paterson, A.; Feltus, F.; Das, K. Linking isoconversional pyrolysis kinetics to compositional characteristics for multiple *Sorghum bicolor* genotypes. *Thermochim. Acta* **2013**, *577*, 46–52. [CrossRef]
85. Müsellim, E.; Tahir, M.H.; Ahmad, M.S.; Ceylan, S. Thermokinetic and TG/DSC-FTIR study of pea waste biomass pyrolysis. *Appl. Therm. Eng.* **2018**, *137*, 54–61. [CrossRef]
86. Sharma, A.; Mohantya, B. Thermal degradation of mango (*Mangifera indica*) wood sawdust in a nitrogen environment: Characterization, kinetics, reaction mechanism, and thermodynamic analysis. *RSC Adv.* **2021**, *11*, 13396–13408. [CrossRef]
87. Phuakpunk, K.; Chalermsinsuwan, B.; Assabumrungrat, S. Pyrolysis kinetic parameters investigation of single and tri-component biomass: Models fitting via comparative model-free methods. *Renew. Energy* **2022**, *182*, 494–507. [CrossRef]
88. Colpani, D.; Santos, V.O.; Araujo, R.O.; Lima, V.M.R.; Tenório, J.A.S.; Coleti, J.; Chaar, J.S.; de Souza, L.K.C. Bioenergy po-tential analysis of Brazil nut biomass residues through pyrolysis. Gas emission, kinetics, and thermodynamic parameters. *Clean. Chem. Eng.* **2022**, *1*, 100002.
89. Poletto, M.; Zattera, A.J.; Santana, R.M. Thermal decomposition of wood: Kinetics and degradation mechanisms. *Bioresour. Technol.* **2012**, *126*, 7–12. [CrossRef]
90. Mallick, D.; Poddar, M.K.; Mahanta, P.; Moholkar, V.S. Discernment of synergism in pyrolysis of biomass blends using thermogravimetric analysis. *Bioresour. Technol.* **2018**, *261*, 294–305. [CrossRef] [PubMed]
91. Vikraman, V.K.; Boopathi, G.; Kumar, D.P.; Mythili, R.; Subramanian, P. Non-isothermal pyrolytic kinetics of milk dust powder using thermogravimetric analysis. *Renew. Energy* **2021**, *180*, 838–849. [CrossRef]
92. Kumar, M.; Mishra, P.K.; Upadhyay, S.N. Thermal degradation of rice husk: Effect of pre-treatment on kinetic and thermo-dynamic parameters. *Fuel* **2020**, *268*, 117164. [CrossRef]
93. Postawa, K.; Fałtynowicz, H.; Szczygieł, J.; Beran, E.; Kułażyński, M. Analyzing the kinetics of waste plant biomass pyrolysis via thermogravimetry modeling and semi-statistical methods. *Bioresour. Technol.* **2022**, *344*, 126181. [CrossRef] [PubMed]
94. Xiao, R.; Yang, W.; Cong, X.; Dong, K.; Xu, J.; Wang, D.; Yang, X. Thermogravimetric analysis and reaction kinetics of lignocellulosic biomass pyrolysis. *Energy* **2020**, *201*, 117537. [CrossRef]
95. Maia, A.A.D.; De Morais, L.C. Kinetic parameters of red pepper waste as biomass to solid biofuel. *Bioresour. Technol.* **2016**, *204*, 157–163. [CrossRef] [PubMed]
96. Sahoo, A.; Kumar, S.; Mohanty, K. Kinetic and thermodynamic analysis of *Putranjiva roxburghii* (putranjiva) and *Cassia fistula* (amaltas) non-edible oilseeds using thermogravimetric analyzer. *Renew. Energy* **2021**, *165*, 261–277. [CrossRef]
97. Hu, L.; Wei, X.-Y.; Guo, X.-H.; Lv, H.-P.; Wang, G.-H. Investigation on the kinetic behavior, thermodynamic and volatile products analysis of chili straw waste pyrolysis. *J. Environ. Chem. Eng.* **2021**, *9*, 105859. [CrossRef]
98. Mishra, A.; Kumari, U.; Turlapati, V.Y.; Siddiqi, H.; Meikap, B. Extensive thermogravimetric and thermo-kinetic study of waste motor oil based on iso-conversional methods. *Energy Convers. Manag.* **2020**, *221*, 113194. [CrossRef]
99. Simões, L.M.S.; Setter, C.; Sousa, N.G.; Cardoso, C.R.; de Oliveira, T.J.P. Biomass to biofuel densification of coconut fibers: Kinetic triplet and thermodynamic evaluation. *Biomass Conv. Bioref.* **2022**, *12*. [CrossRef]
100. Singh, S.; Chakraborty, J.P.; Mondal, M.K. Intrinsic kinetics, thermodynamic parameters and reaction mechanism of non-isothermal degradation of torrefied *Acacia nilotica* using isoconversional methods. *Fuel* **2020**, *259*, 116263. [CrossRef]
101. Muigai, H.H.; Choudhury, B.J.; Kalita, P.; Moholkar, V.S. Co–pyrolysis of biomass blends: Characterization, kinetic and thermodynamic analysis. *Biomass Bioenergy* **2020**, *143*, 105839. [CrossRef]

agriculture

MDPI

Article

Developing and Testing the Air Cooling System of a Combined Climate Control Unit Used in Pig Farming

Ivan Ignatkin [1,*], Sergey Kazantsev [1], Nikolay Shevkun [1], Dmitry Skorokhodov [1], Nikita Serov [1], Aleksei Alipichev [1] and Vladimir Panchenko [2]

[1] Moscow Timiryazev Agricultural Academy, Russian State Agrarian University, Moscow 127550, Russia
[2] Department of Theoretical and Applied Mechanics, Russian University of Transport, Moscow 127994, Russia
* Correspondence: ignatkin@rgau-msha.ru

Abstract: This article presents the results of developing and testing the air-cooling system of a combined climate control unit used in pig farming. The authors have found a water-evaporative system to be the most efficient for cooling the air supply. Cooling systems of this type consume 0.003 kW/kW of electric power to produce 1 kW of cold. Based on the developed mathematical model for water-evaporative cooling in the combined climate control unit, the authors have determined that an air supply with a temperature of 31.2 °C and a relative humidity of 30.4% can be cooled by 8.3 °C when saturated with moisture to a relative humidity of 90.0% (by 11.7 °C at 100%). Experimental studies of the cooling system confirmed the theoretically obtained data.

Keywords: indoor climate; air cooling; water-evaporative systems; sprayed panels; heat recovery units

Citation: Ignatkin, I.; Kazantsev, S.; Shevkun, N.; Skorokhodov, D.; Serov, N.; Alipichev, A.; Panchenko, V. Developing and Testing the Air Cooling System of a Combined Climate Control Unit Used in Pig Farming. *Agriculture* **2023**, *13*, 334. https://doi.org/10.3390/agriculture13020334

Academic Editor: Massimo Cecchini

Received: 22 November 2022
Revised: 11 January 2023
Accepted: 27 January 2023
Published: 30 January 2023

1. Introduction

Pig breeding is a highly productive branch of animal husbandry due to fast-growing pig breeds. Thanks to its nutritional value, availability and extraordinary cooking characteristics, pork meat plays a significant role in the human diet. In Russia, pork meat accounts for 25.6% of the meat resources consumed. The number of pigs in the Russian Federation amounted to 25.9 million by the end of 2021, 90.2% of which were industrially grown [1]. Such enterprises use flow technologies, high animal concentration, and intensive use of breeding stock to achieve maximum productivity. All these factors require a high level of production at all stages. Indeed, any deviation from the optimum production mode inevitably leads to losses.

The indoor climate is determined by a combination of temperature, relative humidity, chemicals, air composition, contaminants, micro-organisms, light, etc. Each of these indicators has a significant impact on animal productivity and must be maintained within strict limits based on the physiological needs of the animals. The most important indicators are temperature and relative humidity [2]. It is reasonable to use these indicators as regulators for the heating and ventilation system.

The body of a pig is covered with very sparse wool. It does not actually protect against external temperature influences. A stable body temperature is maintained by the thermoregulation system, in which the body uses energy to maintain a constant temperature. This energy consumption rate is minimal at an optimum temperature (Figure 1) [3].

Currently, genetics companies have significantly lowered the fat content of pork meat by reducing the thickness of the subcutaneous tissue that acts as natural thermal insulation in pigs. As a result, breeds developed with modern genetic techniques are more sensitive to temperature drops.

Figure 2 shows that a high relative humidity ($\varphi > 75\%$) results in a decrease in pig weight gain (by 20%) and an increase in feed consumption (by 40%) [3].

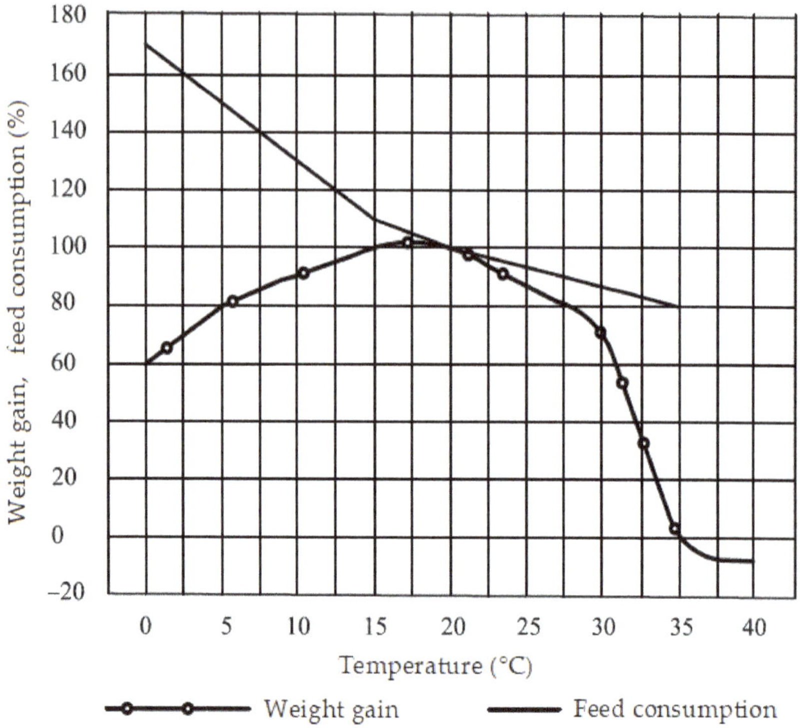

Figure 1. Influence of ambient temperature on the performance of fattened pigs.

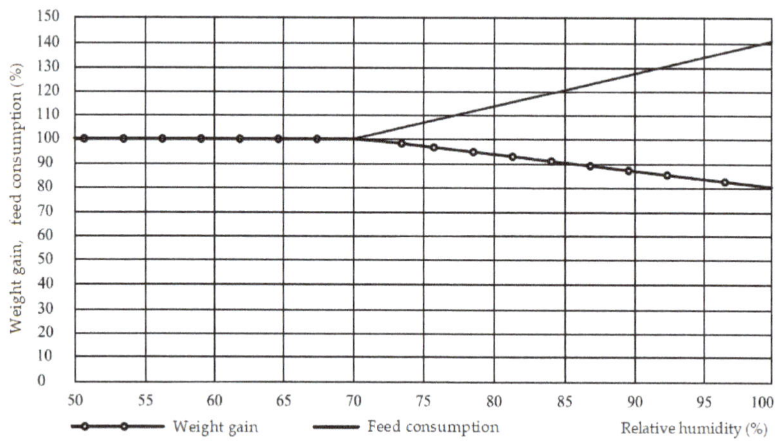

Figure 2. Effect of relative humidity on Large White pig productivity.

Dry air (relative humidity under 50%) also negatively affects the animal's body, causing irritation of the mucous membranes of the eyes and the respiratory tract, decreased local immunity, increased thirst, and consequently reduced appetite and nutrient absorption.

The above materials show the considerable influence of indoor climate on the productivity of animals.

Indoor climate indicators and animal excretion rates are taken by planners as input data from their manuals. It should be noted that the updated norms currently operating in

Russia—"RD-APK 1.10.02.04-12 methodological guidelines on engineering design of pig farms and facilities"—are based on the All-Union Science and Technical Regulations 2–96. The requirements have not changed much as to the emission of heat and harmful gases by animals and indoor climate parameters.

Russian pig breeders have achieved tremendous results in ensuring early maturity and increasing prolificacy [4]. The animal's body has become longer. The ratio of the surface area of a pig to its volume has changed; the thickness of the rump has decreased, which inevitably influences the parameters of its heat exchange with the environment. By comparing domestic and foreign excretion rates and indoor climate requirements, we can see a number of differences [5].

The entry "Nursing sows with piglets" in RD-APK provides data for animals weighing up to 200 kg, although the weight of a typical sow is 200–300 kg. These data are also available in the DIN 18910 standard operating in Germany (Berechnungs- und Planungsgrundlagen fuer das klima in geschlossenen ställen—norms for calculating parameters and indoor climate planning in closed stables (barns)).

The standard indoor climate parameters are also different, as shown in Table 1.

Table 1. Standard indoor climate parameters (data taken from RD APK and DIN 18910).

Age and Gender Group	Weight, kg	Standard Parameters			
		RD-APK 1.10.02.04-12		DIN 18910	
		Indoor Temperature, °C	Indoor Relative Humidity, %	Indoor Temperature, °C	Indoor Relative Humidity, %
Reproductive, barren, gestation sows, boars	Over 50	16	80	16–20	75
Nursing sows with local heating for piglets	Over 100	18	80	20	75
Pigs fattened using the "all empty—all occupied" method	10	24	70	24	70
	20–30	20	80	-	-
	40–50	18	80	-	-
	60–100	16	80	18	70

The data presented in Table 1 show that the requirements are basically the same, but DIN 18910 is stricter with regard to the requirements for relative humidity and temperature. However, the CO_2 concentration requirements are stricter in the RD-APK and amount to 0.2%, compared with 0.3% in DIN 18910. In our opinion, this is due to the fact that in the Soviet Union central heating systems were used everywhere to provide the required concentration of carbon dioxide. However, decentralized heating systems operating on natural gas are more common now due to their lower cost, as described in the thesis written by D. Tikhomirov [6].

Open combustion heat generators direct the flue gases indoors. The main component of natural gas is methane. The products of its combustion include carbon dioxide and water. With these systems, a carbon dioxide concentration of 0.2% is theoretically unachievable in pig houses during the coldest period, but 0.3% is still achievable.

Climate control systems used in crop production [7] and livestock farming aim to achieve normative values for temperature, relative humidity, and pollutant concentration [8]. Temperature is an important indicator of the indoor climate in livestock buildings [9].

In winter, a good supply of heat is required for heating the air supply [10]. Therefore, exhaust air heat recovery systems offer an effective way to reduce heating energy consumption [11,12]. Supply and exhaust air units with heat recovery exchangers utilize the heat in the exhaust air without mixing exhaust and supply air [13], thus supplying clean heated air to the production facilities. There are also ventilation systems with spiral recuperators [14,15] for heat recovery in winter and air cooling in summer.

Regenerative heat exchangers are widely used in the industry, e.g., in transcritical CO_2 heat pumps [16] and in CO_2 heat pump systems with compression/ejection for simultaneous cooling and heating [17].

In decentralized pit-type systems it is important to ensure reliable separation of supply and extract air [18].

However, in summer the challenge is no less important for decreasing the indoor temperature [19]. This is often achieved with individual cooling systems [20].

The importance of energy-saving systems using heat recovery can hardly be overestimated; they can be found in both mobile and stationary equipment [21]. In our opinion, it is reasonable to consider retrofitting supply and exhaust heat recovery units with cooling systems [22]. Such a solution would increase the intensity of equipment use, reducing the payback period.

However, in some cases the use of modular coolers is feasible [23].

This work aims to develop and test an air cooling system in a combined climate control unit used in pig farming.

To achieve this goal, it is necessary to solve the following tasks:

1. To analyze and classify the available air cooling ventilation systems;
2. To justify the construction of the combined climate control unit with the air cooling system;
3. To design a mathematical model of cooling in the climate control unit;
4. To experimentally test the developed system.

Air temperature can be decreased to the required values in different ways.

For pig farms, air cooling methods can be divided into two types: water-evaporative cooling systems and vapor-compression refrigerating units (Figure 3).

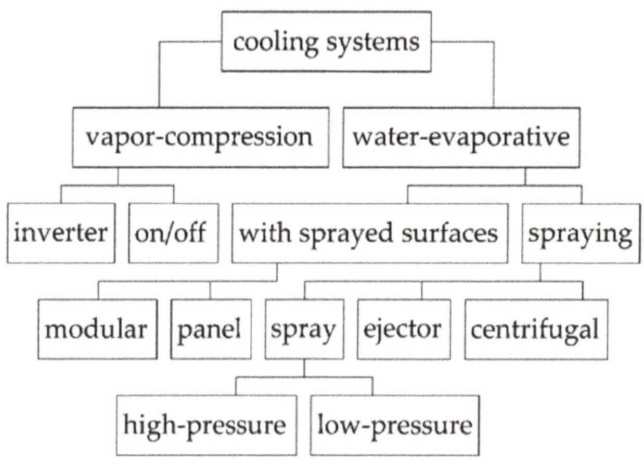

Figure 3. Classification of cooling systems.

The vapor-compression refrigeration unit is a closed hermetic system consisting of an evaporator, a compressor, a heat exchanger condenser, a filter-dryer and a throttle connected by pipelines. The unit is filled with a refrigerant with a boiling point lower than that of the cooled medium. The cooled air comes into contact with the surface of the evaporator and transfers its heat to the refrigerant, turning it into vapor. The compressor forces the refrigerant vapor into the condenser, where the latter is turned into its liquid state. Vapor-compression refrigeration systems are used extensively in industry and civil engineering, but they are rarely used in animal husbandry. However, where strict temperature limits are set, regardless of weather conditions, vapor-compression refrigeration units are the only feasible solution.

With water evaporation cooling, the air is cooled by being blown through a spray chamber or a cartridge package soaked in water and subjected to consecutive heat exchanges. Another option is spraying water into the cooled air stream via nozzles. The evaporated moisture absorbs the heat of vapor formation and the air is cooled.

Water-evaporative cooling is fundamentally simpler and less demanding in terms of maintenance and operating conditions, but it has a number of limitations and its efficiency depends significantly on the temperature and relative humidity of the outdoor air.

In spite of their fundamental similarity, water-evaporative cooling systems are available in a wide range of designs. The most noteworthy are water spray systems and systems with sprayed surfaces. Table 2 presents the results of their comparative assessment.

Table 2. Analysis of cooling system efficiency.

Parameter	Vapor-Compression Refrigerating Units (VCRU)		Water-Evaporative Cooling Systems					
	VCRU On/Off	**Inverter VCRU**	**Low-Pressure Nozzles (2 atm)**	**High-Pressure Nozzles (70 atm)**	**Sprayed Panel**	**Centrifugal Cooling Unit**	**Sprayed Modules**	**Ejector Modules**
Supply air temperature reduction efficiency, °C	-	-	1	3	15	6	8	4.5
Cooling capacity of the system, kW	282	282	34.2	102.6	424.9	170.0	226.6	127.5
Installed electrical capacity of the cooling units, kW	95.9	95.9	1	2.2	1.6	33.4	25.2	66.0
Hourly energy consumption of the cooling units, taking into account irregular use, kWh	95.9	67.1	0.8	2.2	1.2	33.4	25.2	66.0
Set air exchange rate, m^3/h	151,800	151,800	442,750	442,750	366,850	366,850	366,850	366,850
Installed fan power, kW	6.6	6.6	18.9	18.9	15.6	15.6	15.6	15.6
Installed electric power of the system, kW	102.4	102.4	19.9	21.1	17.2	48.9	40.8	81.6
Electrical power consumption for producing 1 kW of cold, kW/kW	0.340	0.238	0.023	0.021	0.003	0.196	0.111	0.518
Refrigeration production efficiency criterion, kW/kW	2.9	2.9	34.2	46.6	265.6	5.1	9.,0	1.9
Water consumption, m^3/h	0	0s	44.2	0.7	3.7	2.5	3.7	0.2
Specific energy consumption per 1000 m^3/h air, W	631.6	442.1	1.8	5.0	3.3	90.9	68.8	179.9

Water is evaporated by absorbing the heat of vapor formation. The energy is thus spent to provide the required air exchange and water supply to the evaporation zone [24]. Ejection and nozzle systems do not directly consume energy for the cooling process. However, their operation requires water supplied at a given flow rate and pressure, which results in electricity costs at the water plant [25]. These costs have been taken into account in estimating the costs of electric power for the production of 1 kW of cold.

The following conditions and assumptions are taken into account when comparing cooling systems:

- vapor-compression cooling systems have unlimited cooling capacity under the operating conditions;
- the estimated air exchange of the water-evaporative cooling systems takes into account equal air exchange for the removal of excess heat and moisture.

Among the water-evaporative cooling systems, systems with sprayed surfaces are the most effective in terms of reducing the air supply temperature. They are also the most energy efficient.

Based on the above and considering the fact that the climate control unit has a large heat exchange surface area, it is advisable to spray water on it to use the effect of water-evaporative cooling of the supply air.

Water-evaporative cooling is widely used in the climate control systems of various production facilities due to its combination of low cost and high efficiency [26]. However, the system's operation is highly dependent on outdoor air parameters [24]. When designing climate control systems, it is important to analyze the main process indicators [27] and to predict the efficiency of water-evaporative cooling in different climate zones. In this connection, it is advisable to design a computational model of the combined unit [28] to link geometric parameters of the cooling unit based on the recuperative heat exchanger, its performance, total aerodynamic resistance of the system and air parameters at the inlet and outlet of the refrigerating unit.

2. Materials and Methods

2.1. Description of the System

The heat-exchanging surfaces are to be cleaned periodically in the process of operation. Consequently, the spraying system performs cleaning and disinfecting functions in addition to cooling [29]. This requires the use of detergent and/or disinfectant solutions. Disinfection can affect both the working surfaces of the unit and the ventilation air, making it possible to disinfect the supply and exhaust air of the serviced area.

The spraying system consists of a pipeline with nozzles, a time switch and a solenoid valve, and is connected to a water supply system for periodic water spraying in the exhaust and supply ducts above the heat exchanger. The sprayed liquid can be used in a flow-through or a cyclic mode. In the latter case, it requires a storage tank, which can take the form of a sump unit [30].

The offered solution ensures a more intensive use of the equipment, helps save on the number of cooling units and provides for the implementation of previously unavailable functions that are extremely important in terms of longevity, operational efficiency and biological safety—thus increasing the economic efficiency of the system as a whole.

The block diagram is shown in Figure 4.

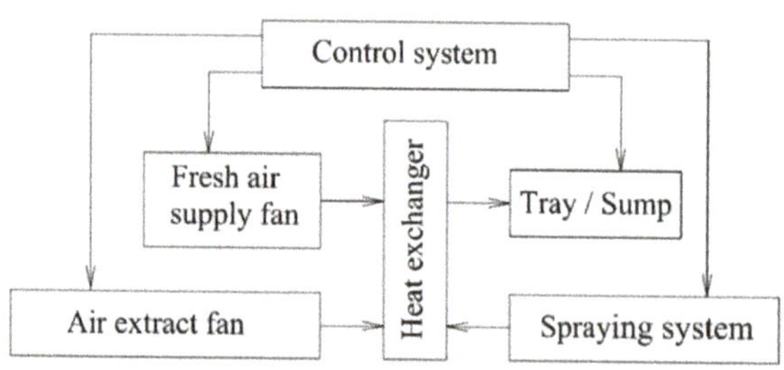

Figure 4. Schematic diagram of the combined climate control unit.

During the warm season, the ventilation system is used to remove excess heat, and the ventilation rate is significantly higher than in winter. In particular, in winter the specific air consumption per 1 kg of live weight of fattened piglets is 0.17 to 0.21 $m^3/kg/h$, and in summertime this value increases to 1.1 to 2.5 $m^3/kg/h$, that is, 5.2 to 14.7 times higher.

In warm periods, air cooling occurs as follows. The indoor air is extracted by separate window or roof fans. The exhaust fan of the unit is reversed, and both channels provide the air inflow. A process diagram of the prototype of the combined climate control unit equipped with the water-evaporative cooling system is shown in Figure 5. The surface of the heat exchanger is sprayed with water via nozzles. On contacting the heat exchanger surface, the water wets it in a film or droplet pattern, creating a large cooling area. The air supply passes through the heat exchanger, comes into contact with the heat exchanger surface, is cooled there and then discharged into the room by the fans.

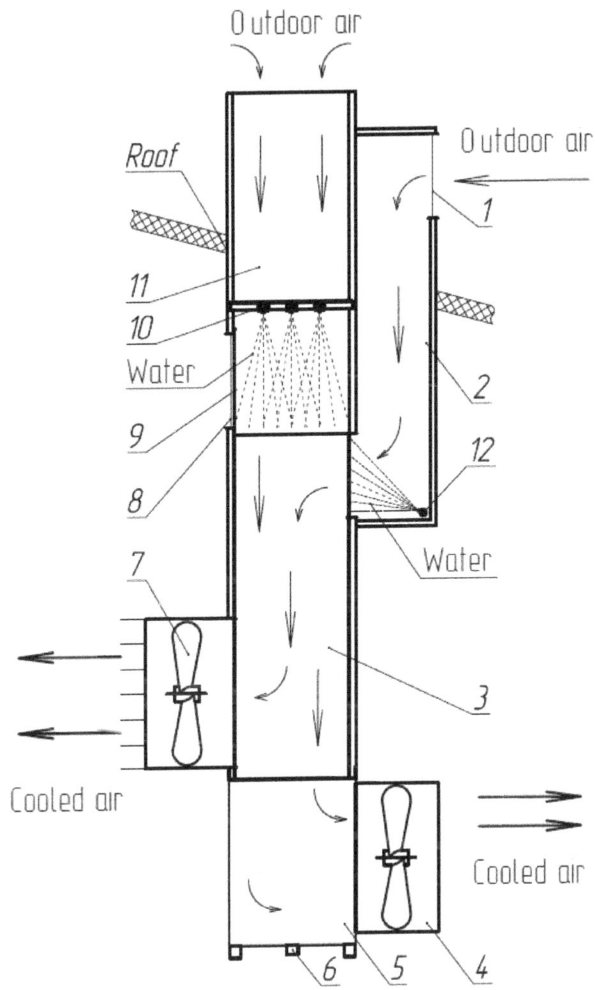

Figure 5. The cooling mode of the combined climate control system: 1—inlet window; 2—inlet duct; 3—heat exchanger; 4—exhaust fan; 5—sump; 6—discharge pipe; 7—supply fan; 8—recirculation opening; 9—recirculation damper; 10—pipeline with nozzles; 11—exhaust duct; 12—pipeline with nozzles.

2.2. Theoretical Research

Modelling of Water-Evaporative Cooling in the Unit

The water-evaporative cooling process in systems with sprayed layers can be represented as a convective heat exchange of the air flow with heat exchanger plates and a simultaneous evaporation of water from the unit surface accompanied by the absorption of vaporization heat.

The finite difference method is widely used to find a numerical solution to heat exchange problems. In a steady-state mode, the system is described by a set of parameters: the surface temperature of the unit (equal to the wet-bulb temperature), temperature and relative humidity of the supply air. The air supply cools as heat and moisture are exchanged, approaching the cooling surface temperature, while its enthalpy (heat content) remains unchanged.

The process is limited by the exposure (the process duration), the relative humidity at the unit outlet (climate requirements) and the outdoor air parameters (the temperature and moisture content), which will determine the physical cooling potential.

For practical calculations, these parameters can be considered conditionally stationary. In the boundary layer, at the cooling surface, the relative humidity is 100%. The temperature of the unit wall is equal to the wet-bulb temperature and is assumed as constant for the considered conditions (Figure 6) [16].

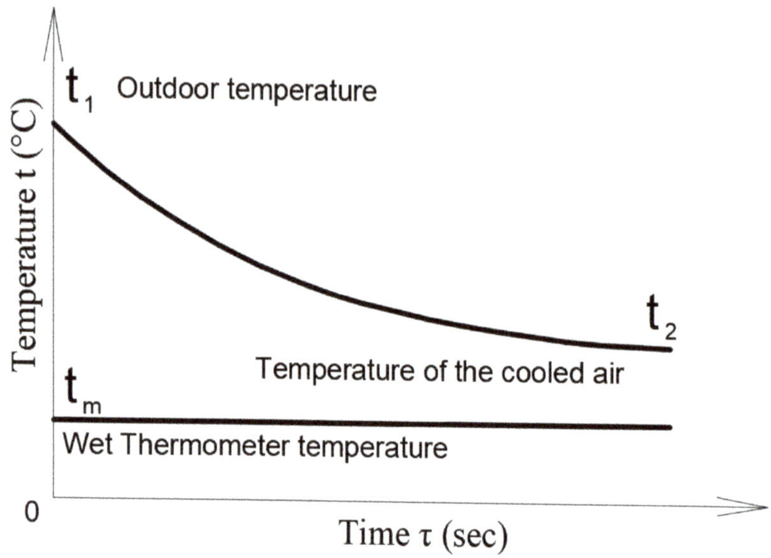

Figure 6. Diagram of water-evaporative air cooling.

The heat and mass transfer process is characterized by equality of heat extracted from the supplied air and the heat of vaporization expended for moisture evaporation from the unit surface [31]. The cooling intensity (heat transfer) of the air supply is determined based on Newton's equation

$$q_{\text{cool}} = S{\cdot}\overline{\alpha}{\cdot}(t_o - t_w), \tag{1}$$

where S—the heat exchange surface area, m^2; $\overline{\alpha}$—the heat transfer coefficient average over the heat exchange surface, W/(m^2·K); t_o—the outdoor air temperature, °C and t_w—the heat exchange surface (wet-bulb) temperature, °C.

The heat transfer coefficient α is the amount of heat transferred through the unit heat transfer surface per unit time at a temperature difference of 1 °C [32].

In engineering calculations, it is expedient to use criterion analysis to determine its value. Thus, based on the aerodynamic similarity criteria it is possible to determine the heat transfer coefficient with high accuracy [33].

$$q_{\text{cool}} = S{\cdot}\overline{\alpha} = Nu\frac{(t_o{\cdot}7{\cdot}10^{-5} + 0.0237)}{d_{e.p}}, \tag{2}$$

where Nu—Nusselt criterion characterizing the correlation of the intensity of heat transfer by convection and heat conduction and $d_{e.p}$—hydraulic diameter of the pipe, m.

In the considered unit, the air ducts are pipes with a rectangular cross-section. The channels of the plates are directed along the main air flow (taken conditionally as vertical), as shown in Figure 7.

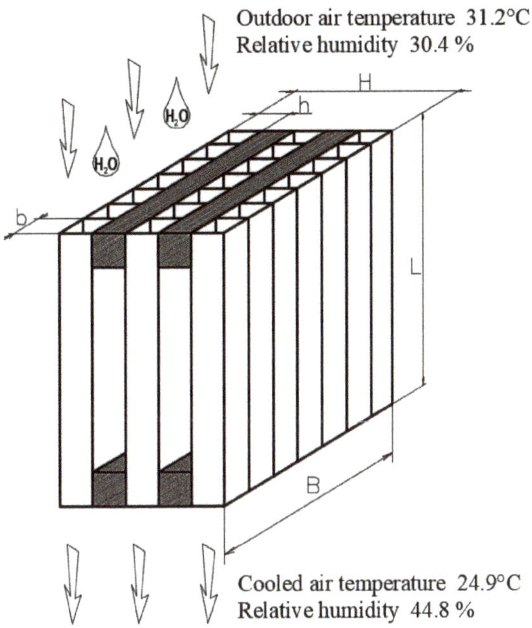

Figure 7. Heat exchanger in cooling mode.

The equivalent diameter for a pipe of rectangular cross-section is determined by the following formula ([32], p. 189):

$$d_{e.p} = 4\frac{A}{P} = 4\frac{b \cdot h}{2 \cdot (b+h)}, \tag{3}$$

where $A = b \cdot h$—cross-section area, m^2; $P = 2 \cdot (b + h)$—perimeter, m; b—width of the tube base of a triangular cross-section, m and h—height of the tube of a triangular cross-section, m.

The Nusselt criterion ($Nu = f\left(\text{Pr}; \text{Re}; \frac{\Delta T}{T}\right)$) is defined by the following formula:

$$Nu = 0.28 \cdot \text{Re}^{0.6} \cdot \overline{\text{Pr}}^{0.36} \cdot \left(\frac{\overline{\text{Pr}}}{\text{Pr}_1}\right)^{0.25}, \tag{4}$$

where Re—the Reynolds criterion value for the air flow in the cooling unit; $\overline{\text{Pr}}$—the average Prandtl criterion value for the air flow in the cooling unit and Pr_c —the Prandtl criterion value for the boundary layer.

The Reynolds criterion value is defined by the formula

$$\text{Re} = \frac{v \cdot d_{e.p}}{\nu_{su}}, \tag{5}$$

where ν_{su}—the average kinematic viscosity of the supply air, determined at an average air temperature in the unit $t_{av} = \frac{t_o + t_w}{2}$.

The Prandtl criterion value for the boundary layer is defined by the following formula:

$$\mathrm{Pr}_l = \frac{\eta_l \cdot c_m \cdot 10^3}{t_w \cdot 7 \cdot 10^{-5} + 0.0237}, \tag{6}$$

where η_l—dynamic viscosity of the layer, Pa·sec and c_m—specific mass heat capacity of the air, J/(kg·K).

The dynamic viscosity of the air in the boundary layer is determined by the following formula:

$$\eta_l = \nu_l \cdot \rho_{su}, \tag{7}$$

where νc—kinematic viscosity of air in the boundary layer, m^2/s and ρ_{su}—air density in the boundary layer, kg/m^3.

The temperature of the air in the boundary layer is asymptotically close to the wall temperature and numerically may be assumed to be equal to the wet-bulb temperature. In this case, the kinematic viscosity of air in the boundary layer can be found using the following formula [34]:

$$\nu_l = \frac{324 - 1.5 \cdot t_w}{10^9} p_0 + 16.81 + 0.048 \cdot t_w \cdot 10^{-6} \cdot 287 \frac{t_w + 273}{p_0}, \tag{8}$$

where t_w—wet-bulb temperature, °C and p_0—atmospheric pressure, Pa.

Air's density largely depends on its temperature and pressure. The present analysis uses the atmospheric pressure, as the working pressure difference in the cooling unit is about 10–30 Pa, which is 0.01–0.03%. This analysis uses approximations of lower accuracy. Taking differential pressure of this order into account would complicate the analysis in the pursuit of imaginary accuracy. The air density of the boundary layer equals

$$\rho_l = \frac{p}{R(t_w + 273)}, \tag{9}$$

where R—gas constant, J/(kg·K).

V.P. Gavrikin and E.A. Kuranov obtained the formula to determine the wet-bulb temperature [35] resulting from the analysis of reference data [30]:

$$t_w = \frac{-7.14 + 0.651 \cdot I}{1 + 9.7 \cdot 10^{-3} \cdot I - 3.12 \cdot 10^{-6} \cdot I^2}, \tag{10}$$

where I—enthalpy (heat content) of the air, kJ/(kg·K).

The results obtained using the formula deviate from the tabulated values within 0.6%, which is quite sufficient for an engineering analysis.

Water-evaporative cooling takes place with the constant heat content, taking into account that enthalpy of humid air depends on the temperature and moisture content. Taking into account the availability of all initial data for outdoor air, it will be convenient to determine its heat content. The enthalpy is to be determined from the formula

$$I = c_{d.a.} \cdot t_o + \frac{(r + c_v \cdot t_o) d_o}{1000}, \tag{11}$$

where $c_{d.a.}$–specific mass heat capacity of dry air, kJ/(kg·K).

On the other hand, in a steady-state process the heat flux can be defined as the vaporization heat of the moisture evaporated in the cooling unit, which is expressed by the following formula:

$$q_{vap} = \frac{W \cdot \overline{\rho}(d_{su} - d_o)(r + c_v \cdot t_w)}{3600 \cdot 10^3}, \tag{12}$$

where W—cooling unit capacity, m^3/h; $\overline{\rho}$—average density of the cooled air, kg/m^3; d_{su}—moisture content of air at the cooling unit outlet, g/kg d.a.; d_o—moisture content of outdoor air, g/kg d.a.; c_v—specific mass heat capacity of dry vapor, kJ/(kg·K) and r—specific heat of water vaporization at 0 °C kJ/kg.

Relationship (12) is of a deterministic nature, and its accuracy is absolute under conditions of initial data reliability. The specific humidity of the air at the unit inlet and outlet is measurable. However, in our case air parameters at the unit outlet are forecast. In a particular case, the expected moisture content of the air at the cooling unit outlet is equal to the one required by the technology. In all other cases, it is different, which is caused by the difference in the physical properties of the interacting media.

The surface may be covered with a continuous stable film (the film-wetting mode) or individual droplets or drips (the droplet-wetting mode), which determines the size of the interaction surface area. When hygroscopic materials are used, in addition to good wettability they have a certain buffer capacity, which can compensate for disturbances in the film mode.

The specific humidity of air is defined by the formula

$$d = 0.77\frac{\varphi}{100}e^{5516.89\left(\frac{1}{273}-\frac{1}{273+t}\right)}, \tag{13}$$

where φ—relative humidity of the air, %; e—the basis of natural logarithm and t—the air temperature, °C.

A similar formula is used to determine the specific humidity of air at the cooling unit outlet.

The water-evaporative cooling process is iso-enthalpic. On L.K. Ramzin's I-d diagram, it appears as a segment connecting points 1 and 2, which characterize the state of moist air before (1) and after (2) entering the cooling unit (Figure 8). It is obvious that water-evaporative cooling is possible only if the following condition is observed: $d_1 < d_2$.

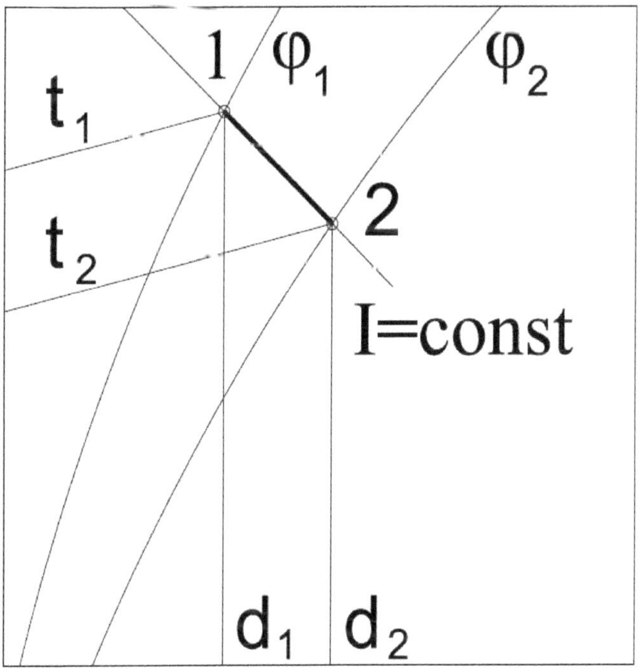

Figure 8. Water-evaporative cooling in the I-d diagram.

The average air velocity in the cooling unit is determined from the formula

$$\overline{\upsilon} = \frac{W}{3600 \cdot \frac{S}{2}}, \tag{14}$$

where W—cooling unit capacity, m^3/h and S—cross-sectional area of the heat exchanger, m^2.

In the process of heat exchange, the air temperature decreases and its density increases proportionally. The average density of the cooled air can be determined from the following formula:

$$\overline{\rho} = \frac{p}{R\left(\frac{t_o + t_w}{2} + 273\right)} \tag{15}$$

The supply air temperature drop is proportional to the heat dissipated from it, provided that it does not exceed q_v (the heat of vaporization, with a known difference in the specific humidity of the air) and can be found from the following system of conditions:

$$\begin{cases} \text{By} q_{\text{cool}} \le q_{vap} t_{su} = \frac{q_{\text{cool}} \cdot 10^3}{0.28 \cdot W \cdot \overline{\rho}} \\ \text{If} q_{\text{cool}} \succ q_{vap} t_{su} = \frac{q_{vap} \cdot 10^3}{0.28 \cdot W \cdot \overline{\rho}} \end{cases} \tag{16}$$

By determining the air supply temperature and taking into account its specific humidity, we can obtain the relative humidity at the cooling unit outlet:

$$\varphi_{su} = \frac{100 \cdot d_{su}}{3.8 \cdot e^{5404.3 \cdot \left(\frac{1}{273} - \frac{1}{273 + t}\right)}}, \tag{17}$$

where d_{su}—moisture content of the air at the cooling unit outlet, g/kg of d.a.

To increase the practical value of the model, it is advisable to determine the minimum necessary water consumption for water-evaporative cooling with a given capacity under specific weather conditions, and the annual water consumption for water-evaporative cooling.

To do this, the wet flow rate for cooling one kilogram of air for specific weather conditions must be determined. Then, the sum of the required air exchange, the specific moisture consumption and the duration of specific weather conditions over a year will give the required value of the annual water consumption for water-evaporative cooling. Weather conditions vary from year to year, so climate data should be used in the design calculations.

The maximum hourly water consumption (m^3/h) for evaporation is determined from the formula

$$Q_w = W_{\max} \cdot \overline{\rho} \cdot \Delta d_{\max} \cdot 10^{-6}, \tag{18}$$

where $W_{\max}$—air exchange rate at maximum design outdoor temperature, m^3/h and $\Delta d_{\max}$—the amount of moisture consumed by one kilogram of supply air during cooling at the maximum estimated outdoor temperature, g/kg of d.a.

Annual water consumption for water-evaporative cooling can be determined from the formula

$$Q_{w.a.} = \sum_{i=1}^{n} W_i \cdot \overline{\rho} \cdot \Delta d_i \cdot \tau_i \cdot 10^{-6}, \tag{19}$$

where W_i—air exchange required at temperature t_i (determined from the balance of humidity and heat), m^3/h; $\Delta d_i = d_{su} - d_o$—amount of moisture consumed by one kilogram of supply air during cooling, g/kg d.a. and τ_i—duration of temperature period t_i, h.

For aggregate calculations, the values of relative humidity can be obtained by linear interpolation of reference points obtained from building regulations (BR) and local building norms (LBN). For more precise calculations it is necessary to use data tables (the relationship between temperature and relative humidity) or diagrams that take into account the duration of specific combinations of temperature and relative humidity for the considered climatic zone. Such data can be found in the works of A.Y. Kreslin and other scientists.

Knowing the required air exchange rate and the permissible total aerodynamic drag or limiting the flow speed with the condition that no droplets of moisture can escape from the sprayed surfaces, we can determine the required cross-sectional area, and, consequently, the required number of units.

2.3. *Experimental Study*

Description of the Experimental Facilities

The climate control system using the combined climate control unit was tested in fattening sector No. 6 on Farm No. 7 of "Firma Mortadel" LLC. Indicators of test conditions are given in Table 3.

Table 3. Parameters and conditions of the tests.

Index	Value
Material of the building	Reinforced concrete
Holding capacity, heads	250
Internal dimension of the building L × W × H, m	14.6 × 14.6 × 2.6
Total floor area, m^2	214.3
Area of the technological section, m^2	200.3
Utilization factor of the floor area of the premises	0.9
Actual number of animals, heads	279
Floor area of the technological section per one animal, m^2/head	0.7
Characteristics of the livestock	
Age and sex group	Fattened pigs
Average weight, kg	70
Climatic conditions	
Region	Vladimir region
Estimated temperature of the coldest five-day period with a probability of 0.98, °C Average monthly relative humidity at 3 p.m. of the warmest month, % temperature, °C	−34
Average monthly relative humidity at 3 p.m. of the coldest month, %	83
Air temperature in summer, °C, probability 0.98	27
Average monthly relative humidity at 3 p.m. of the warmest month, %	57
Indoor climate	
temperature, °C	21
relative humidity, %	60

As a control sector, we used fattening sector No. 5, having 241 pigs with a weight of 30.5 kg (20 animals per pen). The climate maintenance system was designed by FG-CLIMITED, Ontario, CA, USA, using equipment from Ag-Co Products Limited Ontario, CA, USA. The front wall has four extractor fans (one extractor from the manure channel). The air is supplied through supply valves located in the ceiling and distributed in the room at the rate of one valve per box for a total of 12. The automation system is controlled by a temperature sensor installed in the room center at a height of 1.2 m above the floor level.

The experimental sector was additionally equipped with a tested system consisting of three prototypes of a combined climate control system with a water sprinkler cooling system (Figures 9 and 10): No. 1, 2—general exchange ventilation; and No. 3—air extraction from the manure channel and a system for regulating indoor climate parameters according to temperature and relative humidity sensors placed along the passage at a height of 1.2 m. Technical characteristics of the units are presented in Table 4.

Figure 9. View from the pig house roof: 1,2—UT-6000S units (for general exchange ventilation); 3—UT-3000 unit (for air extraction from the manure channel).

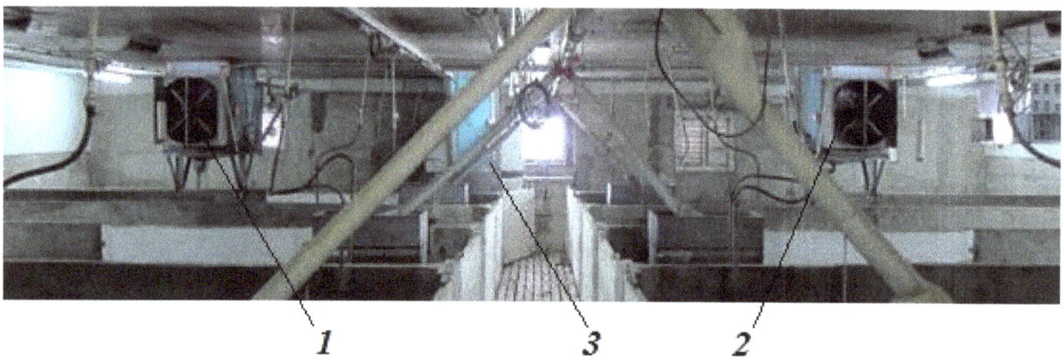

Figure 10. General view of the experimental system of climate control inside the pig house: 1,2—UT-6000C unit (for general exchange ventilation); 3—UT-3000 unit (for air extraction from the manure channel).

Table 4. Technical characteristics of the units.

Designation	Value	
Working designation	UT—6000C	UT—3000
Air capacity, m^3/h	6000	3000
Heat output at −25 °C, kW	50.7	25.4
Installed electric power, kW	1.6	0.8
Type	With defrosting system	Without defrosting system
Note	Reduced sump dimensions	With two supply fans

The summer ventilation system was supplemented with an option to use the climate control units in the cooling mode.

During the test period, the outdoor air temperature ranged from −8 °C to +31 °C, and climate measurements were taken automatically by a climate control computer according to temperature and relative humidity sensors placed as shown in the diagram in Figure 11.

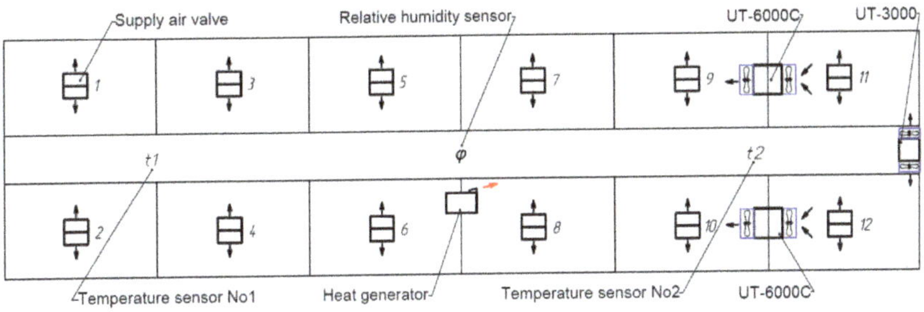

Figure 11. Control point diagram with sensors.

The climate control computer automatically maintains the specified indoor air temperature and relative humidity. Manual climate control is also possible.

As the relative humidity in the room increases, the air exchange rate smoothly increases as well. The air supply is controlled automatically by changing the speed of the fans and the degree of opening of the air inlet flaps. If the room temperature falls below the set point while the excess moisture is being removed, the heating devices are switched on. If the temperature drops further below the set point value, the ventilation level is set to the minimum value (depending on the weight and age of the animals), with the temperature taking priority.

This solution ensures the removal of excess moisture, heat and harmful gases from the house throughout the year.

3. Results

Study of Water-Evaporative Cooling

The problem was solved in Excel 2007 (Microsoft Corporation, Redmond, WA, USA).

Within the variation range of relative humidity from 20 to 100% and outdoor air temperature from 20 to 50 °C; the calculation results are graphically presented in Figure 12.

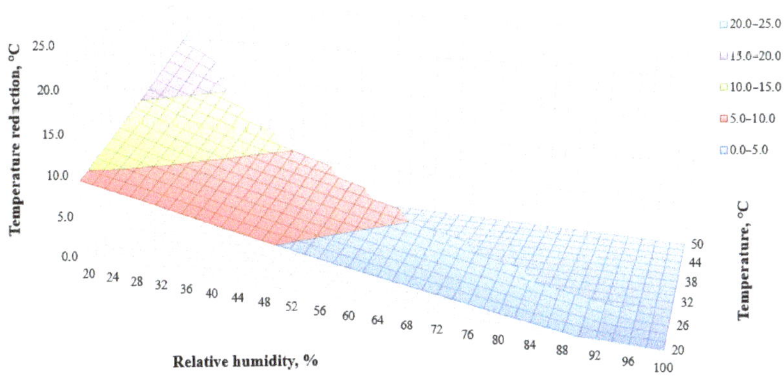

Figure 12. Reduction of air temperature in the unit as a function of the temperature and relative humidity of the outdoor air.

The graph is plotted in a three-dimensional coordinate system, showing the relationship of $\Delta t = f(t, \varphi)$. The result obtained shows that the maximum cooling value is achieved at minimum relative humidity and maximum outdoor air temperature, otherwise the supply air cooling value decreases monotonically.

Over the test period, the outdoor air temperature was 30.4 °C and relative humidity was 31.2%. The test results are shown in Table 5. Figure 13 presents a diagram of the exhaust and supply air temperatures.

Table 5. Test results of the combined climate control unit in the cooling mode.

Air Parameters in Control Points			
Points	**Indoor Temperature, °C**	**Difference between Indoor and Outdoor Temperatures, °C**	**Indoor Relative Humidity, %**
1	29.4	1.8	37.1
2	29.5	1.7	34.4
3	28.8	2.4	34.3
4	28.7	2.5	37.2
5	28.4	2.8	39.2
6	28.6	2.6	38.8
7	27.3	3.9	44.3
8	27.4	3.8	46.2
9	25.3	5.9	45.6
10	24.9	6.3	44.8
11	25.5	5.7	43.9
12	25.3	5.9	42.5
Average:	27.4	3. 8	40.7

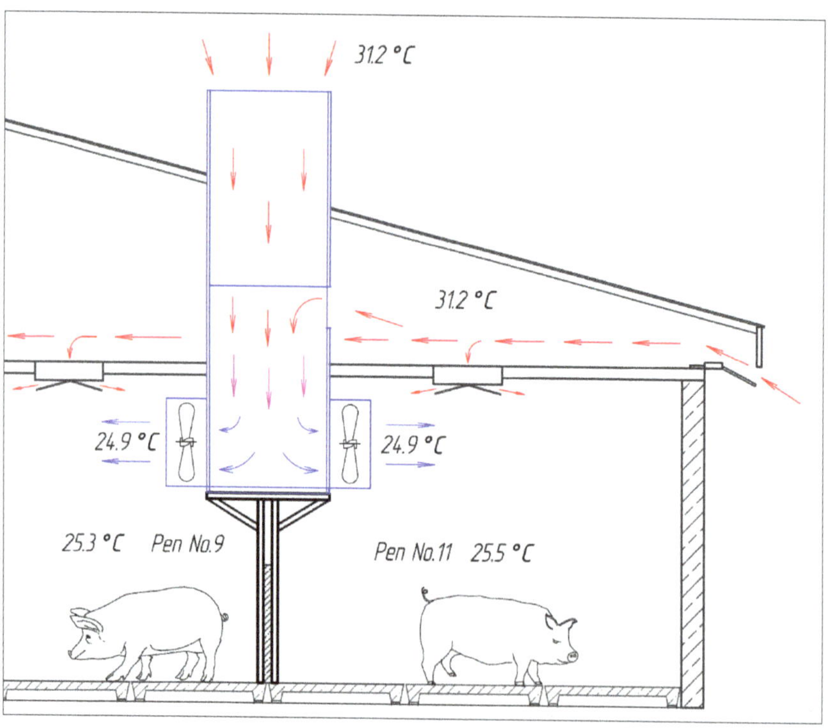

Figure 13. Diagram of the exhaust and air supply temperatures.

Table 5 shows that the maximum value of the air supply temperature reduction was 6.3 °C (in the pens closest to the units). As the distance from the units increased, the indoor air temperature increased. This effect was due to the fact that the share of cooled air (from the units) in the total ventilation system was 49%. As a result, the cooled air mixed with untreated outdoor air from the general ventilation system. Therefore, the average room temperature drop was 3.8 °C.

When analyzing the convergence of the results, we noted some discrepancies with the analytical data. The calculation model shows that supplied air with a temperature of 31.2 °C and a relative humidity of 30.4% can be cooled by 8.3 °C when saturated with moisture to a relative humidity of 90.0% (by 11.7 °C at 100.0% relative humidity).

At small values of the coefficient of spraying heat-exchange plates, we registered reduced values of the air supply temperature as compared to the calculated ones. This effect is caused by the influence of the wettability of the heat exchange surface on the efficiency of water-evaporative cooling. When the water flow rate was increased, the experimental data came close to what was theoretically possible, but this was accompanied by a detachment of moisture droplets.

The stated assumption can be proven by the fact that in sprayed layers made of hygroscopic and well-wetted cellulose, the above effect was not observed in prcatice (under condition of a sufficient water flow rate for water evaporation).

Further research is required to study the influence of the wettability of sprayed surfaces on the efficiency of water-evaporative cooling.

4. Discussion

The problem of ensuring proper climate control in livestock buildings is rather urgent and requires special attention. A considerable amount of research focuses on the development and optimization of climate control systems to increase functionality and improve technological and energy efficiency. The authors aimed to develop and test an air cooling system in a combined climate control system used in pig production. To achieve this goal, we analyzed the existing cooling systems for ventilation air, developed the design of a climate control system and a mathematical model of air cooling, made a prototype and conducted experimental research.

In the course of our research, different methods of air cooling in climate control systems were considered. For livestock farming, it is advisable to use water-evaporative cooling systems with sprayed surfaces. This variant is optimal as these systems are characterized by constructive simplicity, reliability and technological and energy efficiency. As compared to other methods of air cooling, surface-filled water-evaporative cooling systems require a minimum amount of electricity per kW of cooling. This was the main argument for the use of the water-evaporative cooling system with sprayed surfaces in the combined climate control unit.

The authors developed a mathematical model describing air cooling for the offered technical solution. The model considers a heat exchanger consisting of tubes with a rectangular cross-section. The flow of the cooled air and the refrigerant water is direct in the considered heat exchanger. The water sprayed by the nozzles wets the surface of the heat exchanger. This makes it possible to obtain a large cooling surface, which has a positive effect on cooling intensity. Based on the obtained mathematical model, we determined the theoretical values of air temperature reduction depending on outdoor air temperature and humidity.

The developed combined climate control unit with a water-evaporative cooling system was tested on the "Firma Mortadel" Ltd. farm. In the course of testing, measurements of temperature and relative air humidity were taken in the animals' pen according to the conventional procedure, and the test results confirmed expectations. The developed system decreases the air temperature.

The prototypes were tested in the operating heating and ventilation system. The capacity of the three test units was 15,000 m^3/h, which provided 49% of the total capacity of the combined system. Since all the cooling units were concentrated on one side, the cooled air was unevenly distributed inside the space. To reduce this effect, it is advisable to distribute the units evenly over the space and cool the whole air supply.

The efficiency of water-evaporative cooling systems strongly depends on the climatic conditions in the region of operation. The highest efficiency is achieved in hot and arid as well as in temperate climates. Under conditions of high relative humidity—80% and above—the cooling effect tends to zero. As the cooling effect increases, so does water consumption. For example, at an outside air temperature of 40.0 °C and a relative humidity of 20.0%, it is possible to warm air to a temperature of 28.8 °C and ensure a relative humidity of 85.0%. Under these conditions, a unit with a capacity of 6000 m^3 of air per hour will have a water consumption of 23 m^3/h.

5. Conclusions

The analysis of the existing air cooling systems has shown that it is advisable to use water-evaporative cooling systems for livestock houses due to their simplicity, technological and energy efficiency and low requirements for operating conditions.

The comparative analysis of the energy efficiency of cooling systems has shown that systems with sprayed surfaces require a minimum amount of electric power to produce 1 kW of cold (0.003 kW/kW).

The combined climate control system includes a heat exchanger with a large heat exchange surface. Adding a heat exchanger spraying system is therefore sufficient to ensure water-evaporative cooling in the system. This system can also perform cleaning and disinfection of the unit's working surfaces as well as the inlet and outlet air of the serviced area.

To describe the cooling process in the combined climate control unit, the authors have developed a mathematical model that links air parameters at the inlet and outlet of the unit, its geometrical characteristics and performance.

The mathematical model can determine the potential cooling of the air supply and the minimum water consumption required to produce this effect. An air supply with a temperature of 31.2 °C and a relative humidity of 30.4% can be cooled by 8.3 °C when saturated with moisture to a relative humidity of 90.0% (by 11.7 °C at 100.0% humidity).

As the experimental research proved, the air supply was cooled by 6.3 °C (in the pens nearest to the units), which confirms the efficiency of water-evaporative cooling. However, additional research is required to study the influence of the wettability of the sprayed surfaces on the efficiency of water-evaporative cooling.

6. Patents

Heat recovery unit. Ilyin, I.V. and Ignatkin, I.Yu. Patent for the invention No. 2627199 C1 RU, MICROPK A01K 29/09 (2006/01). Applicant and patent holder—Russian State Agrarian University—Moscow Timiryazev Agricultural Academy—No. 2016127599; applied on 8 July 2016; issued on 3 August 2017, Bulletin No. 22-6 p.: ill.

Author Contributions: Software, D.S.; Validation, S.K.; Data curation, N.S. (Nikita Serov); Writing—original draft, I.I.; Writing—review & editing, N.S. (Nikolay Shevkun) and A.A.; Visualization, V.P.; Project administration, I.I. All authors have read and agreed to the published version of the manuscript.

Funding: This article was funded by an intra-university grant from the Russian State Agrarian University—Moscow Timiryazev Agricultural Academy "Food Sovereignty" section for research projects in the field of import substitution within the framework of the "Priority 2030" strategic academic leadership program.

Institutional Review Board Statement: Not applicable.

Informed Consent Statement: Not applicable.

Data Availability Statement: The data presented in this study are available on request from the authors.

Conflicts of Interest: The authors declare no conflict of interest.

References

1. *Agriculture in Russia;* Statistical Digest: Rosstat, Moscow, Russia, 2021; 100p.
2. Petrakis, T.; Kavga, A.; Thomopoulos, V.; Argiriou, A.A. Neural Network Model for Greenhouse Microclimate Predictions. Emerging Agricultural Engineering Sciences, Technologies, and Applications. *Agriculture* **2022**, *12*, 780. [CrossRef]
3. Samarin, G.N. Energy-Saving Technology of Microclimate Formation in Livestock Buildings. Ph.D. Thesis, Moscow State Agroengineering University after V.P. Goryachkin (GMSAU), Moscow, Russia, 2009.
4. Cai, J.; Ding, F.; Hong, Y.; Hu, R. An Impact Analysis of Farmer Field Schools on Hog Productivity: Evidence from China. Agricultural Policy, Food Security, and Rural Development: Innovations and Perspectives. *Agriculture* **2021**, *11*, 972. [CrossRef]
5. American Society of Heating, Refrigerating and Air-Conditioning Engineers (ASHRAE). *ASHRAE Handbook. Heating, Ventilating, and Air-Conditioning Systems and Equipment*, I-P ed.; American Society of Heating, Refrigerating and Air-Conditioning Engineers (ASHRAE): Atlanta, GA, USA, 2008.
6. Tikhomirov, D.A. Energy-Saving Electric Systems and Technical Means for the Heat Supply of the Main Technological Processes in Animal Breeding. Ph.D. Thesis, All-Russian Scientific Research Institute for Electrification of Agriculture (VIESH), Moscow, Russia, 2015.
7. Cevallos, G.; Herrera, M.; Jaimez, R.; Aboukheir, H.; Camacho, O. A Practical Hybrid Control Approach for a Greenhouse Microclimate: A Hardware-in-the-Loop Implementation. Agricultural Systems and Management. *Agriculture* **2022**, *12*, 1916. [CrossRef]
8. Ogunlowo, Q.O.; Akpenpuun, T.D.; Na, W.-H.; Rabiu, A.; Adesanya, M.A.; Addae, K.S.; Kim, H.-T.; Lee, H.-W. Analysis of Heat and Mass Distribution in a Single- and Multi-Span Greenhouse Microclimate. Greenhouse Technology and Management. *Agriculture* **2021**, *11*, 891. [CrossRef]
9. Havelka, Z.; Kunes, R.; Kononets, Y.; Stokes, J.E.; Smutny, L.; Olsan, P.; Kresan, J.; Stehlik, R.; Bartos, P.; Xiao, M.; et al. Technology of Microclimate Regulation in Organic and Energy-Sustainable Livestock Production. Farm Animal Production. *Agriculture* **2022**, *12*, 1563. [CrossRef]
10. Khimenko, A.; Tikhomirov, D.; Trunov, S.; Kuzmichev, A.; Bolshev, V.; Shepovalova, O. Electric Heating System with Thermal Storage Units and Ceiling Fans for Cattle-Breeding Farms. Engineering Innovations in Agriculture. *Agriculture* **2022**, *12*, 1753. [CrossRef]
11. Roulet, C.A.; Heidt, F.D.; Foradini, F.; Pibiri, M.C. Real heat recovery with air handling units. *Energy Build.* **2001**, *33*, 495–502. [CrossRef]
12. Tikhomirov, D.A.; Vasilyev, A.N.; Budnikov, D.; Vasilyev, A.A. Energy-saving automated system for microclimate in agricultural premises with utilization of ventilation air. *Wirel. Netw.* **2019**, *26*, 4921–4928. [CrossRef]
13. Ignatkin, I.Y. Energosberezhenie pri otoplenii v usloviyakh kraynego severa (Energy saving when heating in the conditions of the Far North). *Vestnik NGIEI* **2017**, *1*, 52–58. (In Russian)
14. Adamski, M. Longitudinal flow spiral recuperators in building ventilation systems. *Energy Build.* **2008**, *40*, 1883–1888. [CrossRef]
15. Adamski, M. Ventilation system with spiral recuperator. *Energy Build.* **2010**, *42*, 674–677. [CrossRef]
16. Qin, X.; Wang, D.; Jin, Z.; Wang, J.; Zhang, G.; Li, H. A Comprehensive Investigation on the Effect of Internal Heat Exchanger Based on a Novel Evaluation Method in the Transcritical CO_2 Heat Pump System. *Renew. Energy* **2021**, *178*, 574–586. [CrossRef]
17. Qin, X.; Zhang, Y.; Wang, D.; Chen, J. System development and simulation investigation on a novel compression ejection transcritical CO_2 heat pump system for simultaneous cooling and heating. *Energy Convers. Manag.* **2022**, *259*, 115579. [CrossRef]
18. Ignatkin, I.Y.; Arkhiptsev, A.V.; Stiazhkin, V.I.; Mashoshina, E.V. A method to minimize the intake of exhaust air in a climate control system in livestock premises. In Proceedings of the International Conference on Agricultural Science and Engineering, Michurinsk, Russia, 12–14 April 2021; IOP Conference Series: Earth and Environmental Science. IOP Publishing: Bristol, UK, 2021. [CrossRef]
19. Kim, J.-G.; Lee, I.-B.; Lee, S.-Y.; Jeong, D.-Y.; Choi, Y.-B.; Cho, J.-H.; Kim, R.-W.; Aarnink, A. Development of an Air-Recirculated Ventilation System for a Piglet House, Part 2: Determination of the Optimal Module Combination Using the Numerical Model. Farm Animal Production. *Agriculture* **2022**, *12*, 1533. [CrossRef]
20. Kalinichenko, A.; Havrysh, V.; Hruban, V. Heat Recovery Systems for Agricultural Vehicles: Utilization Ways and Their Efficiency. Advances in Agriculture Mechanization. *Agriculture* **2018**, *8*, 199. [CrossRef]
21. Dieckmann, J. Improving humidity control with energy recovery. *ASHRAE J.* **2008**, *50*, 38–45.
22. El Foujh, Y.; Stabat, P. Adequacy of air-to-air heat recovery ventilation system applied in low energy buildings. *Energy Build.* **2012**, *54*, 29–39.
23. Gubina, I.A.; Gorshkov, A.S. Energy saving in buildings with heat recovery exhaust air. *Constr. Unique Build. Struct.* **2015**, *4*, 209–219.
24. Kazantsev, S.P.; Ignatkin, I.Y. Sistema mikroklimata v svinovodstve s primeneniyem okhladiteley novoy konstruktsii (Climate control system in pig breeding based on cooling units of a new design). *Mekhanizatsiya Elektrif. Sel'skogo Khozyaystva* **2014**, *5*, 18–20. (In Russian)
25. Tikhomirov, D.A.; Tikhomirov, A.M. Improvement and modernization of systems and means of power supply is the most important direction of solving the problems of increasing the energy efficiency of agricultural production. *J. Mach. Equip. Rural Areas* **2017**, *11*, 32–36.

26. Ivanov, Y.G.; Ponizovkin, D.A. Vliyanie parametrov vozdushnoy sredy korovnika na fiziologicheskie pokazateli zhivotnykh (Influence of indoor air parameters on the physiological parameters of animals). *Mekhanizatsiya Elektrif. Sel'skogo Khozyaystva* **2015**, *4*, 18–21. (In Russian)
27. Tikhomirov, D.A. Methodology of calculation heat and energy saving ventilation and heating units used in animal farms. *J. Altern. Energy Ecol.* **2013**, *2*, 125–131.
28. Tikhomirov, D.A. Electrical and thermal calculation of air heater recuperative heat exchanger. *J. Mech. Electrif. Agric.* **2013**, *1*, 15–17.
29. Böger, R.; Rohn, K.; Kemper, N.; Schulz, J. Sodium Hypochlorite Treatment: The Impact on Bacteria and Endotoxin Concentrations in Drinking Water Pipes of a Pig Nursery. Challenges and Perspectives in Pig Farming: Breeding, Husbandry and Management. *Agriculture* **2020**, *10*, 86. [CrossRef]
30. Teploutilizatsionnaya Ustanovka s Adaptivnoy Retsirkulyatsiey (Heat Recovery Unit with Adaptive Recirculation). Russian Federation Patent No. 2016127598, 3 October 2017. (In Russian).
31. Livchak, I.; Naumov, A. *Ventilation of Multi-Storey Residential Buildings*; AVOK-PRESS: Moscow, Russia, 2005.
32. Yudaev, B.N. *Teploperedacha (Heat Transfer)*, 2nd ed.; Vyssh. Shkola: Moscow, Russia, 1981; p. 319. (In Russian)
33. Barkalov, B.V.; Karpis, E.E. *Konditsionirovanie Vozdukha v Promyshlennykh, Obshchestvennykh i Zhilykh Zdaniyakh (Air Conditioning in Industrial, Public and Residential Buildings)*; Stroyizdat: Moscow, Russia, 1971; p. 270. (In Russian)
34. Gavrilkin, V.P.; Kuranov, A.E. Analiticheskoe opredelenie parametrov vlazhnogo vozdukha (Analytical determination of moist air parameters). *Vestnik ASTU* **2007**, *2*, 148–151. (In Russian)
35. Bogdanova, S.N. *Kholodil'naya Tekhnika. Konditsionirovanie Vozdukha. Svoystva Veshchestv (Refrigeration Technology. Air Conditioning. Properties of Substances)*, 4th ed.; SPb GAHPT: St. Petersburg, Russia, 1999. (In Russian)

agriculture

Article

Evaluation of Comparative Field Studies for Root and Onion Harvester with Variable Angle Conveyor

Alexey Dorokhov, Alexander Aksenov, Alexey Sibirev *, Maxim Mosyakov, Nikolay Sazonov and Maria Godyaeva

FSBSI "Federal Scientific Agronomic and Engineering Center VIM", Moscow 109428, Russia
* Correspondence: sibirev2011@yandex.ru; Tel.: +7-964-584-3518

Abstract: The long-term storage of vegetable crops and potatoes in storage, even with a small content of soil and plant impurities in the heap, does not ensure the preservation of the required quality indicators of marketable products. The lack of modern technological foundations for improving the design parameters of machines for harvesting vegetables and potatoes leads to the impossibility of eliminating the loss of root crops, their damage, as well as the high-quality process of cleaning from impurities. This circumstance is due to the fact that modern technologies and technical means of harvesting root crops and onions are not able to provide high-quality marketable products with minimal labor costs. In this regard, a modern separating system of the harvesting machine has been developed, which ensures the variation of technological parameters under changing conditions for the harvesting of root crops and onions. The experimental studies of the developed harvester with an experimental separating system made it possible to ensure the high-quality harvesting of potato and onion tubers with the following parameter values: a completeness of separation of more than 98% and a damage to products up to 1.7%, at a speed of movement of 1.7 m/s for the separating system; a completeness of separation of more than 98% and a product damage of up to 1.1% at a speed of up to 1.0 m/s for the harvester; and a separation completeness of more than 98% and a product damage of up to 1.4% at a commercial product extraction depth of 0.02 m. The results of comparative studies on the quality indicators of the machine for harvesting root crops in the harvesting of potato and onion tubers are presented, indicating the prevailing values of the quality indicators of work in the harvesting of potatoes, depending on the change in the regime indicators of the quality of work.

Keywords: technological process; rod elevator; lifting angle; displacement; experiment; cleaning machine

Citation: Dorokhov, A.; Aksenov, A.; Sibirev, A.; Mosyakov, M.; Sazonov, N.; Godyaeva, M. Evaluation of Comparative Field Studies for Root and Onion Harvester with Variable Angle Conveyor. *Agriculture* **2023**, *13*, 572. https://doi.org/10.3390/agriculture13030572

Academic Editor: Tao Cui

Received: 29 January 2023
Revised: 16 February 2023
Accepted: 23 February 2023
Published: 26 February 2023

1. Introduction

The development of agricultural engineering in the Russian Federation cannot be ensured without the development of machines with digital intelligent systems [1–6].

The quality of work is determined by the optimal solution to the problem of providing the required parameters of the machines, both for sowing, harvesting, and the post-harvest processing of crops [7–9], while maintaining the full potential of seeds and grown products [10–12]. The determining indicator of the quality of crop harvesting is their preservation and the minimum amount of losses and damage to marketable products, with an effective value of cleaning from various impurities, which cannot be achieved without the introduction of modern harvesting machines. The main difference between the harvesting of root crops and onions from the harvesting of other agricultural crops is the increased volume of separated soil and plant impurities, therefore, the separating system of machines for harvesting vegetables and potatoes should have a high-throughput [13,14]. However, the separating system of harvesters, represented by bar elevators, does not allow the full cleaning of commercial products from mechanical impurities, and to eliminate this drawback, separation intensifiers of various types have been introduced into their design. R. Farhadi et. at [15] intensified the process of the separation of potato tubers, the scheme and the general view of which are shown in Figures 1 and 2.

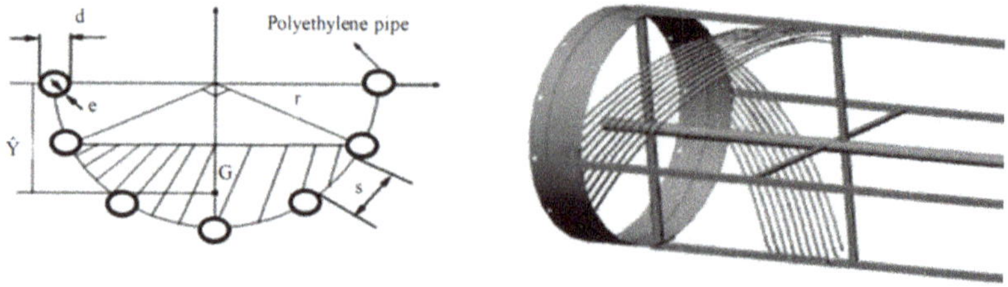

(a) **(b)**

Figure 1. Gradation pipes cross-section at a semicircle of the helix ((**a**) scheme; and (**b**) general form): d—the bar diameter; e—bar thickness; s—bar spacing; r—drum radius; G—center of gravity of the drum; and Ŷ—distance from the center of gravity to the center of the bar.

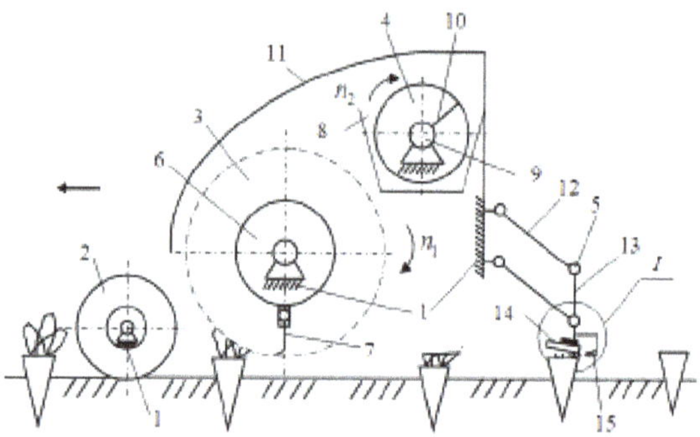

Figure 2. Constructive scheme of the improved machine for haulm cutting: 1—chassis frame; 2—support (reference) wheel; 3—rotor; 4—auger; 5—cutter of heads root crops; 6, 9—drum; 7—knife; 8—chute; 10—spiral; 11—casing; 12—parallelogram hinged suspension; 13—bracket; 14—copier; 15—knife; ←—forward movement.

Figure 1 shows the cross-section of the pipes at a semicircle of the helix. For the determination of the pipe diameter, the following points should be attended to: the semicircle radius and perimeter are 15 cm and πr. The initial distance between the pipes is two cm and the final distance is eight cm. The pipes' numbers of helix semicircle generators are 14 at first, for suitable grading. Anything lower than six pipes does not appear at the last pitch. The pipes' size should exist in industrial productions. This equation should be valid for avoiding semicircle defectiveness.

The various stages of piece construction were done with consideration to the presented matters before parts. Figure 2 shows the total side view of the system. Just as observed in the figure, the helix was rotated by a tractor with use of the multipurpose joint, belt and pulley, gearbox, chain, and sprocket wheel. This power transmission system was applied successfully to the reduction of the speed up to a necessary level, and diminished the rotational speed of 540 on 9 RPM. The main disadvantage of the well-known design of the separating device is the turbulent movement of the tuberous heap, which ensures the distribution of the marketable products of tubers along the diameter of the working body, without the presence of a soil layer, which increases the damage to the potato. A known separating device was developed by Ping Yuan Xiong, Xuan Lin, and Yi Wang [16]. A known machine for harvesting table roots was developed by Storozhuk I.M. and Pankiv V.R., the scheme of which is shown in Figure 2, providing an increase in the quality of the cleaning of commercial products as a result of improving the root adapter [17].

However, the disadvantage of this technical solution is the uneven height of the cut of the tops due to the changes in soil density when the machine moves across the field, which leads to a decrease in the product quality as a result of an additional mechanical impact on the root crop and the impossibility of the further utilization of plant impurities remaining on the surface of the root crop.

Thus, the improvement of the design of the digging working bodies of harvesting machines does not fully allow for the improvement of the quality of the harvesting, therefore, it is necessary to carry out the research and development of the functioning elements of the harvesting machine to clean the commercial products from plant and mechanical impurities.

The purpose of the study is to develop separating devices for harvesters that improve the quality of the separation and reduce the damage to commercial products, which is achieved through the determination of the quality indicators of harvesting with an experimental machine for harvesting potatoes and onions.

2. Materials and Methods

2.1. Bar Elevator with Adjustable Web Angle

The developed bar elevator (RF patent No. 2679734) helps to improve the quality of harvesting, in terms of indicators such as: the damage and completeness of the separation when changing the angle of inclination α_1 of the separating system web, depending on the variation in harvesting conditions [13].

In order to change the vertical position of the bar elevator (1), the weight sensors (2) installed on the harvesting machine (3) are used. (Figure 3).

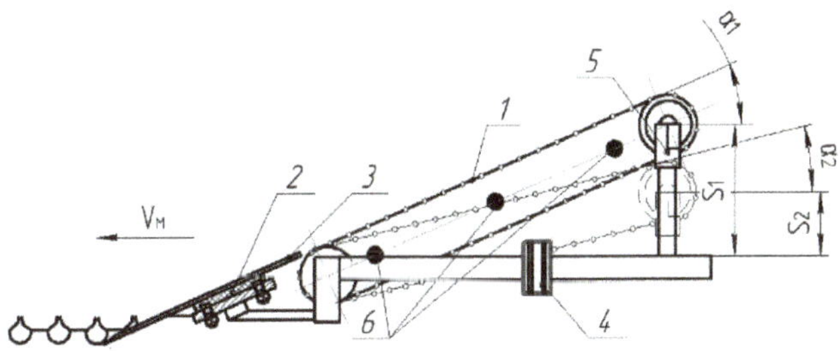

Figure 3. Scheme for changing the angle of inclination of the bar elevator conveyor: 1—bar elevator; 2—weight sensor of the digging share; 3—digging share; 4—microcontroller; 5—electric cylinder; and 6—inertial sensor; ←—forward movement.

The mass control device (2) of the heap entering the separation transmits a control signal to the microcontroller (4), and as a result, the web inclination angle is corrected when the mass of the incoming products changes.

The drive for changing the inclination angle of the web of the bar elevator (1), consisting of electric cylinders (5), ensures the movement of the rod to the required distance when the heap mass changes and, accordingly, the inclination angle of the elevator web (1) changes.

The control of the soil that has passed through the openings of the elevator is provided by inertia sensors (6), located along the length of the elevator (1).

In spite of this, the amount of sieved soil will be characterized by peak amplitude values. Thus, the evaluation of the separation efficiency of the elevator is carried out according to the amplitude values.

If the sieving of the soil on the rod elevator is below the required value set by the microcontroller (4), the angle α_1 of the inclination of the web of the rod elevator (1) is corrected in the set range of values.

Having determined the mass of a heap of root crops and onions on the digging share, the controller with a time delay T, with the movement of the bar elevator, transmits a control signal to move the rod of the electric cylinders.

Field studies on the separating rod elevator with an adjustable tilt angle of the machine for harvesting root crops and onions (Figure 3) were carried out on the fields of the farm "Tsirulev E.P." This was in the Privolzhsky district of the Samara region in 2019 for the harvest of Meduza onion, and also in 2022 for the harvest of the Red Scarlett potato variety at Krasnaya Gorka LLC, in the Kolyshleysky district of the Penza region. The studies were carried out in accordance with STO AIST 8.7-2013 "Machines for harvesting vegetables and melons. Methods for assessing functional indicators".

The scheme of the experiment included the performance of comparative studies on the quality indicators of onion and potato harvesting by a harvester with a developed separating system on a planting area of 0.04 ha.

Before performing studies on the experimental sample of the separating systems, the physical and mechanical properties of the soil were determined, and the quality indicators of the separation of a heap of turnips and potato tubers were also determined. Other factors considered were the soil of the digging plowshare, the forward speed v_K of the movement of the machine for the harvesting of the root crops and onions, as well as the forward speed v_{EL} of the movement of the rod elevator web on the indicators of the quality of harvesting. The assessment of the quality of the technological process of cleaning was carried out according to the following indicators:

- The damage D to the bulbs/tubers;
- The completeness of the separation ν of a heap of onions/tubers [17,18].

The extraction of the marketable products from the soil was carried out at a depth h_L of digging from 0.02 to 0.06 m, with a value of its change equal to 0.1 m.

The step of changing the speed v_K of the harvester varied within 0.2 m/s in the range from 1.0 to 1.8 m/s.

By changing the gear ratio of the separating system, the speed v_{EL} of the rod elevator varied from 1.0 to 1.8 m/s, with a step of 0.2 m/s.

The quality indicators of the machine with an experimental separating system were determined after passing the accounting plot, and the sifted products on the tarpaulin along the entire length of the harvested area were collected.

In the selected heap, its fractional composition was determined: bulbs/tubers, free soil, and the soil associated with marketable products.

Before performing studies on the experimental sample of the separating systems, the physical and mechanical properties of the soil were determined, and the quality indicators of the separation of a heap of turnips and potato tubers were also determined. For conducting research on the site, the following were insured: that there was medium loamy chernozem, that the field relief was even, that the field contour was close to rectangular, and that the length of the rut was 350 m. Other factors considered were the soil of the digging plowshare, the forward speed v_K of the movement of the machine for the harvesting of the root crops and onions, as well as the forward speed v_{EL} of the movement of the rod elevator web on the indicators of the quality of harvesting. The assessment of the quality of the technological process of cleaning was carried out according to the following indicators:

2.2. Bar Elevator with Asymmetric Shakers

Due to the fact that the technology of growing potatoes differs from onion crops in-depth, their patterns of planting, tillage, and, accordingly, the depth of digging during harvesting also differ. This is due to the fact that when onion ripens, more than 1/2 of the bulb is located on the soil surface, which contributes to the ripening, the formation of a good shirt, and facilitates mechanized harvesting while reducing the flow of soil impurities to the separating working bodies, which is a distinctive feature of potato harvesting, since the tuber nest is located at a digging depth of 15–20 cm and, together with marketable products, a large amount of the soil provides impurities.

Therefore, to intensify the process of separating the potato tubers from the soil impurities, separation intensifiers with an asymmetric arrangement of shakers are installed in the separating system of the harvester (Figure 4).

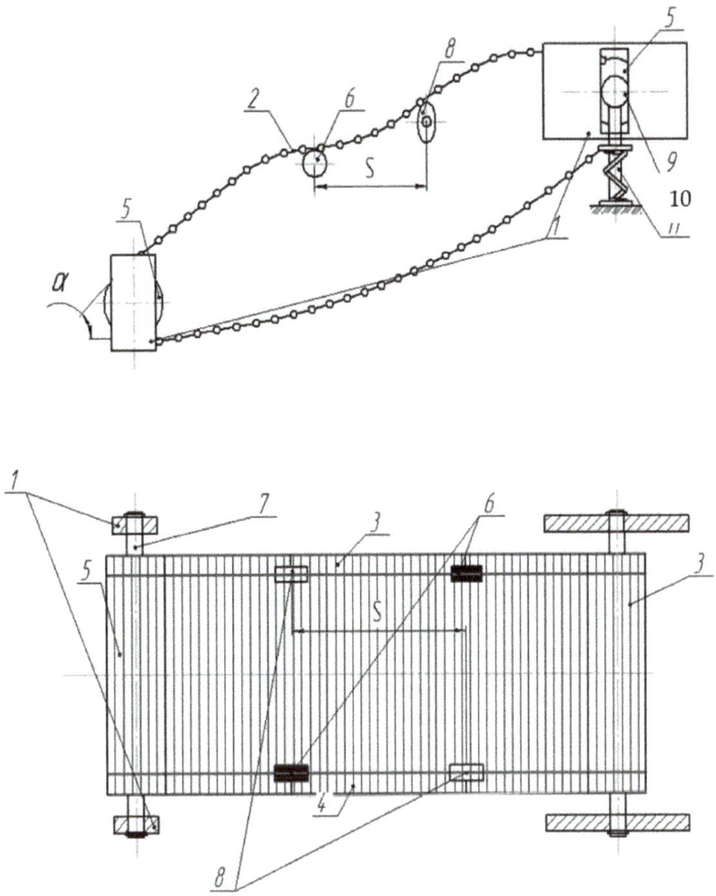

Figure 4. Scheme of separating rod elevator: 1—frame; 2—rod elevator; 3, 4—branches of rod elevator; 5—leading roller; 6—support roller; 7—driven roller; 8—elliptical shaker; 9—driven roller shaft; and 10—the mechanism for regulating vertical movement.

The separating rod elevator with asymmetrically installed passive elliptical shakers (RF patent No. 2638190) and with an adjustable angle of the rod elevator provides a reduction in the damage to and an increase in the quality of the separated products. This is as a result of reducing the effect of the vertical component of the gravity of the roots and bulbs to a minimum, as well as increasing the uniformity of the distribution of a pile of root crops and bulbs along the separating surface, when the angle α of the inclination of the rod elevator changes as a result of the changes in soil and climatic conditions for harvesting root crops, onions, and potatoes [17].

The separating conveyor of the machine for harvesting root crops, onions, and potatoes (Figure 4) contains a bar elevator (2) mounted on frame (1), undersides (3), and branches of a rod elevator (4), of which a leading roller (5) is installed, and support (6) and driven (7) rollers are mounted on a frame (1).

The results of field studies on the developed machine for harvesting potatoes of the Red Scarlett variety obtained dependencies that determined the performance indicators for the completeness of the cleaning and damage to tubers in the conditions of Krasnaya Gorka LLC in 2022.

The studies were carried out after the desiccation of the tops of a potato plant, with its subsequent removal on medium sandy soils at a moisture content along the entire length of the accounting plot, within 18–22%.

When conducting research within the conditions of the above farms to determine the quality indicators of harvesting, with a duration of harvesting an area of 0.1 ha, the bulbs and tubers were sorted into intact and damaged.

The damaged bulbs/tubers included products with slight and severe damage caused by machine harvesting.

Light damage to the bulbs/tubers includes:

- bulbs/tubers, bare up to 1/2 with cracks 1 mm deep, up to 10 mm long.

Severe product damage includes:

- bulbs/tubers more than 1/2 bare, with cracks more than 1 mm deep and more than 10 mm long, with dents more than 10 mm.

The damage to the bulbs/tubers was determined by the formula:

$$P = \frac{G_P}{G_{CT} - G_P} \cdot 100\%, \tag{1}$$

where

G_P—the weight of the damaged standard bulbs/tubers in a heap, kg;
G_{CT}—the mass of the separated bulbs/tubers in a heap, kg.

The completeness of the separation of a heap of onions/tubers was determined by the formula:

$$\nu = \frac{\nu_P^I - \nu_P^K}{\nu_P^I} \cdot 100\%, \tag{2}$$

where

ν_P^I—the mass of the soil impurities in the initial heap, kg;
ν_P^K—the mass of the soil impurities in the container (non-isolated impurities), kg.

The quality of the harvester was determined as follows.

At the beginning of the accounting plot, with the non-stop movement of the harvesting unit, on a signal, a tarpaulin was placed under the separating rod elevator, into which the entire harvested mass was collected.

In the process of passing the plot, a tarpaulin was unwound behind the machine, onto which a heap fell after the separation.

Next, samples were taken from the surface of the tarpaulin from the entire territory of the accounting plot. At the same time, the fractional composition of the heap was determined, which took into account: the bulbs/tubers, the free soil, and the soil associated with the bulbs/tubers.

2.3. Research Processing Methods

The reliability of the obtained data is ensured by the methods of mathematical processing and statistical analysis used for the research results and multivariate analysis, which included the use of licensed mathematical software packages for PC: "Microsoft Excel", "STATISTICA-10.0", "Math CAD 2020".

At the same time, the concepts and elements generally accepted in variation statistics that characterize the variation series were used: the average variation—X, the standard deviation—σ, and the coefficient of variation—ν. Each of the main elements was determined according to the known formulas of variation statistics.

This made it possible to determine the accuracy of the experimental data and establish the acceptable limits within which they are sufficiently reliable.

To determine the number of intervals (K) for the varying values of the parameters of the size–mass characteristics of the tubers, we will use the empirical relationship:

$$K = \sqrt{n}, \tag{3}$$

where n = number of tubers, pcs; and $K = \sqrt{100} = 10$.

The sampling range:

$$R = xmax - xmin, \tag{4}$$

where xmax and xmin are the maximum and minimum values of the investigated feature.

The interval of the investigated feature:

$$D = R/K. \tag{5}$$

3. Results and Discussion

Based on the results of the experimental data processing, graphs were plotted for the dependence of the completeness of separation of the marketable products of bulbs ν_B and tubers ν_T, as well as their damage (D_b and D_t), on the mode and technological parameters of the harvesting machine (h_L and ν_P) and the separating rod elevator (ν_{EL}) with an adjustable angle canvas tilt. The research results are presented in Figures 5–9.

With a change in the depth of digging onions and potato tubers, there is a variation in the completeness of the separation of and damage to marketable products, which is explained by the presence of an additional soil layer between the roots and the working surface of the bar cloth, with an increase in the depth of the digging while reducing the damage and increasing the completeness of the separation, regardless of the harvested crop (Figure 7A,B)

Figure 5. General view of the harvesting machine for root crops and onions, equipped with a separating rod elevator with an adjustable blade inclination angle and a receiving plowshare for digging/picking up root crops and bulbs.

Figure 6. General view of the harvesting unit for harvesting root crops and onions, equipped with a separating rod elevator with an adjustable blade inclination angle: 1—MTZ 1221 tractor; and 2—machine for harvesting root crops and onions.

The greatest damage to 1.8% and 1.4% of the commercial products of onions and potatoes, respectively, is observed with a decrease in the depth of digging to 4 cm, which is explained by the difference in their physical and mechanical properties, and the formation of a more stable peel in the potato tubers than a scaly shirt in bulbs.

The difference between the minimum 1.1% and maximum 1.9% damage to the root crops within the identical values of the studied technological parameter is no more than 0.4%, and is expressed by the correlation dependence:

$$\begin{cases} D_B(h_L) = 1.88 - 0.47 \cdot h_L - 0.07 \cdot h_L^2, \\ D_T(h_L) = 1.39 - 0.85 \cdot h_L + 0.53 \cdot h_L^2. \end{cases} \quad (6)$$

An obvious circumstance is an increase in the completeness of the cleaning of the root crops (Figure 7B), with a decrease in the depth of digging, which is due to a decrease in the supply of soil impurities to the cleaning devices and a simultaneous increase in damage to the marketable products.

The maximum completeness of the cleaning of the bulbs was more than 96%, and is observed at a depth of digging of up to 4 cm, which, when extracting potato tubers from the soil, provides 95.4% of the completeness of the separation.

The minimum value of the completeness of the separation of the root and tuber crops was 91%, and is observed at a depth of 10 cm of digging in the potato tuber nest (which is 91.6% in comparison with the commercial onion products), however, with an increase in the depth of digging to 12 cm, an increase in the completeness of the cleaning is observed by 0.4% and is explained by an increase in the time of the separated mass on the working surface of the rod elevator, in view of increasing its load and reducing the forward speed of its movement. Differences between the completeness of the cleaning of the studied root crops with a change in the depth of digging are observed, on average, up to 0.5%, and are expressed by a correlation dependence:

$$\begin{cases} \nu_B(h_L) = 98.6 - 2.66 \cdot h_L + 0.3 \cdot h_L^2, \\ \nu_T(h_L) = 98.4 - 3.56 \cdot h_L + 0.45 \cdot h_L^2. \end{cases} \quad (7)$$

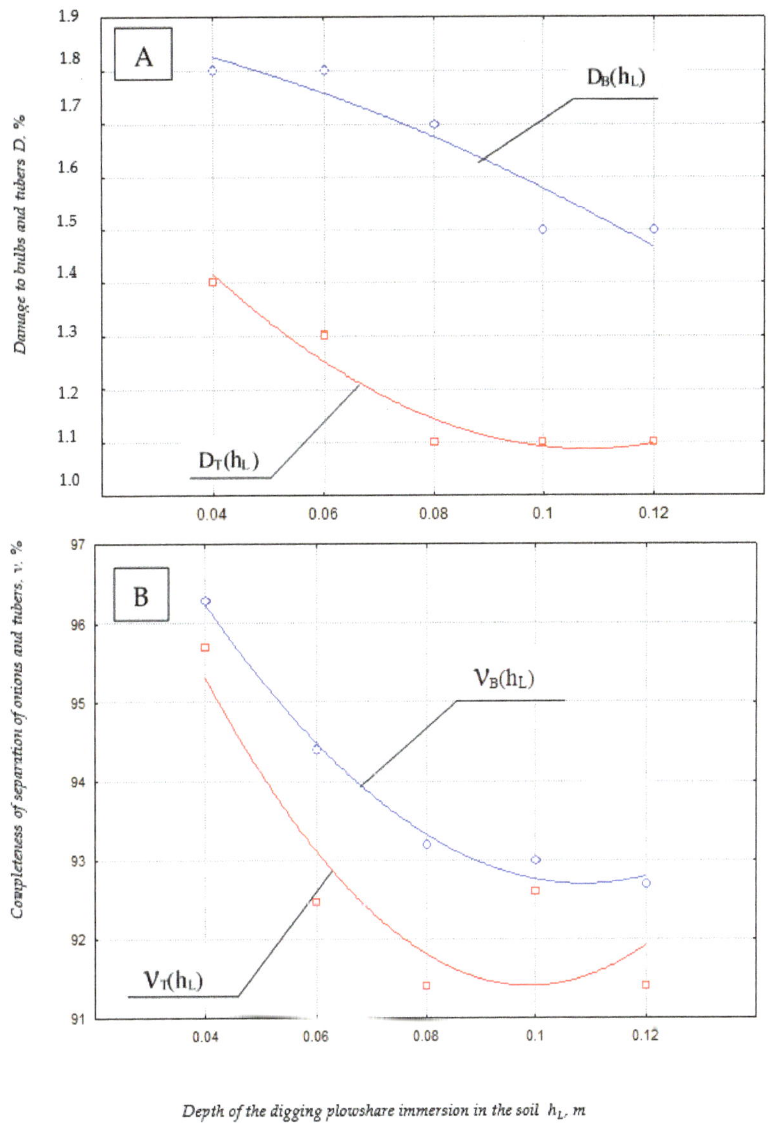

Figure 7. Dependence of: (**A**) damage to bulbs D_B and tubers D_T %; and (**B**) completeness of separation of onions ν_B and tubers ν_T% of the separating rod elevator with an adjustable blade inclination angle from the depth h_L of digging share into the soil.

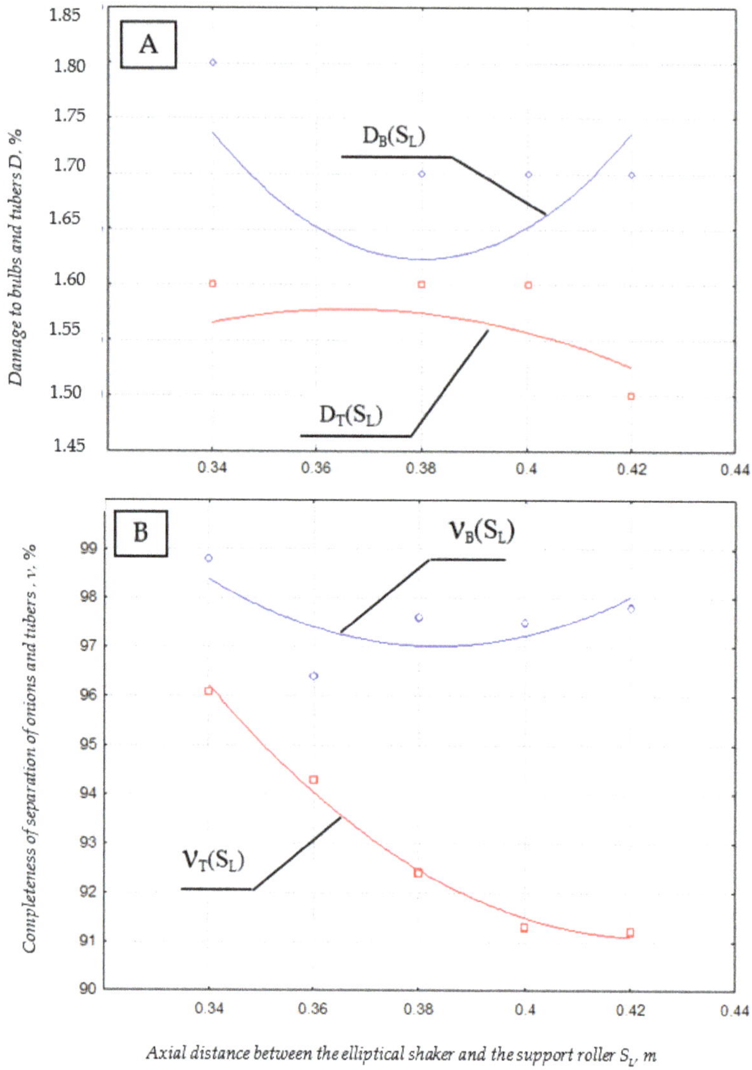

Figure 8. Dependence of: (**A**) damage to bulbs D_B and tubers D_T, %; and (**B**) completeness of separation of onions ν_B and tubers ν_T% of the separating rod elevator with an adjustable blade inclination angle from the center distance S_L between the shakers.

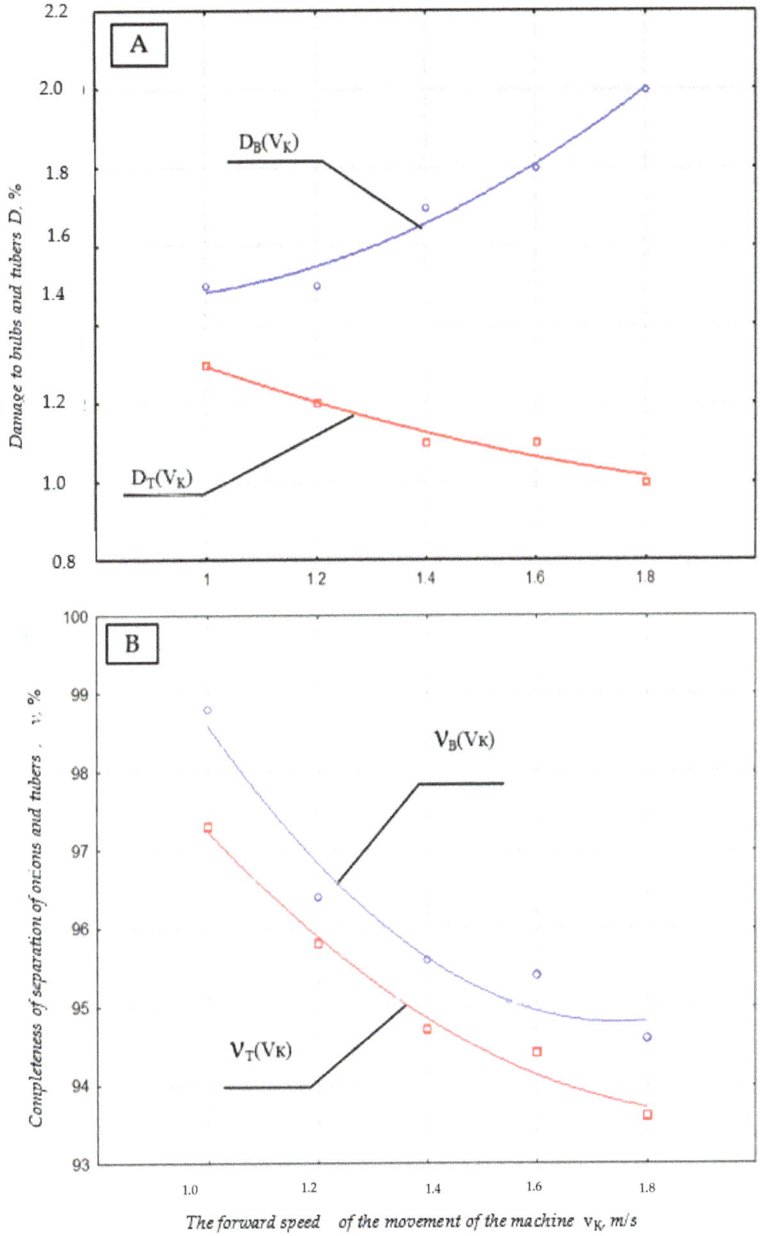

Figure 9. Dependence of: (**A**) damage to bulbs D_L and tubers D_K%; and (**B**) completeness of separation of onions v_B and tubers v_T% of the separating rod elevator with an adjustable blade inclination angle from the forward speed v_K of the machine.

An increase in the quality of the harvesting of potato and onion tubers using a separating rod elevator with asymmetrically installed shakers is ensured by increasing the time for the discrete comparison of impurities with the slotted openings of the working body, as a result of moving not only along the length, but also the width of the rod elevator. At the same time, the maximum amount of the damage to the bulbs is no more than 1.8% at a constant digging depth of 0.1 cm and a forward speed of the rod elevator v_{EL} and machine v_K within 1.0 and 1.2 m/s, respectively. The difference between the maximum and minimum damage to the potato tubers in the studied range of the shaker locations is no more than 0.3%, with a minimum value of 1.52%. In addition, the combination of asymmetrically located shakers on the rod elevator provides a better process of potato tuber separation within 96%, which is 2.3% lower compared to its use in onion harvesting at a technological parameter value of 0.34 cm.

With an increase in the center distance to the maximum allowable value of 0.42 m, a decrease of 91.5% in the completeness of separation is observed in the cleaning of the potato tubers, which is lower than the minimum value of the cleaning of the bulbs by 7%, and is explained by a decrease in the transverse amplitude of the oscillation of the web, due to the connectivity of the potato tubers in the tuber nest of the plant.

The performance indicators are expressed by the correlation dependence:

$$\begin{cases} D_B(S_L) = 1.88 - 0.17 \cdot S_L - 0.02 \cdot S_L^2, \\ D_T(S_L) = 1.54 + 0.03 \cdot S_L - 0.07 \cdot S_L^2. \end{cases} \tag{8}$$

$$\begin{cases} v_{Л}(S_{Л}) = 99.2 - 2.84 \cdot S_L + 0.43 \cdot S_L^2, \\ v_K(S_{Л}) = 98.1 - 3.16 \cdot S_L + 0.35 \cdot S_L^2. \end{cases} \tag{9}$$

The graphic dependences of the change in the quality indicators of the harvesting of potatoes and onions using a rod elevator with an adjustable blade inclination angle on the forward speed v_K of the machine, are shown in Figure 9.

An increase in the studied technological indicator of potato harvesting leads to a decrease in the quality indicators of the minimum values of 1.0% of damage to the tubers, and the 93.8% completeness of separation to the maximum of 1.2% for damage and 97.3% of the completeness of separation.

At the same time, it should be noted that there is a distinctive feature in terms of the damage to onion commercial products, with an increase in the forward speed of the machine from 1.5% to 2.0%, which is explained by an increase in the collisions between the bulbs due to the thickening of their planting, and an increase in the supply of commercial products for separation.

The difference between the minimum values of the damage to bulbs and tubers reaches 1.5% and 1.0%, respectively, with maximum values of 1.2% and 2.0% damage, and is expressed by the equations of parabolic functions:

$$\begin{cases} D_B(v_K) = 1.88 - 0.17 \cdot v_K - 0.02 \cdot v_K^2, \\ D_T(v_K) = 1.54 + 0.03 \cdot v_K - 0.07 \cdot v_K^2. \end{cases} \tag{10}$$

The difference between the quality of the cleaning by groups of commercial products is no more than 2%, with an increased completeness of separation of 97.3% for the harvesting of onions and a minimum value for the harvesting of potatoes of 93.8% with a correlation dependence:

$$\begin{cases} v_B(v_K) = 98.7 - 2.56 \cdot v_K + 0.27 \cdot v_K^2, \\ v_T(v_K) = 98.2 - 1.82 \cdot v_K + 0.15 \cdot v_K^2. \end{cases} \tag{11}$$

Upon analyzing the graphical dependencies (Figure 10), we can say that the maximum damage value is observed for the commercial onion products and is 1.9%, which is 1.7% for potato tubers.

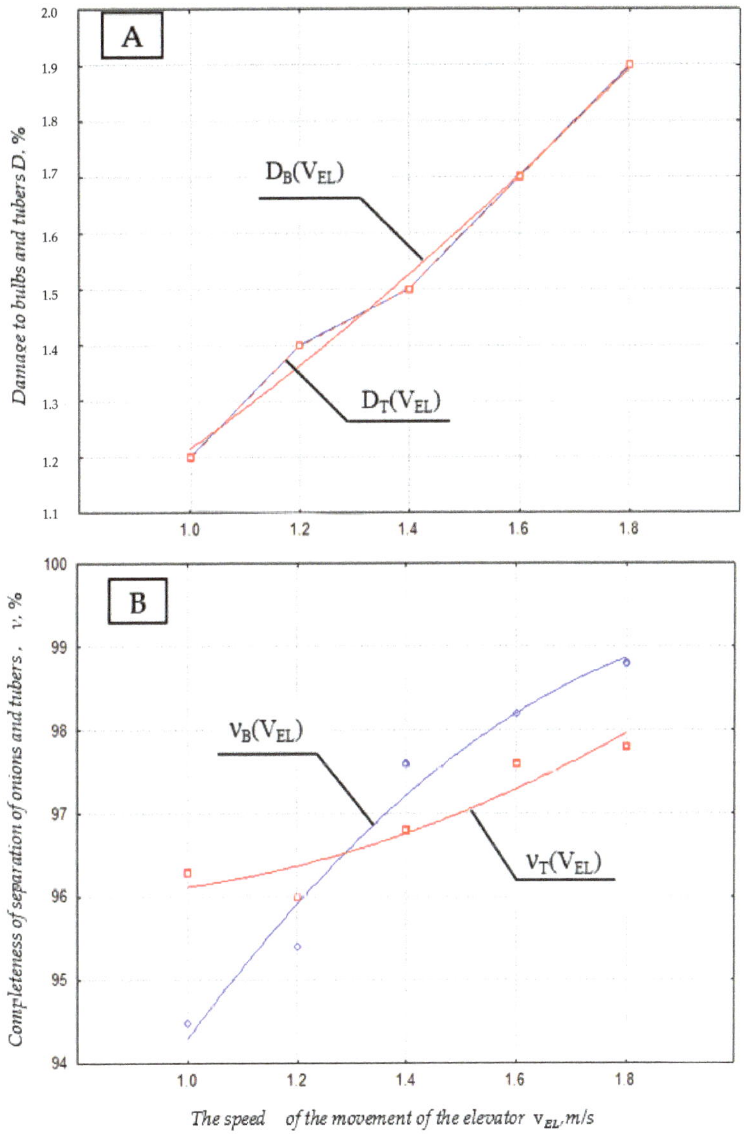

Figure 10. Dependence: (**A**) damage to bulbs D_B and tubers D_T%; and (**B**) completeness of separation of onions ν_B and tubers ν_T% of the separating rod elevator with an adjustable blade inclination angle from the translational speed v_{EL} of the rod elevator.

The minimum damage value of 1.2% is for all the groups of the commercial products, and increases with an increase in the forward speed of the rod elevator in the range of 1.0 to 1.8 m/s. The correlation between the damage to the marketable products of potato tubers and onion bulbs is determined by the expression:

$$D_B,\ D_T(v_{EL}) = 1.08 - 0.12 \cdot v_{EL} + 0.7 \cdot v_{EL}^2. \tag{12}$$

The maximum separation completeness of more than 98.8% is observed in onion harvesting, which is higher than the completeness of the potato tubers' cleaning by 1%, at a forward speed of the rod elevator of 1.8 m/s.

At the minimum value of the studied indicator of the harvesting machine, equal to 1 m/s, the completeness of the cleaning has a minimum value for potato tubers of 96%, which is 2% higher than that for onion harvesting.

In addition, the optimal value of the forward speed of the rod elevator, at which the approximation of the cleaning completeness curves is 96.8% at a speed value of 1.3 m/s, is described by the correlation dependence:

$$\begin{cases} \nu_B(v_{EL}) = 92.4 - 2.08 \cdot v_{EL} - 0.15 \cdot v_{EL}^2, \\ \nu_T(v_{EL}) = 96.2 + 0.31 \cdot v_{EL} + 0.07 \cdot v_{EL}^2. \end{cases} \quad (13)$$

According to the results of the research, it follows that the optimal ratio of the quality indicators of onion harvesting is ensured when crossing curves approximate the completeness of separation at 95.5% and the damage to the bulbs at 1.2% at a forward speed of 1.38 m/s of the machine for the harvesting of root crops and onions.

The reliability of the conducted studies was assessed by the calculated value of the mathematical expectation M(X), and the normal law of the distribution of the damage to the commercial products on a bar elevator is:

$$M(X) = 0.3. \quad (14)$$

Student's distribution quantile:

$$T = qt\left(1 - \frac{\alpha}{2}, \upsilon\right) = 2.012. \quad (15)$$

Calculation of statistics criterion and distribution of a random variable over intervals submitted in Table 1.

Table 1. Calculation of statistics criterion and distribution of a random variable over intervals.

	Observed	Cumulative	Percent	Cumul. %	Expected	Cumulative	Percent	Cumul. %	Observed
2.5	0	0	0	0	0.26	0.26	1.79	1.79	−0.26
2.55	0.5	1	6.66	6.66	0.47	0.74	3.17	4.97	0.52
2.6	1	3	13.33	20.0	0.98	1.73	6.56	11.54	1.01
2.65	1.5	3	0	20.0	1.67	3.41	11.15	22.69	−1.67
2.7	2.5	8	33.33	53.33	2.32	5.73	15.53	38.22	2.67
2.75	2.5	8	0	53.33	2.65	8.39	17.72	55.95	−2.65
2.8	2.5	13	33.33	86.66	2.48	10.88	16.59	72.54	2.51
2.85	2	0	0.00000	86.66	1.91	12.79	12.73	85.27	−1.91
2.9	1	15	13.33	100.0	1.21	13.99	8.01	93.29	0.79
Infinity	1	15	0	100.0	1.0	15.0	6.7	100.0	−1.0

Graphically, the observed and expected frequencies are determined by constructing scatter plots (Figure 11).

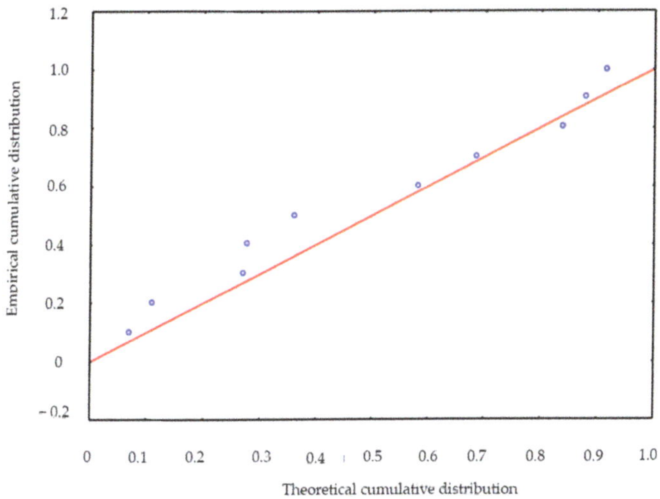

Figure 11. Plot of observed values (empirical cumulative distribution) versus expected frequencies (theoretical cumulative distribution).

The mathematical expectation M(X) of the normal distribution law for the separation of the potato tubers and bulbs on the developed cleaning device:

$$\mathrm{M(X)} = 0.2. \quad (16)$$

Student's distribution quantile:

$$\mathrm{T} = \mathrm{qt}\left(1 - \frac{\alpha}{2}, \upsilon\right) = 2.012. \quad (17)$$

Calculation of statistics criterion and distribution of a random variable over intervals submitted in Table 2.

Table 2. Calculation of statistics criterion and distribution of a random variable over intervals.

	Observed	Cumulative	Percent	Cumul. %	Expected	Cumulative	Percent	Cumul. %	Observed-
95	0	0	0	0	0.16	0.16	1.08	1.08	−0.16
95.5	1	1	6.66	6.66	0.31	0.47	2.06	3.15	0.69
96	1	2	6.66	13.33	0.68	1.15	4.58	7.73	0.31
96.5	0	2	0	13.33	1.26	2.42	8.43	16.16	−1.26
97	0	2	0	13.33	1.93	4.35	12.88	29.04	−1.93
97.5	2.4	8	40	53.33	2.44	6.81	16.32	45.37	3.55
98	2.6	11	20	73.33	2.57	9.38	17.15	62.53	0.42
98.5	0	11	0	73.33	2.24	11.62	14.96	77.49	−2.24
99	3	14	20	93.33	1.62	13.24	10.82	88.31	1.37
99.5	0	14	0	93.33	0.97	14.22	6.49	94.81	−0.97
100	1	15	6.66	100	0.48	14.71	3.23	98.04	0.51
Infinity	0	15	0	100	0.29	15	1.95	100	−0.29

Graphically, the observed and expected frequencies are presented in the scatter plot (Figure 12).

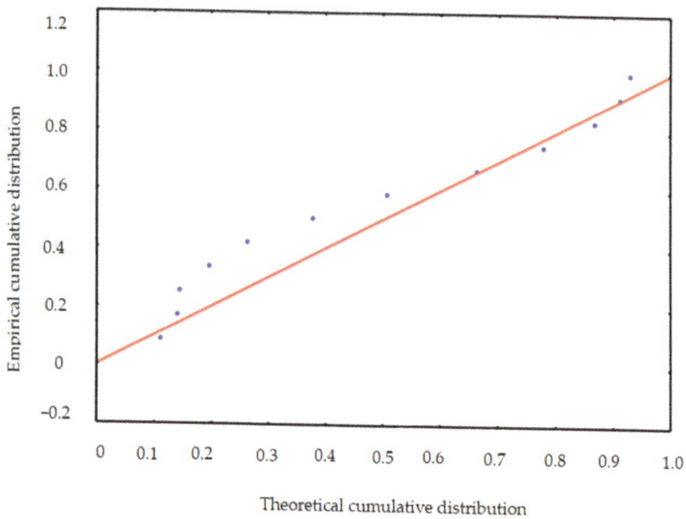

Figure 12. Plot of observed values (empirical cumulative distribution) versus expected frequencies (theoretical cumulative distribution).

The statistical processing of the results of the experimental studies, according to the criteria S_y^2 и S_{LF}^2, are shown in Figure 13.

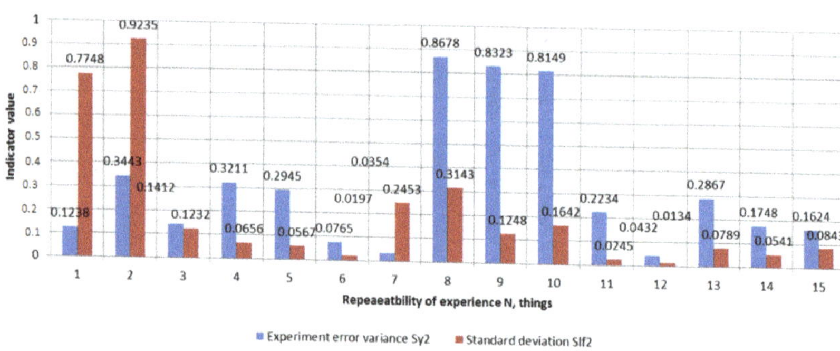

Figure 13. Statistical characteristics of experience error.

The results of the conducted field studies on a machine for harvesting root crops and onions, equipped with a bar elevator with an adjustable blade inclination angle and an asymmetric arrangement of shakers, showed the high-quality performance of the technological process of separating turnips and potato tubers at the optimal values of the parameters: a completeness of separation of more than 98%, and damage to the products of up to 1.7% at a speed of movement of 1.7 m/s of the separating system; a completeness of separation of more than 98%, and s product damage of up to 1.1% at a speed of up to 1.0 m/s of the harvester, and a separation completeness of more than 98%, and product damage of up to 1.4% at a commercial product extraction depth of 0.02 m.

The results obtained from the studies are consistent with the previously known data on the development and testing of machines in terms of the cleaning of the marketable products of root and tuber crops.

The obtained research results are consistent with the previously known directions for improving the design of harvesters.

Research performed by V.P. Khambalkar and colleagues developed an onion harvester that has been helpful in harvesting the onion bulb effectively. The design of the onion harvester greatly depended on the working width, the depth of the operation, and the conveying capacity. The agro-technical data of the crop has been collected from the university field regarding the depth of the onion bulb, the spacing between the plant and the row-to-row distance, the height of crop, and the polar and equator dimensions of the onion bulb. The overall size of the harvester is 1020 × 600 × 300 mm. The width-to-depth ratio for the optimum operation performance is chosen to be 6:1. The depth of the operation according to the agro-technical requirement comes to be 10 cm. The working width of the machine, with respect to the power availability and the crop sowing characteristics, comes to be 60 cm [18].

4. Conclusions

For the manufacture and production of the technical means of a potato harvesting machine, it is recommended to choose technological parameters from experimentally confirmed values that ensure the high-quality performance of the harvesting process.

Improving the design and technological schemes of the machines for harvesting potatoes and onions, on the basis of the conducted research, can further be used in the system for monitoring the quality indicators of the machines for the harvesting and post-harvest processing of potatoes and other agricultural root crops, which will provide a new level of synthesis of the complex technical systems of agricultural production.

The proposed technological and technical solutions can serve as a basis for expanding the research on and practical implementation of Russia's transition to the sustainable development of vegetable production.

Author Contributions: Conceptualization, A.S.; methodology, A.D. and M.G.; software, N.S. and M.G.; validation, A.A.; investigation, A.S.; resources, M.M.; writing—original draft preparation, M.G.; writing—review and editing, A.S.; project administration, A.S.; funding acquisition, N.S. All authors have read and agreed to the published version of the manuscript.

Funding: The research was carried out with the financial support of the Russian Science Foundation of the 2022 contest "Conducting research by scientific groups led by young scientists" of the Presidential Program of research projects implemented by leading scientists, including young scientists No. 22-76-10002.

Institutional Review Board Statement: The authors declare that they have no known competing financial interests or personal relationships that could have appeared to influence the work reported in this paper.

Informed Consent Statement: The authors declare that they have no known competing financial interests or personal relationships that could have appeared to influence the work reported in this paper.

Data Availability Statement: The raw data supporting the conclusions of this article will be made available by the authors, without undue reservation.

Conflicts of Interest: The authors declare that they have no known competing financial interests or personal relationships that could have appeared to influence the work reported in this paper.

References

1. Bentini, M.; Caprara, C.; Martelli, R. Harvesting damage to potato tubers by analysis of impacts recorded with an instrumented sphere. *Biosyst. Eng.* **2006**, *94*, 75–85. Available online: https://www.researchgate.net/publication/222060402_Harvesting_Damage_to_Potato_Tubers_by_Analysis_of_Impacts_recorded_with_an_Instrumented_Sphere (accessed on 7 February 2006). [CrossRef]
2. Akdemir, B.; Ulger, P.; Arin, S. Mechanized panting and harvesting of onion. *Agric. Mech. Asia* **1993**, *24*, 23–26.
3. Mendoza, F.; Lu, R.; Cen, H. Grading of apples based on firmness and soluble solids content using Vis/SWNIR spectroscopy and spectral scattering techniques. *J. Food Eng.* **2014**, *125*, 59–68. Available online: https://www.researchgate.net/publication/259122251_Grading_of_apples_based_on_firmness_and_soluble_solids_content_using_VisSWNIR_spectroscopy_and_spectral_scattering_techniques (accessed on 22 October 2013). [CrossRef]
4. Böl, M.; Seydewitz, R.; Leichsenring, K.; Sewerin, F. A phenomenological model for the inelastic stress–strain response of a potato tuber. *J. Mech. Phys. Solids* **2020**, *137*, 94–103. [CrossRef]

5. Dorokhov, A.S.; Sibirev, A.V.; Aksenov, A.G. Results of field studies on the separation of a heap of onion sets on a bar elevator with asymmetrically installed shakers. *Eng. Technol. Syst.* **2020**, *1*, 133–149. Available online: http://vestnik.mrsu.ru/index.php/ru/articles2/89-20-1/762-10-15507-0236-2910-030-202001-8 (accessed on 31 March 2020).
6. Pasaman, B.; Zakharchuk, V. The determination of the parameters of a ploughshare-rotorting potato digger. Econtechmod. *Int. Q. J.* **2012**, *1*, 43–47. Available online: http://www.journals.pan.pl/dlibra/publication/98931/edition/85237/content (accessed on 28 March 2012).
7. El-Rahman, A.; Magda, M. Development and performance evaluation of a simple grading machine suitable for onion sets. *J. Soil Sci. Agric. Eng. Mansoura Univ.* **2011**, *2*, 213–226. Available online: https://www.researchgate.net/publication/339175331_Performance_Evaluation_of_Onion_Grader (accessed on 10 October 2011). [CrossRef]
8. Khura, T.; Mani, I.; Srivastava, A. Design and development of tractor-drawn onion (Allium cepa) harvester. *Indian J. Agric. Sci.* **2011**, *6*, 528–532. Available online: https://www.researchgate.net/publication/267251851_I_C_A_R_Design_and_development_of_tractor-drawn_onion_Allium_cepa_harvester (accessed on 13 June 2011).
9. Indraja, D.; Ajkhilesh, J.; Vishal, P.; Amarsingh, P.; Ashutosh, D. A Review paper based on Design and Development of An Onion Harvesting Machine. *J. Inf. Comput. Sci.* **2019**, *9*, 333–337. Available online: https://www.researchgate.net/publication/339201506 (accessed on 9 February 2020).
10. Bachche, S. Deliberation on Design Strategies of Automatic Harvesting Systems: A Survey. *Robotics* **2015**, *4*, 194–222. Available online: https://www.researchgate.net/publication/278472968_Deliberation_on_Design_Strategies_of_Automatic_Harvesting_Systems_A_Survey (accessed on 11 June 2015). [CrossRef]
11. Dai, F.; Guo, X.H.; Zhao, W.Y.; Xin, S.L.; Liu, X.L.; Wu, Z.W. Design and experiment of canvas belt combined operation machine for potato digging and plastic film collecting. *Trans. CSAM* **2018**, *3*, 104–113. Available online: https://www.researchgate.net/publication/319244994_Design_and_experiment_of_combined_operation_machine_for_potato_harvesting_and_plastic_film_pneumatic_auxiliary_collecting (accessed on 17 June 2018).
12. Sibirev, A.; Aksenov, A.; Kynev, N.; Sazonov, N. Separating Conveyor of a Machine for Harvesting Root Crops and Bulbs—No. 20181175/25. Patent of the Russian Federation No. 2,679,734, 5 November 2018.
13. Bachche, S.; Oka, K. Modeling and performance testing of end-effector for sweet pepper harvesting robot. *J. Robot. Mechatron.* **2013**, *25*, 705–717. Available online: https://www.researchgate.net/publication/259702640_Design_Modeling_and_Performance_Testing_of_End-Effector_for_Sweet_Pepper_Harvesting_Robot_Hand (accessed on 15 June 2013). [CrossRef]
14. Farhadi, R.; Sakenian, N.; Azizi, P. Design and construction of rotary potato grader. (Part I). *Bulg. J. Agric. Sci.* **2012**, *18*, 304–314. Available online: https://agrojournal.org/18/02-21-12.pdf (accessed on 4 April 2012).
15. Xiong, P.; Lin, X.; Wang, Y.; Shen, S. Simulation of southern potato grader based on adams. *INMATEH—Agric. Eng.* **2016**, *1*, 5–13. Available online: https://www.irjet.net/archives/V4/i10/IRJET-V4I10257.pdf (accessed on 5 January 2016).
16. Storozhuk, I.M.; Pankiv, V.R. Research results of harvesting haulm remnants of root crops. *INMATEH—Agric. Eng.* **2015**, *2*, 101–108. Available online: https://www.researchgate.net/publication/286121902_Research_results_of_harvesting_haulm_remnants_of_root_crops (accessed on 25 January 2015).
17. Sibirev, A.; Aksenov, A.; Kolchin, N.; Ponomarev, A. Separating Conveyor of a Root-Crop Harvester—20171078/14. Patent of the Russian Federation No. 2,638,190, 3 October 2017.
18. Khambalkar, V.; Talokar, A.; Wankhade, K. Design Of Onion Harvester. *Mon. Peer Rev. Open Access Int. e-J.* **2014**, *1*, 11–15. Available online: http://www.yuvaengineers.com/design-of-onion-harvester-ashwini-talokar-kanchan-wankhade-v-p-khambalkar/pdf (accessed on 24 January 2014).

agriculture

MDPI

Article

Design of Device for Optical Luminescent Diagnostic of the Seeds Infected by *Fusarium*

Maksim N. Moskovskiy [1], Mikhail V. Belyakov [1,2,*], Alexey S. Dorokhov [1], Andrey A. Boyko [3], Sergey V. Belousov [4], Oleg V. Noy [5], Anatoly A. Gulyaev [1], Sergey I. Akulov [1], Anastasia Povolotskaya [6] and Igor Yu. Efremenkov [2]

1 Department of Technologies and Equipment for Breeding Works, Federal Scientific Agroengineering Center VIM, 109428 Moscow, Russia
2 Institute of Radio Engineering and Electronics, National Research University MPEI, 111250 Moscow, Russia
3 Don State Technical University, 346780 Rostov-on-Don, Russia
4 Kuban State Agrarian University Named after I.T. Trubilin, 350044 Krasnodar, Russia
5 L.L.C. Rostagroservice, 344012 Rostov-on-Don, Russia
6 St. Petersburg State University, 198504 Saint-Petersburg, Russia
* Correspondence: bmw20100@mail.ru; Tel.: +7-991-688-9404

Abstract: The development and application of optical luminescent methods and devices will help obtain information quickly and objectively about the level of *Fusarium* infection of agricultural plants. For the previously obtained ranges, the spectral characteristics of excitation and luminescence of wheat, barley, and oats of various degrees of infection were measured. The obtained dependences of flows on infection were approximated by linear regression models and relative sensitivities were determined. For wheat and barley, it is advisable to determine the degree of infection by the ratio of flows $\Phi\lambda1/\Phi\lambda2$, which makes it possible to calibrate the measuring device in relative units and increase its sensitivity. A method for determining the degree of infected seeds with *Fusarium* was developed. After the seeds are placed in a light-tight chamber, they are excited by radiation, and photoluminescence is recorded. The electrical signal from the radiation receiver is amplified and processed accounting for previously obtained calibration curves. In the universal device that measures the infection of wheat, barley, and oats seeds, it is necessary to have three radiation sources: 362 nm, 424 nm, and 485 nm. Based on the energy efficiency criteria, optimal LEDs and photodiodes, as well as a microcontroller, switches, operational amplifiers, a display, and other components of the device, were selected.

Keywords: grain plants seeds; *Fusarium*; photoluminescence; linear regression models

Citation: Moskovskiy, M.N.; Belyakov, M.V.; Dorokhov, A.S.; Boyko, A.A.; Belousov, S.V.; Noy, O.V.; Gulyaev, A.A.; Akulov, S.I.; Povolotskaya, A.; Efremenkov, I.Y. Design of Device for Optical Luminescent Diagnostic of the Seeds Infected by *Fusarium*. *Agriculture* **2023**, *13*, 619. https://doi.org/10.3390/agriculture13030619

Academic Editor: Roberto Alves Braga Júnior

Received: 14 December 2022
Revised: 22 February 2023
Accepted: 24 February 2023
Published: 4 March 2023

1. Introduction

One of the main factors which affects yield losses are plant diseases. Plant diseases are dangerous because they are difficult to detect and identify in the early stages. This factor affects crop yields and it can seriously affect the sustainability of the agricultural sector. It is important for agricultural enterprises to detect diseases early in order to control their spread.

The acceptance of operational management decisions depends on the availability of information about the diseases. Traditionally, plant diseases are detected by interpretation of visual symptoms followed by laboratory evaluation [1]. However, these methods require skills and experience in the relevant plant pathology, significant time to complete the diagnosis, and expensive chemicals and equipment. These disadvantages of traditional methods have prompted the development of modern technologies such as machine vision and remote sensing for the detection and identification of plant diseases. These technologies make it possible to assess the disease with greater reliability, accuracy, and speed [2]. These technologies are based on the determination of the optical properties of plants in various spectral ranges. Optical methods such as RGB imaging, multi- and hyperspectral sensors, thermography, or chlorophyll fluorescence show potential for conscious, objective, and reproducible detections for disease detection and quantification of epidemic diseases [3–5].

Using infrared thermography, chlorophyll fluorescence imaging and hyperspectral imaging, *Fusarium* damage to wheat on spikelet scales was monitored. The method made it possible to visualize the temperature difference inside the infected spikelets starting from 5 days. At the same time, the violation of photosynthetic activity was confirmed by the maximum fluorescence yield of spikelets over 5 days. NIR spectroscopy with a reflectance prefix was used to categorize weedy rice from cultivated rice [6]. As a means of assessing microbial contamination and shelf life of leafy green vegetables, spectral analysis of near-IR reflectance (NIR) and transmission (absorption) in the visible region was used [7]. The effectiveness of early detection of three European endemic diseases, wheat septoria, rust, and blotch, has also been analyzed [8].

Monitoring of *Fusarium* damage to wheat on spikelet scales was carried out using infrared thermography, chlorophyll fluorescence imaging, and hyperspectral imaging. The method made it possible to visualize the temperature difference inside the infected spikelets starting from 5 days.

Hyperspectral imaging is used to detect rust blistering in southwestern white pine seedlings [9,10]. The application of hyperspectral imaging for detection of *Fusarium* was evaluated in these study [11]. Hyperspectral images were obtained in the wavelength range of 400–1000 nm. Another study [12] provided the basis for the development of automated and monitoring systems of the myrtle rust detection based on reflectance spectrum sensors. An algorithm for recognizing cucumber diseases based on leaf images using sparse representation classification has been developed [13].

In a coffee leaf rust study [14], researchers obtained data using a Sequoia camera, which produced images with a spatial resolution of 10.6 cm in four spectral bands: green (530–570 nm), red (640–680 nm), far red (730–740 nm), and near infrared (770–810 nm). Researchers used conventional cameras such as the Nikon D80 to image leaf diseases in the study of rice and wheat leaf diseases [15]. The use of low-cost drones equipped with digital cameras as a field phenotyping tool to determine the severity of foliar diseases in a wheat breeding program is being explored [16].

Based on the application of chlorophyll fluorescence and hyperspectral imaging, it is possible to detect *Fusarium* species infection in wheat [17]. To detect brown rust disease in winter wheat, an optical sensor device was developed that excites chlorophyll fluorescence at discrete wavelengths and detects induced emissions [18].

Increasingly, remote sensors are being used to monitor plant health, offering non-destructive spatial detection and quantification of plant diseases at various levels of measurement. When applied on various platforms, these optoelectronic sensors open new possibilities for predicting and responding to stress and plant diseases [19].

Portable NIR reflectance instruments have been used to evaluate feed on farms, including predicting crude proteins, acidic detergent fiber, and others [20–22].

The Leaf Scanner device was developed to analyze the distribution of chlorophyll content in whole leaves based on visible and near-infrared LEDs for visual and near-infrared imaging of leaves [23]. For the same purposes, a crop chlorophyll detector based on an optical sensor with an interference filter was developed in the spectral range of 400–1000 nm [24].

A portable near infrared spectroscopy system has been developed for the rapid measurement of water content in rapeseed leaves. The spectra were collected using an integrated spectrometer from 900 to 1700 nm [25]. Sensors capable of automatically measuring canopy temperature using a thermal imaging camera [26] have been made to non-destructively measure the area of individual leaves outdoors in daylight using an RGB-D sensor, Kinect v2 as part of a handheld device [27], or image processing techniques [28].

A system for estimating biomass from measurements of the vegetation cover of leafy vegetables based on laser sensors has been developed [29]. A portable system has been developed that can take 3D measurements and classify objects based on color and depth images obtained from multiple RGB sensors [30].

A diagnostic tool was proposed to assess the degree of ripeness of oilseed olive in the visible and near infrared ranges, calibrated using image analysis. An RGB image has been received. Spectroscopic analyses were performed using a benchtop FT-NIR and a portable vis/NIR instrument. The desktop device was equipped with a fiber optic probe, and the spectra were recorded in the range of 800–2500 nm [31]. Yang et al. [32] presented a portable touch detector for the degree of sweetness and firmness of kiwi fruit. The detector consisted of a control/processing unit, an LED panel, a driver unit, and a unit for detecting and amplifying the light signal. LEDs of 1000 and 1100 nm were used as the light source.

A reflectance spectrometer using NIR spectra (850–1040 nm) has been proposed to measure protein concentration directly on the combine [33]. At the same time, insufficient attention has been paid to luminescent methods and devices for determining plant infection, which also make it possible to diagnose plant pathologies with high sensitivity but at a lower cost.

Most of the existing methods for assessing biochemical characteristics use destructive chemical analyses, which require time and a lot of labor. In addition, these methods use highly potent chemicals that require special handling and disposal. Currently, various spectroscopic methods are used: reflection spectroscopy, chlorophyll fluorescence spectroscopy, IR thermal imaging, terahertz spectroscopy in the time domain, and hyperspectral imaging have all been used for biochemical studies by focusing on changes in the ratio of chlorophyll content, physical changes, or the aquatic status of plants. The development of reliable, non-invasive, accurate, and effective methods for evaluating breeding material based on physiological, morphological, biochemical, and other indicators that closely correlate with the productivity and stability of plants is relevant.

However, reflective infrared, thermal imaging, terahertz spectroscopy requires expensive high-precision equipment, so luminescent spectroscopy of the ultraviolet and visible range may be an effective alternative. During the excitation and emission of luminescence, a deeper and longer interaction of radiation and biological tissue occurs than during reflection.

The effectiveness of the use of spectral methods has been revealed to assess the contamination during key periods of growth of collected seeds when they are laid for storage.

The spectroscopic diagnostics included FTIR in the mid-IR region, Raman, and luminescence methods. Combination of chemometric tools with FTIR and Raman spectroscopy allowed obtaining approaches based on identified characteristic spectral features which may be used as infection markers. These approaches make it possible to detect the infection on the grain husk. The carotenoid type fungi pigment was identified within the resonance conditions of Raman scattering excitation [34].

The aim of this study is to design a device for optical photoluminescent diagnostics of *Fusarium* infection of seeds of cereal plants. On the basis of previously obtained spectral characteristics, it is necessary to establish the dependence of luminescent fluxes on infection and develop a method for determining the proportion of infected seeds in a sample. For practical implementation of the methodology, based on energy efficiency criteria, components and parts of the device were selected.

2. Materials and Methods

The infection of seeds with *Fusarium* was investigated. Winter wheat "Irishka No. 172", barley "Moskovsky 86", and oats "Salp" were used as seed samples.

The degree of infection of seeds was determined by external signs. Additionally, the chromatography method was used to determine the T-2 toxin by fluorescence in long-wavelength ultraviolet light.

The spectral characteristics of excitation and luminescence were measured using a previously developed technique [35,36]. Statistical processing was carried out, averaged over 250 spectra. The integral absorption capacity N and photoluminescence flux Φ were calculated using the Panorama Pro software package.

3. Results and Discussion

3.1. Obtainment and Analysis of the Spectra

The luminescence spectra $\varphi_l(\lambda)$ were measured based on previously obtained results [35] for wavelengths of excitation maxima λ_e equal to 232, 362, 424, 485, and 528 nm. Figure 1 shows the spectral characteristics of the luminescence of wheat seeds at various λ_e. Researchers [34,37] have suggested that the glow is caused by the luminescence of chlorophyll α. Its presence in infected plants may be associated with infection during the flowering period.

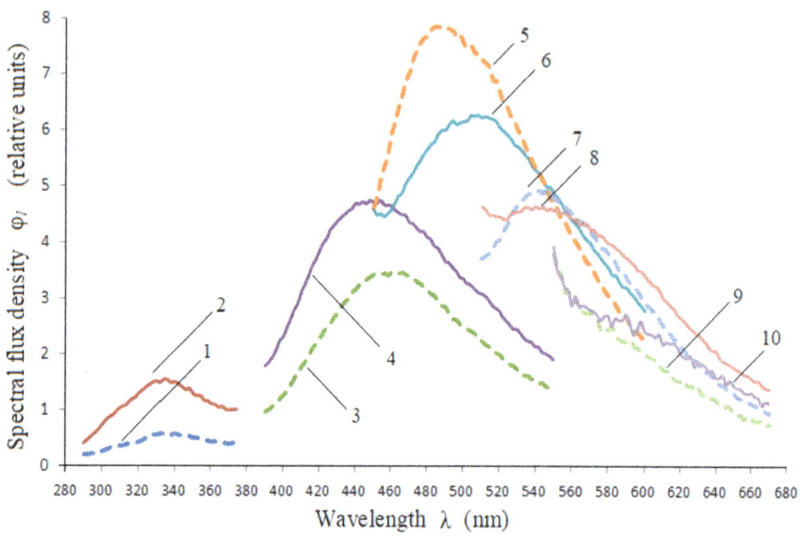

Figure 1. Luminescence spectra of wheat seeds excited by radiation: $\lambda_{e,1}$ = 232 nm: 1—not infected, 2—infected; $\lambda_{e,2}$ = 362 nm: 3—not infected, 4—infected; $\lambda_{e,3}$ = 424 nm: 5—not infected, 6—infected; $\lambda_{e,4}$ = 485 nm: 7—not infected, 8—infected; $\lambda_{e,5}$ = 528 nm: 9—not infected, 10—infected.

Figure 1 shows that all photoluminescence spectra of infected seeds are higher than those of healthy seeds at $\lambda_{e,1}$ = 232 nm, $\lambda_{e,2}$ = 362 nm, $\lambda_{e,5}$ = 528 nm. When excited at $\lambda_{e,3}$ = 424 nm, the spectrum of infected seeds is located lower in the range of 450–542 nm, and at $\lambda_{e,4}$ = 485 nm, the spectra repeatedly intersect.

The calculation of integral flows and the construction of infestation dependencies on the β(Φ) flow are presented in previously published articles [35,36].

3.2. Calculation of Regression Models

The obtained dependencies were approximated by linear regression models.

The relative sensitivity of a change in flux with a change in the degree of infection can be determined by the formula:

$$S_\Phi = 100 \times \left| \frac{\Delta\Phi_\lambda}{\Delta\beta \times \Phi_{\beta=0}} \right| \quad (1)$$

The coefficients of determination and sensitivity are presented in Table 1.

Table 1. Parameters of linear regression models of flow dependences on the degree of infection.

Parameter	λ_e = 232 nm	λ_e = 362 nm	λ_e = 424 nm	λ_e = 485 nm
		Wheat		
R^2	0.94	0.84	0.83	0.91
S_Φ	1.50	0.40	0.13	0.09
		Barley		
R^2	0.94	0.95	0.63	0.29
S_Φ	2.06	1.44	0.29	0.11
		Oats		
R^2	0.50	0.62	0.88	0.97
S_Φ	1.33	0.19	0.31	0.17

For wheat, all dependencies are statistically significant. Approximations for barley at λ_e = 424 nm and λ_e = 485 nm and for oats at λ_e = 232 nm and λ_e = 362 nm are not statistically significant.

The highest sensitivity of wheat during luminescence excitation was at the wavelength of λ_e = 232 nm, similarly for barley. Lower sensitivities were observed when excited by 362 nm wavelength radiation. For oat seeds, the highest sensitivity was observed at λ_e = 424 nm.

It is advisable to determine the degree of infection by the ratio of fluxes $\Phi_{\lambda 1}/\Phi_{\lambda 2}$, which makes it possible to calibrate the measuring device in relative units. It will also help to increase the sensitivity if the wavelength λ_2 is chosen for the falling dependence $\Phi_\lambda(\beta)$. For barley, the highest sensitivity (S_Φ = 2.48) was achieved at the flux ratio Φ_{232}/Φ_{485}, and for wheat (Φ_{232}/Φ_{424}) it was 1.85. The coefficients of determination of the obtained dependencies were 0.99 and 0.98, respectively. For practical problems, inverse dependences of the degree of infection of seeds on their photoluminescence flux $\beta(\Phi_\lambda)$ were obtained.

Using 232 nm radiation to excite photoluminescence has disadvantages. First, there is a relatively low radiation flux (5.8–10.8 times less than at λ_e = 362 nm). Second, there are serious problems with the radiation source: narrow spectrum emitters (for example, LEDs) are available only with a wavelength of at least 250 nm and they are very expensive (from USD 500). In addition, in the shortwave spectrum, UV radiation produces ozone gas and this gas is toxic to humans from the air (sustainable ozone generation occurs when air is irradiated with a wavelength of less than 242 nm). There is not yet sufficient evidence to support widespread use where direct human exposure is expected [38].

Therefore, it is proposed to use excitation by radiation at the wavelength of 362 nm for the practical application of the method for wheat and barley.

Then the calibration equations are as follows, for wheat

$$\beta = 287\frac{\Phi_{362}}{\Phi_{424}} - 137 \quad (2)$$

for barley

$$\beta = 245\frac{\Phi_{362}}{\Phi_{485}} - 57 \quad (3)$$

and for oats

$$\beta = 0.22\,\Phi_{424} - 289 \quad (4)$$

The determination coefficients for Equations (2)–(4) are 0.88, 0.96, and 0.88, respectively.

3.3. *Device of Photoluminescent Diagnostics of Seeds*

Based on the obtained results [36], the method for determining the degree of infection of seeds with *Fusarium* was developed. To implement the method, it is necessary to develop a device. The block diagram of the device is shown in Figure 2.

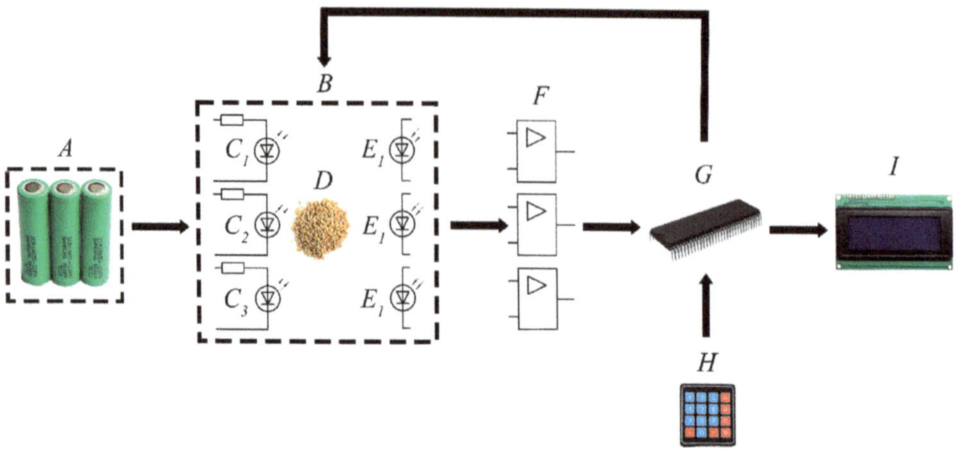

Figure 2. Generalized scheme of the seed contamination sensor. A—power supply, B—light-tight optical unit, C1, C2, C3 are radiation sources with ballast resistors, D are the studied seeds, E1, E2, E3 are radiation receivers, F are operational amplifiers, G—microcontroller, H—keyboard, I—display.

The algorithm of the sensor is as follows:

1. The selected seeds (*D*) are placed in a dark, light-tight chamber (*B*).
2. Photoluminescence is excited for 20 microseconds by two (for wheat or barley) or one radiation source (for oats).
3. Luminescence is detected 0.75–1.0 microseconds after the radiation source is turned off by two photodetectors (for wheat or barley) or one (for oats).
4. The received analog electrical signal from the receivers (photodiodes *E*) is amplified by an amplifier (*F*), converted into a digital signal, and fed to the microcontroller (*G*).
5. On the microcontroller, the degree of infection is calculated taking into account the photo signal and a priori information (calibration Equations (2)–(4)).
6. The received and processed signal is fed to the indicator (*I*). Then a decision is made on subsequent operations with seeds.

3.4. *Justification of the Choice of the Element Base of the Device*

Requirements for an optical device for express diagnostics of grain seeds based on spectral luminescent parameters of grain.

(1) Easy operation and minimal labor intensity.
(2) Minimum permissible measurement error.
(3) The weight of the device is not more than 3.5 kg to ensure maximum operator mobility.
(4) Safety during operation.
(5) Battery life of 5 h.
(6) A light-tight chamber when measuring the seed material should ensure that there is no radiation from the external light background, as well as minimal reflections of the flow from the walls when the LEDs are working.
(7) Convenience of cleaning the light-tight chamber from agricultural crops.
(8) The result of measuring the diagnostic parameter, namely the degree of infection, should be displayed on the indicator for the operator.
(9) Low cost of an express diagnostic device for maximum accessibility to users.

In a universal device that measures the infection of wheat, barley, and oats, it is necessary to have three radiation sources: 362 nm, 424 nm, and 485 nm. The most preferred is the use of LEDs having a narrow spectrum, excellent speed, making it possible to switch sources. The list and parameters of LEDs that can be used to create a device are shown in Table 2.

Table 2. List of radiation sources.

LED Name [Data Source]	Radiation Wavelength λ, nm	Forward Current *I*, mA	Forward Voltage *U*, V	Angle of Radiation φ, °	Flow of Radiation Φ_e, W
VLMU3510-365-130	362	700	4.0	160	0.95
NICHIA NCSU276A	365	700	4.0	160	0.9
NICHIA NVSU233B	365	1400	3.85	160	1.4
NICHIA NCSU033C	365	700	4.0	160	0.75
CREELED424	424	700	4.0	150	0.8
LHUV-0420-0550	420	1000	3.5	150	1.0
LZ4-00UA00-00U6	400	1000	3.9	120	5.45
XPEBBL-L1-0000-00201	485	1000	3.5	140	3.5
MLEBLU-A1-0000-000T01	480	350	3.5	150	0.8
L1CUBLU100000000	485	350	3.5	150	0.034

The numerical criterion for choosing an LED is its effective output during excitation.

$$k_{Is} = \frac{\Phi_{eff}}{\Phi_{full}} = \frac{\int_0^\infty \varphi(\lambda)_{\text{LED}} S(\lambda) d\lambda}{\int_0^\infty \varphi(\lambda)_{\text{LED}} d\lambda}, \quad (5)$$

гд е Φ_{eff}—efficient flux;
Φ_{full}—full flux of exciting radiation;
$S(\lambda)$—spectral sensitivity of seeds—excitation spectrum;
$\varphi(\lambda)_{\text{LED}}$—radiation spectrum of the LED.
The calculation results are presented in Table 3.

Table 3. Results of calculating the effective recoil of radiation sources.

Sources of Radiation	K_{Is}
VLMU3510-365-130	0.98
NICHIA NCSU276A	0.97
NICHIA NVSU233B	0.98
NICHIA NCSU033C	0.98
CREELED424	0.95
LHUV-0420-0550	0.94
LZ4-00UA00 00U6	0.73
XPEBBL-L1-0000-00201	0.90
MLEBLU-A1-0000-000T01	0.93
L1CUBLU100000000	0.92

Secondary selection criterion are the radiation angle, the minimum value of the electric power and the maximum value of the radiation flux. From the analysis of the results of Tables 3 and 4, it follows that the most optimal sources of radiation will be VLMU3510-365-130 LEDs (for λ_e = 362 nm), CREELED424 (for λ_e = 424 nm), and XPEBBL-L1-0000-00201 (for λ_e = 485 nm).

As a radiation receiver, photodiodes are the most optimal due to their effective speed and small overall dimensions. The parameters of some photodiodes are presented in Table 4.

Table 4. Radiation receiver.

Photodiode Name [Data Source]	Spectral Range λ, nm	Sensitivity S, A/W	Dark Current I_d, fA
VEMD5510CF	440–620	0.2	2×10^5
FGAP71	150–550	0.12	4×10^4
Hamamatsu S8265	340–720	0.3	2×10^4
Hamamatsu S1133	320–730	0.4	1×10^4
Hamamatsu S7686	480–660	0.38	2×10^4
BWR21R	420–675	0.009	2×10^7
VBPW34S	430–1100	0.004	3×10^7
VEMD5510CF	440–620	0.2	2×10^5
SFH 2711	470–670	0.001	2×10^5
SLD-70 BG2A	400–700	0.001	1×10^8

The main criterion for choosing a radiation receiver is matching the sensitivity spectrum of the receiver with the photoluminescence spectrum of seeds.

The effective luminescence output for photodiodes in the short-wavelength region of the spectrum is calculated using the formula:

$$k_{ld} = \frac{\Phi_{\text{eff}}}{\Phi_{\text{full}}} = \frac{\int_0^\infty \varphi(\lambda) S_{\text{ph}}(\lambda) d\lambda}{\int_0^\infty \varphi(\lambda) d\lambda}, \tag{6}$$

где $S_{\text{ph}}(\lambda)$—spectral sensitivity of the photodetector;

$\varphi(\lambda)$—photoluminescence spectrum of seeds.

The calculation results are presented in Table 5.

Table 5. Results of calculating the effective luminescence recoil for radiation detectors.

Radiation Receiver	k_{ld}
VEMD5510CF	0.68
FGAP71	0.66
Hamamatsu S8265	0.75
Hamamatsu S1133	0.79
Hamamatsu S7686	0.49
BWR21R	0.87
VBPW34S	0.26
VEMD5510CF	0.76
SFH 2711	0.79
SLD-70 BG2A	0.89

Secondary selection criteria are maximum sensitivity and minimum dark current.

Thus, the optimal receivers are photodiodes VEMD5510CF (for λ_e = 362 nm and λ_e = 485 nm) and BWR21R (for λ_e = 485 nm). Turning on and off the LEDs, photodiodes for the corresponding excitation wavelengths is carried out using the ATmega328P control microcontroller. This microcontroller has a low cost (USD 8), high prevalence, but a low input supply voltage range (1.8–5.0 V). In this regard, it is necessary to use a high-precision voltage converter. The ATmega328P has 14 digital inputs/outputs and 6 analog inputs, which is enough to control selected sources and receivers.

To control the turning on and off of the LEDs, it is necessary to use three analog switches with a high current and voltage load capacity, since the selected LEDs, the parameters of which are presented in Table 3, have an input current of 700–1000 mA and a forward voltage of 3.5–4 V. A single-channel RDC1-S2 N power MOSFET power switch was chosen for such analog switches, which is designed to operate with high voltage, namely with 100 V and a current of 5.6 A. This switch has a low gate voltage threshold of 2 V and is compatible with the selected ATmega328P controller.

It is necessary to use operational amplifiers with high speed and gain, as well as low noise to amplify the electrical signal received from photodiodes. The AD820ANZ was chosen as such an operational amplifier. The AD820ANZ has an output signal slew rate of 3 V/μs, which is sufficient to detect luminescence after the light source is turned off. The operational amplifier is capable of operating with a single supply voltage in the range from 5 V to 36 V.

The output of information on the degree of infection of crops was carried out on the LCD 2004 display with I2C. This display has the 5 V supply voltage, as well as an I2C adapter that provides data exchange between two buses: a parallel LCD bus and an I2C bus. The I2C adapter also has an image contrast adjustment resistor and its own voltage regulator. LCD 2004 with I2C has a sufficient number of familiarity spaces, which allows you to correctly display the name of the culture, calibration Equations (2)–(4), and the degree of infection.

The power source was three NCR18650B batteries with a capacity of 3350 mA·h and a voltage of 3.7 V with the ability to replenish the capacity. The batteries were placed in the battery compartment and connected in series. The ATmega328P microcontroller was designed for power supply up to 5 V; in this regard, it is advisable to use the DC–DC converter MT3608, which will reduce the voltage from the batteries from 11 V to 5 V, and if the voltage drops below 5 V, it will increase to the required values (DC–DC).

The ATmega328P controller has a low output voltage of 3–5 V and a low output current of 50 mA, therefore you must use an additional current converter up to 2 A and voltage from 5 V to power radiation sources, single-channel power switches, operational amplifiers. LM2596 converter, which has an input voltage of 3 to 40 V, an output voltage of 1.5–35 V and an output current of up to 3 A.

The following areas of future research are expected:

- laboratory and field tests of the developed device.
- extension of the application of the developed method to other agricultural plants and other diseases.

A similar spectral method, but based on reflectivity data, was used to detect Fusarium pepper disease [39]. Unfortunately, the creation of a device implementing the method was not reported.

The device proposed in this study, unlike analogs, does not need to analyze the structure of volatile organic compounds [40] and does not require the construction of an image [41,42].

4. Conclusions

The method for determining infestation involves the excitation of seed luminescence and its registration in a light-protective chamber, as well as amplification and processing of the received electrical signal. For instrumental implementation of the method, it is energy efficient to use VLMU3510-365-130, CREELED424, and XPEBBL-L1-0000-00201 LEDs, as well as VEMD5510CF and BWR21R photodiodes.

Author Contributions: Conceptualization, M.V.B. and M.N.M.; methodology, M.V.B.; software, A.P. and I.Y.E.; validation, S.V.B.; formal analysis, O.V.N.; investigation, O.V.N.; resources, A.A.B.; data curation, S.I.A.; writing—original draft preparation, I.Y.E.; writing—review and editing, A.A.G.; visualization, I.Y.E. and A.P.; supervision, M.N.M.; project administration, M.N.M. and A.S.D.; funding acquisition, A.S.D. All authors have read and agreed to the published version of the manuscript.

Funding: This research received no external funding.

Institutional Review Board Statement: Not applicable.

Informed Consent Statement: Not applicable.

Data Availability Statement: The raw data supporting the conclusions of this article will be made available by the authors, without undue reservation.

Conflicts of Interest: The authors declare no conflict of interest. The funders had no role in the design of the study; in the collection, analyses, or interpretation of data; in the writing of the manuscript; or in the decision to publish the results.

References

1. Alemu, K. Detection of Diseases, Identification and Diversity of Viruses: A. *J. Biol. Agric. Healthc.* **2015**, *5*, 132–141.
2. Mohd Ali, M.; Bachik, N.A.; Muhadi, N.A.; Yusof, T.; Gomes, C. Non-destructive techniques of detecting plant diseases: A review. *Physiol. Mol. Plant Pathol.* **2019**, *108*, 101426.
3. Mahlein, A.-K. Plant Disease Detection by Imaging Sensors–Parallels and Specific Demands for Precision Agriculture and Plant Phenotyping. *Plant Dis.* **2016**, *100*, 241–251. [CrossRef] [PubMed]
4. Mahlein, A.-K.; Alisaac, E.A.; Masri, A.; Behmann, J.; Dehne, H.-W.; Oerke, E.-C. Comparison and Combination of Thermal, Fluorescence and Hyperspectral Imaging for Monitoring Fusarium Head Blight of Wheat on Spikelet Scale. *Sensors* **2019**, *19*, 2281. [CrossRef]
5. Mahlein, A.-K.; Kuska, M.T.; Thomas, S.; Wahabzada, M.; Behmann, J.; Rascher, U.; Kersting, K. Quantitative and qualitative phenotyping of disease resistance of crops by hyperspectral sensors: Seamless interlocking of phytopathology, sensors, and machine learning is needed! *Curr. Opin. Plant Biol.* **2019**, *50*, 156–162. [CrossRef] [PubMed]
6. Makmuang, S.; Nootchanat, S.; Ekgasit, S.; Wongravee, K. Non-destructive method for discrimination of weedy rice using near infrared spectroscopy and modified self-organizing maps (SOMs). *Comput. Electron. Agric.* **2021**, *191*, 106522. [CrossRef]
7. Tsakanikas, P.; Fengou, L.-C.; Manthou, E.; Lianou, A.; Panagou, E.Z.; Nychas, G.E. A unified spectra analysis workflow for the assessment of microbial contamination of ready-to-eat green salads: Comparative study and application of non-invasive sensors. *Comput. Electron. Agric.* **2018**, *155*, 212–219. [CrossRef]
8. Johannes, A.; Picon, A.; Alvarez-Gila, A.; Echazarra, J.; Rodriguez-Vaamonde, S.; Díez Navajas, A.; Ortiz-Barredo, A. Automatic plant disease diagnosis using mobile capture devices, applied on a wheat use case. *Comput. Electron. Agric.* **2017**, *138*, 200–209. [CrossRef]
9. Haagsma, M.; Page, G.F.M.; Johnson, J.S.; Still, C.; Waring, K.M.; Sniezko, R.A.; Selker, J.S. Using Hyperspectral Imagery to Detect an Invasive Fungal Pathogen and Symptom Severity in *Pinus strobiformis* Seedlings of Different Genotypes. *Remote Sens.* **2020**, *12*, 4041. [CrossRef]
10. Haagsma, M.; Page, G.F.M.; Johnson, J.S.; Still, C.; Waring, K.M.; Sniezko, R.A.; Selker, J.S. Model selection and timing of acquisition date impacts classification accuracy: A case study using hyperspectral imaging to detect white pine blister rust over time. *Comput. Electron. Agric.* **2021**, *191*, 106555. [CrossRef]
11. Zhang, D.-Y.; Chen, G.; Yin, X.; Hu, R.-J.; Gu, C.-Y.; Pan, Z.-G.; Zhou, X.-G.; Chen, Y. Integrating spectral and image data to detect Fusarium head blight of wheat. *Comput. Electron. Agric.* **2020**, *175*, 105588. [CrossRef]
12. Heim, R.H.J.; Wright, I.J.; Chang, H.C.; Carnegie, A.J.; Pegg, G.S.; Lancaste, E.K.; Falster, D.S.; Oldeland, J. Detecting myrtle rust (Austropuccinia psidii) on lemon myrtle trees using spectral signatures and machine learning. *Plant Pathol.* **2018**, *67*, 1114–1121. [CrossRef]
13. Zhang, S.; Wu, X.; You, Z.; Zhang, L. Leaf image based cucumber disease recognition using sparse representation classification. *Comput. Electron. Agric.* **2017**, *134*, 135–141. [CrossRef]
14. Marin, D.B.; Ferraz, G.A.S.; Santana, L.S.; Barbosa, B.D.S.; Barata, R.A.P.; Osco, L.P.; Ramos, A.P.M.; Guimarães, P.H.S. Detecting coffee leaf rust with UAV-based vegetation indices and decision tree machine learning models. *Comput. Electron. Agric.* **2021**, *190*, 106476. [CrossRef]
15. Jiang, Z.; Dong, Z.; Jiang, W.; Yang, Y. Recognition of rice leaf diseases and wheat leaf diseases based on multi-task deep transfer learning. *Comput. Electron. Agric.* **2021**, *186*, 106184. [CrossRef]
16. Bhandari, M.; Ibrahim, A.M.H.; Xue, Q.; Jung, J.; Chang, A.; Rudd, J.C.; Maeda, M.; Rajan, N.; Neely, H.; Landivar, J. Assessing winter wheat foliage disease severity using aerial imagery acquired from small Unmanned Aerial Vehicle (UAV). *Comput. Electron. Agric.* **2020**, *176*, 105665. [CrossRef]
17. Bauriegel, E.; Herppich, W.B. Hyperspectral and Chlorophyll Fluorescence Imaging for Early Detection of Plant Diseases, with Special Reference to *Fusarium* spec. Infections on Wheat. *Agriculture* **2014**, *4*, 32–57. [CrossRef]
18. Tischler, Y.K.; Thiessen, E.; Hartung, E. Early optical detection of infection with brown rust in winter wheat by chlorophyll fluorescence excitation spectra. *Comput. Electron. Agric.* **2018**, *146*, 77–85. [CrossRef]
19. Fahey, T.; Pham, H.; Gardi, A.; Sabatini, R.; Stefanelli, D.; Goodwin, I.; Lamb, D.W. Active and Passive Electro-Optical Sensors for Health Assessment in Food Crops. *Sensors* **2021**, *21*, 171. [CrossRef]
20. Cherney, J.H.; Digman, M.F.; Cherney, D.J. Handheld NIRS for forage evaluation. *Comput. Electron. Agric.* **2021**, *190*, 106469. [CrossRef]
21. Acosta, J.; Castillo, M.S.; Hodge, G.R. Comparison of benchtop and handheld near-infrared spectroscopy devices to determine forage nutritive value. *Crop Sci.* **2020**, *60*, 3410–3422. [CrossRef]
22. Berzaghi, P.; Cherney, J.H.; Casler, M.D. Prediction performance of portable near infrared reflectance instruments using preprocessed dried, ground forage samples. *Comput. Electron. Agric.* **2021**, *182*, 106013. [CrossRef]
23. Zhang, L.; Wang, L.; Wang, J.; Song, Z.; Rehman, T.U.; Bureetes, T.; Ma, D.; Chen, Z.; Neeno, S.; Jin, J. Leaf Scanner: A portable and low-cost multispectral corn leaf scanning device for precise phenotyping. *Comput. Electron. Agric.* **2019**, *167*, 105069. [CrossRef]

24. Song, D.; Qiao, L.; Gao, D.; Li, S.; Li, M.; Sun, H.; Ma, J. Development of crop chlorophyll detector based on a type of interference filter optical sensor. *Comput. Electron. Agric.* **2021**, *187*, 106260. [CrossRef]
25. Zhou, L.; Zhang, C.; Taha, M.F.; Qiu, Z.; He, Y. Determination of leaf water content with a portable NIRS system based on deep learning and information fusion analysis. *Trans. ASABE* **2021**, *64*, 127–135. [CrossRef]
26. Giménez-Gallego, J.; González-Teruel, J.D.; Soto-Valles, F.; Jiménez-Buendía, M.; Navarro-Hellín, H.; Torres-Sánchez, R. Intelligent thermal image-based sensor for affordable measurement of crop canopy temperature. *Comput. Electron. Agric.* **2021**, *188*, 106319. [CrossRef]
27. Yau, W.K.; Ng, O.-E.; Lee, S.W. Portable device for contactless, non-destructive and in situ outdoor individual leaf area measurement. *Comput. Electron. Agric.* **2021**, *187*, 106278. [CrossRef]
28. Gong, A.; Wu, X.; Qiu, Z.; He, Y. A handheld device for leaf area measurement. *Comput. Electron. Agric.* **2013**, *98*, 74–80. [CrossRef]
29. Buelvas, R.M.; Adamchuk, V.I.; Leksono, E.; Tikasz, P.; Lefsrud, M.; Holoszkiewicz, J. Biomass estimation from canopy measurements for leafy vegetables based on ultrasonic and laser sensors. *Comput. Electron. Agric.* **2019**, *164*, 104896. [CrossRef]
30. Perez, R.M.; Cheein, F.A.; Rosell-Polo, J.R. Flexible system of multiple RGB-D sensors for measuring and classifying fruits in agri-food Industry. *Comput. Electron. Agric.* **2017**, *139*, 231–242. [CrossRef]
31. Tugnolo, A.; Giovenzana, V.; Beghi, R.; Grassi, S.; Alamprese, C.; Casson, A.; Casiraghi, E.; Guidetti, R. A diagnostic visible/near infrared tool for a fully automated olive ripeness evaluation in a view of a simplified optical system. *Comput. Electron. Agric.* **2021**, *180*, 105887. [CrossRef]
32. Yang, B.; Guo, W.; Huang, X.; Du, R.; Liu, Z. A portable, low-cost and sensor-based detector on sweetness and firmness grades of kiwifruit. *Comput. Electron. Agric.* **2020**, *179*, 105831. [CrossRef]
33. Long, D.S.; McCallum, J.D. Adapting a relatively low-cost reflectance spectrometer for on-combine sensing of grain protein concentration. *Comput. Electron. Agric.* **2020**, *174*, 105467. [CrossRef]
34. Pankin, D.; Povolotckaia, A.; Kalinichev, A.; Povolotskiy, A.; Borisov, E.; Moskovskiy, M.; Gulyaev, A.; Lavrov, A.; Izmailov, A. Complex Spectroscopic Study for *Fusarium* Genus Fungi Infection Diagnostics of "Zalp" Cultivar Oat. *Agronomy* **2021**, *11*, 2402. [CrossRef]
35. Bashilov, A.M.; Efremenkov, I.Y.; Belyakov, M.V.; Lavrov, A.V.; Gulyaev, A.A.; Gerasimenko, S.A.; Borzenko, S.I.; Boyko, A.A. Determination of Main Spectral and Luminescent Characteristics of Winter Wheat Seeds Infected with Pathogenic Microflora. *Photonics* **2021**, *8*, 494. [CrossRef]
36. Belyakov, M.V.; Moskovskiy, M.N.; Litvinov, M.A.; Lavrov, A.V.; Khamuev, V.G.; Efremenkov, I.Y.; Gerasimenko, S.A. Method of Optical Diagnostics of Grain Seeds Infected with Fusarium. *Appl. Sci.* **2022**, *12*, 4824. [CrossRef]
37. Pedrós, R.; Moya, I.; Goulas, Y.; Jacquemoud, S. Chlorophyll Fluorescence Emission Spectrum inside a Leaf. *Photochem. Photobiol. Sci.* **2008**, *7*, 498. [CrossRef]
38. Simons, R.M.; Blatchley III, E.R.; Linden, K.G. Far UV-C in the 200-225 nm Range, and Its Potential for Disinfection Applications. Available online: https://www.iuva.org/resources/covid-19/Far%20UV-C%20in%20the%20200%20_%20225%20nm%20range,%20and%20its%20potential%20for%20disinfection%20applications.pdf. (accessed on 30 August 2022).
39. Karadağ, K.; Emin Tenekeci, M.; Taşaltın, R.; Bilgili, A. Detection of pepper fusarium disease using machine learning algorithms based on spectral reflectance. *Sustain. Comput. Inform. Syst.* **2020**, *28*, 100299. [CrossRef]
40. Sanjay, M.; Kalpana, B. Early Mass Diagnosis of Fusarium Wilt in Banana Cultivations using an E-Nose Integrated Autonomous Rover System. *Int. J. Appl. Sci. Biotechnol.* **2017**, *5*, 5–261. [CrossRef]
41. Zhang, D.; Wang, D.; Du, S.; Huang, L.; Zhao, H.; Liang, D.; Gu, C.; Yang, X. A Rapidly Diagnosis and Application System of Fusarium Head Blight Based on Smartphone. In Proceedings of the 8th International Conference on Agro-Geoinformatics (Agro-Geoinformatics), Istanbul, Turkey, 16–19 July 2019; pp. 1–5.
42. Zhang, D.; Wang, Z.; Jin, N.; Gu, C.; Chen, Y.; Huang, Y. Evaluation of Efficacy of Fungicides for Control of Wheat Fusarium Head Blight Based on Digital Imaging. *IEEE Access* **2020**, *8*, 109876–109890. [CrossRef]

agriculture

MDPI

Article

Research on the Time-Dependent Vehicle Routing Problem for Fresh Agricultural Products Based on Customer Value

Daqing Wu [1,2], Jiyu Li [1], Jiye Cui [1] and Dong Hu [3,*]

1 College of Economics and Management, Shanghai Ocean University, Shanghai 201306, China
2 Department of Nanchang Technology, Economic and Technological Development Zone, 901 Yingxiong Dadao, Nanchang 330044, China
3 College of Continuing Education, Shanghai Jianqiao University, Shanghai 202151, China
* Correspondence: pudonglghd@126.com

Abstract: With continuous improvements in people's consumption levels, consumers' demands for safe and fresh agricultural products increase. The increase in the number of vehicles and serious congestion on roads has led to problems, such as the weak timeliness of urban cold chain logistics, high carbon emissions, low customer value and reduced customer satisfaction. In this study, carbon emissions, customer satisfaction, customer value and cost are considered, and an optimization algorithm is established to solve the time-dependent vehicle routing problem in urban cold chain logistics. For road congestion at different time periods during the cold chain distribution process, the segment function is used to express the vehicle speed. According to the characteristics of the model, considering the constraints of the time window and vehicle capacity, an improved NSGA-II algorithm with the local optimization characteristics of the greedy algorithm (G-NSGA-II) is proposed, and the sorting fitness strategy is optimized. In addition, we carry out a series of experiments on existing vehicle routing problem examples and analyze them in a real background to evaluate and prove the effectiveness of the proposed model and algorithm. The experiment results show that the proposed approach effectively reduces the total cost, enhances customer value and promotes the long-term development of logistics companies.

Keywords: green cold chain delivery; fresh agricultural products; customer value; time-dependent road network

Citation: Wu, D.; Li, J.; Cui, J.; Hu, D. Research on the Time-Dependent Vehicle Routing Problem for Fresh Agricultural Products Based on Customer Value. *Agriculture* **2023**, *13*, 681. https://doi.org/10.3390/agriculture13030681

Academic Editors: Vadim Bolshev, Vladimir Panchenko and Alexey Sibirev

Received: 9 February 2023
Revised: 8 March 2023
Accepted: 13 March 2023
Published: 14 March 2023

1. Introduction

Due to the increase in consumer health awareness, the importance of the quality of fresh food in daily life has been highlighted. The huge demand for fresh food has also driven the development of cold chain logistics. According to the prediction of the China Business Industry Research Institute, the market scale of China's cold chain logistics business is likely to exceed RMB 550 billion by 2022 [1]. The International Energy Agency (IEA) reports that 23% of global CO_2 emissions are generated by the transportation sector, in which nearly 75% of emissions is generated by the road sector [2]. Fresh food is mainly distributed using large fuel-guzzling vehicles. The cold logistics industry has gradually become a major energy consumer and carbon emitter due to its rapid development. Therefore, the study of the vehicle routing problem is significant in reducing CO_2 emissions, achieving green logistics and promoting economic development. Through energy conservation and emission reduction, cold chain logistics actively responds to the "14th Five-Year Plan" to achieve the "double carbon goal". It is of great social significance to study the optimization of the cold chain path from a green perspective.

The vehicle routing problem (VRP) was first proposed by Dantzig and Ramser [3]. In order to adapt to social developments and meet the needs of life, models and algorithms are gradually diversifying. The literature can be divided into three categories based on different objective functions; the three categories are as follows:

(1) Research on green VRP (GVRP)

At present, the main low-carbon policies include emission caps, carbon tax, carbon trading and carbon offset. By conducting the study under these four low-carbon policies, Wang et al. [4] found that the carbon tax policy and carbon trading policy achieved a better emission reduction. Many scholars have studied VRP under the carbon tax policy and carbon trading policy. Yao and Zhang [5] proposed a model for minimizing the total cost under the carbon tax system, and then used an improved genetic algorithm to solve it. G. K. Liu et al. [6] constructed a cold chain path optimization model for joint distribution under the carbon tax policy. Through a quantitative analysis of carbon tax, Ning et al. [7] established a model to minimize integrated costs with the carbon tax as a decision variable. To solve this, they proposed a quantum ant colony algorithm based on adaptive rotation angles. Jiang [8] pointed out that carbon tax policies and carbon trading policies are widely used in the global carbon market. Additionally, the carbon trading policy had a more stable effect of reducing carbon emissions than the carbon tax policy. Cheng et al. [9] demonstrated that by setting different policy parameters, carbon trading policies could guide decision makers to choose transport routes with lower emissions. There are also scholars who introduced carbon emissions directly into the model. Ye et al. [10] used spatial block area planning to construct an adaptive network topology for distribution routes. They confirmed that the distribution routes obtained in this way had lower carbon emissions. Ren et al. [11] built a cold chain path optimization model with the purpose of minimizing carbon emissions. They confirmed that incorporating the upper and lower limits of pheromone concentrations into the traditional ant colony algorithm could improve the optimization efficiency. After considering the congestion of cities in different time zones, Xiao et al. [12] used a genetic algorithm with dynamic programming to formulate delivery vehicles' routes and departure times to minimize carbon dioxide emissions. In the study of route optimization for multimodal transport, Demir et al. [13] considered the details of the transport process comprehensively and set carbon emissions as one of the objectives of the model. Sadati et al. [14] introduced a multidepot green vehicle routing problem (MDGVRP), where alternative fuel-powered vehicles were used to serve people.

(2) Research on VRP with customer satisfaction

Due to the perishable and time-sensitive nature of fresh food, service time windows are used to measure customer satisfaction. The degree of customer satisfaction or customer benefits cannot be intuitively indicated in this way. As a result, several scholars have studied satisfaction as a model objective. The model designed by Zulvia et al. [15] for perishable products contains four objectives: minimizing operating costs, minimizing carbon emissions, minimizing cargo damage costs and maximizing service levels, using the fuzzy membership function to represent the service level. They used the multiobjective gradient evolution algorithm (MOGE) to solve the model. According to the characteristics of fresh agricultural products, Wang et al. [16] constructed the penalty cost function based on the time window and used fuzzy logic to calculate satisfaction. Additionally, an improved quantum-behaved particle swarm optimization algorithm (IQPSO) was proposed to solve the problem. Based on delivery time and quality satisfaction, Zhang et al. [17] constructed a time satisfaction function based on a mixed time window and considered the effect of the product deterioration rate on satisfaction. The model was solved using an improved partial elite single-parent genetic algorithm, revealing the changing satisfaction trend with the cost optimization process. Ganji et al. [18] proposed that each customer has a set of time windows with different priorities that represent different values for the customer. Jie et al. [19] used a soft time window to calculate penalty costs. They introduced a mixed-integer nonlinear programming model to minimize the total cost and used three multiobjective metaheuristic algorithms to solve it. Eydi et al. [20] assumed that customers have multiple needs and a time window, prioritizing services with higher demand. Additionally, a multiobjective metaheuristic algorithm was applied to minimize distribution costs and maximize customer satisfaction. Xiao et al. [21] also used a fuzzy membership function to measure customer time satisfaction and established a dual-objective model. They then found fresh food delivery schemes with a nondominated ranking genetic (NSGA-II) algorithm. To minimize energy consumption and maximize customer satisfaction, Ghannadpour et al. [22] categorized customers by priority and used NSGA-II to solve the model.

(3) Research on time-dependent VRP (TDVRP)

The previous cold chain distribution problem is mainly studied for static road networks, which, in real life, are constantly changing. Malandraki [23] first proposed TDVRP considering the different speeds of vehicles in different periods. Some scholars optimized the delivery route under time-dependent road conditions. Jie et al. [19] introduced stochastic factors (traffic congestion and weather changes) into the traditional particle swarm optimization (PSO) algorithm to improve the velocity calculating formula. A hybrid algorithm (HA) combining the sweep algorithm (SA) and improved particle swarm optimization (IPSO) was developed to minimize the total cost of distribution. Due to the different multiple traffic modes, Asgharizadeh et al. [24] proposed an accurate travel-time estimation model by plotting the speed curves of different types of roads. Xu et al. [25] described the relationship between time and velocity in terms of a trigonometric function and developed a novel multiobjective mixed-integer nonlinear programming (MINLP) model. Additionally, they used the improved NSGA-II to solve it. Similarly, Poonthalir et al. [26] set a minimum speed limit and a maximum speed limit to calculate the expected speed using a triangular distribution. They combined the greedy mutation operator with the particle swarm algorithm to reduce cost and fuel consumption. Considering different speeds in different periods, Zhou et al. [27] used the road segment division method to calculate the speed of vehicles in each road segment. Additionally, they designed an improved ant colony algorithm to find the minimum total cost. Under real traffic work, Z. Liu et al. [28] used a continuous function to describe the relation between velocity and time. In the model, they explored the impact of speed, load and road slope on fuel consumption. To solve the model, a hybrid genetic algorithm with a variational neighborhood search was used.

For the VRP models, solving algorithms can be divided into the precise algorithm, traditional heuristic algorithm and metaheuristic algorithm.

1. The precise algorithm is one that can find an optimal solution to a problem. Yu et al. [29] proposed an improved branch-and-price (BAP) algorithm to precisely solve the heterogeneous fleet green vehicle routing problem with time windows (HFGVRPTW). Based on the segmentation of vehicle tours, Lee et al. [30] constructed an extended charging stations network and developed the branch-and-price method on an extended charging station network to solve the problem to optimality. Bruglieri et al. [31] introduced the green vehicle routing problem with capacitated alternative fuel stations (AFSs) and presented two mixed-integer linear programming formulations. Additionally, they described the two variants of the exact cutting planes method for solving the model. The exact solution obtained with the CPLEX solver could be used to test the performance of the heuristic algorithm [32].
2. The core idea of traditional heuristic algorithms is to search locally for a better solution than the current one until no better solution is available. Li et al. [33] constructed a linear programming model with the goal of minimizing the weighted sum of the delivery time and transportation cost. The local search process based on a variety of neighborhood structures and the uppercrossing/solution (US) algorithm were adopted to design an improved iterative local search solution algorithm. Considering the travel time variability, Song et al. [34] constructed a nonlinear and concave objective function. They proposed the augmented Lagrangian relaxation method with a block coordinate descent, linearization and Lagrangian substitution techniques. Yang et al. [35] studied the vehicle routing problem with mixed backhauls and time windows (VRPMBTW), which was solved with dynamic programming in a block nonlinear Gauss–Seidel framework.

3. The metaheuristic algorithm is an improvement on the heuristic algorithm. It is a combination of the stochastic algorithm and local search algorithm. Schermer et al. [36] formulated the vehicle routing problem with drones and en route operations (VRPDERO) as a mixed-integer linear program (MILP), which solved by an algorithm based on the concepts of the variable neighborhood search (VNS) and tabu search (TS). Wang et al. [37] developed a novel adaptive granularity learning-distributed particle swarm optimization (AGLDPSO) with the help of machine learning techniques, including a clustering analysis based on locality-sensitive hashing (LSH) and adaptive granularity control based on logistic regression (LR). Compared with other large-scale optimization algorithms, this algorithm had a better local search ability.

In order to improve the solving efficiency, many scholars adopted hybrid heuristic algorithms, which combine various metaheuristic algorithms together and integrate their different advantages. Wang et al. [38] combined a variant of the ACO (multiant system) with several local search heuristics to improve the solution quality. Wang et al. [39] constructed a biobjective model to minimize the total cost and delivery vehicles, and designed the tabu search algorithm (TS) and nondominated sorting genetic algorithm (NSGA-II) to solve the problem. These algorithms provided the ideas for solutions for the research in this paper.

Based on the models constructed and algorithms used in previous studies, the previous works and the innovations in this study are summarized as follows:

1. From the low-carbon perspective of routing optimization, many scholars have used low-carbon policies in their models, with research mainly focusing on calculating carbon emissions and using a carbon tax to obtain the cost of carbon, while seldom considers carbon trading prices and its spatiotemporal differences. Due to carbon trading policies having a greater advantage in reducing carbon emissions, using carbon trading policies and their uncertainties in studies is conducive to both enriching related research and reducing carbon emissions.
2. Most scholars use soft time windows to calculate penalty costs and customer satisfaction in their models. However, these only consider the impact of the current distribution schemes on customers. Other factors such as the influence and importance of customers also have an impact on the development of enterprises. It is more socially relevant to measure the value of customers by integrating various factors. This study calculates customer value from both current value and potential value.
3. Under time-dependent road networks, there are few studies that have comprehensively considered the perishability of fresh food, customer value and carbon emissions. It is necessary to increase the research in fresh cold chains. This model can not only effectively reduce the distribution cost of fresh products, but also improve customer satisfaction.

Therefore, a green vehicle routing problem with customer value (CV-GVRP) for fresh product distribution is established. The CV-GVRP is an extension of the traditional GVRP model, which adds time-dependent road network conditions, the characteristics of fresh products and customer value. It is a mixed-integer linear programming model with objectives of minimizing the total cost and maximizing customer satisfaction. According to the NP-hard property of the model, a hybrid heuristic algorithm is used to solve the problem. NSGA-II is extensively applied as a multiobjective optimization algorithm. A greedy algorithm combined with the nondominant sorting genetic algorithm II (G-NSGA-II) was designed to improve the local search ability of the traditional algorithm. Comparisons of the results obtained using models with different objective functions for solving the arithmetic examples confirm the superiority of this model. The conclusions provide a reference for determining decisions in the distribution of fresh cold chains.

The remaining sections of this paper are summarized in sequence: Section 2 of this paper describes the research problem and introduces a mathematical model (CV-GVRP) to solve it. In Section 3, an improved algorithm (G-NSGA-II), combining the nondominated ranking genetic algorithm II with the locally greedy algorithm, is proposed. Section 4 is devoted to computing arithmetic examples via the proposed model and algorithm, with the results confirming their performance. Finally, the conclusions and suggestions for future research are discussed in Section 5.

2. Problem Description and Model Formulation

2.1. Problem Description

Fresh cold chain logistics show that distribution networks consist of multiple customer points and distribution centers. After the goods are packed and sorted in the distribution center, they are distributed using multiple refrigerated trucks. In cases where the vehicle speed changes constantly due to traffic conditions, the optimal vehicle travel path can be planned, considering both the delivery cost and customer value, and the delivery service can be provided within the time window specified by the customer.

In order to ensure the feasibility of the model, this paper assumed the following: (1) That all vehicles depart from the same distribution center and return to the distribution center after completing the delivery. (2) The refrigerated vehicles used are the same type and are sufficient to complete the delivery task. (3) The location of the customer, the amount of demand, the time window and their importance, as well as the temperature of the refrigerated vehicle at the opening and closing of the door, are known. (4) The demand at each customer point does not exceed the maximum capacity of the refrigerated vehicle, and each customer is delivered to using only one vehicle. (5) The quality of the fresh products transported is the same and the requirements for the refrigerated temperature are the same. (6) The preservation period and deterioration rate of fresh products are the same and known. (7) The speed of the vehicle in each period is only related to the delay factor of traffic congestion in that period. (8) Positive publicity is given to the enterprise when customer satisfaction is positive.

2.2. Problem Formulation

In this paper, some corresponding parameters and variables were used to construct the model. The variables and definitions are listed in Table 1.

Table 1. Symbols and definitions.

Variable	Definition	Parameter	Definition	Set	Definition
t_i	The time of arrival at the customer i	Q	Maximum load of refrigerated truck	U	Set of all notes including distribution center
t_{ij}	The time it takes to travel from customer i to customer j	v_o	Average travel speed of refrigerated truck	U_c	Set of all customers
d_{ij}	The delivery distance between customer point i and j	C_0	Departure cost of refrigerated truck	K	Set of all vehicles
Q_{ij}	Load capacity from customer i to customer j	C_p	Transportation cost per unit distance of refrigerated truck	U	Set of all time periods
s_i	Time spent serving customer i	θ	Rate of corruption		
$W_i(t_i)$	Customer satisfaction function	R_1	Price per unit of damage		
v_{ij}^p	The velocity from i to j in a time period p	R_2	Profit per unit of fresh food		
ω	Carbon trading price	q_i	Demand of customer i		
Z	The total cost of travel	n_i	The number of potential consumers attracted by the positive publicity of the customer i		

Table 1. *Cont.*

Variable	Definition	Parameter	Definition	Set	Definition
Z_1	The cost of vehicle fixed	(ET_i, LT_i)	Customer's acceptable service time frame		
Z_2	The cost of transportation	$(ET_i^{\alpha}, LT_i^{\beta})$	Customer's expected delivery time frame		
Z_3	The cost of product loss	γ_1	The cost per unit of time waiting for customer i to arrive before ET_i		
Z_4	The penalty cost	γ_2	The cost per unit of time delayed to arrive at customer i after LT_i		
Z_5	The cost of carbon emissions	λ_i	The weight of customer i		
V	Total customer value	φ	Carbon emissions from refrigeration equipment per unit of goods by delivery unit		
g_i	The current value of customer i	ρ_0	Fuel consumption per unit distance at no load		
p_i	The potential value of customer i	ρ^*	Fuel consumption per unit distance at full load		
X_{ij}	0–1 variable: 1 means delivery performed using vehicle k from customer i to customer j, otherwise, it is 0.	ε	Carbon emissions per unit of fuel consumption		
Y_{ik}	0–1 variable: 1 means customer i served using vehicle k, otherwise, it is 0.	ρ^p	The delay coefficient of traffic congestion in time period p		

With the previous comprehensive analysis, the CV-GVRP multiobjective optimization model with a minimum total cost and maximum customer value was formulated. The calculations for Z_1, Z_2, Z_3, Z_4 and Z_5 were explained in the following section. The objective function was as follows:

$$\max V = \sum_{i=1}^{u} \lambda_i (g_i + p_i) \tag{1}$$

$$\begin{aligned}
\min Z &= Z_1 + Z_2 + Z_3 + Z_4 + Z_5 \\
&= C_0 \sum_{k=1}^{k} \sum_{i=1}^{u} X_{0i}^k + \sum_{k=1}^{m} \sum_{i=0}^{u} \sum_{j=0}^{u} d_{ij} C_p X_{ij}^k \\
&+ \sum_{i=0}^{u} \sum_{k=1}^{m} Y_{ik} [q_i (1 - K_1 e^{-\theta t_i}) + Q_{i0} (1 - K_2 e^{-\theta \frac{q_i}{xh}})] R_1 \\
&+ \gamma_1 \sum_{i=1}^{u} \max\{ET_i - t_i, 0\} + \gamma_2 \sum_{i=1}^{u} \max\{t_i - LT_i, 0\} \\
&+ \omega \sum_{k=1}^{m} \sum_{i=0}^{u} \sum_{j=0}^{u} X_{ij}^k d_{ij} \{\varepsilon [\rho_0 + (\rho * - \rho_0) \frac{Q_{ij}}{Q}] + \varphi Q_{ij}\}
\end{aligned} \tag{2}$$

subject to:

$$\sum_{k=1}^{m} q_i Y_{ik} \leq Q, i = 1, 2, 3 \cdots u, k = 1, 2, 3 \cdots m \tag{3}$$

$$\sum_{k=1}^{m} \sum_{j=1}^{u} X_{ij}^k \leq m, i = 0 \tag{4}$$

$$\sum_{k=1}^{m} \sum_{j=1}^{u} X_{ij}^k = 1, i \neq j, i = 0, 1, 2 \cdots u \tag{5}$$

$$\sum_{k=1}^{m}\sum_{i=0}^{u} X_{ij}^{k} = 1, i \neq j, j = 1,2,3\cdots u \tag{6}$$

$$\sum_{k=1}^{m} Y_{ik} = 1, i = 1,2,3\cdots m \tag{7}$$

$$\sum_{j=0}^{u} X_{ij}^{k} = \sum_{j=0}^{u} X_{ji}^{k} \leq 1, i = 0, k = 1,2,3,\cdots m \tag{8}$$

$$t_j = t_i + S_i + t_{ij}, i = 1,2,3\cdots u, j = 1,2,3\cdots u \tag{9}$$

Equations (1) and (2) indicate that the objective functions of the model minimized the total distribution cost and maximized the customer value. Under the constraint of Equation (3), the total order demand assigned to the K-th vehicle could not exceed the loading capacity of the distribution vehicle. Constraint (4) indicated that the number of vehicles involved in the distribution could not exceed the total number of refrigerated vehicles. Each customer point was allowed to be served using only one refrigerated vehicle for one delivery, which was realized with constraints (5) and (6). Constraint (7) indicated that each customer point was served using only one delivery vehicle. Constraint (8) indicated that the refrigerated vehicle departed from the distribution center and returned to the distribution center after completing the transportation task. The time relationship through constraint (9) indicated that the whole distribution process was continuous.

Under a time-dependent road network, the objective function of the CV-GVRP based on the variable price of carbon trading and customer value was to minimize the total cost and maximize the customer value. The total cost contained five components: the vehicle fixed cost, transportation cost, loss cost, penalty cost and carbon emission cost, which were as follows:

2.2.1. Cost Function Composition

(1) Vehicle Fixed Cost

The fixed cost of a vehicle is the cost of purchase, maintenance, labor paid to the driver, etc. It only depends on the number of vehicles performing the task. The function was as follows:

$$Z_1 = C_0 \sum_{k=1}^{m}\sum_{i=1}^{u} X_{0i}^{k} \tag{10}$$

(2) Transportation Cost

In general, the shorter the distance traveled, the lower the cost. Thus, the transportation cost of the vehicle is proportional to the distance traveled. C_p represents the transportation cost within a unit distance. The total transportation cost Z_2 could be expressed as:

$$Z_2 = \sum_{k=1}^{m}\sum_{i=0}^{u}\sum_{j=0}^{u} d_{ij} C_p X_{ij}^{k} \tag{11}$$

(3) Loss Cost

Both the transit period and the load of goods incurred a loss cost. The continuous lifetime function decayed according to an exponential rate. The decay function of fresh products was: $E(t) = E_0 \cdot e^{-\theta t}$ [40]. This formula expresses the quality of the product at different times, where E_0 is the product's initial quality and θ is the sensitivity factor of the product to time. The loss cost of transit without opening the door was as follows:

$$C_g = \sum_{i=0}^{u}\sum_{k=1}^{m} Y_{ik} q_i [1 - K_1 e^{-\theta(t_i - t_0)}] R_1 \tag{12}$$

Temperature changes caused by opening the door during loading could also affect the freshness of the product. The door opening time was determined using the customer service time s_i. The loss cost caused by loading goods while opening the door was:

$$C_l = \sum_{i=0}^{u}\sum_{k=1}^{m} Y_{ik}Q_{i0}(1 - K_2 e^{-\theta S_i})R_1 \tag{13}$$

We could obtain the total loss cost in the whole distribution process as follows:

$$Z_3 = \sum_{i=0}^{u}\sum_{k=1}^{m} Y_{ik}R_1[q_i(1 - K_1 e^{-\theta(t_i - t_0)}) + Q_{i0}(1 - K_2 e^{-\theta S_i})] \tag{14}$$

(4) Penalty Cost

In order to maximize the freshness of our products, we needed to arrive within the time frames set by our customers individually. Otherwise, there would be penalty costs. Suppose that the time window required by the customer was (ET_i, LT_i). We could use t_i to represent the time when the delivery vehicle arrives at customer i. If $ET_i \le t_i \le LT_i$, no additional penalty cost would be incurred. If the goods were not delivered within the time window, the penalty cost Z_4 would be paid. Arriving earlier than the time requested would incur a waiting cost, and arriving later would pay a delayed service fee. The relation was stated as follows:

$$Z_4 = \gamma_1 \sum_{i=1}^{u} \max\{ET_i - t_i, 0\} + \gamma_2 \sum_{i=1}^{u} \max\{t_i - LT_i, 0\} \tag{15}$$

(5) Carbon Emissions Cost

This cost is mainly related to carbon emissions and the price of carbon trading. Carbon emissions mainly come from two sources: one is generated by energy consumption during transportation, and the other is generated by refrigeration equipment.

1. Carbon emissions caused by driving

The amount of CO_2 produced by a vehicle while transporting products is not only related to the distance, but also to the load capacity of the whole process. The fuel consumption per unit distance is different at full load and at no load. The carbon emissions of the travel process from node i to j were given:

$$G_1 = \varepsilon d_{ij}[\rho_0 + (\rho^* - \rho_0)\frac{Q_{ij}}{Q}] \tag{16}$$

2. Carbon emissions from refrigeration

In order to maintain freshness, we used refrigeration equipment in the distribution process. The cumulative carbon emissions from refrigeration between nodes i and j were as follows:

$$G_2 = \varphi d_{ij} Q_{ij} \tag{17}$$

As a result, the total amount of carbon emitted in the whole distribution process was $(G_1 + G_2)$. Due to current carbon trading policies, there are regional variations and significant fluctuations in the price of carbon trading. According to spatial and temporal distribution characteristics of the "carbon K-line", the CV-GVRP calculated the carbon cost through random variables. The price per carbon transaction was $\omega \sim (\overline{\omega} - \hat{\omega}, \overline{\omega} + \hat{\omega})$, and the probability of taking any value in this range was equal.

The carbon emission cost could be expressed as:

$$Z_5 = \omega \sum_{k=1}^{m}\sum_{i=0}^{u}\sum_{j=0}^{u} X_{ij}^{k} d_{ij}\{\varepsilon[\rho_0 + (\rho^* - \rho_0)\frac{Q_{ij}}{Q}] + \varphi Q_{ij}\} \tag{18}$$

2.2.2. Customer Value Measurement Method

The CV-GVRP considered the customer value in two parts: the current value and the potential value. The current customer value depends on the customer demand, and the two are positively correlated. The calculation method was as follows: $g_i = \frac{q_i}{\sum_{i=1}^{u} q_i} q_i R_2$.

The potential value is primarily related to the company's reputation, business strength, technological innovation and other factors, while the quality of the distribution has a more pronounced effect on the company's image. When the distribution is properly optimized, the freshness of the food can improve and customer satisfaction can increase. This motivates more potential consumers and, then, the company gains more potential customer value.

The potential value was calculated using $p_i = W_i n_i q_i R_2$. Customer satisfaction in the CV-GVRP was computed based on the time windows, and the expected time window $\left(ET_i^{\alpha}, LT_i^{\beta}\right)$ was set to be within the specified time window (ET_i, LT_i). Figure 1 shows the customer satisfaction function curve of the model.

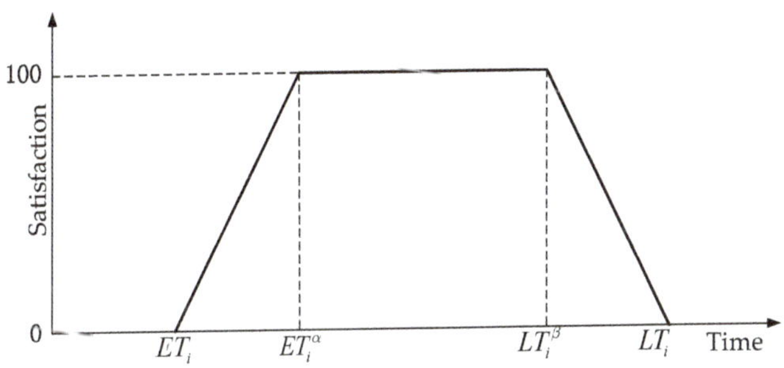

Figure 1. Customer satisfaction curve.

Based on this, the linear change trend of customer satisfaction over time could be expressed with the fuzzy membership function. The time satisfaction function of a customer i was as follows:

$$W_i(t_i) = \begin{cases} \frac{(t_i - ET_i)}{(ET_i^{\alpha} - ET_i)} & t_i \in [ET_i, ET_i^{\alpha}] \\ \frac{(LT_i - t_i)}{(LT_i - LT_i^{\beta})} & t_i \in [LT_i^{\beta}, LT_i] \\ 1 & t_i \in [ET_i^{\alpha}, LT_i^{\beta}] \\ 0 & t_i \notin [ET_i^{u}, LT_i^{\beta}] \end{cases} \tag{19}$$

The calculation of the value of a single customer could be expressed as $\lambda_i(g_i + p_i)$. Giving service priority to customers with high weights λ_i can increase customer satisfaction and customer value, which is helpful for long-term business growth. The value of total customers was as follows:

$$V = \sum_{i=1}^{u_c} \lambda_i (g_i + p_i) \tag{20}$$

2.2.3. Time-Dependent Speed Calculation Method

In real life, road conditions are complex and not static. Thus, vehicle speed in each period was calculated according to formula $v_{ij}^{p} = v_0 / \rho^{p}$. The relationship between speed and time is shown in Figure 2.

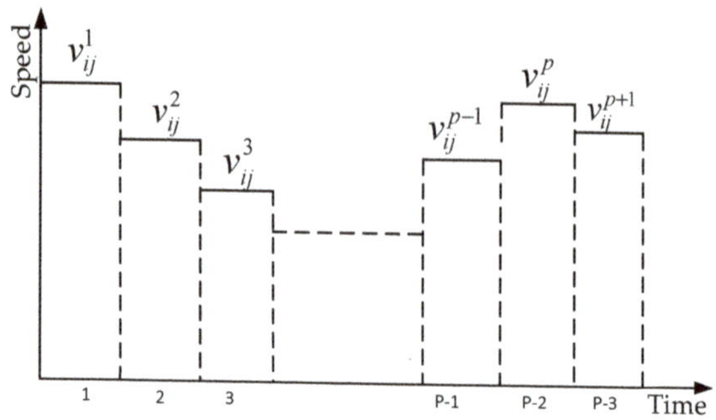

Figure 2. Speed–time relationship.

Ichoua et al. [41] believe that vehicle speed tends to be constant in each subdivided p period. Therefore, the CV-GVRP divided the total time of a day into periods, i.e., $[B_1, F_1]$, $[B_2, F_2]$, …, $[B_p, F_p]$. The speed in each period remained unchanged; the length of the p period is H_p; the driving time in the p period is ; the distance to be driven after the end of p period is L_{ij}^p; meanwhile, $d_{ij} = \sum_{p \in P} d_{ij}^p$ and t_{ij}^p represent the driving time in p period. The total driving time was calculated as follows:

Step 1: First, calculate $d_{ij}^p = L_p v_{ij}^p$. If $d_{ij} < d_{ij}^p$, calculate $t_{ij}^p = d_{ij}/v_{ij}^p$ next to proceed to Step 3, or calculate $L_{ij}^p = d_{ij} - d_{ij}^p$, $t_{ij}^p = L_p$.

Step 2: To calculate the vehicle's driving time in the remaining period, let $\xi = 1$ and the vehicle run in period $p + \xi$, $d_{ij}^{p+\xi} = H_{p+\xi} v_{ij}^{p+\xi}$; if $d_{ij}^{p+\xi} \leq L_{ij}^p$, $t_{ij}^{p+\xi} = H_{p+\xi}$, $L_{ij}^{p+\xi} = L_{ij}^p - d_{ij}^{p+\xi}$ and $\xi = \xi + 1$, then repeat this step. Otherwise, calculate $t_{ij}^{p+\xi} = d_{ij}^{p+\xi-1}/v_{ij}^{p+\xi}$ to proceed to Step 3.

Step 3: The total vehicle travel time $t_{ij} = \sum_{p \in P} t_{ij}^p$ is obtained and the section (i, j) driving time computation is finished.

3. G-NSGA-II Algorithm

3.1. Description of the G-NSGA-II Algorithm

CV-GVRP is a multiobjective model with a comprehensive consideration of enterprise cost and customer value. The nondominated ranking genetic algorithm-II with the elite genetic strategy is widely used in multiobjective optimization and solves the problem exactly [20,39]. The Pareto solution can intuitively reflect the relationship between the two objectives. Using the greedy algorithm to obtain an initial solution can expand the local search and improve the solution quality [32]. To further improve the global search ability and convergence speed of the algorithm, this paper designed an improved NSGA-II (G-NSGA-II) with the characteristics of the local optimization of the greedy algorithm to solve the model.

1. Coding Mechanism

Natural number coding was used to visually describe possible solutions to the problem. A chromosome denotes a feasible distribution solution using client sites as genes of the chromosome, and 0 denotes a distribution center. The chromosome can be expressed as (0, I1, I2..., Ir, 0), which means a vehicle departs from the distribution center (point 0) and services customers I1, I2..., Ir in order. The vehicle then returns to point 0. A schematic of chromosome coding is shown in Figure 3.

2. Population Initialization

Since the greedy algorithm could give an initial solution in a shorter time, this paper combined the greedy algorithm to solve the model through the following steps:

Step 1: First, set the initial path to empty.

Step 2: From the initial node (warehouse), randomly select customer point i and calculate the distance between customer point i and neighboring points. Then, add the customer point j with the shortest delivery distance to the path.

Step 3: Determine whether there are still unserved customer nodes; if so, proceed to the next step, otherwise, terminate the algorithm.

Step 4: Calculate the distance between customer point $i - new$ and the remaining customer points, choose the shortest one and then return to Step3.

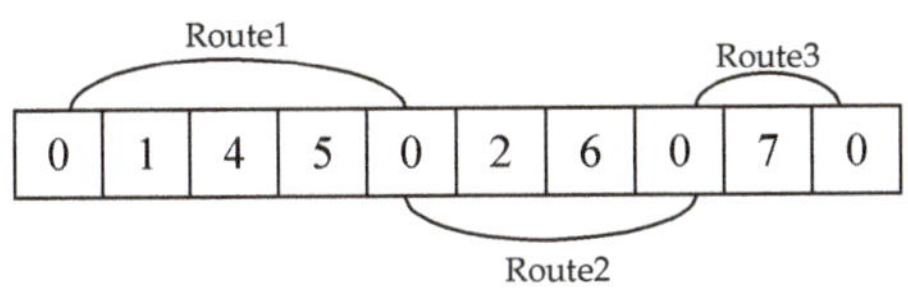

Figure 3. Chromosome coding.

3. Improve the Sorting Fitness Strategy

Figure 4 reflects the density information of individuals. The Pareto ranking of individuals 1, 2, 3, 4, 5 and 6 was one, and the ranking of individuals a, b and c was two. However, the density around individual a was significantly larger than that of b and c. Therefore, to avoid them having the same probability of entering the next generation, the ranking values were combined with the density information around the individuals to distinguish individuals in the same layer. This could improve the diversity of the population distribution without increasing the complexity of the algorithm.

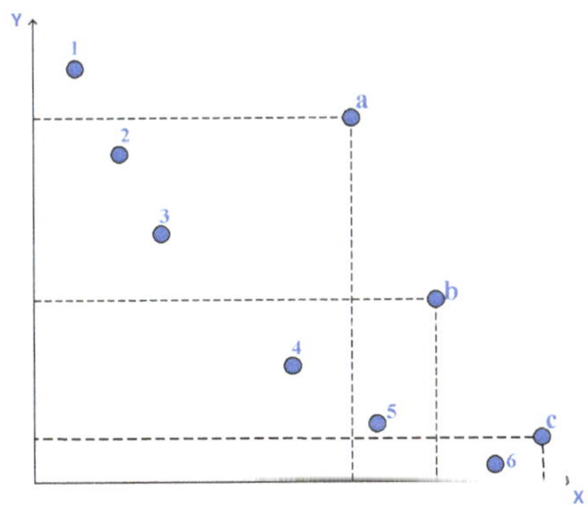

Figure 4. Density information around the individuals.

4. Crowding Calculation

After several iterations to obtain a hierarchy of noninferior solutions, the crowding distance between neighboring individuals was calculated using objective values. The individuals with a more considerable crowding distance in the same layer were retained to ensure the diversity of solutions. $D(i+1)^{fm}$ and $D(i-1)^{fm}$ represent the m-th objective function value of individuals $(i+1)$ and $(i-1)$, respectively, and fm represents the m-th function in the model. $fm^{\max}$ and $fm^{\min}$ are the maximum and minimum values of the m-th objective function values, respectively. The following formula could calculate the crowding degree of individuals:

$$D(i) = D(i) + \frac{D(i+1)^{fm} - D(i-1)^{fm}}{fm^{\max} - fm^{\min}} \tag{21}$$

5. Crossover and Variation

In this paper, we adopted the natural number encoding method and chose the partial-mapped crossover (PMX) for the crossover operation in order to improve the convergence speed of the algorithm. The basic procedure was as follows: 1. Two individuals were randomly identified as crossing nodes in the parent population. 2. The genes of the two crossover nodes were exchanged between the parents. 3. Relationships were mapped based on genes between two crossover nodes, replacing duplicated genes on the same parent to obtain two offspring. The diagram of the PMX is shown in Figure 5.

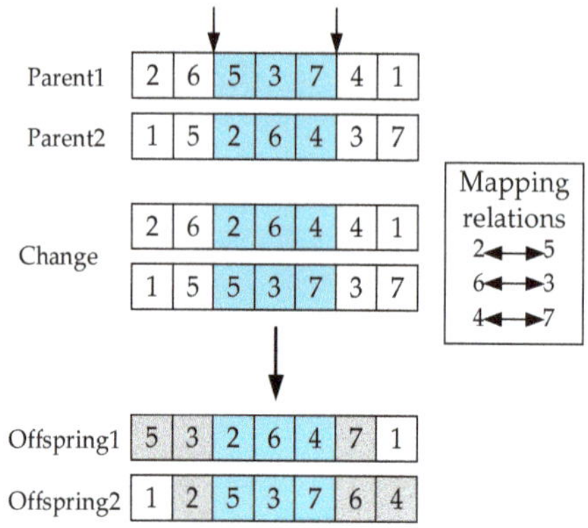

Figure 5. Cross process.

This paper adopted the exchange variation to increase the diversity of individuals in the population and improve the local search ability of the algorithm, in which any two gene points were selected to be swapped. The result of the mutation is shown in Figure 6.

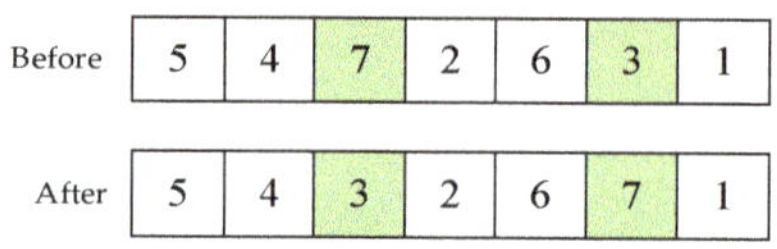

Figure 6. Mutation process.

3.2. Algorithm Process

Figure 7 represents the whole calculation process. The detailed procedure was as follows:

Step 1: To determine the population size, maximum number of iterations, crossover probability and mutation probability, use the greedy algorithm to generate the initial population.

Step 1.1–Step 1.3 are the processes of generating the initial solution:

Step 1.1: Calculate the distance between two customer points and sort them from smallest to largest.

Step 1.2: Determine whether the solution satisfies the subpath condition. If it does, add it to the current path or determine the next path.

1. The path is not closed after being added.
2. After adding, no node should exceed two connecting edges.
3. The total demand of the customer should not exceed the load of refrigeration.
4. It can meet the requirements of the customer's time window.

Step 1.3: Execute the previous step until all subpaths are assigned while both endpoints of each path are connected to the distribution center to form a closed loop.

Step 2: Perform an improved nondominated ranking based on the biobjective function values of the individuals in the initial population. Then, apply a tournament strategy to select and generate new offspring populations through a cyclic crossover and mutation.

Step 3: Combine the offspring and parent populations to generate a new population. Through the elite strategy, compare the improved nondominated ranking value and crowding distance to obtain a better combination of individuals to generate the parent population.

Step 4: Iteratively update the newly generated parent population with genetic manipulation.

Step 5: Judge whether the current iteration number reaches a maximum and output the final result. Otherwise, repeat Step3.

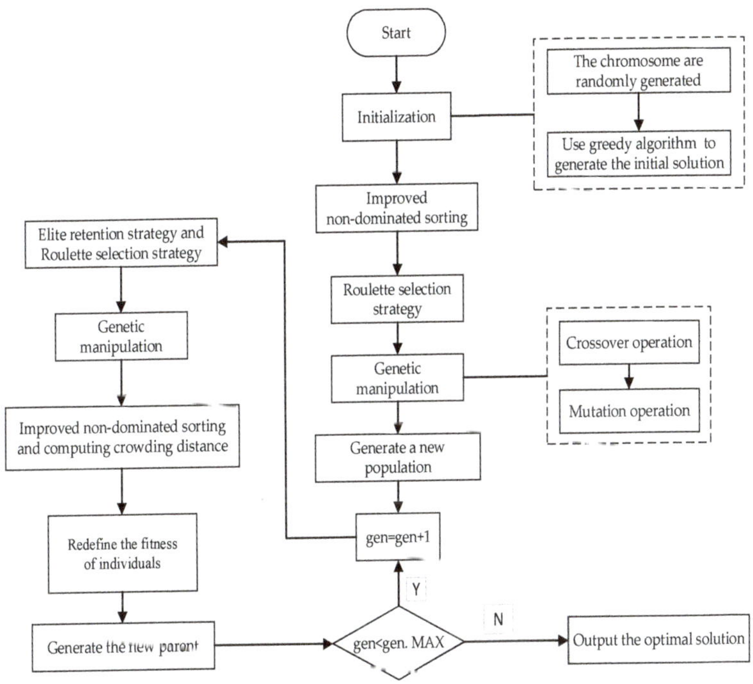

Figure 7. G-NSGA-II algorithm flow chart.

4. Experiments and Results Analysis

In order to further test the performance of the algorithm, the algorithm was compared with the traditional NSGA-II. Experiments were carried out with three classical datasets (Solomon benchmark datasets r101, c101 and rc101 [42]) to test the performance of the algorithms. Customer influence, information dissemination and customer weight information were added to form the small-scale cases (c101_25, r101_25 and rc101_25), the medium-scale cases (c101_50, r101_50 and rc101_50) and the large-scale cases (c101_100, r101_100 and rc101_100), which had 100 customers.

Table 2 shows the values for the model parameter settings. The values of the parameters were obtained from [5,40].

Table 2. Related parameters and values.

Parameter	Value	Parameter	Value
The departure cost of the vehicle C_0C_0 (RMB)	100	Constant value of product deterioration K_1, K_2	(1,1)
The cost of running a vehicle C_p (RMB/Km)	2	Customer weight λ	(1~5)
Spoilage rate of fresh produce θ	0.002	Affect the size	(0~25)
Maximum load of vehicle Q (kg)	500	Customer current value threshold	150
The cost of waiting to arrive early (RMB/h)	2	Customer potential value threshold	20
The penalty cost of late arrival (RMB/h)	3	The price per unit of fresh produce R_1 (RMB)	30
The average speed v_0 (km/h)	50	Profit per unit of fresh R_2 (RMB)	30
Fuel consumption per unit distance when fully loaded ρ^* (L/Km)	0.377	Fuel consumption per unit distance when no load ρ_0 (L/Km)	0.165
CO_2 emission coefficient ε (Kg/L)	2.63	The amount of CO_2 produced by refrigeration per unit distance traveled by goods delivered φ g/(Kg·Km)	0.0066
Benchmark carbon trading price $\overline{\omega}$	30	Range of carbon trading price changes $\hat{\omega}$	(0~30)

Table 3 represents the traffic congestion coefficients for different periods. The parameter values in both tables were from the study by Wang et al. [40]. All case tests were performed on an Intel(R) Core (TM) i5-7200U, 2.50 GHz, 4 GB RAM, Windows 10 (64-bit) computer, and the algorithms used in this paper were programmed on MATLAB R2019b.

Table 3. Traffic congestion factor.

Period of Time	[0,5]	[5,7]	[7,9]	[9,12]	[12,14]	[14,18]	[18,20]	[20,24]
Congestion Delay Coefficient	1	1.2	1.4	1.6	1.5	1.3	1.7	1.2

4.1. Model Solution

There were six customer information points in Table 4. This study used customer point information to solve the CV-GVRP under the G-NSGA-II algorithm, and compared the obtained results with those under the uniparental genetic algorithm. Although the model and algorithm in this paper did not have an advantage in terms of time consumption, they performed better in reducing the total cost and increasing the customer value. The solution results are shown in Table 5.

Table 4. Point-related customer information.

No.	0	1	2	3	4	5
Location	(35,35)	(41,49)	(35,17)	(55,45)	(55,20)	(15,30)
Demand	0	10	7	13	19	26
Acceptable time window	[0.0,24]	[1.8,4.8]	[0.0,2.5]	[0.0,2.0]	[0.5,3.0]	[1.0,4.0]
Optimum time window	[0.0,24]	[2.8,3.8]	[0.5,1.5]	[0.5,1.5]	[1.0,2.0]	[2.0,3.0]
Service time	0	0.3	0.3	0.3	0.3	0.3
Customer importance	1	1	1	1	1.2	1.5
Weight	1	1	1	3	2	4
Scope of influence	0	13	20	15	2	5

Table 5. SAPGA and G-NSGA-II solution results.

Algorithm	Total Cost	Customer Value	Computation Time	Distribution Path
SAPGA	375.3001	378.9227	4.62	0-3-2-4-0 0-5-1-0
G-NSGA-II	322.2635	442.8125	13.44	0-1-0 0-3-4-2-0 0-5-0

4.2. Numerical Benchmark

4.2.1. Comparison between MOPSO, NSGA-II and G-NSGA-II

Using MOPSO, the conventional NSGA-II and G-NSGA-II to solve the nine cases, the parameter settings of the algorithms used in the experiments were as follows: Figures 8–10 show the r101_100, r101_50 and r101_100 Pareto solution results. In this study, the algorithm parameters were set as per the study by Rabiee et al. [43]. The basic parameters of the experiment are shown in Table 6.

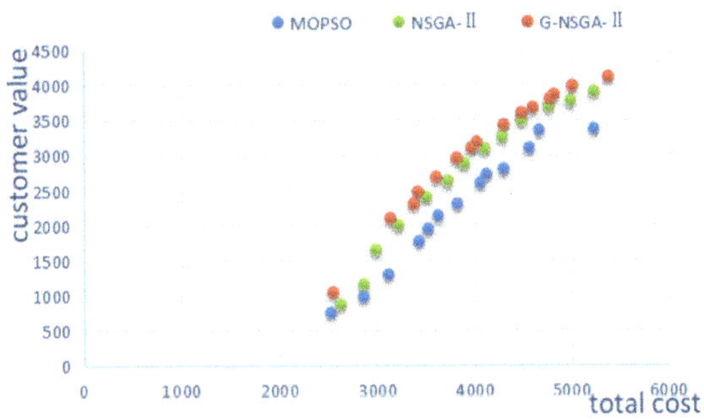

Figure 8. Solution results for r101_100.

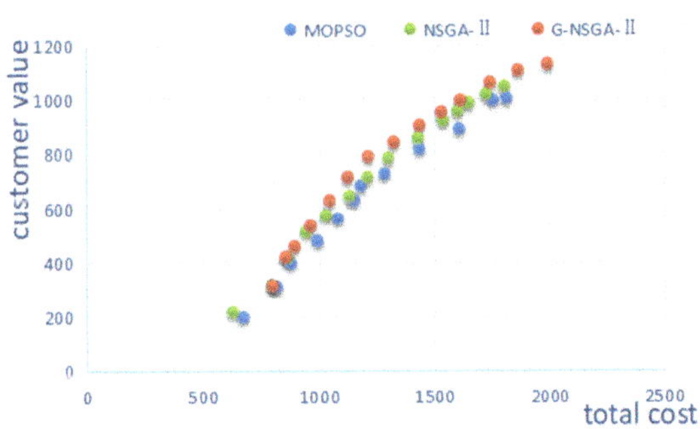

Figure 9. Solution results for r101_50.

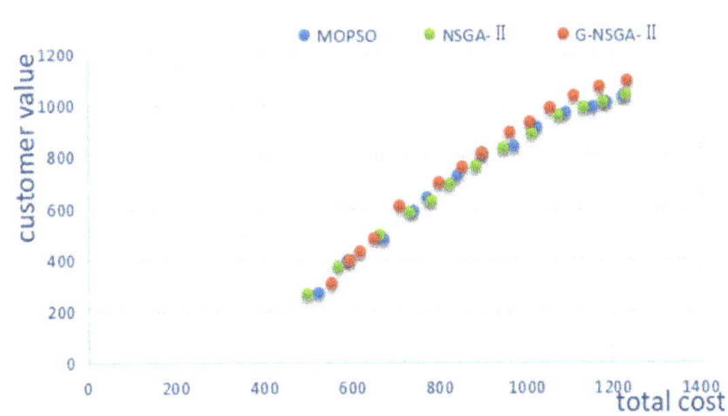

Figure 10. Solution results for r101_25.

Table 6. Parameter settings of each algorithm.

MOPSO		NSGA-II		G-NSGA-II	
Parameter	**Value**	**Parameter**	**Value**	**Parameter**	**Value**
The population number	200	The population number	30	The population number	30
Size of the warehouse	100	Crossover probability	0.9	Crossover probability	0.9
Inertial factor	0.9	Mutation probability	0.1	Mutation probability	0.1
Local velocity factor	1	The number of iterations	500	The number of iterations	500
Global velocity factor	2				
The number of iterations	300				

The algorithm comparison continued using these nine examples, and the following three indicators were used to compare the performance of the algorithms:

1. The number of Pareto solutions (NPS) indicated an algorithm's total number of nondominated solutions.
2. The diversity metric (DM) of solutions was measured with $\sqrt{(\max f_{1i} - \min f_{1i})^2 + (\max f_{2i} - \min f_{2i})^2}$. It determined the diversity of nondominant solutions obtained with each algorithm.
3. The formula $(\sum_{i=1}^{n} \sqrt{f_{1i}{}^2 + f_{2i}{}^2})/n$ was used to calculate the average distance between the Pareto solution and the ideal point (0, 0), which was the mean ideal distance (MID).

For the first two indicators, the higher the value, the better the algorithm performance, while the lower the average ideal distance value, the better the algorithm performance [43]. The comparison results are shown in Table 7.

Table 7. Algorithm performance comparison.

Examples	NPS G-NSGA-II	NPS NSGA-II	NPS MOPSO	DM G-NSGA-II	DM NSGA-II	DM MOPSO	MID G-NSGA-II	MID NSGA-II	MID MOPSO
c101_25	16	15	14	1116.3	889.9	857.3	1400.1	1463.8	1422.7
r101_25	14	13	13	707.4	702.0	645.6	951.4	1150.5	1168.0
rc101_25	15	14	14	455.0	445.1	412.7	1596.3	1582.4	1579.4
c101_50	15	14	14	1352.9	1284.3	1036.5	3112.3	3213.8	3203.9
r101_50	14	12	12	1349.9	1055.2	922.3	1024.6	1447.0	1583.7
rc101_50	15	14	13	1047.1	1049.1	934.9	1907.0	2131.3	2383.2
c101_100	16	14	13	1936.4	1727.5	1504.3	5401.0	6403.8	7799.2
r101_100	15	13	13	2746.8	2469.3	2610.8	4015.9	4412.5	5139.5
rc101_100	16	15	13	1898.8	1727.9	1558.5	5045.8	5144.0	6260.4

In Figure 11, the NPS of G-NSGA-II and NSGA-II was significantly greater than that of MOPSO. The first two were similar in terms of dispersion. However, the median NPS of G-NSGA-II was significantly higher than that of NSGA-II. Figure 12, showing the DM, shows the boxplot of G-NSGA-II, which outperformed the others in dispersion and median. The MID in Figure 13 shows that G-NSGA-II and NSGA-II performed significantly better than MOPSO, with G-NSGA-II having a slightly lower median. Based on the results as previously seen, G-NSGA-II performed better. In summary, the G-NSGA-II algorithm outperformed the traditional NSGA-II and MOPSO ones in terms of convergence and diversity. Its Pareto solutions were more uniformly distributed than those of the traditional NSGA-II. Therefore, it had better results in solving the problem.

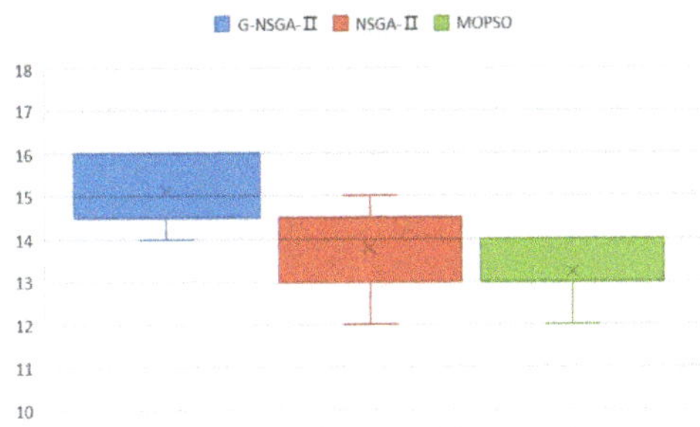

Figure 11. NPS boxplot.

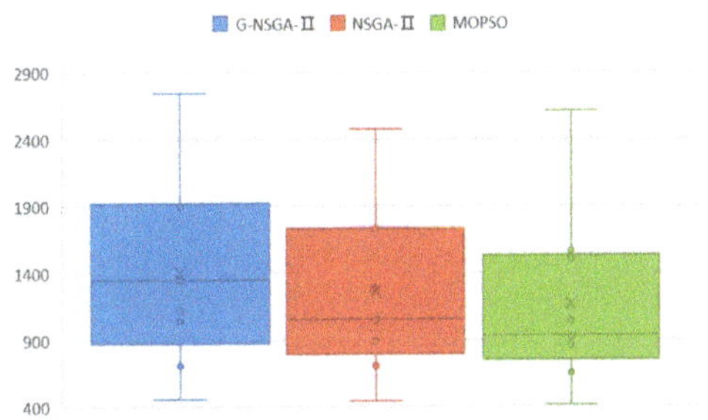

Figure 12. DM boxplot.

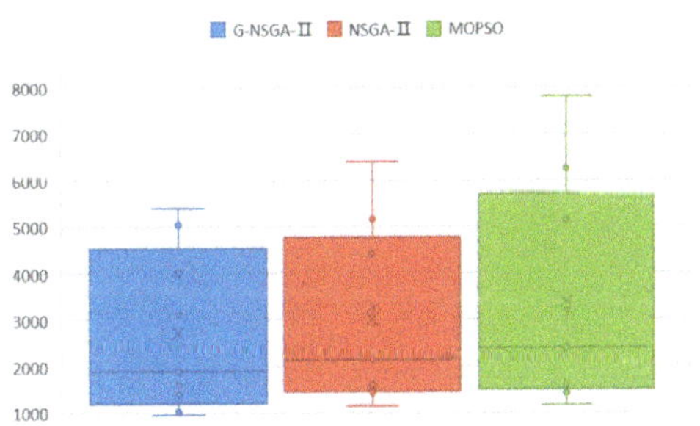

Figure 13. MID boxplot.

4.2.2. Analysis of Carbon Trading Price Sensitivity

Through the analysis of different types of data, it was found that the total cost changed significantly under the influence of variable carbon trading prices, especially in the large-scale examples. The change in price was not a single trend, which was why the carbon trading prices differed from the carbon tax policies. Under its influence, carbon emissions could increase or decrease, so the total cost also fluctuated. Due to firms gaining from selling carbon allowances, the cost could be reduced. Additionally, this behavior could incentivize firms to continue cutting carbon.

The total cost was lower than the situation with a fixed carbon trading price. The changes are shown in Figure 14. Furthermore, it indicated that a proper adjustment in carbon trading prices could motivate enterprises to adjust their distribution schemes. This could reduce the total cost and contribute to declining social carbon emissions from a macroperspective.

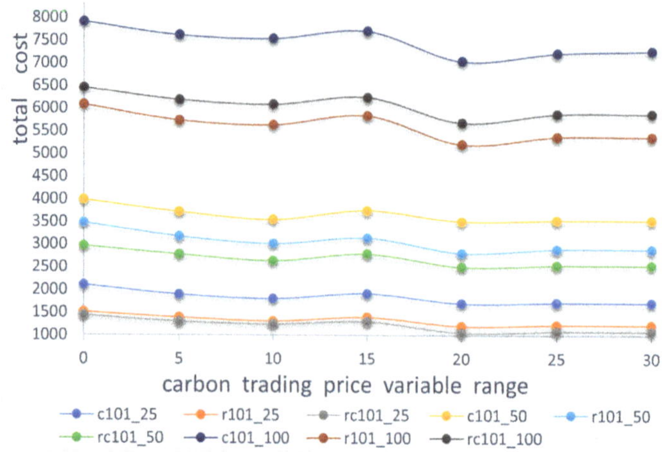

Figure 14. Graph of total cost versus $\hat{\omega}$.

4.2.3. Influence of Time-Dependent Network on Routing Strategy

In cases of different scales, the model in this paper had different improvements. For example, in the small-scale case, the total cost of the CV-GVRP was saved by 6.42%, and the customer value was increased by 15.43% on average. In the medium-scale case, the total cost of the CV-GVRP was saved by 5.4%, and the customer value was increased by 17.68% on average. In the large-scale case, the total cost of the CV-GVRP was saved by 3.75%, and the customer value was increased by 17.40% on average. The results of the comparison are shown in Table 8. The cost savings of the CV-GVRP were more significant in the small-scale cases, and the increase in customer value was more obvious in the medium-scale and large-scale cases. The average increase in the total cost was 5.192%, and the average increase in customer value was 16.838%, which indicated that the CV-GVRP performed better than the model with a static road network in improving customer value. Considering traffic congestion could increase the delivery time and increase the total cost, but the CV-GVRP brought more customer value, which is conducive to the long-term development of the enterprise.

Table 8. Comparison of static road network and CV-GVRP results.

	Static Road Network		CV-GVRP	
	Total Cost	**Customer Value**	**Total Cost**	**Customer Value**
c101_25	1674.5	1271.9	1716.3	1423.1
r101_25	1139.1	907.1	1231.8	1091.9
rc101_25	1197.6	1495.3	1301.0	1705.2
c101_50	3415.3	2115.6	3533.7	2714.9
r101_50	2687.3	2380.7	2893.7	2736.1
rc101_50	2435.0	2512.8	2558.0	2758.3
c101_100	6926.6	3770.9	7299.9	4211.9
r101_100	5171.1	3404.9	5379.1	4106.5
rc101_100	5791.7	5389.1	5898.9	6462.3

4.2.4. Influence of Customer Value on Routing Strategy

The GVRP is a model that does not consider the potential value of customers. It has the same constraints as the CV-GVRP. However, the objective function of the GVRP is customers' total cost and current value, which does not consider the impact of potential value on corporate reputation. Then, the G-NSGA-II was used to solve the GVRP for different cases. In the small-scale case of 25 customers, the total cost of CV-GVRP increased by 7%, the customer value increased by 14% and the satisfaction increased by 26% on average. In the medium-scale case of 50 customers, the total cost increased by 8%, the customer value increased by 14% and the satisfaction increased by 15% on average. In the large-scale case of 100 customers, the total cost increased by 3%, the customer value increased by 21% and the satisfaction increased by 13% on average. Overall, the average increase in the total cost was 3%, the average increase in customer value was 21% and the average increase in satisfaction was 13%. The increase in the total costs was most likely due to the increased delivery distance and time required to serve important customers. Additionally, the improvement in customer satisfaction brought more potential value to the enterprise and increased the total customer value. The results of the comparison are shown in Table 9.

Table 9. Comparison of GVRP and CV-GVRP results.

	GVRP			CV-GVRP		
	Total Cost	**Current Value**	**Satisfaction**	**Total Cost**	**Customer Value**	**Satisfaction**
c101_25	1598.6	1236.3	0.783	1716.3	1423.1	0.963
r101_25	1182.6	939.1	0.726	1231.8	1091.9	0.931
rc101_25	1211.9	1539.1	0.744	1301.0	1705.2	0.937
c101_50	3289.4	2418.7	0.791	3533.7	2714.9	0.916
r101_50	2734.5	2417.2	0.811	2993.7	2736.1	0.927
rc101_50	2373.9	2389.9	0.803	2558.0	2758.3	0.914
c101_100	6999.8	3417.2	0.822	7299.9	4211.9	0.907
r101_100	5266.5	3396.7	0.807	5379.1	4106.5	0.912
rc101_100	5713.6	5467.5	0.793	5898.9	6462.3	0.908

4.3. An instance of Distribution in Shanghai City

A refrigerated logistics company in Shanghai has to deliver to 20 distribution points. The information about customers is shown in Table 10. Assuming that the traffic congestion in the distribution is not considered, the vehicle is driven at a uniform speed of 40km/h and the carbon trading price is fixed at 54 RMB/ton. Moreover, the potential value of the customer is not considered. The solution can be solved with G-NSGA-II and then compared with the results of CV-GVRP. This paper's model considered the variable speed, variable carbon trading price and customer value, resulting in an 8.18% reduction in the total cost, a 23.82% increase in customer value and a 4.05% reduction in time. The results are shown in Table 11. In addition, the number of vehicles used was two less. The two solutions are shown in Figures 15 and 16.

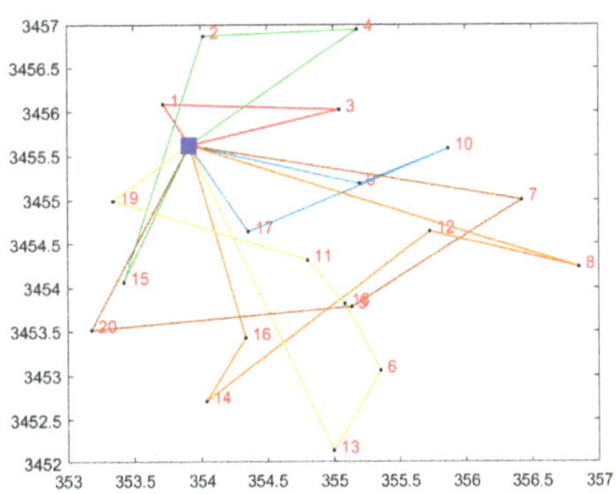

Figure 15. Preoptimization delivery routes.

Table 10. Shanghai Huangpu district distribution point information.

X-axis (km)	*Y*-axis (km)	Time Window (min)	Acceptable Time Window (min)	Demand (t)	Service Time (min)	Importance	Weights	Influences Scope
353.914	3455.623	0–500	0–500	0	0	0	0	0
353.721	3456.089	30–90	30–120	0.80	10	1.2	2	10
354.020	3456.869	60–90	60–120	3.35	20	1.3	2	3
355.047	3456.028	30–90	30–150	2.95	15	1.0	1	2
355.179	3456.943	60–90	60–120	2.40	10	1.0	1	7
355.135	3453.781	70–130	70–160	2.75	20	1.0	1	5
355.351	3453.055	80–110	80–140	3.30	20	1.1	1	8
356.423	3455.002	70–130	70–160	2.80	25	1.2	2	3
356.849	3454.232	60–90	60–120	3.25	20	1.3	3	2
355.197	3455.186	80–140	80–180	2.15	15	1.4	4	4
355.870	3455.585	85–115	85–150	3.05	15	1.4	4	6
354.804	3454.310	80–140	80–170	3.20	25	1.5	5	7
355.729	3454.642	70–100	70–130	3.50	10	1.2	4	8
354.597	3452.140	60–120	60–150	0.55	20	1.3	3	10
354.036	3452.706	80–110	80–150	2.70	20	1.0	2	12
353.423	3454.06	80–140	80–170	1.70	15	1.0	2	13
354.337	3453.427	90–120	90–160	2.25	20	1.0	1	7
354.358	3454.64	90–150	90–180	2.75	25	1.0	2	10
355.082	3453.816	70–100	70–140	1.90	10	1.5	5	15
353.345	3454.992	60–120	60–150	3.15	20	1.3	3	13
353.180	3453.517	80–110	80–140	1.00	15	1.0	2	7

Table 11. Example solution result comparison.

	Pre-Optimization	Optimized
Total cost	1858.5	1706.5
Customer value	1730.0	2142.1
Calculating time	19.6442	18.8483
Delivery route	0→3→1→0 0→9→10→17→0 0→4→2→15→0 0→8→12→14→16→0 0→19→11→18→6→13→0 0→7→5→20→0	0→9→18→19→5→17→0 0→13→1→20→14→15→0 0→2→3→12→6→16→11→0 0→4→8→10→7→0

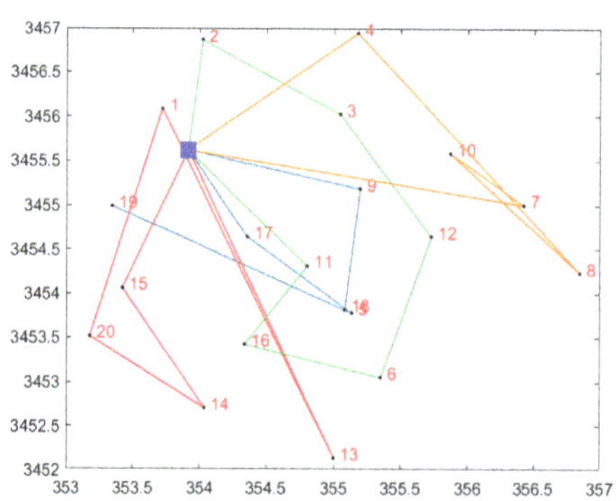

Figure 16. Optimized delivery route conclusion.

4.4. Discussion on the Model and the Algorithm

In general, the experiments focused on two aspects: the algorithm and the model.

To analyze the algorithm performance, six customer points were used as examples to verify the feasibility of the G-NSGA-II algorithm. To further explore the superiority of the G-NSGA-II algorithm, the small–medium–large-scale test sets were constructed using Solomon data. The solution results of the MOPSO, NSGA-II and G-NSGA-II algorithms were compared in convergence and diversity. By plotting boxplots in NPS, DM and MID and comparing their median and dispersion, the G-NSGA-II algorithm outperformed the others.

The CV model was used to analyze the influence of carbon trading uncertainty, time-varying network and customer potential value on the distribution schemes.

1. Variable carbon trading prices were used to reflect the temporal and spatial aspects of carbon trading policies. The experimental results showed that there was no linear relationship between the range of carbon trading price and the total cost, and the reasonable setting of the carbon trading price helped to reduce the cost and achieve a green distribution.
2. The experiments were designed to explore the influence of time-dependent road networks on distribution strategies, and it was found that the total costs, customer values and customer satisfaction would be affected on different scales. For large customer groups, the increase in customer value was more significant.
3. This paper studied the impact of potential value from customers on the distribution strategy. The experimental results showed that considering the potential value of customers could greatly improve customer satisfaction, especially in small groups.

Finally, the CV-GVRP model and the G-NSGA-II algorithm were applied to practical problems, and a planning scheme was proposed for real-life cold chain distributions. Constructing a cold chain distribution model under a carbon trading policy was conducive to reducing carbon emissions and achieving sustainable developments in enterprises. Taking into account the time window and potential value of the customer could help improve the reputation and benefits of the business by providing timely services to important customers.

5. Conclusions

From the perspective of variable carbon trading prices, this paper built a multiobjective optimization model, the CV-GVRP, which aimed to minimize the total cost and maximize customer value, considering both customer value and the time-dependent network. This model was an improvement of the traditional cold chain one, which was mainly applicable to fresh agricultural products with a short shelf life, low temperature and simple requirements for distribution conditions, such as leafy vegetables and fruits. By setting different ranges of carbon trading prices, it was found that flexible carbon trading prices positively affected companies in reducing their carbon emissions. Compared to static networks, the CV-GVRP was found to be advantageous in the total cost savings and customer value growth. Moreover, the CV-GVRP could improve customer satisfaction at a lower cost than models that do not take customer value into account. Small, medium and large-scale examples of c, r and rc, respectively, were constructed using Solomon data. Using the nine examples, G-NSGA-II, NSGA-II and MOPSO were used to solve this model. The values of NPS, DM and MID showed that the G-NSGA-II algorithm performed better than the others.

Future research can incorporate additional factors into the consideration of customer value, such as service quality and product quality. For the calculation of variable speed, only the congestion case was considered. However, the speed limits, emergencies and traffic control also affect vehicle speed in real life. This study considered only the carbon trading policy; in the future, the carbon tax policy and carbon trading policy can be used in coordination to reduce carbon emissions. To better address practical problems, there is still a lot of space to enrich the details of this model.

Author Contributions: D.W.: Conceptualization, Funding Acquisition and Writing—Review, Editing and Original Draft; J.L.: Methodology, Resources, Software and Writing—Original Draft; J.C.: Formal Analysis and Validation; D.H.: Supervision and Project Administration. All authors have read and agreed to the published version of the manuscript.

Funding: This research was funded by the China Education Ministry of Humanities and Social Science Research Youth Fund project (No. 18YJCZH192), the Special Project of National Characteristic Freshwater Fish Industrial Technology System for Construction of Modern Agricultural Industrial Technology System (No. CARS-46). Major project of National Social Science Fund "Research on the development strategy of China's deep blue fishery under the background of accelerating the construction of a marine power" (No. 21 & ZD100).

Institutional Review Board Statement: Not applicable.

Data Availability Statement: All experimental data in this paper came from: https://www.sintef.no/projectweb/top/vrptw/solomon-benchmark/ (accessed on 18 April 2008).

Conflicts of Interest: The authors declare no conflict of interest.

References

1. Cui, F.Z. The combination of point and surface, precise policy to the "14th Five-Year" new journey–Interpretation of the "14th Five-Year" cold chain logistics development plan. *China Logist. Purch.* **2022**, *1*, 24–26. [CrossRef]
2. Zhang, W.; Gajpal, Y.; Appadoo, S.S.; Wei, Q. Multi-depot green vehicle routing problem to minimize carbon emissions. *Sustainability* **2020**, *12*, 3500. [CrossRef]
3. Dantzig, G.B.; Ramser, J.H. The truck dispatching problem. *Manag. Sci.* **1959**, *6*, 80–91. [CrossRef]
4. Wang, C.; Peng, Z.; Xu, X. A Bi-Level Programming Approach to the Location-Routing Problem with Cargo Splitting under Low-Carbon Policies. *Mathematics* **2021**, *9*, 2325. [CrossRef]
5. Yao, Z.; Zhang, Y. Research on cold chain logistics distribution path optimization based on dual perspective of Internet of Things and low carbon. *Ecol. Econ.* **2020**, *36*, 61–66.
6. Liu, G.K.; Hu, J.Y.; Yang, Y.; Xia, S.M.; Lim, M.K. Vehicle routing problem in cold Chain logistics: A joint distribution model with carbon trading mechanisms. *Resour. Conserv. Recy.* **2020**, *156*, 104715. [CrossRef]
7. Ning, T.; Gou, T.; Liu, X.D. Simulation study of fresh agricultural products cold chain logistics strategy considering low carbon constraints. *J. Syst. Simul.* **2022**, *34*, 797–805. [CrossRef]
8. Jiang, J.H. Analysis of Carbon Pricing Mechanism and Proposals on the Improvement of China's Carbon Market under Carbon Neutralization. *Price Theory Pract.* **2022**, *2*, 26–30+90. [CrossRef]
9. Chen, X.Q.; Jin, C.; Yao, Q.G.; Wang, C. Research on Robust Optimization of multimodal transport routing under carbon trading policy. *Chin. J. Manag. Sci.* **2021**, *6*, 82–90. [CrossRef]
10. Ye, P. Vehicle distribution routing optimization considering low carbon emission. *Environ. Sci. Manag.* **2020**, *45*, 38–42.
11. Ren, T.; Chen, Y.; Xiang, Y.C.; Xing, L.N.; Li, S.D. Low-carbon cold chain vehicle routing optimization considering customer satisfaction. *Comput. Integr. Manuf. Syst.* **2020**, *26*, 1108–1117. [CrossRef]
12. Xiao, Y.Y.; Konak, A.A. Genetic algorithm with exact dynamic programming for the green vehicle routing & scheduling problem. *J. Clean. Prod.* **2017**, *167*, 1450–1463. [CrossRef]
13. Demir, E.; Hrušovský, M.; Jammernegg, W.; Van Woensel, T. Green intermodal freight transportation: Bi-objective modelling and analysis. *Int. J. Prod. Res.* **2019**, *57*, 6162–6180. [CrossRef]
14. Sadati, M.E.H.; Catay, B. A hybrid variable neighborhood search approach for the multi-depot green vehicle routing problem. *Transport. Res. E-Log.* **2021**, *149*, 102293. [CrossRef]
15. Zulvia, F.E.; Kuo, R.J.; Nugroho, D.Y. A many-objective gradient evolution algorithm for solving a green vehicle routing problem with time windows and time dependency for perishable products. *J. Clean. Prod.* **2020**, *242*, 118428. [CrossRef]
16. Wang, H.; Li, W.; Zhao, Z.Z.; Wang, Z.F.; Li, M.H.; Li, D.F. Intelligent Distribution of Fresh Agricultural Products in Smart City. *IEEE Trans. Ind. Inform.* **2022**, *18*, 1220–1230. [CrossRef]
17. Zhang, Y.M.; Li, Y.M.; Liu, H.O. Satisfaction-constrained multi-model cold chain logistics VRP optimization research. *Stat. Decis.* **2019**, *35*, 176–181. [CrossRef]
18. Ganji, M.; Kazemipoor, H.; Molana, S.M.H.; Sajadi, S.M. A green multi-objective integrated scheduling of production and distribution with heterogeneous fleet vehicle routing and time windows. *J. Clean. Prod.* **2020**, *259*, 120824. [CrossRef]
19. Jie, K.W.; Liu, S.Y.; Sun, X.J. A hybrid algorithm for time-dependent vehicle routing problem with soft time windows and stochastic factors. *Eng. Appl. Artif. Intell.* **2022**, *109*, 104606. [CrossRef]
20. Eydi, A.; Ghasemi-Nezhad, S.A. A bi-objective vehicle routing problem with time windows and multiple demands. *Ain. Shams. Eng. J.* **2021**, *12*, 2617–2630. [CrossRef]

21. Xiao, Q.; Zhao, H.; Ma, Y. Optimization of cold chain transportation routes in multi-distribution centers. *Packag. Eng.* **2019**, *40*, 116–122. [CrossRef]
22. Ghannadpour, S.F.; Zarrabi, A. Multi-objective heterogeneous vehicle routing and scheduling problem with energy minimizing. *Swarm Evol. Comput.* **2019**, *44*, 728–747. [CrossRef]
23. Malandraki, C.; Daskin, M.S. Time dependent vehicle routing problems: Formulations, properties and heuristic algorithms. *Transp. Sci.* **1992**, *26*, 185–200. [CrossRef]
24. Asgharizadeh, E.; Jooybar, S.; Mahdiraji, H.A.; Garza-Reyes, J.A. A Novel Travel Time Estimation Model for Modeling a Green Time-Dependent Vehicle Routing Problem in Food Supply Chain. *Sustainability* **2022**, *14*, 8633. [CrossRef]
25. Xu, Z.T.; Elomri, A.; Pokharel, S.; Mutlu, F. A model for capacitated green vehicle routing problem with the time-varying vehicle speed and soft time windows. *Comput. Ind. Eng.* **2019**, *137*, 106011. [CrossRef]
26. Poonthalir, G.; Nadarajan, R. A Fuel Efficient Green Vehicle Routing Problem with varying speed constraint (F-GVRP). *Expert. Syst. Appl.* **2018**, *100*, 131–144. [CrossRef]
27. Zhou, X.C.; Lv, Y.; He, C.H.; Liu, C.S.; Yang, K. Multi-vehicle green vehicle routing model and optimization algorithm considering time-varying speed. *Control. Decis.* **2022**, *37*, 473–482. [CrossRef]
28. Liu, Z.Q.; Chen, Y.P.; Li, J.; Zhang, D.Q. Spatiotemporal-Dependent Vehicle Routing Problem Considering Carbon Emissions. *Discret. Dyn. Nat. Soc.* **2021**, *2021*, 9729784. [CrossRef]
29. Yu, Y.; Wang, S.; Wang, J.; Huang, M. A branch-and-price algorithm for the heterogeneous fleet green vehicle routing problem with time windows. *Transp. Res. Pt. B-Methodol.* **2019**, *122*, 511–527. [CrossRef]
30. Lee, C. An exact algorithm for the electric-vehicle routing problem with nonlinear charging time. *J. Oper. Res. Soc.* **2021**, *72*, 1461–1485. [CrossRef]
31. Bruglieri, M.; Mancini, S.; Pisacane, O. The green vehicle routing problem with capacitated alternative fuel stations. *Comput. Oper. Res.* **2019**, *112*, 104759. [CrossRef]
32. He, B.Q.; Li, K.P.; Cheng, X.X. The Research on the Optimization of Courier Companies "Last-Mile" Express Pickup and Delivery Process. *Oper. Res. Manag. Sci.* **2019**, *1*, 27–34.
33. Li, W.L.; Li, K.P.; Tian, Q.N.; Li, X.S. Study on Optimization of retail logistics distribution route considering order release time under outbreak environment. *Chin. Manag. Sci.* **2022**, *9*, 195–205. [CrossRef]
34. Song, M.; Cheng, L. An augmented Lagrangian relaxation method for the mean-standard deviation based vehicle routing problem. *Knowl.-Based Syst.* **2022**, *247*, 108736. [CrossRef]
35. Yang, S.; Ning, L.; Shang, P.; Tong, L.C. Augmented Lagrangian relaxation approach for logistics vehicle routing problem with mixed backhauls and time windows. *Transp. Res. Pt. E-Logist. Transp. Rev.* **2020**, *135*, 101891. [CrossRef]
36. Schermer, D.; Moeini, M.; Wendt, O. A hybrid VNS/Tabu search algorithm for solving the vehicle routing problem with drones and en route operations. *Comput. Oper. Res.* **2019**, *109*, 134–158. [CrossRef]
37. Wang, Z.J.; Zhan, Z.H.; Kwong, S.; Jin, H.; Zhang, J. Adaptive granularity learning distributed particle swarm optimization for large-scale optimization. *IEEE Trans. Cybern.* **2020**, *51*, 1175–1188. [CrossRef]
38. Wang, Y.; Wang, L.; Peng, Z.; Chen, G.; Cai, Z.; Xing, L. A multi ant system based hybrid heuristic algorithm for vehicle routing problem with service time customization. *Swarm Evol. Comput.* **2019**, *50*, 100563. [CrossRef]
39. Wang, Y.; Zhang, J.; Liu, Y.; Xu, M.Z. Optimization of Fresh Goods Multi-Center Vehicle Routing Problem Based on Resource Sharing and Temperature Control. *Chin. J. Manag. Sci.* **2022**, *11*, 272–285. [CrossRef]
40. Wang, N.; Hu, D.W.; Xu, J.; Zhao, J. A time-varying path problem of urban cold chain logistics based on customer value and satisfaction. *Chin. J. Highw.* **2021**, *34*, 297–308. [CrossRef]
41. Ichoua, S.; Gendreau, M.; Potvin, J.Y. Vehicle dispatching with time-dependent travel times. *Eur. J. Oper. Res.* **2003**, *144*, 379–396. [CrossRef]
42. Solomon, M.M. Solomon's VRPTW Benchmark Problems. 1987. Available online: https://www.sintef.no/projectweb/top/vrptw/solomon-benchmark/100-customers/ (accessed on 1 April 1987).
43. Rabiee, M.; Zandieh, M.; Ramezani, P. Bi-objective partial flexible job shop scheduling problem: NSGA-II, NRGA, MOGA and PAES approaches. *Int. J. Prod. Res.* **2012**, *50*, 7327–7342. [CrossRef]

agriculture

MDPI

Article

Exploring the Prospective of Weed *Amaranthus retroflexus* for Biofuel Production through Pyrolysis

Julia Karaeva [1,*], Svetlana Timofeeva [1], Marat Gilfanov [1], Marina Slobozhaninova [1], Olga Sidorkina [1], Ekaterina Luchkina [1], Vladimir Panchenko [2] and Vadim Bolshev [3,*]

1 Institute of Power Engineering and Advanced Technologies, FRC Kazan Scientific Center, Russian Academy of Sciences, 420111 Kazan, Russia
2 Department of Theoretical and Applied Mechanics, Russian University of Transport, 127994 Moscow, Russia
3 Laboratory of Power Supply and Heat Supply, Federal Scientific Agroengineering Center VIM, 109428 Moscow, Russia
* Correspondence: julieenergy@list.ru (J.K.); vadimbolshev@gmail.com (V.B.)

Abstract: *Amaranthus retroflexus* or redroot pigweed is a second generation lignocellulosic fuel. Each biomass sample (leaves, inflorescences and stems) was pyrolyzed in a lab-scale furnace, in a nitrogen atmosphere under non-isothermal conditions at heating rates of 10 °C/min until the furnace temperature reached 550 °C. The pyrolysis characteristics of the three major components were also studied through thermogravimetric analysis. The thermal decomposition of the biomass samples is similar to the process of pyrolysis of lignocellulosic materials and proceeds in three main stages: dehydration, devolatilization, and carbonation. The highest bio-oil yield was obtained for inflorescences (55%) and leaves (45%). Gas chromatography—mass spectrometry analysis was carried out for oil fractions of the pyrolysis liquid from *Amaranthus retroflexus*. The composition of the pyrolysis oil fraction from the leaves had an overbearing aliphatic hydrocarbon nature whereas the oil fraction from inflorescences and stems was composed mainly of oxygen-containing components. The use of *Amaranthus retroflexus* biochars can lead to slag formation in power equipment, so it is advisable to use them to produce composite fuel, for example, mixed with coal. The results would help to better understand the thermal behavior of *Amaranthus retroflexus* biomass and its utilization for fuels or chemicals.

Keywords: biomass; *Amaranthus retroflexus*; pyrolysis; thermogravimetric analysis; bio-oil; biochar

Citation: Karaeva, J.; Timofeeva, S.; Gilfanov, M.; Slobozhaninova, M.; Sidorkina, O.; Luchkina, E.; Panchenko, V.; Bolshev, V. Exploring the Prospective of Weed *Amaranthus retroflexus* for Biofuel Production through Pyrolysis. *Agriculture* **2023**, *13*, 687. https://doi.org/10.3390/agriculture13030687

Academic Editor: Vincenzo Alfano

Received: 3 February 2023
Revised: 13 March 2023
Accepted: 13 March 2023
Published: 15 March 2023

1. Introduction

Second generation biofuels are produced from lignocellulosic biomass derived from agricultural waste, forest waste, municipal and industrial waste, and grass and aquatic plants. All of these types of biomass have one common property; they are formed from non-food resources [1]. *Amaranthus retroflexus* (AR), or redroot pigweed, is a second-generation lignocellulosic biomass. This plant is a fast growing herb of class C4, consisting of 60–70 species [2,3]. AR is a cosmopolitan plant capable of growing in any climatic zone, including the cryolitic zone [4]. The plant can reach 1.5–3 m in height, thus, it is possible to obtain huge biomass resources with little water and fertilizer consumption [2].

Pyrolysis is the most promising technology for thermal utilization, since it allows for the procurement of gaseous, liquid, and solid products, and is also the first stage of all thermochemical processes [5]. The quantity, properties, and application of these three mains products depend on the parameters of the feedstock, the type of reactor and the technological conditions for the implementation of the process (the heating rate, final pyrolysis temperature, pyrolysis atmosphere, etc.) [6].

From a practical point of view, bio-oil and biochar are of the greatest interest, since pyrolytic gas is most often used for its own technological needs. The rich chemical composition of bio-oil allows it to be used as renewable fuels and value-added chemicals [7,8].

Biochar has the following applications: soil amendment, nutrient and microbial carrier, immobilizing agent for remediation of toxic metals and organic contaminants in soil and water, catalyst for industrial applications, porous material for mitigating greenhouse gas emissions and odorous compounds, and feed supplement to improve animal health and nutrient intake [9,10]. Due to its low thermal conductivity, it can also be mixed with soil and used as a thermal backfill [11]. In addition, biochar-derived activated carbon is suitable for batteries and electrode material for supercapacitor applications [12].

There are known studies on the pyrolysis of weeds *Ageratum conyzoides* (goat weed) [13], *Ageratina adenophora* (Crofton weed) [14], *Alternanthera philoxeroides* (alligator weed) [15], *Parthenium hysterophorus* [16], and *Cannabis sativa* [16] to obtain biochar, bio-oil, and biogas. It should be noted that there are studies on the pyrolysis of invasive plants, such as: *Acacia Holosericea* [17], water hyacinth [18], *Prosopis juliflora* [12], and *Eupatorium adenophorum* [19]. However, there are no reports on the utilization of AR as a feedstock for pyrolysis. Therefore, the use of pyrolysis for the use of weeds and invasive plants with the production of bio-oil and biochar can contribute to the formation of a circular economy and increase the profitability of the agricultural industry [20].

For a detailed study of the process of thermochemical conversion, the plant can be divided into fractions. For example, leaves and stems [21], seeds [22–24], flowers, leaves, and stems [25], leaves, hurds, and roots [26], pseudo-stems [27], stem, leaves, fiber, chaff, and seed husks [28], and inflorescences [29,30]. This will make it possible to understand the contribution of each fraction to the material balance of the pyrolysis process. The content of hemicellulose, cellulose, and lignin differs in all constituent parts of the plants, so the quantity and quality of the resulting gaseous, liquid, and solid products varies significantly.

The aim of the work is to study the possibilities of thermal utilization of the AR weed plant, to study in detail the prospects for using its aboveground biomass to obtain new liquid and solid products with high added value. In this research, the authors studied the pyrolytic process of the leaves, stems, and inflorescences. Conventional pyrolysis was performed in combination with a thermogravimetric analysis. Therefore, the present study aims to solve the following problems: (a) determine the material balance of the AR pyrolysis of leaves, inflorescences, and stems of AR; (b) analysis the characteristics of the features of thermal decomposition of leaves, inflorescences, and stems according to TGA data, at a heating rate of 10 °C/min in an inert atmosphere; (c) study of the composition and quality of the chemical composition of the oil fraction of the pyrolysis liquid; and (d) analyze the possibility of using biochar for combustion both as an independent fuel and as part of a mixed fuel, taking into account the parameters of slag formation.

2. Materials and Methods

2.1. Sample Materials

The object of the study was the aerial part of the weed plant AR (Figure 1). The plants were grown in a field with cultivated plants (55°63′ N, 48°73′ E). In addition, they were divided into leaves, inflorescences (partially with seeds), and stems. Leaves and inflorescences were dried at room temperature, crushed, and sieved to obtain a mass with a particle size of less than 5 mm. The stems were crushed immediately, then dried and additionally ground; the particle size also did not exceed 5 mm.

2.2. Physicochemical Characterization

Moisture, ash content, and volatile matter (VM) were measured in accordance with ASTM E1755-01, ASTM E1756-08, GOST R 56881-2016, and GOST 32990-2014. Elemental analysis of the samples was carried out on a EuroEA3000 CHNS analyzer (Eurovector, S.p.A., Milan, Italy). The samples were weighed on a Sartorius CP2P microbalance (Germany) in tin capsules. Callidus 4.1 software was used to evaluate the obtained data.

(a) (b) (c)

Figure 1. Photographs of the dry biomass AR: (**a**) leaves; (**b**) inflorescences; and (**c**) stems.

The oxygen content (O, wt%) was calculated from the difference by Equation (1):

$$O = 100 - H - C - N - Ash, \quad (1)$$

where H, C, N, and Ash are wt% of hydrogen, carbon, nitrogen, and ash content of the fuel, respectively.

Atomic H/C and O/C ratios of AR fractions and their biochars were determined with Equations (2) and (3) [31]:

$$\text{atomic}\frac{H}{C}\text{ratio} = \frac{\text{number of H atoms}}{\text{number of C atoms}} = \frac{H/1}{C/12} \quad (2)$$

$$\text{atomic}\frac{O}{C}\text{ratio} = \frac{\text{number of O atoms}}{\text{number of C atoms}} = \frac{O/16}{C/12} \quad (3)$$

The higher heating value (HHV, MJ/kg) of the leaves, inflorescences, and stems were determined using Equation (4) [32]:

$$HHV_{AR} = 0.3491 \cdot C + 1.1783 \cdot H + 0.1005 \cdot S - 0.1034 \cdot O - 0.0151 \cdot N - 0.0211 \cdot Ash, \quad (4)$$

where C, H, O, N, and S, are, respectively, the carbon, hydrogen, oxygen, nitrogen, and sulfur content of the fuel, wt%.

2.3. *Thermogravimetric Analysis*

The most common method used to study the thermal behavior and thermal stability of fuels is the thermogravimetric analysis. Thermal decomposition data were measured using an STA 449 A1 Jupiter synchronous microthermal analyzer (Netzsch, Selb, Germany). For this, the following experimental parameters were chosen:

- temperature range: from 38 to 1000 °C;
- dynamic inert atmosphere: argon;
- heating rate: 10 °C/min;
- flow rate: 75 mL/min;
- average weight—35 mg;
- atmospheric pressure.

To ensure the repeatability of the experiment with an error of 1.5%, the experimental conditions were repeated at least three times.

2.4. Experimental Pyrolysis Procedure

A laboratory setup was used to study the pyrolysis of the biomass samples. It includes a tubular reactor (Figure 2), in which various organic raw materials can be thermally treated [33,34].

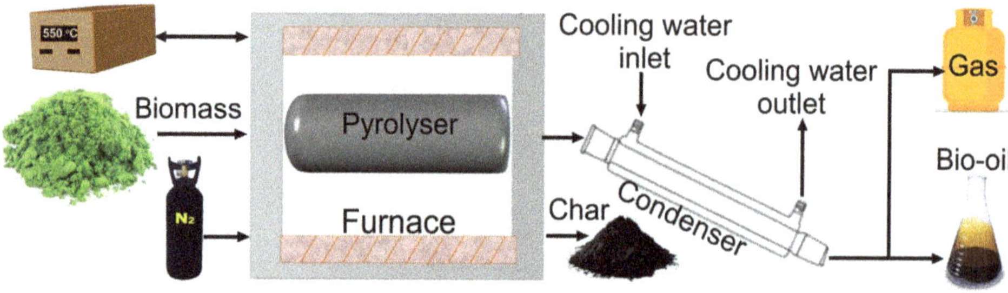

Figure 2. Experimental setup for the study of pyrolysis.

The laboratory setup was preliminarily purged with nitrogen. The prepared biomass sample (weighing about 45 g) was placed in a retort, which was installed in a preheated tubular reactor, which was hermetically sealed with lids. In the reactor, the biomass samples were subjected to a pyrolysis process. The maximum temperature of the pyrolysis process was 550 °C. The heating rate in the experiment was 10 °C/min. As a result of the experiments, three products were obtained: pyrolysis gas, pyrolysis liquid, and solid carbonaceous residue, biochar. Upon completion of the pyrolysis process, the retort was cooled, the solid residue was extracted and its mass yield was determined. Liquid pyrolysis products were collected, and their mass was determined. The gaseous product was determined by the difference in the masses of products from the material balance. Each experiment was repeated at least three times.

2.5. Expanded Measurement Uncertainty

The combined standard uncertainty of the measured value Y (Equation (5)) is obtained according to the law of propagation of uncertainties by summing the squares of the products of the standard uncertainties of all influencing quantities:

$$u(Y) = \sqrt{u^2(m_{ind}) + u^2(m_{device}) + u^2(F_{cor})} \quad (5)$$

The standard measurement uncertainty of m_{ind} is calculated assuming a normal probability distribution using Equation (6):

$$u(m_{ind}) = \sqrt{\frac{\sum_{i=1}^{n} (m_i - m_{ind})^2}{n(n-1)}}, \quad (6)$$

where m_i is the result of the i-th repetition of the weight measurement, m_{ind} is the arithmetic mean of m; n is the number of repeated measurements, $n = 3$.

The uncertainty associated with the value of m_{device} ((Equation (7)) is estimated using the manufacturer's data on the balance. In the laboratory scale passport for a measurement range of up to 50 g, the limits of weighing error are ±0.001 g (Δ). Since the value is given without a confidence level, we accept a rectangular distribution of weighing error values within these limits. The standard uncertainty is estimated according to type B and is:

$$u(m_{device}) = \frac{\Delta}{\sqrt{3}}. \quad (7)$$

The standard uncertainty of the correction factor is calculated from the information on the allowable discrepancy between parallel weight determinations. The given allowable relative discrepancy is $r = 20\%$ and is considered a 95% confidence interval for the difference between two estimates of a quantity distributed according to the normal distribution law. The standard uncertainty of the correction factor will be equal to the standard deviation calculated on the basis of the specified interval, taking into account that the measurement result is taken as the arithmetic mean of the determinations of two parallel samples, according to type B and according to Equation (8):

$$u(F_{cor}) = \frac{r}{100\% \cdot 2.8 \cdot \sqrt{2}}. \quad (8)$$

The expanded uncertainty U is obtained by multiplying the combined standard uncertainty by a coverage factor using Equation (9):

$$U = k \cdot u(Y), \quad (9)$$

where U is the expanded uncertainty, k is the coverage factor ($k = 2$ at a confidence level of approximately 95%, assuming a normal probability distribution of the measure), $u(Y)$ is standard uncertainty.

2.6. GC-MS Analysis of Bio-Oils

Gas chromatography–mass spectrometry (GC–MS) was carried out on a Shimadzu GCMS-QP2010 Ultra chromate mass spectrometer on an HP-5MS column (0.25 μm, 0.32 mm, 30 m) with the following parameters: carrier gas helium "A", temperature injector 300 °C, flow rate through the column 2 mL/min, split mode (10), thermostat temperature program—gradient temperature increasing from 60 to 180 °C in steps of 10 °C/min, then holding for 10 min, then increasing the temperature from 180 to 230 °C in steps of 10 °C/min, then holding for 5 min, then increasing the temperature from 230 to 280 °C in steps of 10 °C/min, then holding for 5 min, then increasing the temperature from 280 to 310 °C in steps of 10 °C/min, then holding for 13 min, and the range of scanned masses: 35–700 m/z. The 1 μL sample was injected into a 10% wt% toluene solution .

2.7. Analysis of Biochars

The chemical analysis of the mineral part was carried out using an EDX-800HS2 energy dispersive fluorescent X-ray spectrometer (Shimadzu, Kyoto, Japan). Measurement conditions: tube: Rh-anode (50 W), voltage: 50 kV, 15 kV, current: auto, atmosphere—air, measured diameter: 10 mm, measurement time: 100 s. The sample was placed on a Mylar film (6 μm thick) in a cuvette for the X-ray fluorescence analysis, after which the cuvette was placed in the instrument and the measurement was taken.

The volatile matter (VM) and ash content were determined according to ASTM D3175-89 and ASTM D3174-04, fixed carbon (FC) was calculated from the difference and calculated by Equation (10):

$$\mathrm{FC} = 100 - \mathrm{VM} - \mathrm{Ash} \quad (10)$$

The HHV (MJ/kg) of the biochars were determined using Equation (11) [35]:

$$\mathrm{HHV_{biochar}} = (15.59 \cdot \mathrm{VM} + 35.36 \cdot \mathrm{FC} - 0.78 \cdot \mathrm{Ash})/100 \quad (11)$$

The basic-acid ratio B/A (Equation (12)), slag viscosity index S_R (Equation (13)), and fouling index F_u (Equation (14)) were used to determine the possibility of contamination of the heating surfaces of the power equipment when using biochar as a fuel [36]:

$$B/A = (\mathrm{Fe_2O_3}+\mathrm{CaO}+\mathrm{MgO}+\mathrm{Na_2O}+\mathrm{K_2O}+\mathrm{P_2O_5})/(\mathrm{SiO_2}+\mathrm{Al_2O_3}+\mathrm{TiO_2}) \quad (12)$$

$$S_R = 100 \cdot \mathrm{SiO_2}/(\mathrm{SiO_2}+\mathrm{Fe_2O_3}+\mathrm{CaO}+\mathrm{MgO}) \quad (13)$$

$$F_u = (B/A)/(Na_2O + K_2O) \quad (14)$$

3. Results and Discussion

3.1. Results of Proximate and Ultimate Analyses

Table 1 presents the results of the determination of the physicochemical characteristics of the AR biomass samples. Proximate and ultimate analysis results indicate the potential use of the studied samples in thermochemical conversion processes. In the structure of all parts of the plant, a high content of volatile components was found, 68.3 to 77.9 wt%. This indicates a high reactivity of the material, as well as the fact that they can be converted into pyrolysis products with a large amount of gaseous and liquid components. The samples had a moisture content of about 7 wt%, which did not exceed the permissible limit of 10 wt%. A higher moisture content can lead to an increase in drying costs due to the need for additional thermal energy and a decrease in the efficiency of the thermal conversion of biomass [37].

Table 1. Results of the proximate and ultimate analyses of the AR biomass samples.

Parameter	Leaves	Inflorescences	Stems
Moisture, wt%	7 ± 0.01	7 ± 0.01	7 ± 0.01
VM, wt%	68.3 ± 0.4	74.2 ± 0.7	77.9 ± 0.8
Ash, wt%	23 ± 0.02	8 ± 0.01	11 ± 0.01
FC, wt%	8.7	17.8	11.1
HHV_{AR}, MJ/kg	21.0	25.6	24.1
C, wt%	34.6 ± 0.22	40.9 ± 0.12	35.4 ± 0.09
H, wt%	5.14 ± 0.09	6.34 ± 0.03	6.43 ± 0.13
N, wt%	4.04 ± 0.05	5.45 ± 0.09	3.97 ± 0.1
O, wt%	33.2 ± 0.17	39.3	43.2 ± 0.18
H/C	1.8	1.9	2.2
O/C	0.7	0.7	0.9

The elemental composition of all samples and their ash content are within the limits characteristic of lignocellulosic materials [38–40]. High ash content is considered a problem in thermochemical conversion processes because it can cause fouling or aggregation, as well as lead to some disposal problems, lower energy conversion rates, and ultimately higher recycling costs. AR leaves are characterized by the highest ash content. Given that this plant was grown on unprepared soil and without fertilizing, the increased ash content in the leaves can be explained by the biological characteristics of the plant. The HHV of the samples was determined from the results of the elemental analysis. For leaves, it was the minimum value equal to 21 MJ/kg; the maximum value was 25.6 MJ/kg for AR stems. The high HHV values in this study are comparable to those for various municipal solid wastes [41–43].

3.2. Thermal Degradation Analysis

The thermogravimetric (TG) and differential thermogravimetric (DTG) curves of the AR samples are shown in Figure 3. Mass loss during thermal decomposition was recorded in an inert medium at a temperature of 10 °C/min.

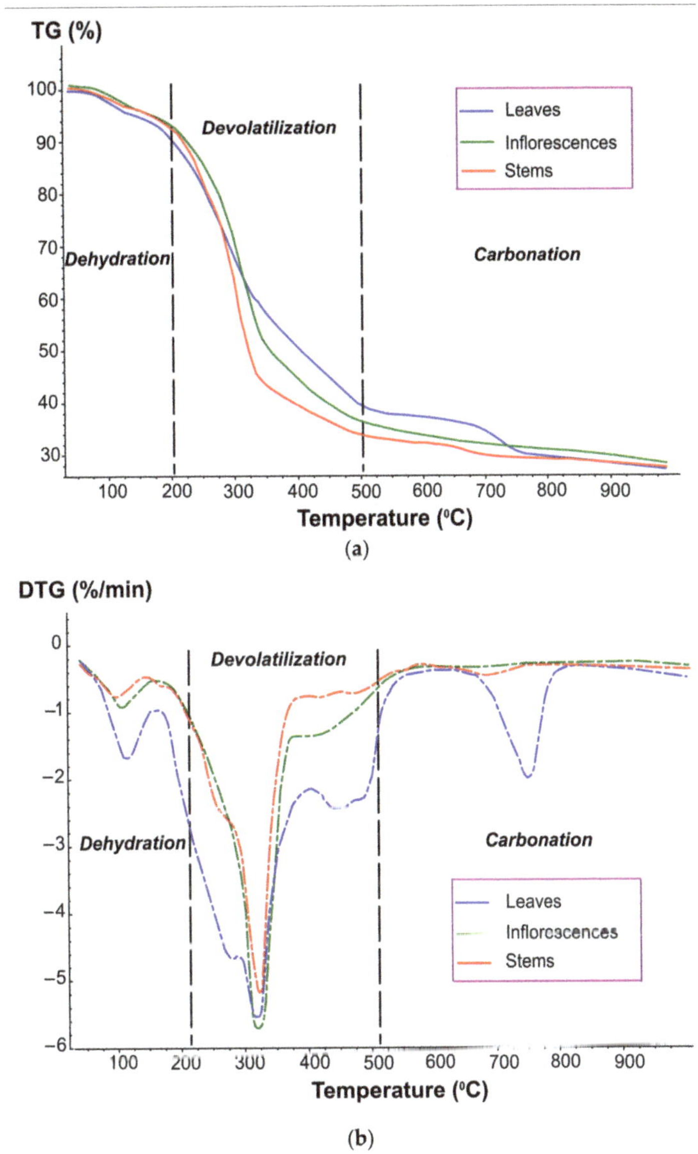

Figure 3. TG (**a**) and DTG (**b**) curves of the AR biomass at a heating rate of 10 °C/min.

The thermal decomposition of the three biomass samples studied is similar to the process of pyrolysis of lignocellulosic materials and proceeds in three main stages, which are characterized by the processes of dehydration, devolatilization, and carbonation (Table 2). The dehydration stage starts from 38 °C and proceeds on average up to 200 °C for the studied samples. Mass loss at this stage is considered to be the removal of free moisture, accompanied by evaporation from the sample surface upon heating, as well as chemically bound moisture. In addition to evaporation of the moisture, a slight release of volatile components is possible at this stage [44–46]. The average mass loss at this stage is 7.7%. At this stage, small peaks are found on the DTG curves. The first peak on the DTG curve is associated with the removal of free moisture at a temperature of about 100 ° C.

Table 2. Main stages of thermal decomposition.

Sample	Heating Rate (°C/min)	Pyrolysis Stage	Starting Temperature (°C)	Ending Temperature (°C)
Leaves	10	Dehydration	38	200
		devolatilization	200	502
		carbonation	502	1000
Inflorescences	10	dehydration	38	216
		devolatilization	216	512
		carbonation	512	1000
Stems	10	dehydration	38	190
		devolatilization	190	520
		carbonation	520	1000

The main stage of thermal decomposition for all samples starts at about 200 °C and ends in the temperature range of 502–520 °C. The processes that occur at this stage are associated with the release of volatile components. Significant mass loss is reflected as a peak in the DTG curve, which is due to complex thermochemical reactions during the conversion of biomass organic matter. In addition, the peak shown on the DTG curve characterizes the maximum decomposition rate, which is due to the breakdown of hemicellulose and cellulose [47]. As is known, the thermal decomposition of hemicellulose occurs in the temperature range of 180 to 300 °C, and cellulose from 300 to 480 °C [48,49]. Hemicellulose is composed of short-chain heteropolysaccharides and has an amorphous and branched structure. Monosaccharides are the main functional groups of hemicellulose with a small amount of uronic acids and acetyl groups [50]. The behavior of hemicellulose during pyrolysis is largely reflected by the characteristics of these building blocks during the thermal conversion process. Cellulose is a linear macromolecular polysaccharide consisting of a long chain of glucose units linked by β-1,4-glycosidic bonds [50]. The decomposition of cellulose is carried out by the depolymerization of various chemical bonds with the formation of carbon monoxide and dioxide, and carbon residues, as well as by the formation of bonds at high temperature to obtain liquid pyrolysis components.

Peaks in the DTG curves at the second stage indicated the decomposition of hemicellulose, which decomposes in the temperature range of 200–340 °C. Peaks at temperatures of 320–450 °C confirmed the thermal decomposition of cellulose. However, the temperature range changed as a result of the different biochemical compositions of the samples studied. For example, during the pyrolysis of AR stems, the maxima on the DTG curves shift, probably because of the high content of hemicellulose, which decomposes before cellulose.

At the carbonation stage, the processes of thermal cracking and dehydrogenation, as well as the decomposition of the solid carbonaceous residue and inorganic substances in its composition, take place [51]. The average mass loss in the carbonation stage is 8.7% for the studied samples. The main process at this stage is the thermal conversion of the complex structure of lignin, which decomposes almost throughout the entire temperature range from 190 to 900 °C. Unlike the carbohydrate structure of cellulose and hemicellulose, lignin has an aromatic matrix that gives strength and rigidity to the cell walls. Cellulose and hemicellulose have been shown to rapidly decompose over a short temperature range, while lignin slowly decomposes over a wider temperature range up to the end temperature of the experiment with the maximum formation of solid carbonaceous residues [52–54]. At temperatures above 680 °C, apparent shoulder peaks appeared in the DTG curves, which is due to the decomposition of inorganic components with low thermal stability [55,56]. At temperatures above 750 °C, practically no mass loss was observed, and the average residual weight as a result of heat treatment was 27.5% (Table 3).

Table 3. TG-data on changes in mass.

Sample	Mass Loss, %			Residual Mass, %
	Dehydration	Devolatilization	Carbonation	
Leaves	8.6	51.8	12.5	27.1
Inflorescences	8.5	57.1	8.3	26.7
Stems	6.1	59.4	5.6	28.9
Average, %	7.7	56.1	8.7	27.5

The content of hemicellulose, cellulose, and lignin in biomass affects the yield of pyrolysis products [57]. A high content of cellulose and hemicellulose contributes to the production of bio-oil, while a higher concentration of lignin results in more biochar produced. The structural complexity and stability of lignin make it difficult to destroy it during pyrolysis, leading to a higher yield of biochar [58].

3.3. Product Distribution and Yields

As a result of the study of the pyrolysis process at a temperature of 550 °C, three products were obtained: gas, bio-oil, and biochar (Table 4). Generally, the yields of liquid, solid and gaseous products are 50–70 wt%, 13–25 wt%, and 12–15 wt%, respectively [59]. A similar trend is typical for the results obtained. For all samples, the mass fraction of bio-oil among the pyrolysis products was the largest. In a number of studies, it has already been shown that the pyrolysis temperature from 450 to 550 °C contributes to the production of liquid products [35,60]. The yield of pyrolysis products when using AR leaves is similar to pine, AR inflorescences are comparable to *Acacia cincinnata* trunk, and AR stems are similar to agricultural biomass residues [61].

Table 4. Distribution of product pyrolysis yields and uncertainty analysis.

Sample	Pyrolysis Product	Mass, g	Mass Share, %	Expanded Uncertainty
Leaves	bio-oil	20.3	45.2	0.285
	biochar	17.5	38.9	0.587
	gas	7.1	15.9	0.836
Inflorescences	bio-oil	24.9	55.3	0.385
	biochar	14.5	32.2	0.473
	gas	5.6	12.5	0.203
Stems	bio-oil	17.7	39.3	0.428
	biochar	16.6	36.9	0.539
	gas	10.7	23.8	0.710

The results obtained in this study were compared with experimental data on the pyrolysis of plant biomass. Table 5 presents data on the products of thermal decomposition obtained during the pyrolysis of various biomass, its parts, and the plant as a whole. Experiments in which the process temperature was 500–600 °C and the heating rate varied from 5 to 50 °C/min were considered as pyrolysis conditions.

Table 5. Distribution of pyrolysis products obtained from various raw materials.

Biomass	Type of Pyrolysis	Pyrolysis Products, wt%			Reference
		Bio-Oil	Gas	Biochar	
Ageratum conyzoides	without catalyst	30	43	23	[13]
Crofton weed	catalytic	27–29	45–49	25–28	[14]
Alternanthera philoxeroides	without catalyst	45	22	33	[15]
Eupatorium adenophorum	catalytic	29–32	36–41	28–33	[19]
Acacia holosericea	without catalyst	33–38	33–37	26–34	[17]
Mixture of discarded vegetables and fruits	catalytic	35	22	30	[20]
Cortaderia selloana	without catalyst	19–34	45–62	18–27	[21]
Banana pseudo-stem (*Musa acuminate*)	without catalyst	28	32	42	[27]
Agricultural biomass residues	without catalyst	32	32	33	[61]
Acacia holosericea trunk	without catalyst	47	22	31	[61]
Amaranthus retroflexus	without catalyst	39.3–55.3	12.5–23.8	32.2–36.9	This study

A large yield of pyrolysis liquid was observed during pyrolysis of *Alternanthera philoxeroides* biomass [15]. The authors of [15] conclude that with an increase in temperature from 350 °C to 450 °C, the bio-oil yield and the yield of gas increase while the yield of biochar decreases. An increase in temperature leads to an increase in the gas yield and a decrease in the yield of pyrolysis liquid, which is associated with the secondary cracking of pyrolysis vapors at a higher pyrolysis temperature. During the pyrolysis of the banana pseudo-stem, a high mass yield of biochar was observed [21]. During the thermochemical conversion of the mixture of discarded vegetables and fruits at 500 °C, the maximum yield of bio-oil was observed, with an increase in temperature, its yield decreased [20]. During catalytic pyrolysis, there was a tendency to increase the yield of pyrolysis liquid and pyrolysis gases [14,19,20].

The heating rate during pyrolysis also affects the product yield. With fast pyrolysis, the yield of liquid products is higher, as this technology is characterized by high heating rates [17,20]. The analysis of studies showed that the temperature of 600 °C turned out to be optimal for further research on optimization due to its highest yields [15,17,20,27]. A comparative analysis showed that the results obtained in this work are comparable with the values obtained by other authors. The difference in values is explained by the composition of the feedstock, the influence of the parameters of the pyrolysis process and the influence of catalysts.

3.4. Composition of Bio-Oils

The composition of pyrolysis products is dominated by bio-oil or pyrolysis liquid, which is associated with the rich chemical composition of AR biomass. It contains at least 15 substances of phenolic nature, pectin substances, organic acids, tannins, amino acids, flavonoids, and other chemical components [62]. Furthermore, a significant content of hemicellulose can contribute to a higher yield of pyrolysis liquid [63]. Bio-oil consists of an aqueous fraction and an oil fraction (Figure 4). The highest yield of the aqueous fraction was obtained during the pyrolysis of AR stems, which may be due to the high content of holocellulose in them. The high content of the aqueous fraction in all bio-oil samples is due to lignin oligomers present in biomass due to the presence of hydrophilic polar functional groups [64].

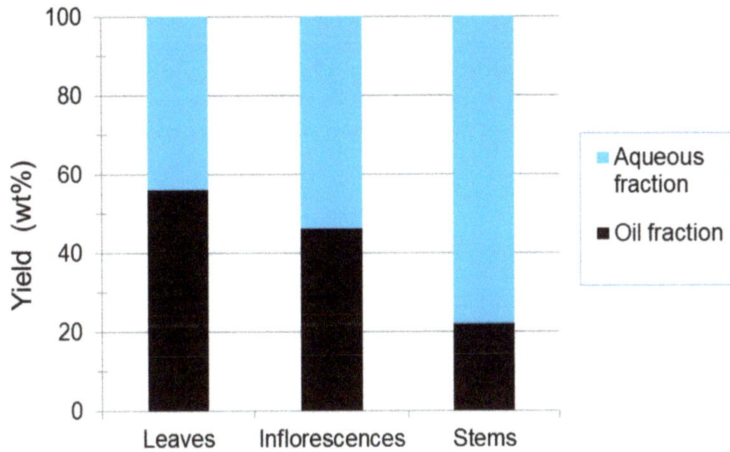

Figure 4. The content of the fractions in the pyrolysis liquid.

The oil fraction can be used as a fuel directly or converted into a high-quality fuel or chemical material. While the aqueous phase cannot be used directly as a fuel. As a result of the GC–MS analysis of the oil fractions of the pyrolysis liquid, their chemical composition was determined. It varies significantly depending on the part of the plant and is a complex mixture of organic compounds that contains hundreds of chemicals in total [65].

In the oil fraction from AR leaves, mass spectra of 72 substances of organic nature were determined, which were combined into four groups (Figure 5).

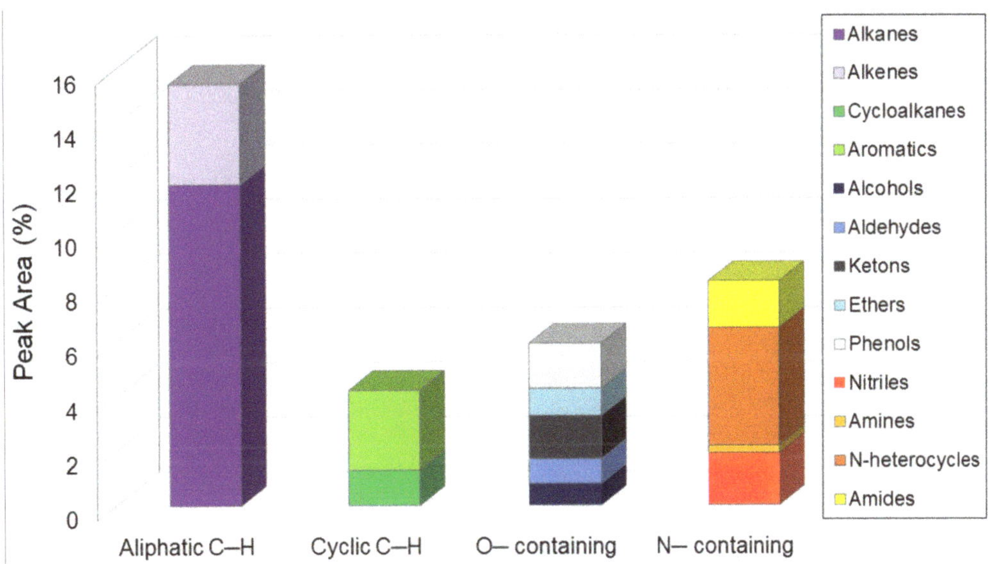

Figure 5. Chemical composition of the oil fraction from leaves.

In total, 33.8% of the total composition of the oil fraction was identified. The majority of all identified compounds are aliphatic hydrocarbons (15.5%), of which 11.8% are paraffinic hydrocarbons (C_{11}–C_{31}) and 3.7 % are olefinic hydrocarbons (C_{12}–C23). This carbon distribution is consistent with some bio-oils that can be used as liquid fuels in terms of carbon distribution [66,67]. A small part (4.23%) consists of cyclic hydrocarbons. The group of oxygen-containing components includes alcohols (0.79%), aldehydes (0.91%), ketones (1.58%), ethers (1.02%), and phenols (1.64%). Nitrogen-containing compounds account for 8.2%, more than half of which (4.34%) are heterocyclic in nature. The slight presence of seeds in the samples can lead to an increase in the content of hydrocarbons in bio-oil [68].

A GC–MS analysis of the oil fraction from AR inflorescences was also carried out, as a result of which the mass spectra of 80 organic substances were obtained (Figure 6). In total, 38.8% of the total composition of the liquid was identified.

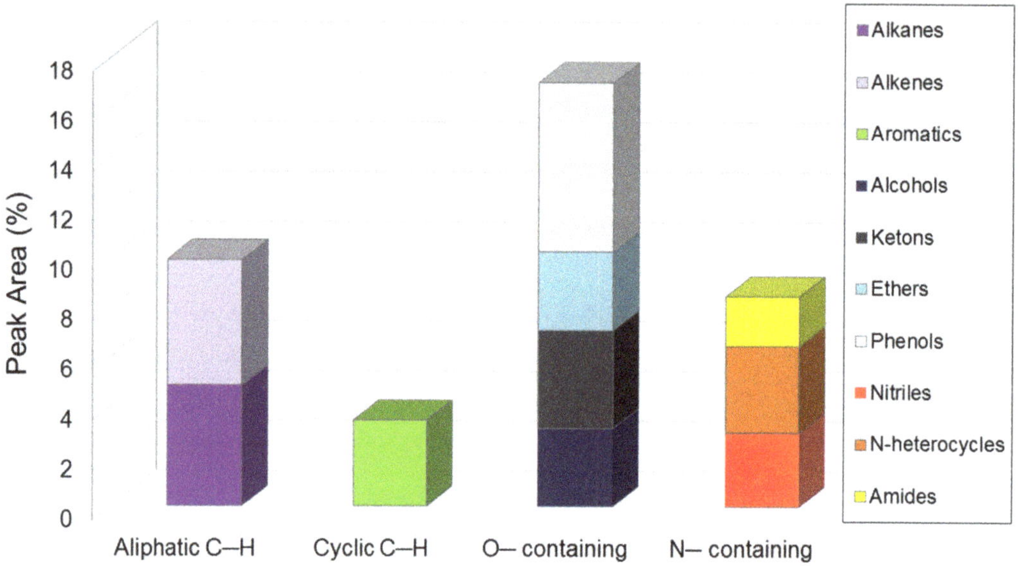

Figure 6. Chemical composition of the oil fraction from inflorescences.

There were marked changes in the composition of the oil fraction obtained from inflorescences relative to the composition of the oil fraction obtained from leaves. The main mass fraction of all identified compounds here is represented by an oxygen-containing group of substances (17%): alcohols (3.12%), ketones (3.94%), ethers (3.17%), and phenols (6.77%). At the same time, the proportion of phenols increased markedly. The content of aliphatic hydrocarbons decreased to 9.88%, mainly due to a decrease in the content of paraffinic hydrocarbons to 4.84%. The fractions of hydrocarbons also changed: paraffinic (C_{16}–C_{44}) and olefinic (C_{14}–C_{30}), respectively. Cyclic hydrocarbons make up 3.44%. Nitrogen-containing compounds account for 8.43%, of which less than half are heterocycles (3.48%).

The chemical composition of the pyrolysis liquid from the AR stems is mainly relatively simple and is represented by phenolic compounds (Figure 7). The data obtained are consistent with [4]. Phenolic compounds in bio-oil are a typical conversion product of lignocellulosic biomass, obtained mainly from the decomposition of lignin [69].

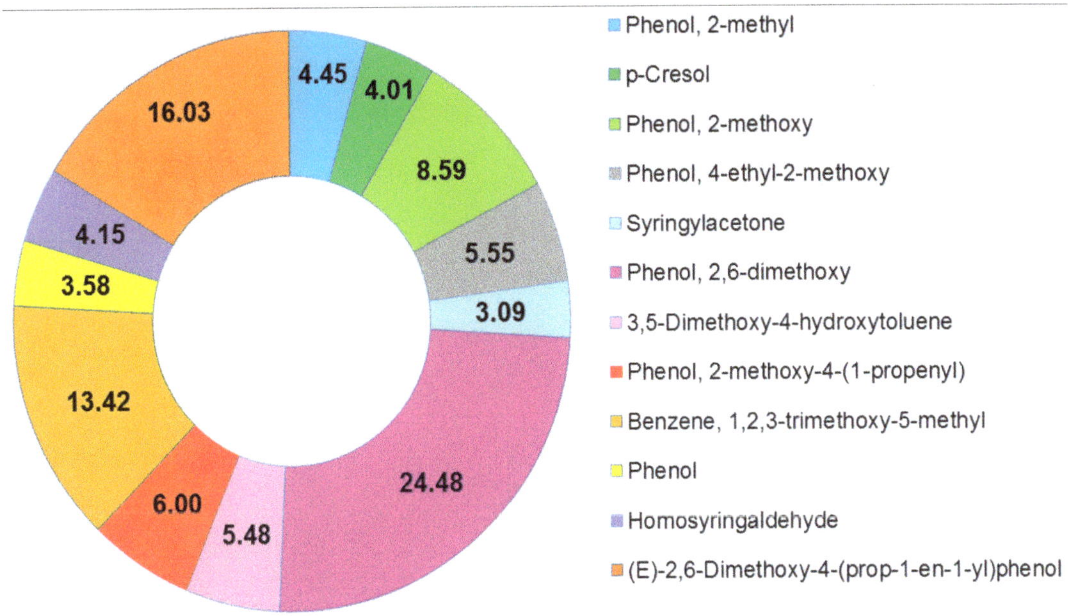

Figure 7. Chemical composition of the oil fraction of the stems.

Mass spectra of 12 organic substances were obtained. Phenolic compounds from AR stems have great potential to produce valuable chemical compounds.

3.5. Composition of Biochars

Biochar is a form of stable carbon and is a viable option for minimizing climate change and carbon footprint reduction. The pyrolysis process significantly increases the carbon content in the feedstock and converts it into stable aromatic compounds, increasing their number and density. The main characteristics of biochars obtained from AR are presented in Table 6. The highest amount of fixed carbon is found in samples from inflorescences, and the lowest from leaves. The ash content of the studied biochars varies from 31.9 to 46.8%. Leaves are characterized by the maximum ash content, and inflorescences are characterized by the minimum. The higher calorific value naturally decreases with the growth of the ash residue. The comparison of biochars obtained from inflorescences of AR weed and cultivated plants [70] showed that they have a similar composition; they contain a lot of calcium and potassium.

The study of the mineral part of biochar is a top priority when considering it as a fuel for power plants. Since most of the chemical elements contained in the samples have high melting and boiling points, after the pyrolysis process, they are concentrated in a solid carbonaceous residue. In the case of intensive slag formation processes, there is a high probability of not only unscheduled equipment shutdowns but also its failure due to the accumulation of uncontrolled amounts or forms of deposits. To predict the slag-forming and polluting properties of biochars, we determined the values of B/A, S_R, *and* Fu.

The optimal value of B/A, at which the fuel will not be slagged, should be less than 0.6. The value of this parameter indicates that the biochar ash of AR will be highly prone to slagging. The slag viscosity index S_R (>72) corresponds to a high viscosity and therefore a low tendency to slagging. The optimal value of the fouling index F_u should be less than 0.6. The parameters obtained indicate extreme slag formation in power equipment when using AR biochars.

Table 6. Results of the proximate analysis, oxide compositions of biochars, and slagging parameters.

Parameter	Leaves	Inflorescences	Stems
VM, wt%	35.1 ± 0.6	19.7 ± 0.7	20.4 ± 0.6
Ash, wt%	46.8 ± 0.4	23.3 ± 0.5	31.9 ± 0.5
FC, wt%	18.1 ± 0.4	57.0 ± 0.6	47.7 ± 0.5
HHV, MJ/kg	11.5	23.0	19.8
CaO, wt%	49.9	27.9	22.6
K_2O, wt%	27.6	51.1	64.2
MgO, wt%	8.3	3.34	3.6
P_2O_5, wt%	5.58	5.65	3.84
SO_3, wt%	4.58	2.68	1.64
SiO_2, wt%	2.51	4.21	0.54
Cl, wt%	0.75	3.89	3.11
Fe_2O_3, wt%	0.4	0.81	0.28
MnO, wt%	0.15	0.13	-
Br, wt%	0.08	0.02	0.01
SrO, wt%	0.08	-	0.02
ZnO, wt%	0.05	0.06	-
CuO, wt%	0.04	0.25	0.09
B/A	36.5	21.1	176
S_R	4.11	11.6	1.99
F_u	1009	1076	11,291

A promising solution to the problem is the use of composite fuel from a mixture of coal and biochar [71]. When using Krasnogorskiy coal (south of west Siberia) [72] in a mixture with biochars, one can obtain high-quality fuel with optimal B/A and S_R values (Figure 8).

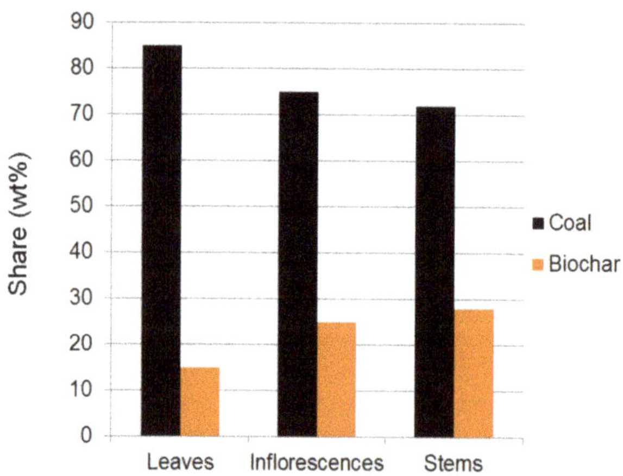

Figure 8. Composite fuel composition.

If biochar is planned to be used without mixing with high-quality fuel, then it is advisable to use it as a soil additive. Adding biochar to the soil not only captures carbon, but also reduces greenhouse gas emissions from the soil, is a promising option for managing agricultural waste, increases soil resilience, reduces the need for fertilizers, and also has a number of other environmental benefits [73]. Thus, the use of biochar promotes carbon sequestration [74]. Mild conditions of thermochemical conversion (at temperatures below 600 °C) make it possible to avoid the melting of ash and retain nutrients in a form accessible to microorganisms and plants [75]. The removal of biomass from the biocenosis should not lead to the disruption of the biogeochemical cycle of substances, so it is advisable to use biochars as soil additives. This use of biochar can also improve soil fertility by increasing its cation exchange and water retention capacity, as well as microbial activity [73,76,77].

A study [78] showed that biochar increased soil pH and improved its electrical conductivity, aggregate stability, water retention, and micronutrient content. In this case, it is necessary to take into account: soil type, seasonality, application rates, and characteristics of cultivated crops. In the study [76], it was noted that, depending on the feedstock and pyrolysis conditions, biochars have the most favorable effect on acidic and neutral soils. The authors of [79,80] proved that the introduction of biochar is of great importance for clay, sandy, and shale soils. The high content of calcium, potassium, magnesium, and phosphorus suggests that AR-derived biochars can not only improve soil structure, but also provide plants with essential nutrients. Thus, AR weed has a promising potential as a raw material for the production of soil additives.

4. Conclusions and Future Perspectives

In this study, an analysis was made of the possibilities of the thermal utilization of the aboveground biomass of the AR weed plant. The results of the proximate and ultimate analysis confirm the potential use of the studied samples in thermochemical conversion processes. TGA showed that the thermal decomposition of the studied AR biomass is similar to the process of pyrolysis of lignocellulosic materials and proceeds in three main stages, which are characterized by the processes of drying, release of volatile components, and carbonation. As a result of the study of the pyrolysis process in a laboratory setting at a temperature of 550 °C, three products were obtained: gas, bio-oil, and biochar. The composition of pyrolysis products is dominated by bio-oil or pyrolysis liquid, which is associated with the rich chemical composition of AR biomass. The main part of the oil fraction of AR leaves is aliphatic hydrocarbons (15.5%). The pyrolysis liquid from the inflorescences is represented by oxygen-containing compounds (17%). The chemical composition of the liquid product from AR stems is relatively simple and mainly by phenolic compounds. The results obtained allow us to conclude that the oil fractions of the pyrolysis liquid of AR parts can be used both as a fuel and as a source of valuable chemicals. An analysis of the characteristics and microelement composition of biochars showed that they can be used to produce composite fuel, for example, mixed with coal. It should be noted that all obtained biochars have a high potential as a raw material for the production of soil additives. The results of the study made it possible to study in detail the potential of using AR from its aboveground biomass to produce new liquid and solid products with high added value.

Author Contributions: Conceptualization, J.K.; methodology, J.K.; software, S.T. and V.P.; validation, V.B. and V.P.; formal analysis, O.S. and E.L.; investigation, S.T., M.G., M.S., O.S., and E.L.; resources, J.K., V.B., and V.P.; data curation, J.K., S.T., V.B., and V.P.; writing—original draft preparation, S.T., M.S., O.S., and E.L.; writing—review and editing, J.K. and V.B.; visualization, S.T; supervision, V.B. and V.P.; project administration, J.K.; funding acquisition, V.B. and V.P.; All authors have read and agreed to the published version of the manuscript.

Funding: This research received no external funding.

Institutional Review Board Statement: Not applicable.

Data Availability Statement: Not applicable.

Conflicts of Interest: The authors declare no conflict of interest.

Nomenclature

List of abbreviations and symbols	
Letters of the Latin alphabet	
C	carbon atom
H	hydrogen atom
C-H	hydrocarbons
Abbreviations	
AR	*Amaranthus retroflexus*
ASTM	American Society for Testing and Materials
B/A	basic-acid ratio
C	carbon content (wt%)
°C	degree Celsius
DTG	differential thermogravimetric
FC	fixed carbon (wt%)
F_u	fouling index
g	gram
GC–MS	Gas chromatography—mass spectrometry
H	hydrogen content (wt%)
HHV	higher heating value (MJ/kg)
k	coverage factor
kV	kilovolt
m	weight (g)
m	meter
mg	milligram
MJ/kg	megajoule per kilogram
mL/min	milliliter/minute
mm	millimeter
n	number of repeated measurements
N	nitrogen content (wt%)
O	oxygen content (wt%)
r	given allowable relative discrepancy
s	second
S	sulfur content
S_R	slag viscosity index
TG	thermogravimetric
VM	volatile matter (wt%)
W	watt
µm	micrometer
u	uncertainty
Greek symbols	
Δ	limit of weighing error
Superscripts and subscripts	
AR	*Amaranthus retroflexus*
i	sequence number of the experiment

References

1. Martínez-Gutiérrez, E. Biogas production from different lignocellulosic biomass sources: Advances and perspectives. *3 Biotech* **2018**, *8*, 233. [CrossRef] [PubMed]
2. Madadi, M.; Wang, Y.; Xu, C.; Liu, P.; Wang, Y.; Xia, T.; Tu, Y.; Lin, X.; Song, B.; Yang, X.; et al. Using Amaranthus green proteins as universal biosurfactant and biosorbent for effective enzymatic degradation of diverse lignocellulose residues and efficient multiple trace metals remediation of farming lands. *J. Hazard. Mater.* **2021**, *406*, 124727. [CrossRef]
3. Izunoglu, S.; Kus, E. Thermogravimetric Evaluation for the Pyrolysis Process of Pellets Produced from Quinoa and Amaranth Harvest Residues. *J. Agric.* **2022**, *5*, 64–73. [CrossRef]
4. Karaeva, J.V.; Timofeeva, S.S.; Kovalev, A.A.; Kovalev, D.A.; Gilfanov, M.F.; Grigoriev, V.S.; Litti, Y.V. Co-pyrolysis of agricultural waste and estimation of the applicability of pyrolysis in the integrated technology of biorenewable hydrogen production. *Int. J. Hydrog. Energy* **2022**, *47*, 11787–11798. [CrossRef]

5. Zhong, Y.; Ding, Y.; Jiang, G.; Lu, K.; Li, C. Comparison of Artificial Neural Networks and kinetic inverse modeling to predict biomass pyrolysis behavior. *J. Anal. Appl. Pyrolysis* **2023**, *169*, 105802. [CrossRef]
6. Pachón Gómez, E.M.; Domínguez, R.E.; López, D.A.; Téllez, J.F.; Marino, M.D.; Almada, N.; Gange, J.M.; Moyano, E.L. Chicken litter: A waste or a source of chemicals? Fast pyrolysis and hydrothermal conversion as alternatives in the valorisation of poultry waste. *J. Anal. Appl. Pyrolysis* **2023**, *169*, 105796. [CrossRef]
7. Machado, H.; Cristino Ana, F.; Orišková, S. Bio-Oil: The Next-Generation Source of Chemicals Rui Galhano dos Santos. *Reactions* **2022**, *3*, 118–137. [CrossRef]
8. Sánchez-Borrego, F.; Álvarez-Mateos, J.P. Biodiesel and Other Value-Added Products from Bio-Oil Obtained from Agrifood Waste. *Processes* **2021**, *9*, 797. [CrossRef]
9. Yang, L.; Deng, Y.; Shu, Z.; Chen, Q.; Yang, H.; Tan, X. Application of Invasive Plants as Biochar Precursors in the Field of Environment and Energy Storage. *Front. Environ. Sci. Sec. Toxicol. Pollut. Environ.* **2022**, *10*, 902915. [CrossRef]
10. Bolan, N.; Hoang, S.A.; Beiyuan, J.; Gupta, S.; Hou, D.; Karakoti, A.; Joseph, S.; Jung, S.; Kim, K.-H.; Kirkham, M.B.; et al. Multifunctional applications of biochar beyond carbon storage. *Int. Mater. Rev.* **2022**, *67*, 150–200. [CrossRef]
11. Patwa, D.; Muigai, H.H.; Ravi, K.; Sreedeep, S.; Kalita, P. A Novel Application of Biochar Produced from Invasive Weeds and Industrial Waste in Thermal Backfill for Crude Oil Industries. *Waste Biomass Valor* **2022**, *13*, 3025–3042. [CrossRef]
12. Raj, F.R.M.S.; Boopathi, G.; Kalpana, D.; Jaya, N.V.; Pandurangan, A. Sustainable development through restoration of Prosopis juliflora species into activated carbon as electrode material for supercapacitors. *Diam. Relat. Mater.* **2022**, *121*, 108767. [CrossRef]
13. Bhattacharjee, N.; Biswas, A.B. Pyrolysis of Ageratum conyzoides (goat weed). *J. Therm. Anal. Calorim.* **2020**, *139*, 1515–1536. [CrossRef]
14. Lin, L.; Cheng, S. Catalytic pyrolysis of crofton weed: Comparison of their pyrolysis product and preliminary economic analysis. *Env. Prog Sustain. Energy* **2022**, *41*, e13742. [CrossRef]
15. Bhattacharjee, N.; Biswas, A.B. Pyrolysis of Alternanthera philoxeroides (alligator weed): Effect of pyrolysis parameter on product yield and characterization of liquid product and bio char. *J. Energy Inst.* **2018**, *91*, 605–618. [CrossRef]
16. Tahir, N.; Tahir, M.N.; Alam, M.; Yi, W.; Zhang, Q. Exploring the prospective of weeds (*Cannabis sativa* L., *Parthenium hysterophorus* L.) for biofuel production through nanocatalytic (Co, Ni) gasification. *Biotechnol. Biofuels* **2020**, *13*, 148. [CrossRef]
17. Reza, M.S.; Ahmed, A.; Caesarendra, W.; Abu Bakar, M.S.; Shams, S.; Saidur, R.; Aslfattahi, N.; Azad, A.K. Acacia Holosericea: An Invasive Species for Bio-char, Bio-oil, and Biogas Production. *Bioengineering* **2019**, *6*, 33. [CrossRef]
18. Huang, H.; Liu, J.; Liu, H.; Evrendilek, F.; Buyukada, M. Pyrolysis of water hyacinth biomass parts: Bioenergy, gas emissions, and by-products using TG-FTIR and Py-GC/MS analyses. *Energy Convers. Manag.* **2020**, *207*, 112552. [CrossRef]
19. Zeng, K.; Yan, H.; Xia, H.; Zhang, L.; Zhang, Q. Catalytic pyrolysis of Eupatorium adenophorum by sodium salt. *J. Mater. Cycles Waste Manag.* **2021**, *23*, 1626–1635. [CrossRef]
20. Patra, B.R.; Nanda, S.; Dalai, A.K.; Meda, V. Slow pyrolysis of agro-food wastes and physicochemical characterization of biofuel products. *Chemosphere* **2021**, *285*, 131431. [CrossRef]
21. Pérez, A.; Ruiz, B.; Fuente, E.; Calvo, L.F.; Paniagua, S. Pyrolysis technology for Cortaderia selloana invasive species. Prospects in the biomass energy sector. *Renew. Energy* **2021**, *169*, 178–190. [CrossRef]
22. Altantzis, A.I.; Kallistridis, N.C.; Stavropoulos, G.; Zabaniotou, A. Peach Seeds Pyrolysis Integrated into a Zero Waste Biorefinery: An Experimental Study. *Circ. Econ. Sust.* **2022**, *2*, 351–382. [CrossRef]
23. Onorevoli, B.; Machado, M.E.; Polidoro, A.d.S.; Corbelini, V.A.; Caramão, E.B.; Jacques, R.A. Pyrolysis of Residual Tobacco Seeds: Characterization of Nitrogen Compounds in Bio-oil Using Comprehensive Two-Dimensional Gas Chromatography with Mass Spectrometry Detection. *Energy Fuels* **2017**, *31*, 9402–9407. [CrossRef]
24. Santos, R.M.; Santos, A.O.; Sussuchi, E.M.; Nascimento, J.S.; Lima, Á.S.; Freitas, L.S. Pyrolysis of mangaba seed: Production and characterization of bio-oil. *Bioresour. Technol.* **2015**, *196*, 43–48. [CrossRef]
25. Xue, T.; Zhang, X.; Jin, J.; Zhang, J.; Li, C.; Fu, Q. The Pyrolysis Characteristics and Kinetics Analysis on Typical Thermoplastic Artificial Green Plants. In Proceedings of the 2019 9th International Conference on Fire Science and Fire Protection Engineering (ICFSFPE), Chengdu, China, 18–20 October 2019; pp. 1–8. [CrossRef]
26. Salami, A.; Heikkinen, J.; Tomppo, L.; Hyttinen, M.; Kekäläinen, T.; Jänis, J.; Vepsäläinen, J.; Lappalainen, R. A Comparative Study of Pyrolysis Liquids by Slow Pyrolysis of Industrial Hemp Leaves, Hurds and Roots. *Molecules* **2021**, *26*, 3167. [CrossRef]
27. Taib, R.M.; Abdullah, N.; Aziz, N.S.M. Bio-oil derived from banana pseudo-stem via fast pyrolysis process. *Biomass Bioenergy* **2021**, *148*, 106034. [CrossRef]
28. Zvicevicius, E.; Ziura, K.; Cerniauskiene, Z.; Tilvikiene, V. Evaluating potential of organic carbon production from fibre hemp biomass. In Proceedings of the 20th International Scientific Conference Engineering for Rural Development, Jelgava, Latvia, 26–28 May 2021. [CrossRef]
29. Farias, R.O.; Lins, P.V.S.; Gabriel, R.; David, A.; Silva, R.; Luísa, L.; Carvalho, M.; Albuquerque, A.A.; Carvalho, S.H.V.; Soletti, J.I.; et al. Pyrolysis of Coconut Inflorescence Wastes: Production, Effects of Parameters, Characterization and Optimization of Phenolic-Rich Bio-Oil. *Int. J. Environ. Res.* **2022**, *16*, 15. [CrossRef]
30. Saravanan, A.; Senthil Kumar, P.; Jeevanantham, S.; Karishma, S.; Yaashikaa, P.R. Modeling analysis on the effective elimination of toxic pollutant from aquatic environment using pyrolysis assisted palmyra palm male inflorescence. *Environ. Res.* **2021**, *197*, 111146. [CrossRef]

31. Marrot, L.; Candelier, K.; Valette, J.; Lanvin, C.; Horvat, B.; Legan, L.; DeVallance, D.B. Valorization of Hemp Stalk Waste Through Thermochemical Conversion for Energy and Electrical Applications. *Waste Biomass Valor* **2022**, *13*, 2267–2285. [CrossRef]
32. Channiwala, S.A.; Parikh, P.P. A unified correlation for estimating HHV of solid, liquid and gaseous fuels. *Fuel* **2002**, *81*, 1051–1063. [CrossRef]
33. Karaeva, J.V.; Timofeeva, S.S.; Islamova, S.I.; Gerasimov, A.V. Pyrolysis kinetics of new bioenergy feedstock from anaerobic digestate of agro-waste by thermogravimetric analysis. *J. Environ. Chem. Eng.* **2022**, *10*, 3. [CrossRef]
34. Karaeva, J.V.; Timofeeva, S.S.; Bashkirov, V.N.; Bulygina, K.S. Thermochemical processing of digestate from biogas plant for recycling dairy manure and biomass. *Biomass Conv. Bioref.* **2021**, *13*, 685–695. [CrossRef]
35. Bai, X.; Wang, G.; Zhu, Z.; Cai, C.; Wang, Z.; Wang, D. Investigation of improving the yields and qualities of pyrolysis products with combination rod-milled and torrefaction pretreatment. *Renew. Energy* **2020**, *151*, 446–453. [CrossRef]
36. Pronobis, M. Evaluation of the influence of biomass co-combustion on boiler furnace slagging by means of fusibility correlations. *Biomass Bioenergy* **2005**, *28*, 375–383. [CrossRef]
37. Bhattacharjee, N.; Biswas, A.B. Physicochemical analysis and kinetic study of orange bagasse at higher heating rates. *Fuel* **2020**, *271*, 117642. [CrossRef]
38. Suntivarakorn, R.; Treedet, W.; Singbua, P.; Teeramaetawat, N. Fast pyrolysis from Napier grass for pyrolysis oil production by using circulating Fluidized Bed Reactor: Improvement of pyrolysis system and production cost. *Energy Rep.* **2018**, *4*, 565–575. [CrossRef]
39. Xiao, R.; Yang, W.; Cong, X.; Dong, K.; Xu, J.; Wang, D.; Yang, X. Thermogravimetric analysis and reaction kinetics of lignocellulosic biomass pyrolysis. *Energy* **2020**, *201*, 117537. [CrossRef]
40. Mierzwa-Hersztek, M.; Gondek, K.; Jewiarz, M.; Dziedzic, K. Assessment of energy parameters of biomass and biochars, leachability of heavy metals and phytotoxicity of their ashes. *J. Mater. Cycles Waste Manag.* **2019**, *21*, 786–800. [CrossRef]
41. Özbay, G.; ve Özçifçi, A. Vacuum pyrolysis of woody biomass to bio-oil production. *Politek. Derg.* **2021**, *24*, 1257–1261. [CrossRef]
42. AlDayyat, E.A.; Saidan, M.N.; Al-Hamamre, Z.; Al-Addous, M.; Alkasrawi, M. Pyrolysis of Solid Waste for Bio-Oil and Char Production in Refugees' Camp: A Case Study. *Energies* **2021**, *14*, 3861. [CrossRef]
43. Ben, H.; Wu, F.; Wu, Z.; Han, G.; Jiang, W.; Ragauskas, A.J. A Comprehensive Characterization of Pyrolysis Oil from Softwood Barks. *Polymers* **2019**, *11*, 1387. [CrossRef] [PubMed]
44. Mishra, R.K.; Mohanty, K. Kinetic analysis and pyrolysis behavior of low-value waste lignocellulosic biomass for its bioenergy potential using thermogravimetric analyzer. *Mater. Sci. Energy Technol.* **2021**, *4*, 136–147. [CrossRef]
45. Ermolaev, D.V.; Timofeeva, S.S.; Islamova, S.I.; Bulygina, K.S.; Gilfanov, M.F. A comprehensive study of thermotechnical and thermogravimetric properties of peat for power generation. *Biomass Conv. Bioref.* **2019**, *9*, 767–774. [CrossRef]
46. EL-Sayed, S.A.; Mostafa, M.E. Kinetics, thermodynamics, and combustion characteristics of Poinciana pods using TG/DTG/DTA techniques. *Biomass Conv. Bioref.* **2021**, *11*, 1–25. [CrossRef]
47. Li, L.; Rowbotham, J.S.; Greenwell, C.H.; Dyer, P.W. An introduction to pyrolysis and catalytic pyrolysis: Versatile techniques for biomass conversion. In *New and Future Developments in Catalysis: Catalytic Biomass Conversion*; Elsevier: Amsterdam, The Netherlands, 2013; pp. 173–208. [CrossRef]
48. El-Sayed, S.A.; Mostafa, M.E. Thermal pyrolysis and kinetic parameter determination of mango leaves using common and new proposed parallel kinetic models. *RSC Adv.* **2020**, *10*, 18160–18179. [CrossRef]
49. Saikia, R.; Chutia, R.S.; Kataki, R.; Pant, K.K. Perennial grass (*Arundo donax* L) as a feedstock for thermo-chemical conversion to energy and materials. *Bioresour. Technol.* **2015**, *188*, 265–272. [CrossRef]
50. Wang, S.; Dai, G.; Yang, H.; Luo, Z. Lignocellulosic biomass pyrolysis mechanism: A state-of-the-art review. *Prog. Energy Combust. Sci.* **2017**, *62*, 33–86. [CrossRef]
51. Atienza-Martínez, M.; Ábrego, J.; Gea, G.; Marías, F. Pyrolysis of dairy cattle manure: Evolution of char characteristics. *J. Anal. Appl. Pyrolysis* **2020**, *145*, 104724. [CrossRef]
52. Cai, W.; Liu, Q.; Shen, D.; Wang, J. Py-GC/MS analysis on product distribution of two-staged biomass pyrolysis. *J. Anal. Appl. Pyrolysis.* **2019**, *138*, 62–69. [CrossRef]
53. Burhenne, L.; Messmer, J.; Aicher, T.; Laborie, M.-P. The effect of the biomass components lignin, cellulose and hemicellulose on TGA and fixed bed pyrolysis. *J. Anal. Appl. Pyrolysis* **2013**, *101*, 177–184. [CrossRef]
54. Yu, J.; Paterson, N.; Blamey, J.; Millan, M. Cellulose, xylan and lignin interactions during pyrolysis of lignocellulosic biomass. *Fuel* **2017**, *191*, 140–149. [CrossRef]
55. López-González, D.; Fernandez-Lopez, M.; Valverde, J.L.; Sanchez-Silva, L. Thermogravimetric-mass spectrometric analysis on combustion of lignocellulosic biomass. *Bioresour. Technol.* **2013**, *143*, 562–574. [CrossRef]
56. Fernandez-Lopez, M.; Parascanu, M.M.; López-González, D.; Soreanu, G.; Avalos-Ramírez, A.; Sanchez, P.; Valverde, J.L.; Sanchez-Silva, L. Catalytic and non catalyyic pyrolysis of biologically treated manure. *Environ. Eng. Manag. J.* **2015**, *14*, 349–355. [CrossRef]
57. Yogalakshmi, K.N.; Poornima, D.T.; Sivashanmugam, P.; Kavitha, S.; Yukesh, K.R.; Sunita, S.; Adish Kumar, S.; Kumar, G.; Rajesh Banu, J. Lignocellulosic biomass-based pyrolysis: A comprehensive review. *Chemosphere* **2022**, *286*, 131824. [CrossRef]
58. Hoang, A.T.; Ong, H.C.; Rizwanul Fattah, I.M.; Chong, C.T.; Cheng, C.K.; Sakthivel, R.; Ok, Y.S. Progress on the lignocellulosic biomass pyrolysis for biofuel production toward environmental sustainability. *Fuel Process. Technol.* **2021**, *223*, 106997. [CrossRef]

59. Hu, X.; Gholizadeh, M. Biomass pyrolysis: A review of the process development and challenges from initial researches up to the commercialisation stage. *J. Energy Chem.* **2019**, *39*, 109–143. [CrossRef]
60. Collard, F.-X.; Blin, J. A review on pyrolysis of biomass constituents: Mechanisms and composition of the products obtained from the conversion of cellulose, hemicelluloses and lignin. *Renew. Sustain. Energy Rev.* **2014**, *38*, 594–608. [CrossRef]
61. Vuppaladadiyam, A.K.; Vuppaladadiyam, S.S.V.; Sahoo, A.; Murugavelh, S.; Anthony, E.; Bhaskar, T.; Zheng, Y.; Zhao, M.; Duan, H.; Zhao, Y.; et al. Bio-oil and biochar from the pyrolytic conversion of biomass: A current and future perspective on the trade-off between economic, environmental, and technical indicators. *Sci. Total Environ.* **2023**, *857*, 159155. [CrossRef]
62. Soshnikova, O.V.; Yatsuk, V.Y. The investigations of the chemical composition of *Amaranthus retroflexus* L. *Russ. Med. Biol. Her.* **2010**, *18*, 135–141. [CrossRef]
63. Palamanit, A.; Khongphakdi, P.; Tirawanichakul, Y.; Phusunti, N. Investigation of yields and qualities of pyrolysis products obtained from oil palm biomass using an agitated bed pyrolysis reactor. *Biofuel Res. J.* **2019**, *6*, 1065–1079. [CrossRef]
64. Dhyani, V.; Bhaskar, T. A comprehensive review on the pyrolysis of lignocellulosic biomass. *Renew. Energy* **2018**, *129*, 695–716. [CrossRef]
65. Mantilla, S.V.; Manrique, Á.M.; Gauthier-Maradei, P. Characterization of Phenolic Compounds Contained in Bio-Oil Stemming from Agricultural Biomass Wastes. *JPC J. Planar. Chromat.* **2016**, *29*, 361–365. [CrossRef]
66. Apaydin-Varol, E.; Pütün, E.; Pütün, A.E. Slow pyrolysis of pistachio shell. *Fuel* **2007**, *86*, 1892–1899. [CrossRef]
67. Gerçel, H.F. The Effect of a Sweeping Gas Flow Rate on the Fast Pyrolysis of Biomass. *Energy Sources* **2002**, *24*, 7. [CrossRef]
68. Garg, R.; Anand, N.; Kumar, D. Pyrolysis of babool seeds (*Acacia nilotica*) in a fixed bed reactor and bio-oil characterization. *Renew. Energy* **2016**, *96*, 167–171. [CrossRef]
69. Khuenkaeo, N.; Phromphithak, S.; Onsree, T.; Naqvi, S.R.; Tippayawong, N. Production and characterization of bio-oils from fast pyrolysis of tobacco processing wastes in an ablative reactor under vacuum. *PLoS ONE* **2021**, *16*, 7. [CrossRef]
70. Karaeva, J.; Timofeeva, S.; Islamova, S.; Bulygina, K.; Aliev, F.; Panchenko, V.; Bolshev, V. Pyrolysis of Amaranth Inflorescence Wastes: Bioenergy Potential, Biochar and Hydrocarbon Rich Bio-Oil Production. *Agriculture* **2023**, *13*, 260. [CrossRef]
71. Chen, X.; Gao, X.; Wu, H. Characterization of Ashes from Co-Firing Biochar with Coal under Pulverized-Fuel Conditions. *ACS Eng. Au.* **2022**, *2*, 397–405. [CrossRef]
72. Podbaronova, J. Assessment of Coals from Russia and Countries of Former Soviet Union for Utility Fluidized Bed Boilers. Master's Thesis, Lappeenranta University of Technology, Lappeenranta, Finland, 2010. Available online: https://urn.fi/URN:NBN:fi-fe201006212081 (accessed on 12 March 2023).
73. Li, Z.; Unzué-Belmonte, D.; Cornelis, J.T.; Linden, C.V.; Struyf, E.; Ronsse, F.; Delvaux, B. Effects of phytolithic rice-Straw biochar, soil bu_ering capacity and pH on silicon bioavailability. *Plant Soil.* **2019**, *438*, 187–203. [CrossRef]
74. Thomazini, A.; Spokas, K.; Hall, K.; Ippolito, J.; Lentz, R.; Novak, J. GHG impacts of biochar: Predictabilityfor the same biochar. *Agric. Ecosyst. Environ.* **2015**, *207*, 183–191. [CrossRef]
75. Schaffer, S.; Pröll, T.; Afif, R.A.; Pfeifer, C. A mass- and energy balance-based process modelling study for the pyrolysis of cotton stalks with char utilization for sustainable soil enhancement and carbon storage. *Biomass Bioenergy* **2019**, *120*, 281–290. [CrossRef]
76. Wijitkosum, S.; Jiwnok, P. Elemental Composition of Biochar Obtained from Agricultural Waste for Soil Amendment and Carbon Sequestration. *Appl. Sci.* **2019**, *9*, 3980. [CrossRef]
77. Laird, D.A.; Fleming, P.; Wang, B.Q.; Horton, R.; Karlen, D. Biochar impact on nutrient leaching from a Midwestern agricultural soil. *Geoderma* **2010**, *158*, 436–442. [CrossRef]
78. Randolph, P.; Bansode, R.R.; Hassan, O.A.; Rehrah, D.; Ravella, R.; Reddy, M.R.; Watts, D.W.; Novak, J.M.; Ahmedna, M. Effect of biochars produced from solid organic municipal waste on soil quality parameters. *J. Environ. Manag.* **2017**, *192*, 271–280. [CrossRef]
79. Subedi, R.; Taupe, N.; Pelissetti, S.; Petruzzelli, L.; Bertora, C.; Leahy, J.J.; Grignani, C. Greenhouse gas emissions and soil properties following amendment with manure-derived biochars: Influence of pyrolysis temperature and feedstock type. *J. Environ. Manag.* **2016**, *166*, 73–83. [CrossRef]
80. Yuan, J.-H.; Xu, R.-K. Effects of biochars generated from crop residues on chemical properties of acid soils from tropical and subtropical China. *Soil Res.* **2012**, *50*, 570–578. [CrossRef]

agriculture

Article

Design and Experiments of a Double-Cutterbar Combine Header Used in Wheat Combine Harvesters

Linghe Yuan, Mingming Lan, Xun He, Wenhe Wei, Wanzhang Wang * and Zhe Qu

College of Mechanical and Electrical Engineering, Henan Agricultural University, Zhengzhou 450002, China; linghe_y@126.com (L.Y.); lanming@henau.edu.cn (M.L.); hexun@henau.edu.cn (X.H.); wenhe@stu.henau.edu.cn (W.W.); quzhe071171@henau.edu.cn (Z.Q.)

* Correspondence: wangwz@henau.edu.cn

Abstract: To solve the problems of congestion and increased power consumption of wheat combine harvesters (WCHs) caused by excessive feed rate, this paper proposes a method to reduce the feed rate by decreasing the feed length of the stalk and designs a double-cutterbar combine header (DCH). Using the threshing test bench and taking the feed rate, the feed length of the stalk, and the speed of the tangential threshing rotor as the influencing factors and the conveying time as the index, the influence of different parameters on the conveying performance was analyzed. The optimal parameters were obtained: the feed rate was 8 kg/s, the feed length of the stalk was 380 mm, the speed of the cutting drum was 554 r/min, and the conveying time was 8.089 s. The optimized parameter combination was tested and verified, and the test results show that the relative error with the predicted value was 0.198%, proving the reliability of the optimized parameters. The critical components of the DCH were designed, the movement process of the profiling mechanism was simulated using ADAMS software, and the structural dimensions of the profiling mechanism were determined. The field performance test of the WCH with a DCH was carried out. The results showed that the loss rate and stubble height met the operation quality requirements. At the same operation speed, the fuel consumption was 11.2% less than that of the WCH with a conventional header, providing a technical reference for the efficient harvest of the WCHs.

Keywords: WCH; DCH; feed length of the stalk; conveying performance; fuel consumption

Citation: Yuan, L.; Lan, M.; He, X.; Wei, W.; Wang, W.; Qu, Z. Design and Experiments of a Double-Cutterbar Combine Header Used in Wheat Combine Harvesters. *Agriculture* **2023**, *13*, 817. https://doi.org/10.3390/agriculture13040817

Academic Editors: Vadim Bolshev, Vladimir Panchenko and Alexey Sibirev

Received: 8 March 2023
Revised: 30 March 2023
Accepted: 30 March 2023
Published: 31 March 2023

1. Introduction

When harvesting, the feed rate of WCHs constantly changes due to factors including header height, harvesting speed, crop moisture content, and density [1–4]. Farmers have recently used plant growth regulators in the wheat plantation process to improve wheat yield, which leads to the high moisture content of wheat stalks during harvest. Compared with the low moisture content of wheat stalks, the feed rate of the WCHs will increase at the same harvest speed, which further increases the workload imposed on the operating parts of the WCHs and even causes blockages. Methods such as elevating the header or reducing the cutting width and operating speed of WCHs are usually adopted to adjust the feed rate to avoid blockages. However, elevating the header will increase the stubble height, which affects the operation quality and is not conducive to subsequent crop planting [5]. At the same time, reducing the cutting width and operating speed will reduce the operational efficiency and increase the harvest loss [6].

As an essential part of the WCH, the header significantly impacts the harvest quality [7]. Most WCHs in China use rigid screw conveyor headers and a mechanical transmission system. Large-scale WCHs in other countries mostly use flexible transmission belt headers with high intelligence. Many scholars have also researched the automatic control of header operation parameters in recent years. Through automatic adjustment of header height [8–14], reel height and rotation speed [15–17], and screw conveyor clearance [18], problems such as combined harvester congestion and increased harvest loss caused by changes in feed rate can be mitigated. Still, it is only partially suitable for China's wheat harvest operation. In their research on the structure of the combine header, Li et al. [19] optimized the structure parameters of the header frame. Xie et al. [20] designed a belt conveyor header for the soybean combine harvester to solve the problems of uneven feeding and congestion. Zhang et al. [21] designed a double crank plane five-bar reel mechanism to solve issues such as easy winding and hanging straws of the reel. Qing et al. [7] designed a reel with improved tine trajectory for harvesting oilseed rape to reduce harvest loss. Van et al. [22] designed a screw conveyor based on a cam mechanism for retracting the fingers to solve the problem of a dead zone between the screw conveyor and inclined conveyor and to improve conveying performance.

The factor limiting the performance of the combine harvester is its ability to handle a large amount of stalk and separate the grain from it. To reduce the number of stalks entering the combine harvester, Shelbourne Company designed the stripper header, which only combs the grains off the stalks and feeds them into the combine harvester, which can effectively reduce the feed rate but results in a high harvest loss [23,24]. Wang [25] once proposed to feed only the head of grain into the combine harvester and designed a secondary cutting and directional conveying device. The test results show that reducing the feed length can effectively reduce power consumption and improve operation efficiency, but its adaptability could be better. Dai et al. [26] designed a plot wheat seed harvester, which realized the harvest of wheat spike and effectively reduced the load of the threshing and cleaning system. The above research shows that harvesting only the upper part of the wheat plant is feasible. This paper proposes a method to reduce the feed rate by reducing the feed length of the wheat stalk and determines the optimal feed length of the wheat stalk through bench tests. The DCH of the WCH is designed to solve the problems of congestion and increased power consumption.

2. Materials and Methods

2.1. Structure and Working Principles of the DCH

A DCH was designed based on the horizontal screw conveyor header of WCHs, which has two layers of cutterbars, as shown in Figure 1a. The DCH is mainly composed of the frame, a reel, a screw conveyor, upper cutterbars, lower cutterbars, wobble boxes, a profiling mechanism, and a transmission system. Among them, the profiling mechanism consists of a fixed beam, support arms, a height-limiting connecting rod, a swing block, and the ground wheel of the upper cutter. The swing block, support arm, and connecting rod installed with a height-limiting block form a crank rocker mechanism, as shown in Figure 1b. The fixed beam is welded to the support arms on both sides, and the upper cutter and ground wheel are installed on the fixed beam. In the case of uneven terrain, the profiling mechanism can enable the lower cutterbars to fluctuate with the change in wheat field terrain, which ensures consistency of cutting height and protects the cutterbars from damage to some extent.

Before an operation, a hydraulic height adjustment handle on the DCH is used to lower the header. In lowering the header, the ground wheel is the first to come into contact with the ground. The height of the lower cutterbars no longer undergoes significant fluctuation, while the upper cutterbars continue to descend to the set height. A schematic representation of the DCH in service is shown in Figure 2. The plant height of wheat is L_1; the upper cutterbars first cut the upper part of the wheat stalks, which can then be pulled into the screw conveyor under the effect of the reel. The cut part with a stalk length of L_2 is fed into the combine harvester; the middle part of the wheat stalks can be cut by the lower cutterbars and then scattered across the field. The length of the central part of wheat stalks is L_3. The uncut part of the wheat stalks remains on the ground, forming stubble with a height of h. The fixed scale of the oil cylinder on the DCH can be adjusted to tune the length of stalks cut by the upper cutterbars to adapt to harvesting wheat plants with different heights.

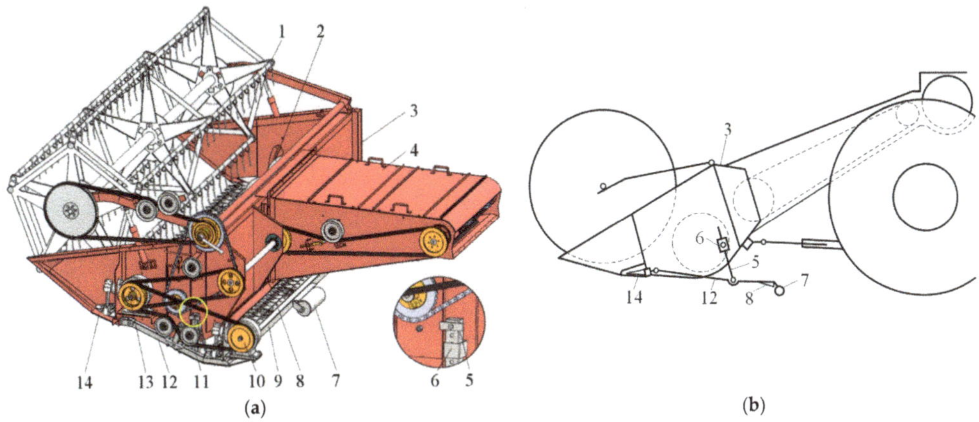

Figure 1. Structure of the DCH. (**a**) Three-dimensional structural diagram of the DCH; (**b**) Structural diagram of the profiling mechanism. 1. Reel; 2. Screw conveyor; 3. Frame; 4. Inclined conveyor; 5. Height-limiting connecting rod; 6. Swing block; 7. Ground wheel; 8. Lower cutterbar; 9. Fixed beam; 10. Wobble box of the lower cutterbars; 11. Belt expansion device; 12. Support arm; 13. Wobble box of the upper cutterbars; 14. Upper cutterbar.

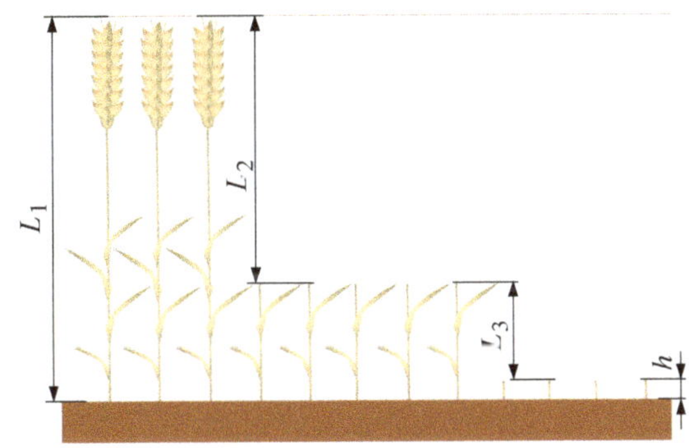

Figure 2. Operation diagram of the DCH.

2.2. Conveying Performance Bench Test

When using the DCHs to harvest wheat, the length of the wheat stalk cut by the header and fed into the combine does not match the working parameters of the relevant DCH parts, which adversely affects the conveying performance. Therefore, to reduce the probability of congestion, a test was carried out on the threshing test bench to determine the best stalk feed length and working parameters by exploring the influence of the parameters of WCHs on the conveying performance, which can provide the basis for the structural design of the DCH.

2.2.1. Test Equipment

The structure of the threshing test bench consists of a frame, a longitudinal axial flow threshing and separating device, a tangential flow threshing and separating device, an inclined conveyor, a screw conveyor, a material conveying platform, and a grain collecting box, as shown in Figure 3. The tangential flow threshing and separating device, inclined conveyor, and screw conveyor bring wheat stalks into the longitudinal axial flow threshing and separating device. In contrast, the tangential flow threshing and separating device completes some of the threshing tasks. Moreover, a 37 kW three-phase asynchronous motor and its supporting frequency converter are used simultaneously to drive the tangential flow threshing rotor, the inclined conveyor, and the screw conveyor and adjust their speeds. The longitudinal axial flow threshing rotor imposes a significant workload; a 75 kW three-phase asynchronous motor and its supporting frequency converter are exclusively used for driving the threshing rotor, for which the angular velocity is also adjusted. The material conveying platform is 10 m long and feeds wheat stalks into the harvester. The frequency converter is used to adjust the rotational speed of the asynchronous motor to change the conveying velocity, thereby controlling the feed rate.

Figure 3. Structural view of the threshing test bench. 1. Frame; 2. Longitudinal axial flow threshing and separating device; 3. Tangential flow threshing and separating device; 4. Material conveying platform; 5. Inclined conveyor; 6. Screw conveyor; 7. Motor; 8. Grain collection boxes; 9. Frequency converter.

2.2.2. Experimental Materials and Methods

The test wheat cultivar was Zhoumai 32, and its average plant height was 743 mm. The test used manually harvested wheat and a cutting height of 50 to 80 mm. The grain moisture content was identified as 14.2%.

Combining this with the requirements of Equipment for harvesting—Combine harvesters—Test Procedure (GB /T 8097-2008), conveying time was taken as the testing indicator for the performance of conveying. The conveying time refers to when the wheat is first conveyed by the conveying platform to the screw conveyor, then passes through the inclined conveyor and the tangential flow threshing rotor, and finally arrives at the longitudinal axial flow threshing rotor. An observation window is set on the side sealing plate of the tangential threshing device, and the NORPIX FR-1000 high-speed digital camera is used to photograph the movement of materials in the transition area of the tangential axial threshing device, as shown in Figure 4. In the test, the acquisition frame rate of the high-speed digital camera was set to 200 frames per second, and the image recording began when turning on the conveying platform. The time taken for the material conveying platform to deliver the wheat to the screw conveyor is t_1. After the tests ended, Image-Pro Premier image analysis software was used to analyze the collected images. To avoid a small number of stalks remaining in the header frame and inclined conveyor during the later stage of material conveying from affecting the judgment of the end time, based on less than or equal to five stalks in five consecutive frames of images, the time corresponding to the first frame of the image was selected as the end time t_2 [27], as shown in Figure 5. The results of the experiments have shown that the number of images with less than or equal to five stalks in succession is less than 10 frames, so the error in judging the conveying time of high-speed photographic images is less than 0.05 s. The conveying time t is:

$$t = t_2 - t_1 \quad (1)$$

Figure 4. Conveying performance experiment site. 1. Observation window; 2. High-speed digital camera; 3. Light source; 4. Computer.

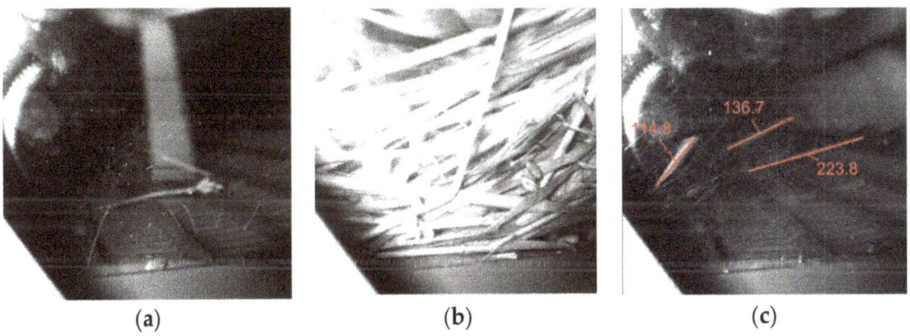

Figure 5. Conveying time collection. (**a**) Image at the beginning of the experiment; (**b**) Image during the stalk conveying; (**c**) Image at the end of the stalk conveying.

Three factors and three levels of the Box–Behnken response surface analysis methods were used in the test [28]. The test factors, such as feed rate, feed length of stalk, and speed of the tangential threshing rotor, were expressed as A, B, and C, respectively, and the conveying time was the test indicator, which was expressed as Y. The coding table of test factors is shown in Table 1.

Table 1. Codes of experiment factors.

Codes	Factors		
	Feed Rate A (kg/s)	Feed Length of Stalk B (mm)	Speed of the Tangential Threshing Rotor C (r/min)
−1	8	300	500
0	9	450	550
1	10	600	600

Adjust the speed of the axial threshing rotor to 900 r/min and the speed of the conveying bench to 1 m/s. Cut the harvested wheat stalks and keep the upper part of the stalks at 300 mm, 450 mm, and 600 mm, respectively. Weigh the materials with different lengths of 24 kg, 27 kg, and 30 kg by an electronic scale and lay them evenly on the rear 3 m conveying platform with the wheat ears toward the threshing test bench to achieve feeding speeds of 8 kg/s, 9 kg/s, and 10 kg/s.

2.3. *Design of Critical Components of the DCH*

2.3.1. Reel

A reel is installed at the front of the DCH, which picks up and conveys wheat stalks into the cutting device. In this process, the reel also supports the wheat stalks being cut. The reel can help the cutting device fulfill the cutting operation while pushing harvested wheat stalks into the screw conveyor to avoid cut stalks piling up at the front of the header. As a cam-action reel exerts a strong stalk-lifting force and has a small impact on cropped ear-heads, they are primarily used in rice and WCHs. The structure consists of central spoke wheels, a central rotatable shaft, tine bars, tines, an eccentric spoke wheel, and cranks, as shown in Figure 6. The eccentric spoke wheel, cranks, and central spoke wheel constitute a parallel four-bar linkage mechanism that allows the tines to maintain a well-adjusted dip angle. The diameter of the reel D_1 is 1000 mm, and the eccentricity of an eccentric spoke wheel and central spoke wheel e_1 is 72 mm.

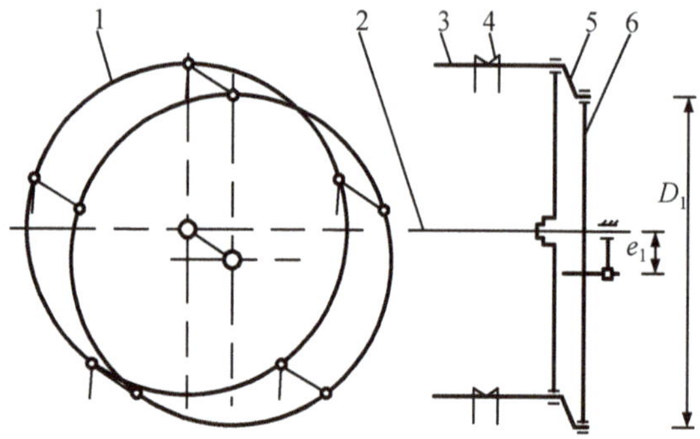

Figure 6. Structural diagram of a cam-action reel. 1. Central spoke wheels; 2. Central rotatable shaft; 3. Tine bar; 4. Tine; 5. Crank; 6. Eccentric spoke wheel.

2.3.2. Screw Conveyor

The structure of the screw conveyor is composed of a cylinder welded with left- and right-handed spiral blades and retractable fingers (Figure 7). The retractable fingers are installed at the screw cylinder, and 16 retractable fingers (with four retractable fingers in each group) are hinged abreast on the retractable finger shaft. The retractable fingers are riveted to a crank and a fixed shaft. The eccentric distance between the center of the retractable fingers and the screw cylinder is e_2. To avoid the winding of wheat stalks, the perimeter of the screw cylinder must be larger than the length of wheat stalks that have entered the DCH [29]. Generally, the diameter D_2 of the screw cylinder is 300 mm. To improve the conveying performance of the spiral conveyor on short wheat stalks, the outer diameter of the spiral blades D_3 is 500 mm, and the screw pitch S is 460 mm.

The screw cylinder, when rotating, drives the retractable fingers to also rotate. As the screw cylinder and retractable fingers are not concentric, the retractable fingers undergo telescopic motion relative to the screw cylinder surface. When the retractable fingers turn back, they should retract into the screw cylinder, but their protrusion of 10 mm remaining outside the screw cylinder should be ensured to avoid the wear-out of the ends of the retractable fingers. When the retractable fingers turn forward, they should stretch to 40–50 mm outside the spiral blades within the screw cylinder in a bid to ensure their clamping capability [30]. Thus, we obtain,

$$e_2 = \frac{D_3 - D_2}{4} + (15 \sim 20) \tag{2}$$

$$L = \frac{D_2}{2} + 10 + e_2 \tag{3}$$

where e_2 is set to 70 mm and the length of retractable fingers L is 280 mm.

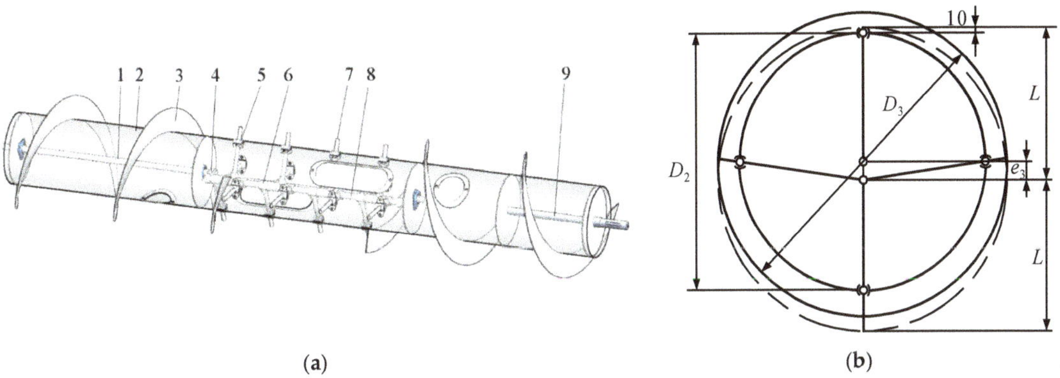

Figure 7. Structural diagram of a screw conveyor. (**a**) Overall structure; (**b**) Schematic diagram of retractable fingers. 1. Fixed central axis; 2. Screw cylinder; 3. Spiral blade; 4. Crank; 5. Sleeve; 6. Finger seat; 7. Finger; 8. Finger shaft; 9. Conveyor driving shaft.

2.3.3. Parameters of the Header Triangle

The triangle of the header refers to a particular space composed of three main working parts: a reel, a cutterbar, and a screw conveyor, as shown in Figure 8. Compared with the conventional header, the wheat stalks cut and transported by the DCH are short in length. The triangle is too large, so the crops quickly accumulate between the cutter and the screw conveyor. When the crop is stacked to a certain amount, it will be grabbed by the blades of the screw conveyor, resulting in uneven transportation, feeding, and even blocking; if the triangle area is too small, the harvest loss will increase. The vertical distance H_1 between the central rotatable shaft of the reel and the upper cutterbar, the horizontal distance L_4 between the screw conveyor center and the upper cutterbar beam, and the forward displacement b of the central rotatable shaft (taking the central rotatable shaft directly above the upper cutterbar as the central position) affect the size of the triangle. L_4 is usually 350–500 mm. We selected L_4 as 450 mm. To make the reel tine shaft have a backward horizontal speed above the cutter, the maximum forward displacement of the reel main shaft $b_{\max}$ is [30]:

$$b_{\max} = \frac{D_1}{2\lambda\sqrt{\lambda^2 - 1}} \tag{4}$$

In Formula (4), λ is the ratio of the circumferential speed of the reel to the operating speed of the combine harvester, which is generally 1.5–1.7 [30]. Taking λ as 1.5, the b_{max} is 298 mm. The reel can move back correctly to increase its ability to push the stalk, but the distance from the screw conveyor when the reel is adjusted to the final position δ_1 should be greater than 25 mm.

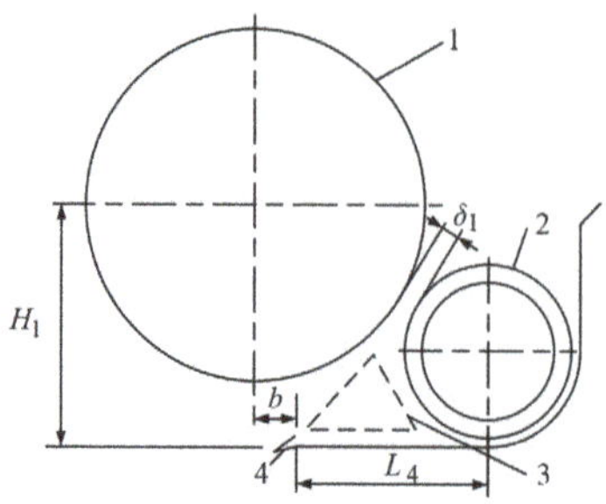

Figure 8. Diagrammatic sketch of a header triangle. 1. Reel; 2. Screw conveyor; 3. Header triangle; 4. Cutterbar.

Combined with the best straw feed length determined by the bench test, the minimum height of H_1 is 753 mm, calculated according to Formula (5) [30].

$$H_1 \geq \frac{D_1}{2} + \frac{2(L_1 - L_3 - h)}{3} = \frac{D_1}{2} + \frac{2L_2}{3} \tag{5}$$

2.3.4. Profiling Mechanism of the Lower Cutterbar

As shown in Figure 1, the support arm, height-limiting connecting rod, swing block, and ground wheels of the profiling mechanism for the lower cutterbar constitute the swing block of the crank. When harvesting, the greater the distance between the upper cutterbar and the ground of the wheat field, the larger the length of the height-limiting connecting rod. However, due to the influences of the DCH's structure and transmission system, the height-limiting connecting rod should be a suitable length.

Statistics indicate that the average plant height of the main wheat varieties in Henan Province, China, is 775 mm. The maximum distance between the ground and upper cutterbar is preliminarily determined to be 400 mm to meet the optimum feed length of stalks. Meanwhile, the lower cutterbar can fluctuate by up to 50 mm with the ground height. The L_{BA} was determined using ADAMS software to simulate and study two processes: (1) the motions of the components with the DCH descending and (2) their motions with a change in ground height during harvest. Create a ground with 50 mm bumps and depressions using 3D software, assemble it with a simplified model of the DCH according to actual working conditions, import the model into ADAMS software, merge the upper cutterbars, header frame, and inclined conveyor frame, merge the lower cutterbars and fixed beam, and add motion constraints to the moving parts. During the simulation of header descent, the motion constraints are shown in Table 2. Establish a height sensor between the upper cutterbars and the ground to stop the header from descending after the upper cutterbars reach the set height. The simulation model is shown in Figure 9. During the harvesting simulation, adjust the model to make the ground wheel contact the ground; change constraint 1 in Table 2 to a translational joint; and add a driver to achieve the DCH movement.

Table 2. Motion constraints.

NO.	Components	Type of Motion Constraint
1	Inclined conveyor—ground	Revolute joint
2	Header fame—support arm	Revolute joint
3	Header fame—swing block	Revolute joint
4	Support arm—height limiting connecting rod	Revolute joint
5	Height limiting connecting rod—swing block	Translational joint
6	Support arm—Fixed beam	Fixed joint
7	Fixed beam—ground wheel	Revolve joint
8	Height-limiting block—swing block	Contact force
9	Ground—ground wheel	Contact force

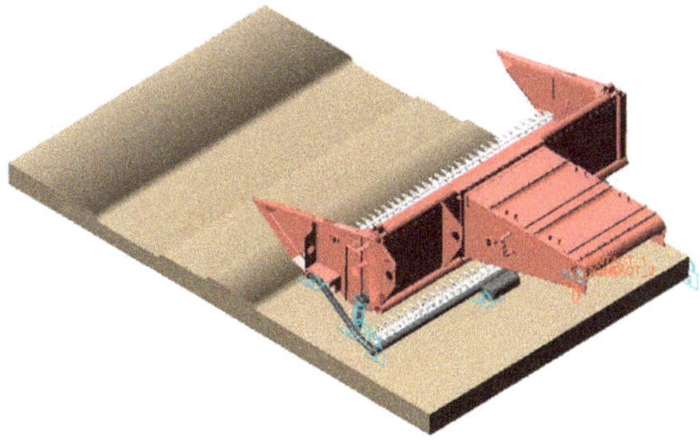

Figure 9. Simulation model.

2.3.5. Transmission System

The transmission system of DCH is shown in Figure 10. The power of the engine is transmitted to the DCH through the belt and chain drive. Then the driving shaft of the DCH can transfer power through the belt drive transmission to the wobble box of the upper cutterbars. The wobble boxes of the upper and lower cutterbars can transfer power through the belt drive transmission to power both the upper and lower movable cutters. Meanwhile, the driving shaft of the DCH drives the rotation of the screw conveyor by way of a chain drive transmission. The transmission occurs between the driving shaft and the reel to drive the rotation of the reel through the chain and belt drive transmission. The transmission ratio is shown in Table 3.

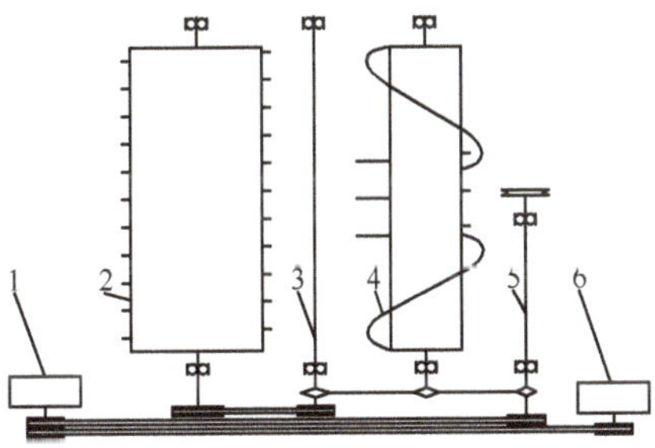

Figure 10. Schematic diagram of the DCH transmission system. 1. Wobble box of the lower cutterbars; 2. Reel; 3. Intermediate shaft of the reel; 4. Screw conveyor; 5. Driving shaft; 6. Wobble box of the upper cutterbars.

Table 3. Transmission ratio of the DCH.

No.	Components	Transmission Ratio
1	Driving shaft—Screw conveyor—Intermediate shaft of the reel	13:35:56
2	Intermediate shaft of the reel—Reel	20:(5~9)
3	Driving shaft—Wobbler of the lower cutterbars	1:1
4	Wobble box of the lower cutterbars—Wobble box of the upper cutterbars	1:1

2.4. Field Experiment

2.4.1. Experimental Arrangement

On 30 May 2022, Xinjiang-9B type WCHs equipped with the newly designed DCH and conventional header were subjected to in-situ operating tests in Xingyang, Henan Province,

China. Field surveys were made according to the requirements of relevant standards. The wheat cultivar was Zhoumai 32, with an average plant height of 743 mm and a grain moisture content of 14.5%. The ratio of stalk to grain was 1.18, and testing conditions met the requirements [31]. The suitable field plot should satisfy the criteria, including that the measured zone should be 50 m long, there must be a stable zone 20 m in front of the measured zone, and a harvester parking site no less than 15 m behind the measured zone; the stubble height is 100 to 120 mm [32]. In the experiment, the combine harvesters worked along the lengthwise direction of field plots at a speed of 5 km/h.

2.4.2. Data Collection and Processing

A 1 m^2 sampling zone was selected in the representative zone along the advancing direction of the combine harvesters at each sampling site. In the sampling zones, all grains and ear-heads were collected; thereafter, once threshed and cleaned, they were weighed. Based on Equations (6) and (7), the harvesting loss rates were calculated, respectively, and then the average value of the harvest loss rates in the five sampling zones was determined [33].

$$F_j = \frac{W_{sh} - W_z}{W_{ch}} \times 100 \tag{6}$$

$$F = \frac{\sum F_j}{5} \tag{7}$$

where F_j is the loss rate of the sampling point, %; W_{sh} is the mass of grain loss per square meter, g/m^2; W_z is the mass of grains naturally falling per square meter, g/m^2; W_{ch} is the mass of grain per square meter, g/m^2; and F is the average harvest loss rate, %.

The stubble height was measured using the five-point sampling method applied in the experimental field. The stubble heights of three subpoints were determined by measuring at the left, middle, and right sides of the horizontal direction of the harvesting range in each point. The average value of the three subpoints was taken as the stubble height at that point, taking the mean average value of the five points.

Before each experiment, the same amount of diesel oil was added to the fuel tank. After completing the test, the diesel oil remaining in the fuel tank was discharged and weighed. According to Equation (8), the fuel consumption per unit area Q could be calculated. Two groups of tests were repeated three times before taking the mean average value.

$$Q = 10,000 \times \frac{q_1 - q_2}{B \times L_4} \tag{8}$$

In the formula, q_1 is the weight of fuel in the fuel tank before a test, kg; q_2 is the weight of fuel in the fuel tank at the end of a test, kg; B is the cutting width of the WCH, m; and L_4 is the length of the measured zone, m.

3. Results and Discussion

3.1. Conveying Performance Test Analysis

3.1.1. Regression Analysis

The experimental results are shown in Table 4. Using Design-Expert 8.0.6 software to regress and fit the experimental results, we obtained the regression mathematical model of conveying time Y:

$$\begin{aligned} Y = {} & 0.212A^2 + 5.463 \times 10^{-6}B^2 + 1.057 \times 10^{-4}C^2 + 6.25 \times 10^{-4}AB - 1.65 \times 10^{-3}AC \\ & -1.75 \times 10^{-5}BC - 2.181A + 9.5 \times 10^{-4}B - 0.0983C + 41.932 \end{aligned} \tag{9}$$

The determination coefficient R^2 of the regression equation was 0.998, indicating a high degree of fitting; the regression model was analyzed by variance, and the results are shown in Table 5. It can be observed that the model had a value of $p < 0.0001$, indicating that the regression equation is significant and can describe the relationship between each factor and response value; the lack of fit was$p = 0.1882 > 0.05$, showing that the residual item is not significant, and there are no other main factors affecting the results, so the regression model was established. $p < 0.01$ was set for *A*, *B*, *C*, *BC*, A^2,B^2, and C^2, which have a very significant effect on the results; $p < 0.05$ was set for *AB* and *AC*, indicating that they have a significant effect on the results; and $p > 0.05$ was set for factors and interaction terms, which have no significant effect on the results. The order of significance of each factor on the screening efficiency, from large to small, was feed rate, feed length of the stalk, and speed of the tangential threshing rotor.

Table 4. Results of the conveying performance test.

No.	Factors			Conveying Time Y (s)
	Feed Rate *A* (kg/s)	Feed Length of Stalk *B* (mm)	Speed of the Tangential Threshing Rotor *C* (r/min)	
1	−1	0	−1	8.53
2	1	0	−1	10.735
3	0	−1	−1	9.17
4	1	0	1	10.15
5	1	−1	0	9.91
6	1	1	0	10.625
7	0	0	0	8.975
8	0	0	0	8.945
9	−1	1	0	8.465
10	0	1	1	9.235
11	−1	−1	0	8.125
12	−1	0	1	8.275
13	0	1	−1	10.025
14	0	−1	1	8.905
15	0	0	0	8.92

Table 5. Variance analysis of the conveying time.

Source	Sum of Squares	df	F Value	*p*-Value
Model	9.68	9	426.24	<0.0001 **
A	8.05	1	3444.5	<0.0001 **
B	0.63	1	268.37	<0.0001 **
C	0.45	1	192.07	<0.0001 **
AB	0.035	1	15.04	0.0117 *
BC	0.027	1	11.65	0.0190 *
AC	0.069	1	29.48	0.0029 **
A^2	0.17	1	70.78	0.0001 **
B^2	0.056	1	23.87	0.0045 **
C^2	0.26	1	110.25	0.0001 **
Residual	0.012	5		
Lack of Fit	0.010	3	4.47	0.1882
Pure Error	1.517×10^{-3}	2		
Cor Total	8.99	14		

Note: $p < 0.01$ (extremely significant, **); $p < 0.05$ (significant, *).

According to the regression equation, the influence of the interaction of factors on the results is shown in Figure 11. When the speed of the tangential threshing rotor is low, the conveying time decreases with the shorter feed length of the stalk. The shorter the length of the stalk, the better the passing performance in the transition area between the tangential threshing rotor and the longitudinal axial flow rotor, and the shorter the conveying time. When the speed of the tangential threshing rotor is high, the conveying time decreases first and then increases with the shortening of the feed length of the stalk. The conveying time decreases first and then increases with the increasing speed of the tangential threshing. This is because the conveying capacity increases when the rotating speed of the cutting drum increases; if the speed of the tangential threshing rotor is too high, the stalks are not fed in time by the axial threshing rotor, and there is congestion in the transition between the axial flow rotor and the tangential threshing rotor, causing the conveying time to increase.

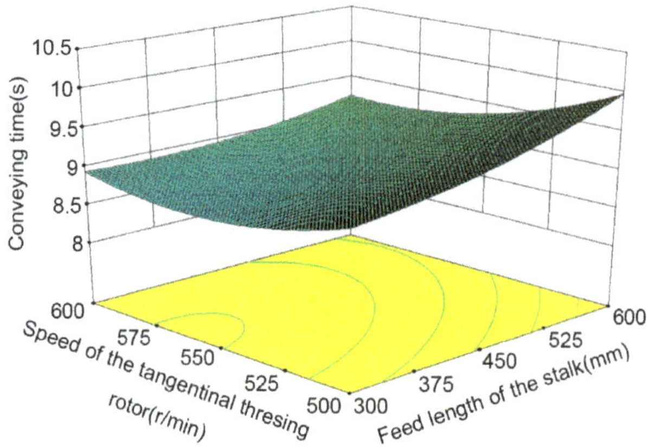

Figure 11. Effect of the interaction of factors on the conveying time.

3.1.2. Parameter Optimization and Verification

In order to achieve better material-conveying performance, the optimization module was used to optimize the regression model and set the solution target as the minimum response value [34]. The optimal parameter combination was obtained as follows: feed rate 8.06 kg/s, feed length of the stalk 377.19 mm, speed of the tangential threshing rotor 553.84 r/min. The optimal parameter combination was rounded as follows: feed rate of 8 kg/s, feed length of the stalk 380 mm, speed of the tangential threshing rotor 554 r/min, and conveying time 8.089 s. The optimized parameter combination was tested and verified. The average value of the five tests was taken, the conveying time was 8.105 s, and the relative error with the predicted value was 0.198%, verifying the reliability of the optimized parameters.

3.2. Simulation Analysis of the Profiling Mechanism of the Lower Cutterbar

The changes in the upper and lower cutterbars, as well as the varying L_{BA} while the DCH descended, are displayed in Figure 12. At the beginning of the simulation, under the effect of gravity, the height-limiting block and the swing block make contact with each other; no relative sliding occurs between the height-limiting connecting rod and the swing block; and the lower and upper cutterbars descend simultaneously. The simulation time is 0.185 s. In the profiling mechanism, the ground wheel makes contact with the ground, the height of the lower cutterbars no longer shows a large range of change, and the header continues to descend. Meanwhile, due to the supporting effect of the ground surface on the ground wheel, the height-limiting connecting rod begins to move upwards along the swing block, and the height of the lower cutterbars slowly drops, accompanied by a decrease in L_{BA}; the simulation ends when the height of the upper cutterbars reaches 400 mm. In this case, the upper and lower cutterbars no longer change positions. The height of the lower cutterbars is 100 mm, and L_{BA} is decreased from 328.6 mm to 280.2 mm.

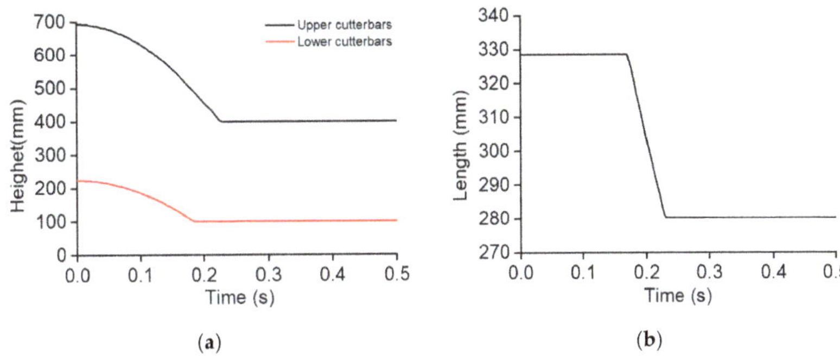

Figure 12. Motions of the components with the DCH descending. (**a**) Height change of the upper and lower cutterbars; (**b**) Length change of the L_{BA}.

In the simulation of the motion of components with the change in ground heights during harvesting, the changes in upper cutterbars and lower cutterbars, as well as varying L_{BA}, are displayed in Figure 13. The height of the upper cutterbars is kept unchanged, while the height of the lower cutterbars, when a hump is encountered on the ground wheel, increases and decreases in the case of a concave ground profile, finally realizing the function of the profiling mechanism. According to Figure 13b, in this process, L_{BA} is shown to change within the range of 254.7 mm to 322.1 mm. By combining the change in L_{BA} during the descent of the header with the dimensions of the swing block and height-limiting block, the length of the height-limiting connecting rod of the profiling mechanism is found to be 365 mm.

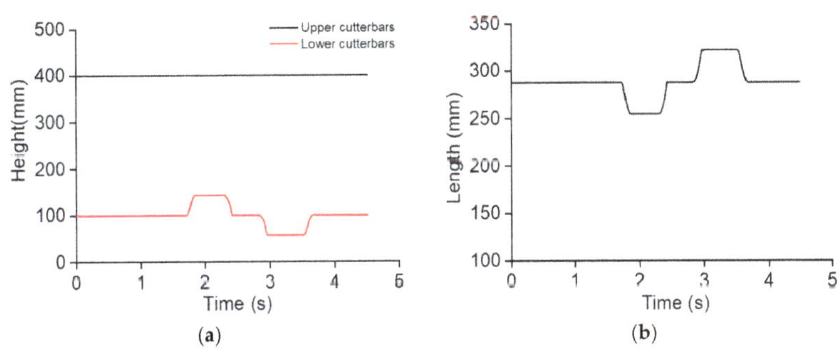

Figure 13. Motions of the components during harvesting. (**a**) Height change of the upper and lower cutterbars; (**b**) Length change of the L_{BA}.

3.3. Field Experiment Analysis

As shown in Figure 14, the loss rate of using the conventional header wheat combine is 0.97%, and the loss rate of using the DCH is 1.05%, but less than 1.2%, which conforms to the relevant national operation quality requirements. The feed rate is reduced for the DCH, and the operating parameters of the threshing and separation device and the cleaning device do not match, increasing the loss rate. Therefore, in the follow-up study, it is also necessary to research the impact of the change in the feed length of the stalk on threshing separation and cleaning performance.

As shown in Table 6, the stubble height of the DCH is 11.02 cm, lower than the stubble height of the conventional header, and it meets the relevant operation quality requirements. The variation coefficient of the stubble height of the DCH is higher than that of the conventional header, indicating that the consistency of the stubble height of the DCH is lower than that of the conventional header. The main reason is that after the upper

9. Brune, M. Method and Device for Height Control of Headers. Europe Patent No. EP1813144A1, 1 August 2007.
10. Hunt, C.D.; Pankal, M. Header Height Control for Combine Harvester. World Patent NO. WO2021242867A1, 2 December 2021.
11. Zhuang, X.; Li, Y. Header height control strategy of harvester based on robust feedback linearization. *Trans. Chin. Soc. Agric. Mach.* **2020**, *51*, 123–130.
12. Xie, Y.; Alleyne, G.; Greer, A.; Deneault, D. Fundamental limits in combine harvester header height control. *J. Dyn. Syst. Meas. Control* **2013**, *135*, 034503. [CrossRef]
13. Daniel, M. Header Height Control of Combine Harvester via Robust Feedback Linearization. Master's Thesis, Iowa State University, Ames, Iowa, 2016.
14. Lopes, G.T.; Magalhaes, P.S.G.; Nobrega, E.G.O. Optimal header height control system for combine harvesters. *Biosyst. Eng.* **2002**, *81*, 261–272. [CrossRef]
15. Baumgarten, J.; Wilken, A.; Neitemeier, D.; Bormann, B.; Spiekermann, S.; Irmer, D. Cutting Table Length Adaptation. European Patent NO. EP3858129A1, 8 August 2020.
16. Chen, J.; Wang, S.; Lian, Y. Design and test of header parameter keys electric control adjusting device for rice and wheat combined harvester. *Trans. Chin. Soc. Agric. Eng.* **2018**, *34*, 19–26.
17. Honeyman, F.; Dreyer, D.; Glade, M. Auto Reel Height. World Patent NO. WO2019234539A1, 12 December 2019.
18. Li, H.; Wan, X.; Xu, Y.; Jiang, Y.; Liao; Q. Clearance adaptive adjusting mechanism for header screw conveyor of rape combine harvester. *Trans. Chin. Soc. Agric. Mach.* **2017**, *48*, 115–122.
19. Li, Y.; Li, Y.; Xu, L.; Hu, B.; Wang, R. Structural parameter optimization of combine harvester cutting bench. *Trans. Chin. Soc. Agric. Eng.* **2014**, *30*, 30–37.
20. Xie, H. Design and Experiment of Belt Conveyor Header for Soybean Combine Harvester. Master's Thesis, Shandong University of Technology, Zibo, China, 2019.
21. Zhang, M.; Jin, M.; Wang, G.; Liang, S.; Wu, C. Design and test of double crank planar five-bar reel in rape windrower. *Trans. Chin. Soc. Agric. Mach.* **2022**, *53*, 115–122.
22. Van, V.S.; Vandergucht, S. Agricultural Machine Equipped with Cam Mechanism for Gathering Crop Material. Europe Patent No. EP3064053B1, 13 December 2017.
23. Henry, W.B.; Nielsen, D.C.; Vigil, M.F.; Calderon, F.J.; West, M.S. Proso millet yield and residue mass following direct harvest with a stripper-header. *Agronomy* **2008**, *100*, 580–584. [CrossRef]
24. Shelbourne, K.; Pakenham, S. Crop Strippers and Stripper Toothing. Europe Patent No. EP0976315B1, 7 April 2004.
25. Wang, Y. Research on apparatus with two cutters and conveyers in two ways for the combine harvester, *Trans. Chin. Soc. Agric. Mach.* **1995**, *26*, 84–89.
26. Dai, F.; Zhao, W.; Han, Z.; Li, X.; Gao, A.; Liu, X. Improvement and experiment on 4gx-100 type wheat harvester for breeding plots. *Trans. Chin. Soc. Agric. Mach.* **2016**, *47*, 196–202.
27. Wang, W.; Liu, W.; Yuan, L.; Qu, Z.; Zhang, H.; Zhou, Z. Simulation and experiment of single longitudin axial material movement and establishment of wheat plants model. *Trans. Chin. Soc. Agric. Mach.* **2020**, *51*, 170–180.
28. Zhang, H.; Chen, B.; Li, Z.; Zhu, C.; Jin, E.; Qu, Z. Design and simulation analysis of a reverse flexible harvesting device for fresh corn. *Agriculture* **2022**, *12*, 1953. [CrossRef]
29. Fu, J.; Zhang, G.; Xie, G.; Wang, Y.; Gao, Y.; Zhou, Y. Development of double-channel feeding harvester for ratoon rice. *Trans. Chin. Soc. Agric. Eng.* **2020**, *36*, 11–20.
30. Chinese Academy of Agricultural Mechanization Sciences. *Agricultural Machinery Design Manual*, 1st ed.; China Agricultural Science and Technology Press: Beijing, China, 2007; pp. 891–904.
31. Technical Specification for Quality Evaluation of Grain Combine Harvesters. *NY/T 2090-2011*; China Standards Press: Beijing, China, 2011.
32. Whole-Feed Combine Harvester-Evaluation Index and Measurement Methods for Fuel Consumption. *GB/T 29002-2012*; China Standards Press: Beijing, China, 2012.
33. The Work Quality of Grain (Wheat) Combine Harvester. *NY/T 995-2006*; China Standards Press: Beijing, China, 2006.
34. Xia, Q.; Zhang, W.; Qi, B.; Wang, Y. Design and experimental study on a new horizontal rotary precision seed metering device for hybrid rice. *Agriculture* **2023**, *13*, 158. [CrossRef]

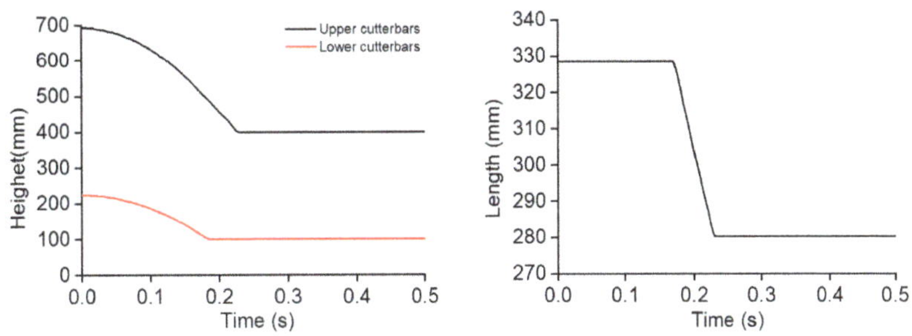

(**a**) (**b**)

Figure 12. Motions of the components with the DCH descending. (**a**) Height change of the upper and lower cutterbars; (**b**) Length change of the L_{BA}.

In the simulation of the motion of components with the change in ground heights during harvesting, the changes in upper cutterbars and lower cutterbars, as well as varying L_{BA}, are displayed in Figure 13. The height of the upper cutterbars is kept unchanged, while the height of the lower cutterbars, when a hump is encountered on the ground wheel, increases and decreases in the case of a concave ground profile, finally realizing the function of the profiling mechanism. According to Figure 13b, in this process, L_{BA} is shown to change within the range of 254.7 mm to 322.1 mm. By combining the change in L_{BA} during the descent of the header with the dimensions of the swing block and height-limiting block, the length of the height-limiting connecting rod of the profiling mechanism is found to be 365 mm.

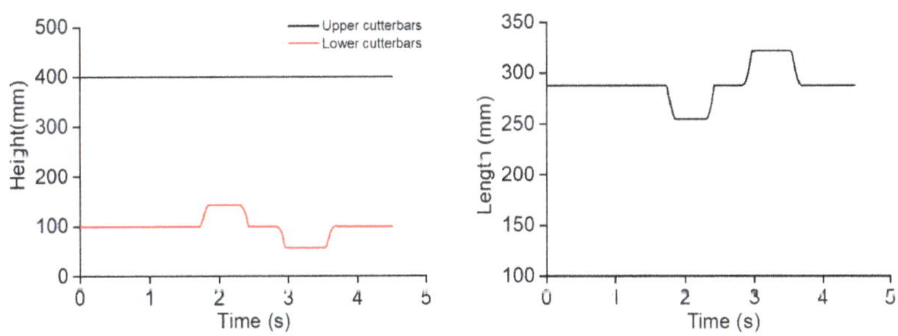

(**a**) (**b**)

Figure 13. Motions of the components during harvesting. (**a**) Height change of the upper and lower cutterbars; (**b**) Length change of the L_{BA}.

3.3. Field Experiment Analysis

As shown in Figure 14, the loss rate of using the conventional header wheat combine is 0.97%, and the loss rate of using the DCH is 1.05%, but less than 1.2%, which conforms to the relevant national operation quality requirements. The feed rate is reduced for the DCH, and the operating parameters of the threshing and separation device and the cleaning device do not match, increasing the loss rate. Therefore, in the follow-up study, it is also necessary to research the impact of the change in the feed length of the stalk on threshing separation and cleaning performance.

As shown in Table 6, the stubble height of the DCH is 11.02 cm, lower than the stubble height of the conventional header, and it meets the relevant operation quality requirements. The variation coefficient of the stubble height of the DCH is higher than that of the conventional header, indicating that the consistency of the stubble height of the DCH is lower than that of the conventional header. The main reason is that after the upper cutterbars of the DCH cut the upper stalks, the remaining stalks are tilted by the thrust of the header plate before being cut, resulting in uneven cutting after the lower cutterbars cut.

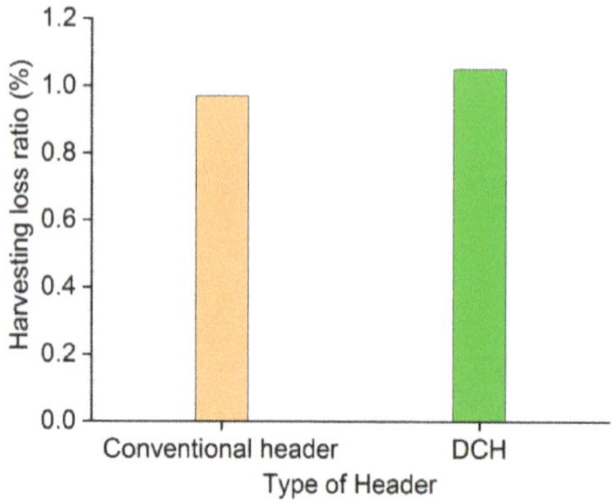

Figure 14. Harvesting loss rate of the WCH using different headers.

Table 6. Measurement results of the stubble height.

Type of Header	Stubble Height(cm)	Coefficient of Variation (%)
DCH	11.02	3.17
Conventional header	11.44	1.81

As shown in Figure 15, under the same operating speed, the fuel consumption per unit area of the WCH with a conventional header is 33.9 kg/hm^2, and the fuel consumption per unit area of the WCH with a DCH is 30.1 kg/hm^2, which reduces the fuel consumption by 11.2%. Compared with the conventional header, the DCH has two layers of cutterbars, and the power consumption of the header is increased. However, at the same operating speed, the length of the stalk fed into the combine becomes shorter, the feed rate decreases, and the power consumption of the conveying, threshing, and cleaning systems decreases, thus reducing the fuel consumption per unit area.

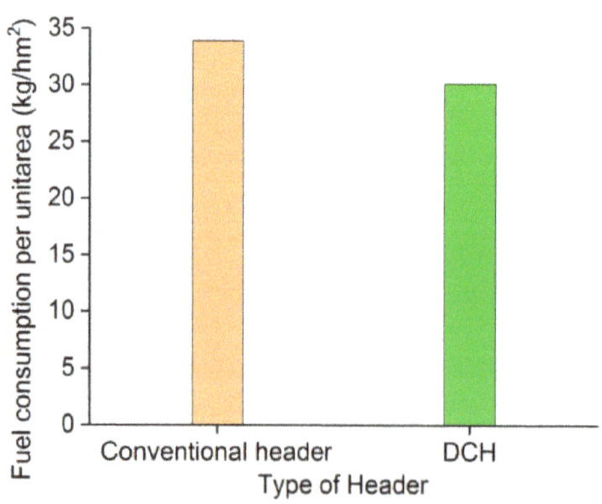

Figure 15. Fuel consumption per unit area of the WCH using different headers.

4. Conclusions

(1) In this paper, a DCH was designed that can only cut the upper part of the wheat plant into the WCH. By reducing the feed length of the stalk, the problems of the wheat

combine, such as congestion and high power consumption, are solved, providing a technical reference for the efficient harvest of the wheat combine.

(2) The effects of feed rate, feed length of the stalk, and speed of the tangential threshing rotor on conveying time were studied using a threshing test bench and optimized to obtain the optimal parameter combination of the test factors: the feed rate was 8 kg/s, the feed length of the stalk was 380 mm, the speed of the tangential threshing rotor was 554 r/min, and the conveying time was 8.089 s. The optimized parameter combination was tested and verified, and the test results show that the relative error with the predicted value was 0.198%, proving the optimized parameters' reliability.

(3) Combined with the best feed length of the stalk determined by the bench test, the critical components of the DCH were designed. The movement process of the profiling mechanism was simulated using ADAMS software, which verifies the correctness of the structural design, and the length of the height-limiting connecting rod was determined to be 365 mm.

(4) The field performance experiment of the DCH of the WCH was carried out, and the average harvest loss rate was 1.05% and the average stubble height was 11.02 cm, which all met the relevant operation quality requirements; at the same operating speed, the fuel consumption was 11.2% less than that of the combine harvester with a conventional header.

Author Contributions: Conceptualization, L.Y. and M.L.; methodology, X.H. and W.W. (Wanzhang Wang); investigation, L.Y., W.W. (Wanzhang Wang) and X.H.; writing—original draft preparation, L.Y., M.L. and X.H.; writing—review and editing, L.Y. and W.W. (Wanzhang Wang); visualization, W.W. (Wenhe Wei) and Z.Q.; funding acquisition, M.L. and W.W. (Wanzhang Wang). All authors have read and agreed to the published version of the manuscript.

Funding: This research was funded by China Agriculture Research System (CARS-03), Henan Province Science and Technology Projects (222102110235), and Henan Province Science and Technology Projects (232102110271).

Institutional Review Board Statement: Not applicable.

Informed Consent Statement: Not applicable.

Data Availability Statement: The data used to support the findings of this study are available from the corresponding author upon request.

Acknowledgments: The authors would like to thank Zhengzhou Zhonglian Harvesting Machinery Co., Ltd., as well as gratefully appreciate the reviewers who provided helpful suggestions for this manuscript.

Conflicts of Interest: The authors declare no conflict of interest. There is no conflict of interest with Zhengzhou Zhonglian Harvesting Machinery Co., Ltd.

References

1. Fan, C.; Zhang, D.; Yang, L.; Cui, T.; He, X.; Zhao, H. Development and performance evaluation of the electric-hydraulic concave clearance control system based on maize feed rate monitoring. *Int. J. Agric. Biol. Eng.* **2022**, *15*, 156–164. [CrossRef]
2. Shamilah, A.M.; Darius, E.P. Actual field speed of rice combine harvester and its influence on grain loss in Malaysian paddy field. *J. Saudi Soc. Agric. Sci.* **2020**, *19*, 422–425.
3. Sun, Y.F.; Liu, R.J.; Zhang, M.; Li, M.Z.; Zhang, Z.Q.; Li, H. Design of feed rate monitoring system and estimation method for yield. *Comput. Electron. Agric.* **2022**, *201*, 107322. [CrossRef]
4. Zhang, Y.; Chen, D.; Yin, Y.; Wang, X.; Wang, S. Experimental study of feed rate related factors of combine harvester based on grey correlation. *IFAC-PapersOnLine* **2018**, *51*, 402–407. [CrossRef]
5. Quick, G.R. Laboratory Analysis of the Combine Header. *Trans. ASAE* **1973**, *16*, 5–12. [CrossRef]
6. Nik, M.A.E.; Khademolhosseini, N.; Abbaspour-Fard, M.H.; Mahdinia, A.; Alami-Saied, K. Optimum utilisation of low-capacity combine harvesters in high-yielding wheat farms using multi-criteria decision making. *Biosyst. Eng.* **2009**, *103*, 382–388. [CrossRef]
7. Qing, Y.; Li, Y.; Yang, Y.; Xu, L.; Ma, Z. Development and experiments on reel with improved tine trajectory for harvesting oilseed rape. *Biosyst. Eng.* **2021**, *206*, 19–31. [CrossRef]
8. Yang, R.; Wang, Z.; Shang, S.; Zhang, J.; Qing, Y.; Zha, X. The design and experimentation of EVPIVS-PID harvesters' header height control system based on sensor ground profiling monitoring. *Agriculture* **2022**, *12*, 282. [CrossRef]

9. Brune, M. Method and Device for Height Control of Headers. Europe Patent No. EP1813144A1, 1 August 2007.
10. Hunt, C.D.; Pankal, M. Header Height Control for Combine Harvester. World Patent NO. WO2021242867A1, 2 December 2021.
11. Zhuang, X.; Li, Y. Header height control strategy of harvester based on robust feedback linearization. *Trans. Chin. Soc. Agric. Mach.* **2020**, *51*, 123–130.
12. Xie, Y.; Alleyne, G.; Greer, A.; Deneault, D. Fundamental limits in combine harvester header height control. *J. Dyn. Syst. Meas. Control* **2013**, *135*, 034503. [CrossRef]
13. Daniel, M. Header Height Control of Combine Harvester via Robust Feedback Linearization. Master's Thesis, Iowa State University, Ames, Iowa, 2016.
14. Lopes, G.T.; Magalhaes, P.S.G.; Nobrega, E.G.O. Optimal header height control system for combine harvesters. *Biosyst. Eng.* **2002**, *81*, 261–272. [CrossRef]
15. Baumgarten, J.; Wilken, A.; Neitemeier, D.; Bormann, B.; Spiekermann, S.; Irmer, D. Cutting Table Length Adaptation. European Patent NO. EP3858129A1, 8 August 2020.
16. Chen, J.; Wang, S.; Lian, Y. Design and test of header parameter keys electric control adjusting device for rice and wheat combined harvester. *Trans. Chin. Soc. Agric. Eng.* **2018**, *34*, 19–26.
17. Honeyman, F.; Dreyer, D.; Glade, M. Auto Reel Height. World Patent NO. WO2019234539A1, 12 December 2019.
18. Li, H.; Wan, X.; Xu, Y.; Jiang, Y.; Liao; Q. Clearance adaptive adjusting mechanism for header screw conveyor of rape combine harvester. *Trans. Chin. Soc. Agric. Mach.* **2017**, *48*, 115–122.
19. Li, Y.; Li, Y.; Xu, L.; Hu, B.; Wang, R. Structural parameter optimization of combine harvester cutting bench. *Trans. Chin. Soc. Agric. Eng.* **2014**, *30*, 30–37.
20. Xie, H. Design and Experiment of Belt Conveyor Header for Soybean Combine Harvester. Master's Thesis, Shandong University of Technology, Zibo, China, 2019.
21. Zhang, M.; Jin, M.; Wang, G.; Liang, S.; Wu, C. Design and test of double crank planar five-bar reel in rape windrower. *Trans. Chin. Soc. Agric. Mach.* **2022**, *53*, 115–122.
22. Van, V.S.; Vandergucht, S. Agricultural Machine Equipped with Cam Mechanism for Gathering Crop Material. Europe Patent No. EP3064053B1, 13 December 2017.
23. Henry, W.B.; Nielsen, D.C.; Vigil, M.F.; Calderon, F.J.; West, M.S. Proso millet yield and residue mass following direct harvest with a stripper-header. *Agronomy* **2008**, *100*, 580–584. [CrossRef]
24. Shelbourne, K.; Pakenham, S. Crop Strippers and Stripper Toothing. Europe Patent No. EP0976315B1, 7 April 2004.
25. Wang, Y. Research on apparatus with two cutters and conveyers in two ways for the combine harvester, *Trans. Chin. Soc. Agric. Mach.* **1995**, *26*, 84–89.
26. Dai, F.; Zhao, W.; Han, Z.; Li, X.; Gao, A.; Liu, X. Improvement and experiment on 4gx-100 type wheat harvester for breeding plots. *Trans. Chin. Soc. Agric. Mach.* **2016**, *47*, 196–202.
27. Wang, W.; Liu, W.; Yuan, L.; Qu, Z.; Zhang, H.; Zhou, Z. Simulation and experiment of single longitudin axial material movement and establishment of wheat plants model. *Trans. Chin. Soc. Agric. Mach.* **2020**, *51*, 170–180.
28. Zhang, H.; Chen, B.; Li, Z.; Zhu, C.; Jin, E.; Qu, Z. Design and simulation analysis of a reverse flexible harvesting device for fresh corn. *Agriculture* **2022**, *12*, 1953. [CrossRef]
29. Fu, J.; Zhang, G.; Xie, G.; Wang, Y.; Gao, Y.; Zhou, Y. Development of double-channel feeding harvester for ratoon rice. *Trans. Chin. Soc. Agric. Eng.* **2020**, *36*, 11–20.
30. Chinese Academy of Agricultural Mechanization Sciences. *Agricultural Machinery Design Manual*, 1st ed.; China Agricultural Science and Technology Press: Beijing, China, 2007; pp. 891–904.
31. Technical Specification for Quality Evaluation of Grain Combine Harvesters. *NY/T 2090-2011*; China Standards Press: Beijing, China, 2011.
32. Whole-Feed Combine Harvester-Evaluation Index and Measurement Methods for Fuel Consumption. *GB/T 29002-2012*; China Standards Press: Beijing, China, 2012.
33. The Work Quality of Grain (Wheat) Combine Harvester. *NY/T 995-2006*; China Standards Press: Beijing, China, 2006.
34. Xia, Q.; Zhang, W.; Qi, B.; Wang, Y. Design and experimental study on a new horizontal rotary precision seed metering device for hybrid rice. *Agriculture* **2023**, *13*, 158. [CrossRef]

agriculture

MDPI

Article

Influence of Factors Determining Weeds' Plant Tissue Reaction to the Electric Pulse Damage Impact

Igor Yudaev [1], Yuliia Daus [1], Vladimir Panchenko [2] and Vadim Bolshev [3,*]

1 Energy Department, Kuban State Agrarian University, Krasnodar 350044, Russia; etsh1965@mail.ru (I.Y.); daus_yulia@list.ru (Y.D.)
2 Department of Theoretical and Applied Mechanics, Russian University of Transport, Moscow 127994, Russia
3 Laboratory of Power Supply and Heat Supply, Federal Scientific Agroengineering Center VIM, Moscow 109428, Russia
* Correspondence: vadimbolshev@gmail.com

Abstract: Due to the emerging danger to the life of animals and people, today there is a turn to safe technologies for controlling weeds by physical methods, both from the point of view of ecology and food safety, which include the destruction of plants using an electric current, in particular, high-voltage electrical pulses. The purpose of the study presented in the article is to identify and evaluate the effect of high-voltage electrical pulses on the irreversible damage to the intracellular structures of the plant tissue of weeds and unwanted grasses during their electric weed control, characterizing and evaluating the parameters and modes associated with such processing. Experimental studies were carried out using a laboratory experimental setup that consists of a pulse voltage generator, a control circuit for a spherical forming spark gap, and schemes for measuring the electrical resistance of the plant tissue of the weed sample. The lesion level made it possible to control the depth of irreversible damage to the internal structure of the plant tissue of weeds by measuring its tolerance (the conductivity of the tissue increased with increasing damage to the cellular components of the tissue).The irreversible damage to the plant tissues of weeds for weeds of various biological groups, which is characterized by reaching the value of at least 4.0–7.5 degrees of damage to their tissues, can be acted on them with high-voltage electrical pulses in the treated tissue of an electric field intensity of at least 3.74 kV/cm, while ensuring specific processing electric energy for the reliable processing of weeds: for *Euphórbia virgáta*, thise quals 5.2…17.5 J/cm^3; for *Amaránthus retrofléxus*, it is 3.5…7.7 J/cm^3; for *Cirsium arvense*, it is 2.7…10.9 J/cm^3;for *Sónchus arvénsis*, it is 3.7…15.8 J/cm^3; and for *Lactúca tatárica*, it is 3.3…8.1 J/cm^3.

Keywords: high-voltage electrical pulse; processing; plant tissue; irreversible damage; intracellular structure

Citation: Yudaev, I.; Daus, Y.; Panchenko, V.; Bolshev, V. Influence of Factors Determining Weeds' Plant Tissue Reaction to the Electric Pulse Damage Impact. *Agriculture* **2023**, *13*, 1099. https://doi.org/10.3390/agriculture13051099

Academic Editor: Jacopo Bacenetti

Received: 4 April 2023
Revised: 10 May 2023
Accepted: 12 May 2023
Published: 21 May 2023

1. Introduction

According to the Food and Agriculture Organization of the United Nations (FAO), potential crop losses are 13.8% from pests, 11.6% from diseases, and 9.5% from weeds. Every year, weeds cause great damage to the yield and quality of agricultural crops, so the average percentage of crop losses in different countries is estimated over the past decade: for spring wheat, it ranges from 7.9% to 47.0%; for winter wheat it ranges from 2.9% to 34.4%; and the average global loss of total production of corn is about 37%, of peanuts it is 36%, of soybeans it is 31%, etc. [1,2].Currently, serious attention is being paid to reducing environmentally unsafe impacts in the crop industry of the agrarian sector, especially when growing organic products. One of the main technological operations in agriculture is weed control, which is implemented mainly through the use of artificially synthesized herbicides. The most popular herbicide in the world today is glyphosate. In 2015, the International Agency for Research on Cancer (IARC) stated that glyphosate is the active chemical in "Roundup" and is "probably carcinogenic". Food and Agriculture Organization of the

United Nations (FAO), the European Chemicals Agency, and the European Food Safety Authority (EFSA) concluded that it is "unlikely to pose a carcinogenic risk" to humans when eating foods grown with it [3]. In 2022, the European Union considered a possible phasing out of the use of designated glyphosate pesticides. That is why today, from the point of view of ecology requirements and food safety, there is a turn to safe effective technologies for controlling weeds by various physical methods, including the destruction of plants using electric currents, in particular, high-voltage electrical pulses [4].

The application of electric impact to control weeds and unwanted vegetation has been known since the 1800s, but there are still a lot of unresolved issues, such as, first of all, what the mechanism of the processes that lead to plant death is; whether there is a certain consistency in the sensitivity of plant tissues of weeds to electrical impact; how the lesion levels of plant tissue can be assessed, using methods that are understandable to all with a simple means of control; how the change in the electrically conductive properties of plant tissue depends on the parameters of the electrical impact, etc. Comparison with the electrical impact types, direct current, sinusoidal voltage, and electrical pulses, made it possible to give priority to high voltage pulses, since it is the processing of this type of electric action that allows deeper irreversible lesions to plant tissue and at the same time, a smaller amount of electrical energy is spent on the process [4,5].

To consider the processes occurring in plant tissues during the destruction of weeds using high-voltage electrical pulses [6–8], it is necessary to study the sensitivity of the processed plant tissue to the electrical effects and assess the lesion level of the normal functioning of its intracellular components depending on the main damaging factors of electrical action—the electric field intensity and the quantity of energy absorbed by the processed tissue [5,9–12].

In the global practice of scientific and applied research on the study of the effect of high-voltage electrical pulses on the intracellular structure of plant objects and materials, research works carried out in the interests of the processing industry in the production of food and semi-finished products are the most studied, in particular, the processing of fruits and vegetables, for example, citrus and gourds, grapes, apples, cherries, sugar beets, etc. [13–17]. The main objective of this research was to study the effect of electrical pulses on the internal structure of plant materials for the purpose of irreversible damage (destruction) and subsequent release of intracellular fluid for additional juice yield, drying acceleration, non-thermal pasteurization of products, etc. [18–20]. As a quantitative indicator characterizing the "destruction" of intercellular and intracellular structures of plant tissues, research scientists use such an indicator as the degree of disintegration [9,21–23], the value and change of which directly depend on the electric field intensity, duration of exposure, pulse form, and the quantity of energy absorbed by the processed object [9,10,12,24–29].

Weeds cause serious harm in semi-arid and arid regions of the Russian Federation, one of which is the Lower Volga region. An essential feature of all methods of tillage in the region is that each operation aimed at combating weeds must ensure not only the complete destruction of weeds, but also the preservation of soil moisture, the lack of which adversely affects the development of cultivated plants. In the fight against weeds, mechanical and chemical methods are widely used. Cultivation is the most effective since, in addition to the destruction of weeds, the soil is loosened, as a result of which metabolic processes in it are improved, which positively affects the growth and development of crops. Weeds themselves are cut by the operating bodies of machines for surface tillage. However, this method is not always effective, since when the field is infested with root shoot weeds, the main weeds of the soil areas of the Lower Volga region, after processing by the operating bodies of cultivators, after some time, new, large foci appear. Root weeds propagate vegetatively, and the most effective is to extract them from the soil together with the root, i.e., combing. However, in the modern system of agricultural machines, there are no tools that allow combing out weeds from the soil along with the root in various phases and periods of their development.

In this paper, the sensitivity of plant tissue of weeds to damaging electric pulse effects was studied. This article did not set a detailed consideration of the technology of electric pulse weed control, design a unit for its implementation, and conduct an economic comparison with known control methods. However, answering the question of respected reviewers, we can say that the introduction of electric pulse weed control into the technology of preparing sown areas (care for fallow lands) for growing crops makes it possible to increase the efficiency of weed control in terms of human labor costs by 62% compared to chemical treatment, and it is practically comparable with cultivation: in terms of direct energy costs, by 83% compared to62%, and in terms of the cost of manufacturing technical means for each method, by 35% compared to41%, respectively. Electric pulse fallow weed control can reduce the total energy costs, for example, when cultivating winter wheat on an area of 100 hectares by 12.89% compared to chemical weed control and by 5.13% compared to conventional cultivation.

The research hypothesis is as follows: during electric pulse processing, the intracellular components of weed plant tissues are primarily damaged, and their change (the ratio of the number of damaged and not yet destroyed cells) should determine the change in the electrically conductive properties of the processed tissues. The depth of processing will be determined by the parameters of the electric pulse impact: the amount of energy absorbed by the plant tissue and the strength of the electric field in it.

The authors of the paper proposed assessing the susceptibility of plant tissue of weeds to damaging electrical effects using such a quantitative indicator as its lesion level, which is understood as the value Sn. The numerical value of this indicator is determined by the ratio of the total electrical tolerance of the plant tissue or its active component, recorded before the start of processing, to the tolerance of the same tissue, which changes during processing, at the particular measuring current frequency, in this case at the frequency of f = 10 kHz [5,24,30,31].

The purpose of the study is to identify and evaluate the effect of high-voltage electrical pulses on the irreversible damage to the intracellular structures of the plant tissue of weeds and unwanted grasses during their electric weed control, characterizing and evaluating the parameters and modes associated with such processing.

2. Materials and Methods

Laboratory experimental setup. Experimental studies were carried out in accordance with the procedures of the experimental setup (Figure 1), which were described in detail in articles [30,31]. The laboratory experimental setup consists of three principal components (blocks): (1) pulsed voltage generator (IVG); (2) control circuit for a spherical forming spark gap; and (3) schemes for measuring the electrical tolerance of the plant tissue of the weed sample. Each of the listed blocks performs its functional purpose.

The experimental setup is powered from a single-phase alternating voltage grid, the required voltage value of which was set at the circuit terminals using a laboratory autotransformer. A digital voltmeter was used to visually control the value of the supply voltage. The standard AT-50 apparatus was used as a source of high direct voltage, at the output of which voltage up to 50 kV can be obtained.

High-voltage capacitors were used as the storage capacities of the discharge circuit of the pulsed voltage generator (GVP). To obtain different values of electrical energy, the capacity of the discharge circuit required for processing weed samples was changed by switching the number of capacitors into batteries. A high-voltage resistor was used as a charging current limiter.

An important role in the GVP is played by spark gaps, which act as a switch and switch capacitors from the charge mode to the discharge mode. The stability of the amplitude of the output pulses of the generator depends on their stable operation.

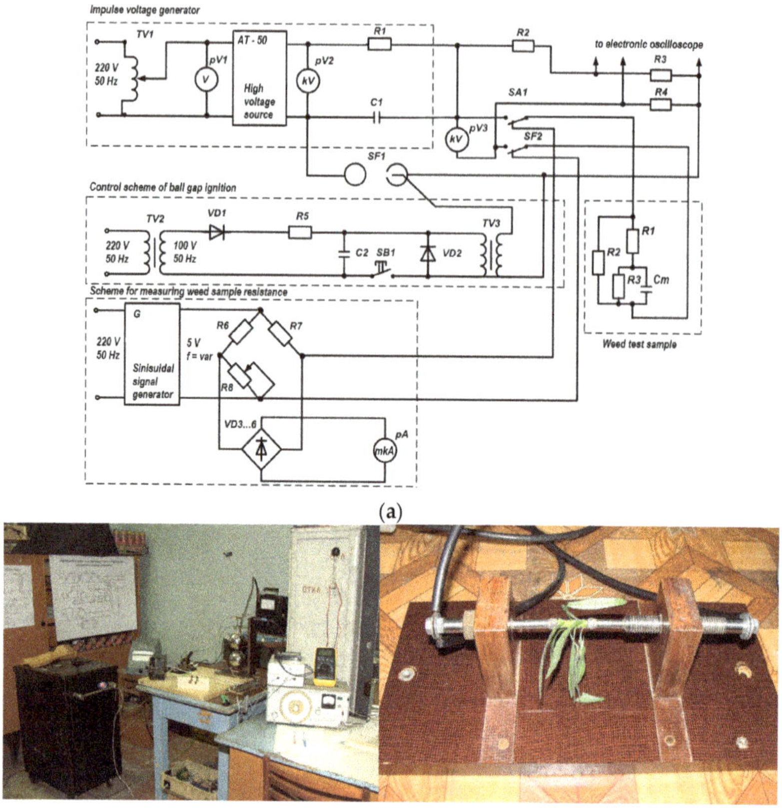

(a)

(b)

Figure 1. Schematic diagram of the laboratory experimental setup (**a**) and the appearance of the laboratory experimental setup (**b**).

To control the moment of discharge in the GVP in order to synchronize its operation with the recording oscilloscope, as well as to calculate the number of pulses affecting the plant tissue of the studied weeds, a controlled spark gap was used. In the laboratory experimental setup, a standard adjustable spark gap was used, which was assembled on the basis of two spherical electrodes with a diameter of 100 mm, the surfaces of which were chrome-plated.

The moment when the starting pulse was supplied corresponded to the end of the charging time of the GVP capacitors. The ball gap control circuit was assembled using a *TV3* pulse transformer. The principle of the circuit operation is as follows: the voltage U is supplied from the secondary winding of the step-down standard transformer *TV2*, and the capacitance *C2* is charged through the resistance *R5*. After the SB1 button was switched on (ignition coil), the capacitance was discharged to the primary winding of the *TV3* pulse transformer, and the current flowing through it at the moment of switching contributed to the occurrence of a high-voltage pulse in the secondary winding. The created high voltage pierced the gap between the main and igniting electrodes of the lower ball of the main discharge gap *SF1*, which was ionized, and the main discharge occurred in it, as a result of which the energy stored in the storage tanks *C1* of the GVP was applied to the plant tissue sample. Diode *VD2*, connected in parallel with the primary winding of the ignition coil, acted as a shunt to prevent the *TV3* pulse transformer from failing. The ignition circuit was mounted in a separate housing, and high voltage was supplied to the spherical spark gap using a high-voltage wire.

Methodology for preparing and conducting this study is as follows. The fragment of the weed plant tissue carved with a surgical scalpel from the plant was placed in the interelectrode space in the working spark gap *SF2* of the discharge grid. The electrodes played the role of both working and measuring electrodes and were spring-loaded stainless-steel rods at the ends of which grooves or blackened platinum mesh segments were used, to which felt pads were attached. To reduce the "plant tissue-electrode" contact resistance and to ensure high-quality contact with plant samples, when using felt pads, the contact surface of the overlays was moistened with 5% NaCl solution after each experiment. Galvanic isolation between the discharge circuit of the GVP and the measuring circuit was carried out using a three-position switch.

The species composition of weeds that are typical for the five-year period of research and found everywhere in crops and fallows in the Lower Volga region is diverse. The dominant types of weeds are (1) dicotyledonous perennial root weeds: *Sónchus arvénsis*, which fill43–54% of weedy areas; *Convōlvulus arvēnsis* fills 26–32%; *Cirsium arvense* fills 12–23%; *Euphórbia virgate* fills 13–19%; and *Lactúca tatárica* fills 4–8%; (2) dicotyledonous annual weeds: *Oxýbasis úrbica, Chenopódium úrbicum*, etc. fills 16 ... 23%; *Amaránthus retrofléxus*, etc. fills 15–30%; *Thláspi, Barbaréa, Raphanus raphanistrum*, etc.–3 ... 9%, etc.

Experiments were carried out with weeds, which are most often found on the lands of agricultural fields of the Lower Volga region. These are annual weeds: *Amaranthus retroflexus, Oxýbasis úrbica, Cánnabis ruderális, Fallópia convólvulus*, and *Xánthium strumárium*, as well as perennial, root shoot weeds, *Cirsium arvense, Sónchus arvénsis, Euphórbia virgáta, Lactúca tatárica*, and *Convōlvulus arvēnsis*.

Weeds grew in the fields of the All-Russian Research Institute of Irrigated Agriculture, on rain-fed fallow lands, with an average or even high degree of infestation by weeds of the same predominant type, which were approximately in the same phase of development.

The studied weeds were in the following phases (periods of development): the beginning of budding, flowering, and the beginning of fruiting. To study the response and sensitivity of the plant tissue of weeds to electrical pulses, experiments were carried out not with the whole plant, but with a separate part—a certain volume of the leaf stem part, the area with the root neck, and part of the root system. The size of the studied part of the weed plant was taken into account in the form of its certain volume, for which the recorded energy absorbed by the plant tissue was recalculated in the form of its specific value (J/cm^3).

Whole weeds, together with the root system extracted from the soil, were dug up in the experimental field, housed in a portable container, and delivered to the laboratory, which was located within walking distance of 100–150 m. In the laboratory on the preparation machine (fixed pair of scalpels), fragment of the plant tissue of the stem, root collar, and root was cut out and placed in the processing chamber between spring-loaded immersion electrodes. After measuring the tolerance of the studied area of the plant, the electrodes were disconnected from the measuring complex and connected to the discharge circuit, in which the plant sample was electro pulsed. Then, the reverse switching followed, and the tolerance of the plant tissue was measured after the electrical action. The impact on the plant sample was continued until subsequent exposures did not lead to the change in the tolerance of the plant tissue of the sample, that is, when the level of irreversible damage to plant tissue reached its maximum value.

Methodology for processing research results and its justification is as follows. The lesion level of the plant tissue damage was found as the ratio of the tolerance of undamaged (living) tissue to the tolerance of damaged tissue.

To establish the dependence of the lesion level of the plant tissue of weeds on the electric field intensity in it and on the quantity of energy absorbed by it, the analytical expression that relates the lesion level to the intensity of the influencing factor and the duration of its action was applied. The applicability of the proposed approach to the analysis of the irreversible electrical pulse damage process to the intracellular structures of weed plant tissues was previously substantiated in the study of the susceptibility of sunflower and tobacco plant tissues to electric pulse exposure [5,30,32]:

$$S = k_s \cdot E^n \cdot t^q \tag{1}$$

where $E = U_0/l_{pm}$ is the average electric field intensity in the plant tissue of the studied weed area, which is characterized by the length l_{pm}, coinciding with the length of the path of the processing current, and the cross-sectional diameter of the processed sample averaged over this length d_{pm}, kV/m; U_0 is initial voltage or voltage on the output capacitor of the discharge circuit applied to the studied sample of plant tissue, kV; t is time period of the electrical action or the duration of the acting pulse, s; n is exponent degree n = const, determining the dependence of the electric field intensity change in plant tissue on the lesion level; q is exponent degree q = const, which determines the dependence of the exposure time to plant tissue on the lesion level; and k_s is proportionality factor, $(\text{cm/kV})^n \cdot \text{s}^{-1}$.

The distinctive point in the application of Equation (1) for the conducted analysis should consider the fact that electric pulse weed control is a process of longitudinal electrical processing of weed tissues, and not their processing across, as described in [5,30,32]. In the analytical Equation (1), to make analysis more convenient, the exposure time t should be considered to the power $q = 1$, which corresponds to the main linear section of the dependence $S_n = f(E, t)$, characterizing that part of it when the transition from the end of the initial phase of processing to its completion is clearly distinguished. Such an assumption is quite possible, since the studied plant tissues of weeds were affected by single pulses that had the same duration in each series of experiments:

$$S = k_s \cdot E^n \cdot t = k_s \cdot \left(\frac{R_m \cdot i}{l_m}\right)^n \cdot t = k_s \cdot \left(\frac{R_m^{n-1} \cdot i^{n-2}}{l_m^n}\right) \cdot W = k_s \cdot \left(\frac{\rho_m}{l_m \cdot S_m}\right) \cdot m \cdot W_1 \cdot {}^{n-2} \tag{2}$$

or

$$S = k_s \cdot \left(\frac{R_m^{n-1} \cdot i^{n-2}}{l_m^n}\right) \cdot m \cdot W_1 = k_s \cdot \left(\frac{R_m \cdot C_k}{2}\right) \cdot m \cdot {}^n \tag{3}$$

where i is the average current value through the plant tissue, A; $R_{\text{PT}} = (\rho_{\text{PT}} \cdot l_{\text{PT}})/S_{\text{PT}}$ is tolerance of the investigated area of plant tissue with electrical resistivity ρ_{PT}, length l_{PT}, and cross-sectional area S_{PT}, Ohm; $W_1 = i^2 \cdot R_{\text{PT}} \cdot t = 0.5k_{\text{p}} \cdot C_k U_0^2 = (0.45\text{–}0.475) \cdot C_k U_0^2$ is single pulse energy, J; k_{p} is the degree of discharge of the capacitor of the discharge circuit (in terms of voltage), usually taken as equal to $k_{\text{p}} = 0.9\text{–}0.95$; C_k is storage capacity of the discharge circuit, F; and m is number of acting pulses.

Elements of planning experiments are as follows. The required reproducibility of measurements depends on the standard of measurement and the desired reliability of the result; therefore, based on the accepted values of the specified reliability of the result and the marginal error, the numbers of reproducibility in the experiments were3–5 measurements.

The five-year research period was determined by the fact that the growth and development of plants depend on various soil and climatic conditions, which can change significantly from year to year, and therefore all experiments during this five-year period were carried out according to the same methodology with the same repetition on plants growing on the same soil area, in the same periods of development and the same tillage period, so that it is possible to determine the variability of the studied parameters to determine the possible deviation of the obtained results. The results deviated from year to year, and according to the results of the entire experiment, this deviation ranged from average data to 10–15%. As a result, it was concluded that the obtained data on the value of the lethal dose, the specific energy of processing, and the value of the lesion level to plant tissue are reliable and do not change above the established margin of error over the years.

The character of all obtained dependencies is the same for all studied weeds. The difference is only in quantitative indicators (different lesion levels with the same influencing factors). Therefore, the paper presents the dependencies of the most resistant weed that requires more energy consumption—*Euphorbia virgáta*—as a typical example.

3. Results of the Research

Preliminary analysis of Equations (1) and (2) shows that the lesion level of plant tissues S_n of weeds depends on the electric field intensity E in them, and the nature of the dependence is determined by the value of the certain exponent n, as well as from the total quantity of damaging energy W, which is determined by the product of the number of influencing pulses m and the power of the single pulse W_1.

The quantitative lesion level and the nature of the dependence behavior $S_n = f(E)$ (Figure 2) is determined by the value of the exponent n, which was previously found empirically for plant tissues to be as follows: for tobacco, $n = 1.26$; for sunflower, $n = 1.5$; for "aloe" plant, $n = 0.75$; and for apple pulp, $n = 2.45$ [5,30]. Here, it is necessary to note an obviously traceable fact—considering the electrical processing of various plant objects, it can be noticed that the n exponent depends on the number of influencing impulses m or the energy W supplied to the processing object. Therefore, the found values of the n indicator should be characterized as averaged for a specific process of electric pulse processing of the specific plant material.

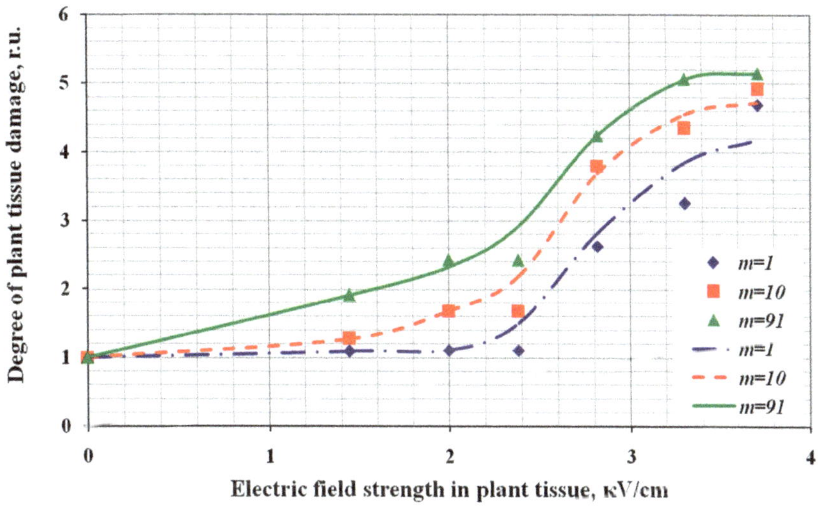

Figure 2. Dependencies of the lesion level of the plant tissue of the root system of *Cirsium arvense* on the electric field intensity in the tissue at different numbers of acting pulses (m). The IVG was characterized by the discharge circuit capacitance of 500 pF and inductance of 110 μH

To confirm the above preliminary hypothesis about the different sensitivity to electric pulse lesions to the plant tissue of weeds, the experimental results were grouped, and experimental dependencies $Sn = f(E)$ were built on them for weeds of different biological groups [32], some of which, as showing the most qualitatively and theoretically the justified nature of the behavior (1–3), are presented in Figure 3.

From the point of view of optimizing the technological process of electric pulse destruction of weeds, it is interesting to analyze the influence on the lesion level S_n of the initial voltage of the discharge circuit U_0 and, consequently, as mentioned earlier, the electric field strength in their plant tissue. To elucidate the essence of this influence, a set of experimental studies were carried out, in which the capacitance of the circuit C_k remained constant, and the processing voltage U_0 changed. As a result of these studies, a set of dependencies $S_n = f(m)$ was obtained (Figures 4 and 5).

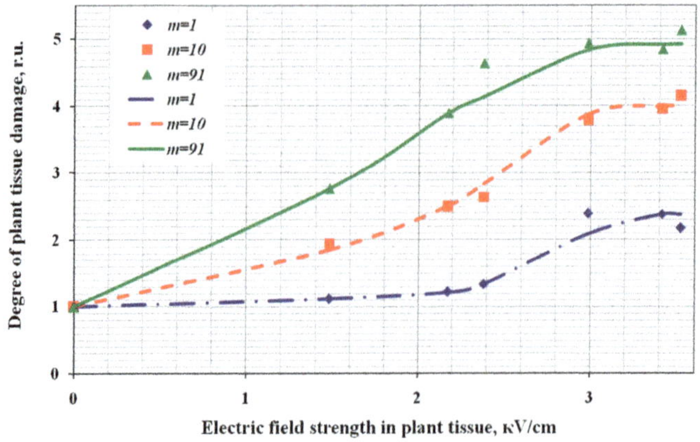

(a)

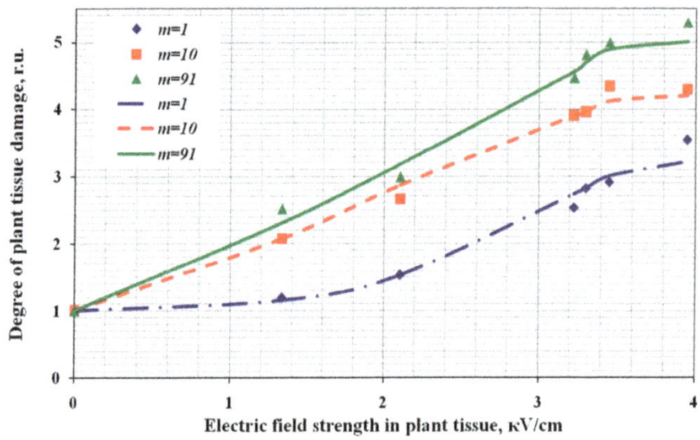

(b)

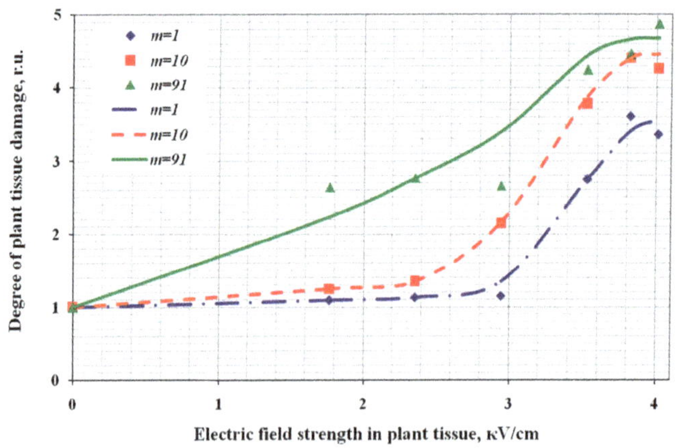

(c)

Figure 3. Dependencies of the lesion level of the plant tissue of the root system of weeds: (**a**) *Euphórbia virgáta;* (**b**) *Cirsium arvense*); and (**c**) *Amaránthus retrofléxus* from the electric field intensity in the plant tissue with a different number of influencing pulses (*m*). The IVG was characterized by the discharge circuit capacitance equal to 1000 pF and inductance of 110 μH.

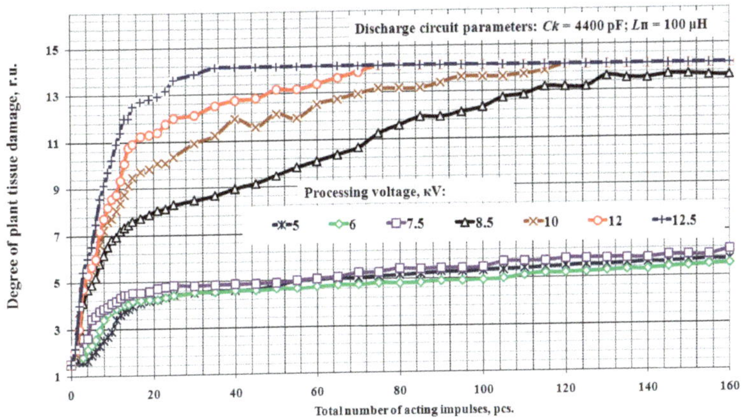

Figure 4. Dependencies of the plant tissue lesion level of the stem of *Euphórbia virgáta* on the total number of acting pulses at diverse values of the processing voltage, provided that the initial resistance values of the processed areas are different.

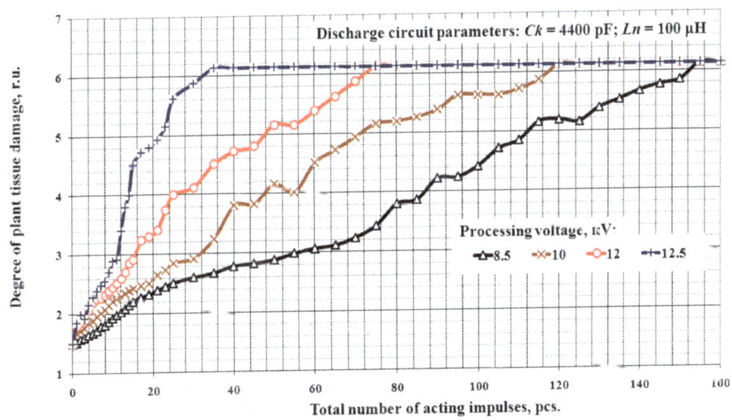

Figure 5. Dependencies of the plant tissue lesion level of the stem of *Euphórbia virgáta* on the total number of acting pulses at diverse values of the processing voltage, provided that the initial resistance values of the processed areas are the same.

In the studied example, in the first series of experiments (Figure 4), samples of plant tissue of the stem of *Euphorbia virgáta* were subjected to electrical processing, while the samples were cut from an adult plant of approximately the same period of development and recorded from the electrical resistance, which was of different values for each of them. In the considered example, samples of plant tissue of the stalk of *Euphórbia virgáta* were subjected to electrical treatment, the studied fragments of which were affected by electrical pulses with unit energy of W_1 = 0.055; 0.079; 0.124; 0.159; 0.22; 0.317; and 0.344 J, and the discharge circuit of the pulse voltage generator had the following parameters: C_k = 4400 10–12 F; L_p = 100 10–6 H; and U_0 = 5; 6; 7.5; 8.5; 10; 12; and 12.5 kV.

To confirm the assumption that if the studied samples of weeds had the same initial resistance, then in the process of electrical processing, their plant tissue could reach the same limiting lesion level, appropriate experiments were carried out. Electric influences were exerted on the plant tissue of fragments of the stems of *Euphórbia virgáta*, with equal initial resistances. Electrical pulse energy W_1 = 0.16; 0.22; 0.32; and 0.35 J (Figure 5). Discharge circuit parameters are C_k = 4400 10–12 F; L_p = 100 10–6 H; and U_0 = 8.5; 10; 12; and 12.5 kV.

4. Discussion of the Results and Analysis

The numerical values of the index n were determined by the calculation method on the main linearly increasing section of the dependence $S_n = k_s \cdot E^n \cdot t$. To conduct this, the graphic dependencies $S_n = f(E)$ were rebuilt on the logarithmic scale (Figure 2), and the n index was determined as the value of the slope tangent of the straight sections of the dependencies $S_n = f(E)$ to the abscissa axis. Each of the dependencies presented in Figure 3 was built for a specific number of acting electrical pulses m at the constant value of the capacitance of the capacitor bank of the discharge circuit C_k and the corresponding value of its inductance L_k.

Based on the results of the experiment and conducted calculations for various plots of different biological species of weeds growing in the Lower Volga region, the average values of the n index were determined and presented in Table 1 [32].

Table 1. Averaged *n* index values for various weed types.

Weed Type (in Latin)	Average *n* Index Valuesfor Various Weeds and Plant Tissues of Plots			
	Stem	Plot with Root Neck	Root	Average Value
Cirsium arvense	1.15	1.24	1.22	1.20
Sónchus arvénsis	1.19	1.23	1.25	1.22
Euphórbia virgáta	1.06	1.21	1.14	1.14
Lactúca tatárica	1.17	1.25	1.26	1.23
Amaránthus retrofléxus	1.22	1.40	1.21	1.28
Oxýbasis úrbica, Chenopódium úrbicum	1.30	-	1.27	1.29
Average value	1.18	1.27	1.23	1.23

The different values of the index *n* presented in Table 1, in our opinion, are explained by the different sensitivity of the tissues of the studied weed areas of different biological groups to the impact on them of high-voltage electrical pulses.

In Figure 6, the constructed graphical dependencies on the logarithmic scale are straight lines—the dependence of the n index value on the number of influencing pulses *m* or on the energy absorbed by the plant tissue is the same, since the impacts were organized by pulses with the same energy in each of them. Even a cursory analysis of the results presented in Figure 6 shows that the more acting pulses (energy) are supplied to the plant sample, the greater the value of the index *n*. As an example, consider the plant tissue of the root of *Euphorbia virgáta*; at $m = 10$, the exponent is $n = 0.63$, and at $m = 1000$, its value increases to $n = 1.796$.

To analyze the difference insensitivity for various tissues, the results were grouped, and the experimental dependencies $S_n = f(E)$ were plotted for weeds of different biological groups [32]. These dependencies (Figure 4) have a pronounced S-shaped character, which is determined by the presence in the initial section in the curves with the slight increase in the lesion level and the value of tension in the plant tissue of the section, which is often linear, with a very high rate of change in the lesion level, as well as of the section in which, with the increase in the value of the electric field intensity, the significant change in the lesion level is no longer observed.

The analysis of the character of the curves presented in Figure 4 makes it possible to draw the following conclusion: the plant tissue of the annual weed root system *Amaránthus retrofléxus*, when exposed to single pulses, reaches greater damage than the tissues of the perennial weeds *Euphórbia virgáta* and *Cirsium arvense*. For $m = 91$, the lesion level reaches one value, and the behavior of the dependencies becomes very similar to each other.

In the electrical treatment of weeds, characterized by the flow of current along plant tissues, the applied initial voltage of the discharge circuit U_0 determines the value of the electric field intensity E in them, at which the lesion level of the plant tissue reaches its maximum value $S_n = S_{nmax}$, and therefore its value is of research interest. The creation of such tension in the plant tissue of weeds leads to their death. To find these tension values, a separate series of experiments were carried out with samples from different plots of different weed species (Table 2). At the same time, it should be noted that plants of different years of research and periods of development, growing on different soil plots, have differences in damaging values of tension, with different expenditures of a minimum specific quantity of energy to damage the internal structure of weeds.

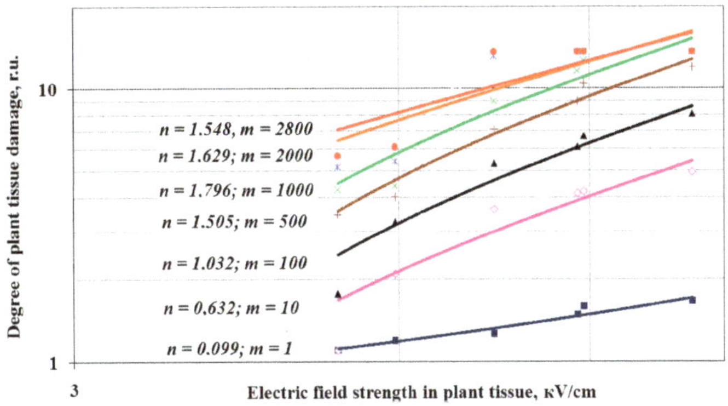

Figure 6. Dependencies of the lesion level of the root system plant tissue of *Euphórbia virgáta* on the electric field intensity in the tissue for diverse number of influencing pulses (m), plotted on the logarithmic scale. The IVG was characterized by the discharge circuit capacitance of 4400 pF and the inductance of 100 μH.

Table 2. The average electric field intensity and the maximum values of the lesion level of plant tissue in various areas of weeds.

Weed Type	Weed Area	Limit Lesion Level S_{nmax}, r.u.		Electric Field Intensity in Plant Tissue E, kV/cm	
		By Areas	Average	By Areas	Average
Cirsium arvense	stem	4.79	4.71	3.55	3.74
	passage	5.25		3.82	
	root	4.09		3.84	
Sónchus arvénsis	stem	7.71	7.51	3.52	3.66
	root	7.32		3.81	
Euphórbia virgáta	stem	4.60	4.59	3.68	3.53
	passage	5.39		3.37	
	root	3.77		3.55	
Oxýbasis úrbica, Chenopódium úrbicum	stem	7.38	7.31	3.41	3.55
	passage	6.89		3.53	
	root	7.65		3.72	
Amaránthus retrofléxus	stem	5.19	5.56	3.45	3.65
	passage	6.39		3.76	
	root	5.09		3.73	

According to the results of almost all experimental studies, the dependence of the lesion level S_n of the weed plant tissue on the number of acting pulses m–$S_n = f(m)$ or on the energy W–$S_n = f(W)$ absorbed by the plant tissue of weeds, which is the same, was plotted for variable values of the parameter bit circuit: U_0 = var or C_k = var. These dependencies have a clearly manifested S-shaped character, which manifests itself in three main areas: (1) in which the lesion level (S_n) changes slightly with the summing up of a certain number of high-voltage pulses; (2) on which there is a sharp increase in the lesion level (S_n) with the increase in the number of acting pulses or with an increase in the energy absorbed by the weed plant tissue (m or W); and (3) on which the lesion level (S_n), having reached its maximum value $S_n = S_{nmax}$, does not change. The steepness index of the second section or the growth rate of the lesion level (S_n) characterizes the sensitivity of the processed plant tissue to damaging electrical effects. The steeper the change in the behavior of the dependence $S_n = f(W)$ in this area, the more susceptible the plant tissue of the weed, in our opinion, to electro pulse exposure. From the point of view of ongoing violations of the normal existence of intracellular structures of plant tissue processes, the limit value S_{nmax} is an indicator that determines the greatest damage to the treated tissue when exposed to pulses with the same dose of electrical power. From an energy point of view, when the limiting value of lesion level to the plant tissue is reached, the electrical effect on the plant should end, since further processing of the weeds does not increase the damage, but only leads to an excessive consumption of electrical power. Additionally, this quantity of electrical power should be considered a lethal or damaging dose of energy for a particular species and part (leafy part or root system) of the treated weed.

Equation (2) determines the proportional rise in the S_n value resulting in a damaging energy increase. The quantity of this energy supplied to the plant by one acting pulse is affected by the circuit capacitance value C_k and the initial voltage U_0 supplied to the processed object. Subsequent influencing pulses result in irreversible lesion accumulation and disruption of its normal functioning in the plant tissue, reaching a critical lesion level value. The speed of reaching the limit values of the lesion level S_{nmax} is affected by the power of one acting pulse.

Of practical interest is information on the acting energy doses that result in the maximum value of the plant tissue lesion level, since the identification of the values of these doses allows the evaluation of the total energy costs and the design of technological requirements for the studied technology. The total impact energy value and its specific indicator (ratio of total impact energy to the volume of the processed plant sample) are used for analysis.

To assess the energy impact, the number of influencing pulses m with the specific energy of one pulse W_{1pulse} must be considered. The W_{1pulse} values were obtained by experimental studies performed with weeds growing in the Lower Volga region; the results were processed and are presented in Table 3 [32].

Table 3. The values of the total impact of the specific electrical power on the plant tissue of various areas of weeds and the limiting lesion level to this tissue.

Weed Type	Weed Plant Area	The Limiting Lesion Level of Plant Tissue, *Snmax*, r.u.		Specific Processing Electric Energy *Wsp*, J/cm^3	
		By Areas	Average	By Areas	Average
Cirsium arvense	stem	5.6…7.4	5.4…7.0	2.4…8.1	2.7…10.9
	passage	5.2…6.3		2.8…10.3	
	root	5.3…7.3		2.9…14.3	
Sónchus arvénsis	stem	4.8…7.7	4.7…7.5	3.3…15.2	3.7…15.8
	root	4.6…7.3		4.1…16.4	
Euphórbia virgáta	stem	4.8…4.7	4.3…4.5	4.9…24.7	5.2…17.5
	passage	3.9…4.5		5.6…12.7	
	root	4.3…4.3		5.1…15.2	
Lactúca tatárica	stem	5.6…7.4	5.5…7.3	3.8…6.8	3.3…8.1
	passage	5.8…6.9		3.3…8.4	
	root	5.0…7.7		2.6…8.7	
Amaránthus retrofléxus)	stem	4.1…7.1	4.0…6.8	3.4…8.2	3.5…7.7
	passage	4.4…6.7		3.1…7.2	
	root	3.6…6.6		3.9…7.8	

The results presented in Table 3 were obtained during five years of research. The scatter of the results is caused by diverse parameters of discharge circuits for each unit: in one case, the processing was carried out at voltages of 3–7.5 kV and 5–12.5 kV, respectively; in addition, capacitance values of the capacitors' discharge circuit differ.

In the case when samples of weeds with different values of initial resistance were processed, the limiting lesion level rose with the increase in the applied initial voltage U_0 (Figure 5). At the same time, the dependence was quite clearly traced: with the rise in the energy of the acting pulses, the rate of reaching the limiting value of the lesion level also increased.

The character of the obtained experimental dependencies (Figure 6) fully confirmed the above hypothesis: assuming that the studied samples of weeds would have the same initial resistance (even despite the difference in their biometric parameters, and the difference in periods and phases of development), then in the process of electrical processing, their plant tissue could reach the same limiting lesion level. The difference would be observed only in the rate of increase in the lesion level: the less sensitive the plant tissue is to electrical action, the slower the achievement of the limit value of the lesion level occurs, and the greater the amount of energy consumed for this. A slightly increased energy consumption to achieve the limit value of the lesion level S_{nmax} can be explained by the fact that the studied samples of weeds were in the phase of development of the "beginning of fruiting" and therefore had a slightly higher resistance value than in the previous experiment [11,29].

The obtained results on the study of the dependence of irreversible lesion level on the plant tissue of weeds are quite closely comparable with similar works on the use of an electric pulsed field in the food and processing industry [9,16,33,34]. When studying the sensitivity of weeds, this kind of analysis has not yet been carried out, and therefore it is not possible to compare the obtained results with data from similar studies. In the case comparing it with the work on food technology, it can be concluded that the nature of the dependencies is identical, but it is not correct to compare the obtained quantitative indicators that affect the change in the tissue level lesions, since in these experiments the technological tasks were often the extraction of juice or the removal of excess liquid, and in the electrical treatment of weeds, the main task was the need to irreversibly lesion and lead to the death of unwanted plants.

Assuming that the studied samples of weeds would have the same initial resistance (even despite the difference in their biometric parameters, and the difference in periods and phases of development), then in the process of electrical treatment, their plant tissue could reach the same limit lesion level. The difference would be observed only in the rise of the lesion level: the less sensitive the plant tissue is to electrical action, the slower the achievement of the limit value of the lesion level occurs, and the more energy is spent for this.

5. Conclusions

The lesion level of the intracellular structures of the plant tissue of the root system and the leaf–stem part of weeds of various biological groups increases with the increase in the value of the electric field intensity in their tissue, which reaches its limiting (maximum) value at E = 3.1…3.8 kV/cm.

The greatest increase in the dependence $S_n = f(E)$ begins with the values of tension, for example, in the tissues of the root system of weeds: in *Euphorbia vine,* it is 2.2 kV/cm; in *Cirsium arvense,* it is 2, 0 kV/cm; and in *Amaránthus retrofléxus,* it is 2.9 kV/cm.

Exponent n characterizes the sensitivity of plant tissue of weeds to electrical damage. Having analyzed the values of the exponent n obtained by calculation on the basis of experimental results, it can be stated that the most sensitive plant tissue is for *Amaránthus retrofléxus* and *Oxýbasis úrbica*, and *Chenopódium úrbicum*: their average indexes are n = 1.28 and n = 1.29, respectively. The least sensitive plant tissues are in *Euphórbia virgáta*, n = 1.14. Plant tissues of *Euphorbia virgáta* are the most resistant to electrical damage, since the numerical values of the exponent n have the lowest values.

With the increase in the electrical power supplied to the weed plant, both by a single pulse and a series of pulses, the lesion level of the internal structures of the weed plant tissue increases. For reliable, irreversible damage to the intracellular structures of the plant tissues of the root system of the studied weeds, it is necessary to create an electric field intensity of at least 3.8 kV/cm in them during electric pulse treatment, and at least 3.7 kV/cm is needed to damage the aerial part. The plant tissue of the stem and root part of such weeds as *Cirsium arvense* and *Sónchus arvénsis*, as well as *Lactúca tatárica*, is most sensitive to electrical impacts, since the lesion level increases more intensively with each influencing pulse and reaches its limiting (maximum) value earlier than in plant tissues of the weed *Euphórbia virgáta* with equal energy impacts.

For reliable, irreversible damage to the intracellular structures of plant tissues of the studied weeds, it is necessary to spend at least 17.5 J/cm^3 of specific electrical power, and to achieve the value of the limiting (maximum) lesion level, at least 7.5 J/cm^3 is needed.

The speed of reaching the maximum value of the lesion level of the plant tissue of weeds depends on the voltage of the discharge circuit U_0—the larger this value, the faster the plant tissue reaches the limiting lesion level S_{nmax}.

The obtained results of the research allow us to discuss the clearly existing dependence of the irreversible lesion level to the plant tissue of weeds on the parameters of the electrical pulse impact. The sensitivity of the tissue to the effects of high-voltage electrical pulses on plants suggests that the implementation in practice of such a process of electrical processing of weeds will have a clear technological effect—the death of weeds and unwanted vegetation. At the same time, the specific values of electric energy and electric field intensity found to achieve the maximum lesion level in the plant tissue of weeds will allow us to evaluate the parameters of the technical device, such as the output parameters of the high-voltage pulse formation unit, and in general the parameters of the electric energy source.

The continuation of the study of the processes during electric impulse weed control of soil areas should be the coordination of the obtained specific indicators and parameters of the real electrical equipment of the electric pulse cultivator. At the same time, special attention should be paid to the study of the processes in case of lesions to weeds that occur when the unit and its electrode system are moved in relation to the weeds themselves in space and time; that is, to evaluate the dependence of the change in the lesion level to plant tissue on the parameters of electric pulse processing in dynamics.

The issue of using the obtained results for designing a control system for the modes and parameters of electric pulse processing deserves a separate discussion. The choice of the lesion level to the plant tissue of weeds, in this case, is the most acceptable option since, technically, it is most simple to implement a resistance measurement carried out at the same frequency but performed at different time intervals during processing to compare the result obtained with predetermined values of the limiting level processing of weeds prevailing on the soil massif, and form a control command to continue or stop the supply of high-voltage pulses to the system for supplying electrical energy to plants.

Author Contributions: Conceptualization, I.Y. and Y.D.; methodology, I.Y.; validation, Y.D., V.P. and V.B.; investigation, I.Y.; data curation, I.Y. and Y.D.; writing—original draft preparation, I.Y., V.P. and V.B.; writing—review and editing, I.Y.; visualization, V.B.; supervision, I.Y. All authors have read and agreed to the published version of the manuscript.

Funding: This research received no external funding.

Data Availability Statement: Data presented in this study are available on fair request to the corresponding author.

Conflicts of Interest: The authors declare no conflict of interest.

References

1. El-Metwally, I.M.; Saudy, H.S.; Elewa, T.A. Natural Plant By-Products and Mulching Materials to Suppress Weeds and Improve Sugar Beet (*Beta vulgaris* L.) Yield and Quality. *J. Soil Sci. Plant Nutr.* **2022**, *22*, 5217–5230. [CrossRef]
2. Saudy, H.S.; El–Metwally, I.M. Effect of Irrigation, Nitrogen Sources, and Metribuzin on Performance of Maize and Its Weeds. *Commun. Soil Sci. Plant Anal.* **2022**, *54*, 22–35. [CrossRef]
3. Ojelade, B.S.; Durowoju, O.S.; Adesoye, P.O.; Gibb, S.W.; Ekosse, G.-I. Review of Glyphosate-Based Herbicide and Aminomethylphosphonic Acid (AMPA): Environmental and Health Impacts. *Environ. Health Impacts. Appl. Sci.* **2022**, *12*, 8789. [CrossRef]

4. Baev, V.I.; Yudaev, I.V. Efficiency estimation of type of the electrical exposure on plants at their processing. *AD ALTA J. Interdiscip. Res.* **2018**, *8*, 252–257.
5. Baev, V.I.; Yudaev, I.V.; Petrukhin, V.A.; Baev, I.V.; Prokofyev, P.V.; Armianov, N.K. Electrotechnology as one of the most advanced branches in the agricultural production development. In *Handbook of Research on Renewable Energy and Electric Resources for Sustainable Rural Development*; Kharchenko, V., Vasant, P., Eds.; IGI Global: Hershey, PA, USA, 2018; pp. 283–310.
6. Vincent, C.; Panneton, B.; Fleurat-Lessard, F. *Electrical Weed Control: Theory and Applications, in Physical Control Methods in Plant Protection*; Springer: Berlin/Heidelberg, Germany, 2001; 321p.
7. Sahin, H.; Yalinkilic, M. Using Electric Current as a Weed Control Method. *Eur. J. Eng. Res. Sci.* **2017**, *2*, 59–64. [CrossRef]
8. Stankovic, M.V.; Cvijanovic, M.; Dukic, V. Ecological importance of electrical devices innovative in the process of anti *Ambrosia artemisiifolia* L. *Econ. Agric.* **2016**, 3, 861–870. [CrossRef]
9. Angersbach, A.; Heinz, V.; Knorr, D. Electrophysiological model of intact and processed plant tissues: Cell disintegration criteria. *Biotechnol. Prog.* **1999**, *15*, 753–762. [CrossRef]
10. Asavasanti, S.; Ristenpart, W.; Stroeve, P.; Barrett, D.M. Permeabilization of plant tissues by monopolar pulsed electric fields: Effect of frequency. *J. Food Sci.* **2011**, *76*, 98–111. [CrossRef] [PubMed]
11. Knorr, D.; Angersbach, A. Impact of High Electric Field Pulses on Plant Membrane Permeabilization. *Trends Food Sci. Technol.* **1998**, *9*, 185–191. [CrossRef]
12. Asavasanti, S.; Ersus, S.; Ristenpart, W.; Stroeve, P.; Barrett, D.M. Critical electric field strengths of onion tissues treated by Pulsed electric fields. *J. Food Sci.* **2010**, *75*, 433–443. [CrossRef]
13. Lebovka, N.I.; Bazhal, M.I.; Vorobiev, E.I. Estimation of characteristic damage time of food materials in pulsed-electric fields. *J. Food Eng.* **2002**, *54*, 337–346. [CrossRef]
14. Mohamed, M.E.; Eissa, A.H.A. Pulsed Electric Fields for Food Processing Technology. In *Structure and Function of Food Engineering*; IntechOpen: London, UK, 2012; pp. 275–306.
15. Bajgai, T.R.; Hashinaga, F. High electric field drying of Japanese radish. *Dry. Technol.* **2001**, *19*, 2291–2302. [CrossRef]
16. Barba, F.J.; Parniakov, O.; Pereira, S.A.; Wiktor, A.; Grimi, N.; Boussetta, N.; Saraiva, J.A.; Raso, J.; Martin-Belloso, O.; Witrowa-Rajchert, D.; et al. Current applications and new opportunities for the use of pulsed electric fields in food science and industry. *Food Res. Int.* **2015**, *77*, 773–798. [CrossRef]
17. Picart, L.; Cheftel, J.C. Pulsed electric fields. In *Food Preservation Techniques*; Zeuthen, P., Bogh-Sorensen, L., Eds.; CRC Press: Boca Raton, FL, USA, 2003; pp. 57–68.
18. Visockis, M.; Bobinaitė, R.; Ruzgys, P.; Barakauskas, J.; Markevičius, V.; Viškelis, P.; Šatkauskas, S. Assessment of plant tissue disintegration degree and its related implications in the pulsed electric field (PEF)—Assisted aqueous extraction of betalains from the fresh red beetroot. *Innov. Food Sci. Emerg. Technol.* **2021**, *73*, 1–10. [CrossRef]
19. Zderic, A.; Zondervan, E.; Meuldijk, J. Breakage of Cellular Tissue by Pulsed Electric Field: Extraction of Polyphenols from Fresh Tea Leaves. *Chem. Eng. Trans.* **2013**, *32*, 1795–1800.
20. Wiktor, A.; Sledz, M.; Nowacka, M.; Rybak, K.; Chudoba, T.; Lojkowski, W.; Witrowa-Rajchert, D. The impact of pulsed electric field treatment on selected bioactive compound content and color of plant tissue. *Innov. Food Sci. Emerg. Technol.* **2015**, *30*, 69–78. [CrossRef]
21. Barbosa Canovas, G.V.; Zhang, H.Q. (Eds.) *Pulsed Electric Fields in Food Processing: Fundamental Aspects and Applications*; CRC Press: Boca Raton, FL, USA, 2019; 289p.
22. Yudaev, I.; Kokurin, R.; Daus, Y.; Ivanov, D. Automatizing process of electropulse treatment of plant raw materials. *IOP Conf. Ser. Earth Environ. Sci.* **2022**, *996*, 012022. [CrossRef]
23. Mañas, P.; Barsotti, L.; Cheftel, J.C. Microbial inactivation by pulsed electric fields in a batch treatment chamber. Effects of some electrical parameters and food constituents. *Innov. Food Sci. Emerg. Technol.* **2000**, 2, 239–249. [CrossRef]
24. Yudaev, I.V.; Daus, Y.V.; Kokurin, R.G. Substantiation of criteria and methods for estimating efficiency of the electric pulse process of plant material. *IOP Conf. Ser. Earth Environ. Sci.* **2020**, *488*, 012055. [CrossRef]
25. Lebovka, N.I.; Praporscic, I.; Vorobiev, E.I. Enhanced expression of juice from soft vegetable tissues by pulsed electric fields: Consolidation stages analysis. *J. Food Eng.* **2003**, *59*, 309–317. [CrossRef]
26. Lebovka, N.I.; Shynkaryk, N.V.; Vorobiev, E. Pulsed electric field enhanced drying of potato tissue. *J. Food Eng.* **2007**, *78*, 606–613. [CrossRef]
27. De Vito, F.; Ferrari, G.; Lebovka, N.I.; Shynkaryk, N.V.; Vorobiev, E. Pulse duration and efficiency of soft cellular tissue disintegration by pulsed electric fields. *Food Bioprocess Technol* **2008**, *1*, 307–313. [CrossRef]
28. Raso, J.; Heinz, V.; Alvarez, I.; Toepfl, S. *Pulsed Electric Fields Technology for the Food Industry: Fundamentals and Applications*; Springer: Berlin/Heidelberg, Germany, 2021; 561p.
29. Vorobiev, E.; Lebovka, N. Pulsed-electric fields induced effects in plan tissues: Fundamental aspects and perspectives of applications. In *Electrotechnologies for Extraction from Food Plants and Biomaterials*; Springer: New York, NY, USA, 2009; pp. 39–81.
30. Yudaev, I.V.; Baev, V.I.; Brenina, T.P.; Eliseev, D.S. *Sornyerasteniyakakob"ektelektricheskoipropolki: Biologicheskieosobennosti i elektrofizicheskiesvoistva (Weeds as an Object of Electric Weeding: Biological Features and Electro-Physical Properties)*; Stanitsa-2: Volgograd, Russia, 2004.
31. Yudaev, I.V. Analysis of variation in circuit parameters for substitution of weed plant tissue under electric pulse action. *Surf. Eng. Appl. Electrochem.* **2019**, *55*, 219–224. [CrossRef]

32. Yudaev, I.; Daus, Y.; Kokurin, R.; Prokofyev, P.; Gamaga, V.; Armianov, N. Electrotechnology as one of the most advanced branches in the agricultural production development. In *Advanced Agro-Engineering Technologies for Rural Business Development*; Kharchenko, V., Vasant, P., Eds.; IGI Global: Hershey, PA, USA, 2019; pp. 396–430.
33. Bauer, M.V.; Marx, C.; Bauer, F.V.; Flury, D.M.; Ripken, T.; Streit, B. Thermal weed control technologies for conservation agriculture—A review. *Weed Res.* **2020**, *60*, 241–250. [CrossRef]
34. Hill, K.; Ostermeier, R.; Töpfl, S.; Heinz, V. Pulsed Electric Fields in the Potato Industry. In *Pulsed Electric Fields Technology for the Food Industry*; Raso, J., Heinz, V., Alvarez, I., Toepfl, S., Eds.; Food Engineering Series; Springer: Cham, Switzerland, 2022.

agriculture

Article

High-Efficiency Photovoltaic Equipment for Agriculture Power Supply

Olga Shepovalova *, Andrey Izmailov, Yakov Lobachevsky and Alexey Dorokhov

Federal Scientific Agroengineering Center VIM, 1-St Institutskiy Proezd, 5, 109428 Moscow, Russia
* Correspondence: shepovalovaolga@mail.ru; Tel.: +7-906-092-95-60

Abstract: Developing an energy supply based on resources whose use does not spoil the noosphere and the creation of such energy supply of efficient equipment whose operation does not cause any damage to nature and man is an urgent task. The need for such an approach is especially relevant and noticeable in agriculture. This article presents the final results of complex studies of new PV devices and PV systems based on them. Considered in the article are the best solutions we propose to improve PV equipment and make it more attractive for agricultural consumers. The developed vertical and planar high-voltage multijunction silicon PV cells and PV modules on their basis are presented. The first type of modules have a maximum power point voltage of up to 1000 V, specific power of up to 0.245 ± 0.01 W/cm^2, and efficiency of up to 25.3% under a concentration ratio range of 10–100 suns. The samples of the second module type (60,156.75 × 156.75 mm PV cells) have an open-circuit voltage of 439.7 V, a short-circuit current of 0.933 A, and a maximum power of 348 W. Additionally, two types of newly designed solar energy concentrators are described in this article: one-dimensional double-wing concentrator ensuring low Fresnel optical losses and multi-zone parabolotoric microconcentrator with the uniform radiation distribution in the focal region, as well as modules based on these concentrators and the developed PV cells. For PV modules, the maximum power degradation is 0.2–0.24% per year in a wet ammonia environment. For concentrating PV modules, this degradation is 0.22–0.37% per year. This article sets out the principles of increasing the efficiency of PV systems by increasing the level of systematization and expanding the boundaries of PV systems. The thus-created PV systems satisfy 30–50% more consumer needs. Thanks to a higher output voltage and other specific features of the developed modules, PV system loss decreased by 12–15%, and maintenance losses also decreased.

Keywords: green energy supply; agricultural electrification; high-efficiency photovoltaic equipment; complex energy supply systems

Citation: Shepovalova, O.; Izmailov, A.; Lobachevsky, Y.; Dorokhov, A. High-Efficiency Photovoltaic Equipment for Agriculture Power Supply. *Agriculture* **2023**, *13*, 1234. https://doi.org/10.3390/agriculture13061234

Academic Editor: Massimo Cecchini

Received: 19 December 2022
Revised: 22 May 2023
Accepted: 25 May 2023
Published: 12 June 2023

1. Introduction

One of the main ways to overcome the negative impacts of modern industrial civilization is the transition to maximum harmonization with nature, which is especially topical and noticeable in the countryside [1]. The energy supply based on the resources whose use does not spoil the noosphere and the creation of such energy supply of efficient equipment whose operation does not cause any damage to nature and man is one of the mechanisms of environmental restoration. The resource for this is, first of all, solar energy [2–6].

The transition to the use of renewable energy resources is the most obvious and justifiable in agriculture. Photovoltaic (PV) systems are one of the best equipment options for specified purposes, especially considering the development of building-integrated photovoltaics (BIPV) and agrivoltaics (AgriPV).

For agriculture and rural areas, the use of photovoltaic solar energy conversion is especially topical because of the following factors [7–9]:

(1) Agriculture has the most opportunities for implementing the optimal variants of PV systems;
(2) Rural territories generally suffer from a greater constant shortage of electric power;
(3) In power supply to agricultural consumers, the inefficiency of improperly designed systems is more obvious, and such systems have a more substantial impact on life activity.

At the same time, the power supply to agriculture imposes special requirements for PV equipment, namely requirements associated with the impact of aggressive environments, e.g., ammonia, thermal and humidity conditions, etc., as well as special requirements for safety, ecological compatibility, and aesthetics. Agriculture is more sensitive to changes (increases) in the equipment cost.

In order to implement power supply in rural areas, first of all, through the application of solar energy resources and expansion of the scale of agricultural power supply with the use of PV systems, it is necessary to increase the efficiency of PV systems and make them more attractive for consumers. At present, the solutions to this problem are as follows:

(1) Efficiency enhancement of solar energy conversion is achieved through improvement in the design of PV devices and manufacturing technology (i.e., the efficiency of PV cells, concentrators, and modules), reduction in the degradation of PV devices' characteristics over time, and the enhancement of devices' resistance to environmental factors. This ensures an increase in the energy output and a minimization of the occupied space [10–14].
(2) A combination of functions (i.e., BIPV, AgriPV, etc.) ensures that there will be no need for a separate space on the land's surface for placing PV modules, in addition to cost reduction and improved aesthetics [15,16].
(3) An increase in the level of systematization, expansion of the boundaries of PV systems, and their development as complex systems ensure a reduction in losses and a higher level of consumer satisfaction [17].
(4) The placement of PV systems at the consumption site ensures the organization of energy supply based on an analogy with nature, a reduction in costs, and additional opportunities for the implementation of the first three factors mentioned above [7].
(5) An increase in the diversification of efficient PV equipment for various operational conditions enhances the PV systems' versatility and expands their application possibilities.

The overall purpose of our many years of research is to increase PV equipment efficiency at all levels [18], including PV cells, PV modules, and PV systems, thus considerably increasing the possibilities of using PV systems in agriculture.

At present, more than 90% of industrial-grade PV cells are PV cells fabricated on a silicon wafer with a *p–n* junction unit, which is due to their substantial advantages over any other types of PV cells (e.g., relatively low cost, accessibility and abundance of silicon, acceptable output characteristics and properties of such PV cells, and well-proven technology) [10,11,13,19].

Homogeneous semiconductor PV cells on the basis of just one *p–n* junction can only generate voltages limited by the potential barrier height on the junction. For silicon and gallium arsenide (i.e., the best technologically investigated PV materials), this voltage is approximately 0.6 V to 0.9 V under nonconcentrated solar radiation. At the same time, most typical electronic devices require considerably greater voltage levels for their operation. When using single-junction PV cells for obtaining the required values of voltage, one has to connect a sufficiently large number of cells/modules in series, which results in power loss due to contact resistance and non–uniformity in the parameters of individual cells/modules. Additionally, it leads to a reduction in efficiency and sturdiness against external factors. The typical planar crystalline silicon PV modules (60 mono-Si PV cells with single *p–n* junction and with a size of 156.75 × 156.75 mm) have an open-circuit voltage ($V_{o.c}$) of 44 V and a short-circuit current ($I_{s.c}$) of 9.5 A, and their voltage and current at the maximum power point are 36.3 V and 8.8 A, respectively. The required voltage and power values of a PV array based on traditional PV modules are usually provided at sufficiently high DC values. Reducing these values can significantly reduce ohmic losses and therefore will both increase the efficiency and reliability of PV systems and reduce costs. Moreover, such PV modules occupy quite a great area [10–13,19–22]. However, in agriculture, limitations generally exist for the placement of PV modules and, as mentioned above, have more stringent requirements for immunity to environmental factors and cost.

Based on this, the first task of our research was to develop high-voltage silicon PV cells that would enable a significant increase in the output voltage of PV equipment and overcome other disadvantages of conventional PV modules that limit the PV equipment used for agriculture power supply.

Earlier on, research was pursued to develop high-voltage vertical multi-junction Si PV cells for space applications [23]. These cells had acceptable efficiency only when the concentration was 1000 suns or higher, and they were not convenient for terrestrial applications. More recent studies also refer to cells operating under high concentrations (efficiency of up to 19% under 2480 suns), which have also not been applied in terrestrial photovoltaics [24]. Terrestrial high-voltage vertical multi-junction Si PV cells suitable for mass application were, for the first time, proposed and developed in the course of the studies summarized in this paper.

Considering the fact that mass technologies are oriented toward the production of planar PV modules and also that the components of PV systems (mass design solutions), such as PV plants, BIPV, and AgriPV, are designed on the basis of the conventional planar PV modules, planar high-voltage silicon PV cells are required. Making such cells was a new idea. Our ideas and theoretical studies were presented in [25–30]. Besides increasing the output voltage, these cells had to accomplish the objective of creating high-voltage PV modules with dimensions matching the standard dimensions of the mass-produced crystalline Si PV modules. Additionally, such modules should be manufactured with the maximum use of existing equipment.

The higher the concentration of the solar energy arriving at radiation receivers (RRs) is, the higher the photoelectric conversion efficiency is. The use of concentrators enables an increase in the irradiance of the active surface of PV cells and therefore increases the efficiency of PV modules/PV systems [11,14,31,32]. A review of the current state of concentrator applications in photovoltaics is presented in [32–36]. Two types of solar radiation concentrators were of interest in terms of developing highly efficient concentrator photovoltaic (CPV) modules: focon-type concentrators and concentrators containing a symmetrical reflecting surface (RS) fabricated as a chute (compound parabolic concentrators). Concentrators of these types have the following important advantages, compared with other types of concentrators: (1) a long daily operation period within the parametric angle of the concentrator without tracking the sun; (2) the possibility of using diffuse radiation, thus increasing the amount of the total irradiance coming to the concentrator aperture, by up to 20%. The mirror RS on the current technological horizon may have a reflection coefficient of 99% or more, which sets this mirror concentrator apart from lens-based ones.

The main disadvantage of concentrators manufactured in the form of solids of revolution, including focon types, and other toroids, as well as concentrators based on Fresnel lenses, is a nonuniform distribution of irradiance intensity over the active surface of RRs. This results in a decrease in efficiency because PV converters are most efficient when irradiance is uniform. Compound parabolic concentrators have uniform radiation density in the rectangular focal region. The main problem of such concentrators is that the rays reflected from the part of RS located close to the focal plane fall on the active surface of RRs at high angles. This results in a sharp increase in the coefficient of radiation reflected from the active surface of RRs (Fresnel losses) [14,36–39].

Moreover, all of the operated efficient PV systems based on CPV modules are bulky and have incomparably larger dimensions and weights than those based on traditional planar PV modules. They require complex and insufficiently reliable tracking systems. This, to a large extent, brings to naught their advantage of a higher performance factor. In addition, naturally, this makes them hardly applicable in agricultural power supply systems.

Free from these defects are CPV modules based on microconcentrators whose dimensions are comparable to those of conventional planar PV modules. The best results in implementing microconcentrators are presented in [40–42]. There is a considerable problem associated with the unevenness of RR illumination while using mirror concentrators, which may be reduced by optimizing the form of RS, and there is also a problem of inherent losses caused by absorption when Fresnel lenses are used [36,39].

In certain situations, the use of toroidal concentrators is the most effective. In [43], we presented a series of newly developed semitoroidal concentrators.

Another task of our research was to remove/minimize the previously mentioned defects of focon-type concentrators and compound parabolic concentrators and to design efficient CPV modules on the basis of the developed concentrators and high-voltage multi-junction cells.

The aim of this article was to present the final generic research results that yielded real practical outcomes, starting from scientific ideas to operating devices and installations. Schematically, the reviewed studies are outlined in Figure 1. The results presented in the paper include two types of high-voltage PV cells (Section 2), optimized solar concentrator structures (Section 3), and four types of PV modules on their basis (Sections 2 and 3). As well, this article presents research on the development of more efficient PV systems by increasing the level of systematization and expanding the boundaries of PV systems. The final efficiency of such systems is evaluated by the degree of consumer need satisfaction (Section 4). To date, the results of our theoretical studies related to PV cells and concentrators have mainly been published. The research results of fabricated PV cells, concentrators, and modules on their basis, are published for the first time in this article. For the time being, quite a high correlation between theoretical calculations and the operating characteristics of real instruments was achieved, which enabled us to present the relevant results in this manuscript.

When proposing the ideas related to the development of PV devices (Sections 2 and 3), along with efficiency improvement, a task was set to minimize the costs at the stage of the introduction of devices into production (by minimizing changes in the manufacturing technology) and at the stage of the creation of PV systems based on the developed PV and CPV modules (by minimizing changes in the technology and design of PV systems and mating structures). These aspects were indispensably taken into account when analyzing and choosing the optimal options. The solutions to this task included the choice of silicon as the basic material for PV cells, the creation of planar high-voltage cells whose dimensions correspond to those of conventional crystalline silicon PV cells, and CPV modules whose dimensions correspond to those of traditional planar PV modules.

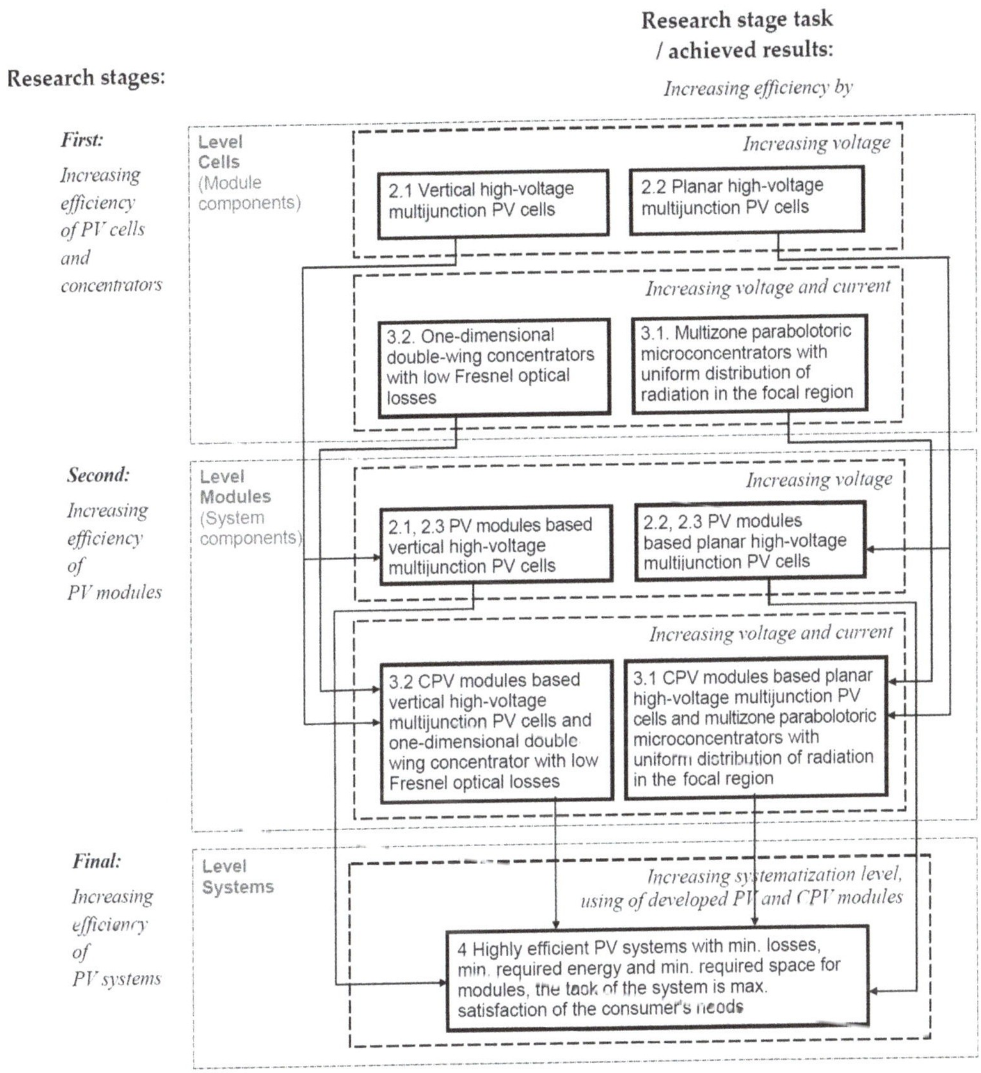

Figure 1. General scheme of research (numbers correspond to the section/subsection number in this article).

2. Increasing Energy Production by Increasing Voltage—Improvement in the Design of Photoelectric Cells

2.1. Vertical High-Voltage Multi-Junction Silicon PV Cells

Three technologies have been used to form multi-layer structures: soldering unit cells using solder plugs with diffused layers, epitaxy, and breakdown [44]. Currently, PV cells are manufactured via plug soldering with subsequent cutting into size-defined cells and the application of antireflection coating with nanoclusters to the end face (active surface) [45,46]. Figure 2a,b show the structure of a typical vertical high-voltage PV cell and a PV module composed of these cells. As one can see from Figure 2, both the cells and modules of this design are much more space-saving than those manufactured using traditional planar technology. The cell voltage is equal to the sum of the voltages of microcells (0.58–0.62 V per microcell). The efficiency achieved for the designed cells under single-fold radiation is comparable to that of conventional silicon PV cells.

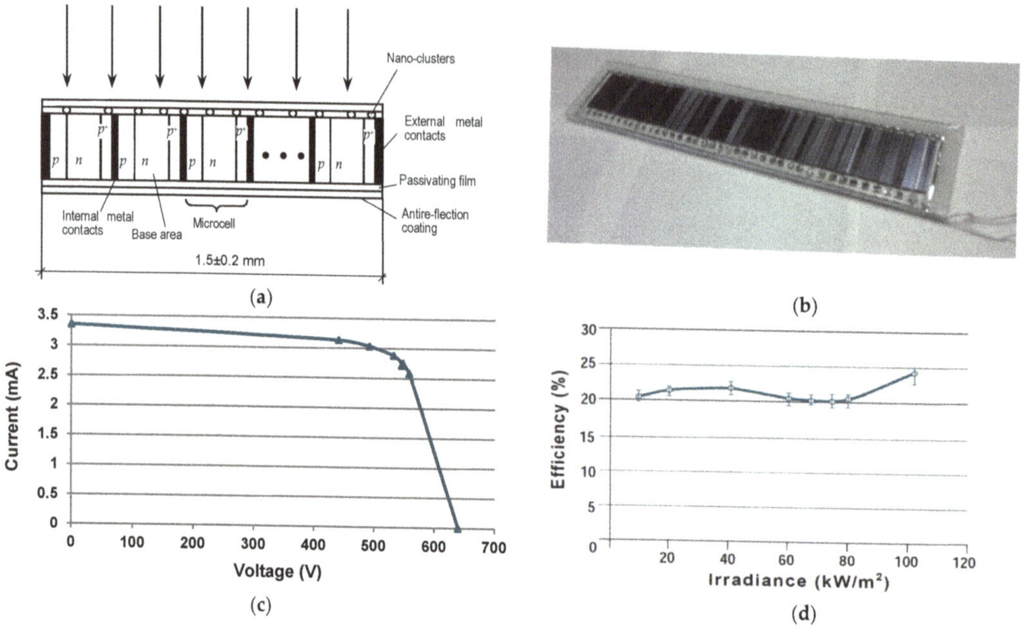

Figure 2. (**a**) PV cell structure; (**b**) PV module based on vertical multi-junction PV cells, 40 60 × 10 mm PV cells; (**c**) *I–V* curve of developed PV modules, 10 suns; (**d**) dependence of efficiency on irradiance.

PV cells have universality in relation to the radiation level, i.e., in contrast to traditional silicon planar PV cells, their voltage does not decrease as concentration increases. Moreover, they have low sensitivity to the radiation incidence angle.

PV modules based on vertical high-voltage multi-junction silicon PV cells have a voltage rate of up to 1000 ± 15 V at the maximum power point. For solar radiation concentration ratios ranging from 10 to 100 suns, the specific power of such modules is up to 0.245 ± 0.01 W/cm^2, and their efficiency is up to 25.3%. The modules are encapsulated with polysiloxane gels (see Section 2.3), which facilitates the extension of their expected service life up to about 50 years. The operating range of their ambient temperature is from –60 to 110 °C. The module efficiency, which is 18–22%, remains the same when the temperature rises up to 60 °C, which simplifies their cooling system while working with concentrators. The production process does not require the use of silver, screen printing, photolithography, or any other time-consuming procedure or expensive materials. Figure 2c shows the *I–V* curve of the developed PV module. The test conditions were as follows: direct normal irradiance of 10 × 1000 W/m^2, cell temperature of 25 °C, and spectral irradiance distribution of direct normal AM1.5. Figure 2d shows the PV module's efficiency dependence on the level of active surface irradiance [46].

The operating voltage of 1000 ± 15 V enables the use of these PV modules with transformerless inverters and connecting them to high-voltage direct-current 110–500 kV power lines without the need for converting substations. To obtain the operating voltage of 1000 V with the use of conventional planar PV modules, it would be necessary to connect more than 1800 150 × 150 mm planar PV cells in series, in which case the total length of the modules would exceed 300 m. It is also important that the operating voltage of one module can be equal to the accepted nominal DC voltage of large photovoltaic systems (system voltage), for example, solar power plants such as AgriPV systems.

The main problem involves the reproducibility of characteristics (wide spread of parameter values). Thus far, it has been impossible to achieve parameter values in the required narrow interval with the change from batch to batch.

2.2. Planar High-Voltage Multi-Junction Silicon PV Cells

The currently developed planar high-voltage silicon PV cells are multi-junction homogeneous PV structures [25–30]. The doped layers form planar diodes $n^+–p–p^+$ ($p^+–n–n^+$; $n–p–p^+$; $p–n–n^+$) or $n–p$ microstructures (unit cells) connected in series in the direction of radiation propagation. Figure 3 shows the design options of planar high-voltage PV cells, while Table 1 presents the advantages, design specificity, and technological features borrowed from PV cells of various types in order to develop planar high-voltage multi-junction PV cells.

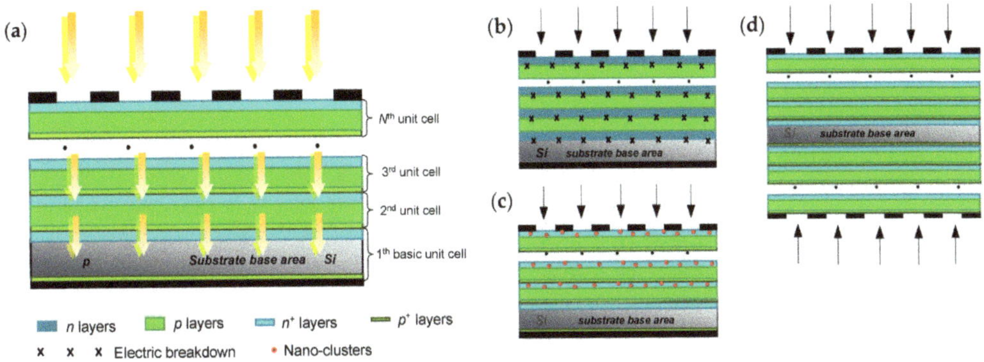

Figure 3. (**a**) High-voltage planar PV cells with basic option; (**b**) structure manufactured using the electric breakdown technology; (**c**) structure with nanoclusters; (**d**) bifacial structure.

Table 1. Advantages, specificity of design, and technologies borrowed from various types of PV cells to develop high-voltage planar multi-junction PV cells.

PV Cell Type	Advantages, Specificity of Design, and Technology
Conventional planar	Dimensions Basic complete processing lines Consequent designs Preferable direction of radiation adsorption
Silicon multi-junction	High output voltage Ability to operate under concentrated solar radiation Low electric losses
III-V-based multi-junction	Multilayer design Layer preparation technology, Interconnection via tunnel junctions Layer thickness
Silicon-based thin film	Layer thickness

The small (≤10 μm) thickness of layers and, consequently, unit cells make them practically transparent considering solar radiation. Each unit cell receives radiation that consecutively passed through the previous semiconductor layers. The thickness of the base unit cell area does not exceed the diffusion length of minority charge carriers.

The theoretical and most of the experimental parts of this research were carried out using the basic option of the design, as shown in Figure 3a. The basic option of PV cells was fabricated through the use of epitaxy by transferring epitaxial layers to the basic unit cell (silicon wafer with a $p–n$ junction formed via diffusion). The first basic unit cell ensures the mechanical strength of the entire structure. As the p^+ and n^+ layers are heavily doped, a conductive contact occurs between them owing to the quantum-mechanical effect of tunneling through the $p^+–n^+$ junction. The dimensions of PV cells are determined by the dimensions of the wafer and correspond to the dimensions of traditional crystalline Si PV cells.

The thickness of p, n, and n^+ layers is from 12 nm to 9.8 μm, while that of p^+-type layers is from 10 nm to 120 nm. The concentration of active impurity in thin heavily doped p^+-type layers is more than 10^{19} cm^{-3}. In general, the thickness of layers increases with the distance from the active surfaces, which enables the production of PV cells with the required output parameters and their nonsignificant distribution with a wide tolerance for layer identity. The number of layers is determined based on equipment capability and the specifics of the layer formation process, as well as the expediency of using solar cells with a specific number of layers in the PV modules.

To avoid losses in the circuit, each unit cell must operate under radiation at its optimal point of the volt–ampere characteristic, which means the equality of the photogenerated currents flowing through the unit cells:

$$\begin{cases} \int\limits_0^\infty \frac{d\Phi}{d\omega} \cdot e^{-\alpha \sum\limits_{k=i+2}^{N} d_k} \cdot \left(1 - e^{-\alpha d_{i+1}}\right) \cdot d\omega = \int\limits_0^\infty \frac{d\Phi}{d\omega} \cdot e^{-\alpha \sum\limits_{k=i+1}^{N} d_k} \cdot \left(1 - e^{-\alpha d_i}\right) \cdot d\omega; \; i = 2,3,\ldots,N-1 \\ \int\limits_0^\infty \frac{d\Phi}{d\omega} \cdot e^{-\alpha \sum\limits_{k=i+1}^{N} d_k} \cdot \left(1 - e^{-\alpha d_i}\right) \cdot d\omega = \int\limits_0^\infty \frac{d\Phi}{d\omega} \cdot e^{-\alpha \sum\limits_{k=2}^{N} d_k} \cdot Q_i(\omega) \cdot d\omega; \quad i = 2,3,\ldots,N \end{cases} \tag{1}$$

where Φ is the incident photon flux; α is the radiation absorption coefficient depending on photon frequency ω or wavelength λ and on the nature of the semiconductor; d_k and d_i are the thickness of the k-th or i-th unit cell, respectively; $Q_i(\omega)$ is the spectral charge carriers collection efficiency for the p–n junction in the i-th unit cell; and N is the number of unit cells in an entire PV cell.

The requirement of Equation (1) is met by determining the optimal values of the thickness of each unit cell as follows:

$$d_i = \frac{1}{\alpha} \ln \frac{1 - e^{-2\alpha\delta} + Q_i e^{-\alpha\delta}\left(1 - e^{-2\alpha\delta(i-1)}\right)}{1 - e^{-2\alpha\delta} + Q_i e^{-\alpha\delta}\left(1 - e^{-2\alpha\delta(i-2)}\right)}; \; i = 2,3,\ldots,N \tag{2}$$

where $\alpha\delta$ is the complex optical–technological parameter of the PV cell.

The density of the photocurrent generated in the i-th unit cell is subject to light absorption in the previous layers.

$$J_{\text{ph}i} = q\Phi_0 S_i(\alpha, \xi); \; i = 1, 2, \ldots, N \tag{3}$$

where q is the absolute value of the electron's charge, Φ_0 is the incident photon flux density, S_i is the spectral charge carriers collection efficiency (photoresponse) of the i-th unit cell, and ξ is the angle of radiation incidence onto the active surface.

Volt–ampere characteristic is expressed as follows:

$$V = \frac{AkT}{q} \sum_{i=1}^{N} \ln\left(\frac{\left(\frac{q\Phi_0 \cdot Q_i}{1+(N-1)\cdot Q_i} - J_{\text{ph}}\right)}{J_{0i}} + 1 \right) \tag{4}$$

where A is the parameter of characteristic curvature, k is the Boltzmann constant, T is the cell temperature, and J_{0i} is the reverse dark current density in the i-th unit cell.

The collection efficiencies of spectral charge carriers of unit cells are determined as follows:

$$\begin{gathered} S_1(\alpha, \xi) = S_b(\alpha, \xi) = \cos\xi \cdot e^{-\frac{\alpha}{\cos\vartheta}\left(\sum\limits_{k=2}^{N} d_k + 2\delta(N-1)\right)} \cdot Q_b(\alpha, \vartheta) \\ S_i(\alpha, \xi) = \cos\xi \cdot e^{-\frac{\alpha}{\cos\vartheta}\left(\sum\limits_{k=i+1}^{N} d_k + 2\delta(N-i) + \delta\right)} \left(1 - e^{-\frac{\alpha}{\cos\vartheta} d_i}\right); \; i = 2, \ldots, N-1 \\ S_N(\alpha, \xi) = \cos\xi \cdot e^{-\frac{\alpha}{\cos\vartheta}\delta} \cdot \left(1 - e^{-\frac{\alpha}{\cos\vartheta} d_N}\right) \end{gathered} \tag{5}$$

where Q_b is the collection efficiency of the charge carriers in the basic cell, ϑ is the refraction angle in a semiconductor structure, and δ is the thickness of the tunnel layer in the unit cells (the difference between the tunnel layers of unit cells can be neglected).

Angle ϑ is determined using the relation $\sin\xi = n \cdot \sin\vartheta$, where n is the radiation refraction index in the semiconductor. Angle ξ is equal to 0° when the rays are perpendicular to the active surface. The reverse dark current density in the i-th unit cell, given the uniform technology used in the fabrication of unit cells, does not depend on the unit cell number, i.e., $J_{0i} = J_0$, as evident from Equation (1).

The plots showing the main characteristics of the developed high-voltage multi-junction PV cell with the basic structure (see Figure 3a) and different unit cell numbers are presented in Figure 4. The characteristics of the final experimental samples quite accurately correspond to the results of computer simulations and calculations.

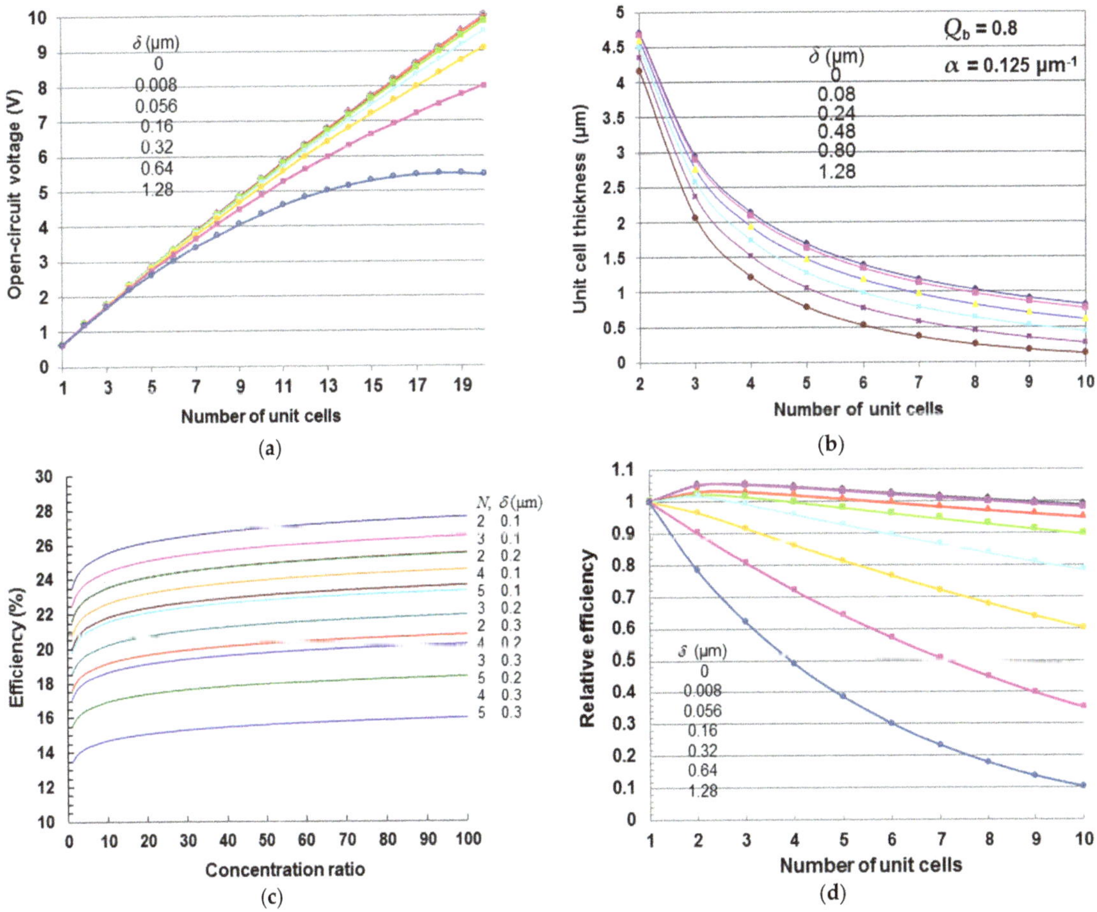

Figure 4. (**a**) Open-circuit voltage of high-voltage planar PV cells under solar radiation, $a = 0.125\ \mu m^{-1}$, A = 2, $J_0 = 10^{-7}$ A/cm^2, and $q\Phi = 45$ mA/cm^2; (**b**) dependence of the optimum unit cell thickness on it number; (**c**) dependence of efficiency on solar radiation concentration; (**d**) relative efficiency, $a = 0.125\ \mu m^{-1}$, A = 2, $J_0 = 10^{-7}$ A/cm^2, and $q\Phi = 45$ mA/cm^2.

For the experiments, PV cells (see Figure 3a) were originally fabricated using the basic design option with the number of unit cells N from 1 (only the basic unit cell without epitaxy layers) to 20 (see Figure 4a). For technological reasons, more detailed research was carried out to determine the characteristics of unit cells using samples with $N = 2$, $N = 5$, and $N = 10$.

The dependence of the open-circuit voltage of PV cells on N is shown in Figure 4a, for various values of tunnel layer thickness δ for $\alpha = 0.125\ \mu m^{-1}$, $Q_b = 0.8$, a radiation intensity equal to that of solar radiation ($q\Phi = 45\ mA/cm^2$), and for other parameters corresponding to the recombination current mechanism (A = 2; $J_0 = 10^{-7}\ A/cm^2$). The open-circuit voltage increases monotonically with N (growing almost linearly for small values of N), weakly depending on the values of J_0 and Q. In the asymptotic limit, for $N \cdot Q >> 1$, it becomes dependent on N, to an extent that is weaker than the linear law and does not change for any value of Q.

As seen from Equation (2), the optimal thickness d_i decreases in inverse proportion to the radiation absorption coefficient α, logarithmically reduces as the unit cell number i increases, and increases as the collection efficiency of charge carriers Q_b increases in the first basic unity cell. The dependence of the optimal thickness of a unit cell on its number for various values of δ is shown in Figure 4b. PV cells may operate under concentrated radiation without a noticeable decrease in voltage, and they have low sensitivity to the solar radiation incidence angle. The efficiency of PV cells does not exceed the maximum efficiency of the basic unit cell only for small values of the tunnel layer thickness δ. For a given value of radiation absorption coefficient α, this thickness, layer number N, and the relevant value of open-circuit voltage must not exceed the corresponding limiting values. For example, for $\alpha = 0.125\ \mu m^{-1}$, there must be $\delta \leq 0.16\ \mu m$, $N \leq 9$, and $V_{o.c} \leq 4.5$ V. A sharp drop in PV cell efficiency was observed if $\delta \geq 0.32\ \mu m$.

Experimental samples of PV modules were manufactured from 60 PV cells with a size of 156.75 × 156.75 mm with $N = 10$. The PV cells were connected in a series–parallel comprising two strings of 30 cells each or in series. The modules were encapsulated using polysiloxane gels (see Section 2.3). Figure 5a shows the module during testing. Additionally, Figure 5b shows the averaged *I–V* curve of the developed modules under the standard test conditions (STC) (cell temperature of 25 °C, irradiance of 1 sun, i.e., 1000 W/m^2, and spectral irradiance distribution of AM1.5).

(a)

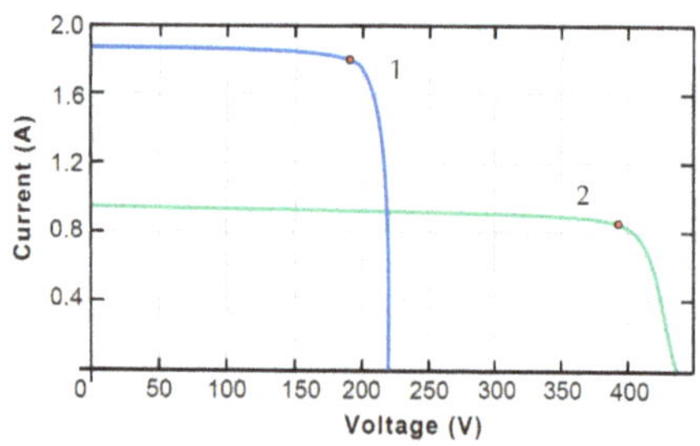

(b)

Figure 5. (**a**) PV module with planar high-voltage multi-junction silicon PV cells ($N = 10$) on the test bench; (**b**) *I–V* curve of the developed modules at STC, $N = 10$ (curve 1—two strings of 30 cells each and curve 2—1 strings of 60 cells).

2.3. *Performance Deterioration Reduction and External Impact Hardening—Improvement in PV Module Manufacturing Technology*

Planar crystalline Si terrestrial PV modules manufactured using the conventional lamination technology lose up to 15–20% of their maximum power at STC after 20 years of operation in tropical climates and after 25 years of operation in temperate climates. One of the reasons for it is the degradation of optical polymeric encapsulants such as ethylene vinyl acetate (EVA) and other plastics induced by ultraviolet radiation and high temperatures [47].

High-voltage cell modules considered in Sections 2.1 and 2.2 were manufactured using polysiloxane gels instead of conventional EVA as the filling material [46,48]. The replacement of the standard encapsulant with EVA and the conventional lamination process with a silicon composition that is poured as liquid and then hardens and turns into polysiloxane gels considerably slows down the performance deterioration of such modules in time. The process reaction is as follows:

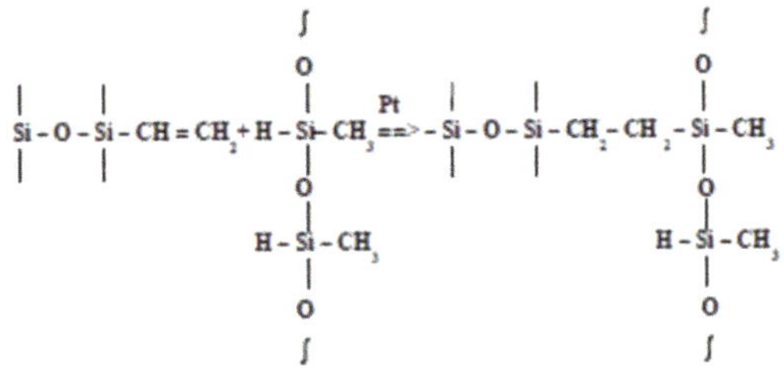

According to the laboratory test data (tests in compliance with requirements of IEC 61215-2:2021, IEC 62716:2013, IEC 61701:2020, and IEC/TS 63126:2020 [49–52]), the module's ultimate power degradation at STC was approximately 15% for 50 years. Such a result was achieved thanks to a considerably lower corrosiveness of silicon gels. Moreover, the initial output power of the modules increased thanks to higher gel transparency and a decrease in the operating temperature of PV cells. PV modules are fire-safe and have increased resistance to operation in environments with a high concentration of ammonia and in tropical and marine climates.

The comparative characteristics of modules encapsulated with polysiloxane gels and conventional silicon modules filled with EVA are shown in Table 2.

Table 2. Comparison of the characteristics of PV modules with two types of filling materials [48].

Parameter	Encapsulant	
	Polysiloxane Gels	**EVA**
Degradation of maximum power at STC (%) per year for 25 years	0.2–0.3 about 7.5	0.6–2.0 15 ÷ 20
Operating temperature (°C)	−60 to +110	−40 to +85
UV radiation durability	high	low
Refraction index	1.406	1.482
Transparency for solar radiation (%)		
λ = 360 nm	90	8
λ = 400 nm	92	62
λ = 600 ÷ 1000 nm	93	91
Corrosive agents		
during encapsulation	none	acetic acid
during operation	none	acetic acid
Mechanical stress		
during encapsulation	no	yes
during operation	no	yes
Modulus of elongation (N/mm^2)	0.006	10.0
Linear coefficient of thermal expansion ($10^{-4} \cdot K^{-1}$)	2.5	4.0

The encapsulation technology ensures a reduction in costs thanks to higher operational characteristics of modules (higher resistance to aggressive environments, e.g., in environments with the presence of ammonia in the ambient air, high humidity, etc., which are characteristic of various agricultural facilities), along with a considerable decrease in the degradation of these characteristics.

3. Increasing Energy Production by Increasing Voltage and Current per Unit Area of Cell Active Surface—Concentrator Design Improvement

In the course of our studies, we sought to optimize the form of the RS of both focon-type and compound parabolic concentrators. In the case of focon-type concentrators, it was important to solve the problem of nonuniform radiation distribution over the RR's operating surface, while in the case of compound parabolic concentrators, we had to minimize Fresnel losses. As a result, two types of effective CPV modules were built: the first based on focon-type concentrators and planar high-voltage cells, and the second based on compound parabolic concentrators and vertical high-voltage cells. The designs were based on the conditions that the concentrators should have such dimensions that would lead to comparable dimensions of the resulting CPV modules to those of conventional crystalline silicon PV modules, and that CPV modules could be used in individual power supply systems. Additionally, the design of CPV modules should match the components of mass-produced PV systems as much as possible.

3.1. Multi-Zone Parabolotoric Microconcentrators with Uniform Radiation Distribution in the Focal Region

The reflecting surface of the developed multi-zone solar radiation microconcentrators was formed via the rotation of a certain curve (generatrix) around a symmetry axis that does not cross this curve and passes through the focal spot center. A microconcentrator with an internal mirror RS operates on the principle of collecting the reflected rays into a circular focal region.

Figure 6 shows examples of a single-zone concentrator and a five-zone concentrator, as well as the outlay of a finished concentrator with high-voltage planar PV cells (see Section 2.2).

The RS of a microconcentrator is composed of several zones lying one above another. The RS generatrix constitutes several smoothly merging parabolas each of which forms a separate zone. Each zone has the form of an axisymmetric solid with a symmetry axis that coincides with that of the concentrator. Each zone of the RS ensures a homogeneous reflected radiation distribution over the entire focal region, and the number of reflected rays incident on each point of the concentrator's focal region is equal to the number of zones (see Figure 6d). At the same time, the form of the RS meets the condition of minimizing the incident angle for rays reaching the RR's active surface.

The general system of equations describing the RS profile is as follows:

$$\begin{cases} \left(\frac{dx_i}{dy_i}\right)^2 + \frac{2y_i}{x_i - z_i} \cdot \frac{dx_i}{dy_i} = 1;\ i = 1, 2, \ldots, N \\ x(y_i - 0) = x(y_i + 0);\ x_i \in [r_i; R_i];\ R_i = r_{i+1} \text{ at } i = 1, 2, \ldots, N-1 \\ x_1 = r_1 \text{ at } y_1 = 0 \\ \lim\limits_{y \to y_i - 0} \frac{dx}{dy} = \lim\limits_{y \to y_i + 0} \frac{dx}{dy} \\ \sum\limits_{i=1}^{N} \left| \frac{x_i dx_i}{z_i dz} \right| = \sum\limits_{i=1}^{N} K_i \\ K = 1 + \sum\limits_{i=1}^{N} K_i;\ K_i = \frac{R_i^2}{r_i^2};\ K = \frac{R_N^2}{r_1^2} \end{cases} \quad (6)$$

where, as shown in Figure 6a, x_i is the current position of the point of ray arrival at the i-th zone on the X axis; x_i is equal to the radius of a section of the RS by the plane that is perpendicular to the concentrator's rotation axis and passes through the point y_i ($r_i \leq x_i \leq R_i$); y_i is the current position of the point of ray arrival at the i-th zone on Y axis coinciding with the concentrator's rotation axis; z_i is the position of the point of arrival of the ray reflected from the i-th zone at the focal spot on the X axis ($-r_i \leq z_i \leq r_i$); K_i is the concentration ratio in the focal region ensured by the i-th zone of the RS; K is the total concentration ratio in the focal region; N is the number of zones; and r_i and R_i are the radiuses of the smallest and largest sections of the i-th zone's RS, which are perpendicular to the concentrator's rotation axis.

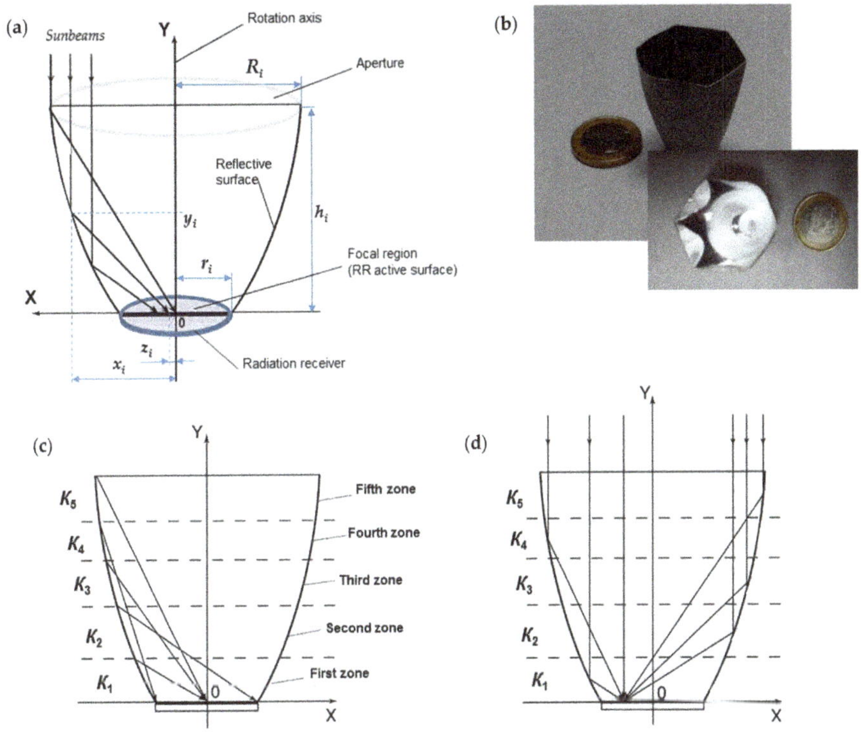

Figure 6. (**a**) Example of a single zone design; (**b**) five-zone microconcentrator with planar high-voltage PV cells corresponding to Figure 6c,d; (**c**) beam path diagram for the uppermost points of each zone, in the five-zone microconcentrator (the first zone corresponds to Figure 6a); (**d**) beam path diagram, for beams falling onto one point of RR, in the five-zone microconcentrator shown in Figure 6c.

The equations for various options of the microconcentrator design are defined in [53].

There are four options for single zones in accordance with the four possible variants of ray paths in the concentrator [53]. The microconcentrator RS is formed via the alternation of four variants of single zones. For instance, the concentrator shown in Figure 6b–d begins from a zone that is similar to that shown in Figure 6a. Then, three more zones are added where the reflected ray consequently arrives from the uppermost point of the zone at the RS center or periphery. Additionally, the fifth zone is the same as the first zone but with different values of r, R, and h. Each zone of the microconcentrator exposes the focal plane to light with concentration ratios K_1, K_2, K_3, K_4, and K_5, respectively. Generally, the concentration ratio of each zone K_i is distinct.

The greater the number of the zone is, the lower the reflection from the RR's active surface, and the higher the efficiency; however, this also results in a larger concentrator height.

The concentrator size depends on the size of the focal region/active surface of the RR. The smaller the focal region is, the smaller the concentrator's dimensions would be. Its size may be reduced to a very small size by using the developed PV cells, which are quite serviceable with millimeter dimensions and may be determined based on the convenience of assembly of the relevant module. The weight characteristics will be only determined by the concentrator wall thickness. The concentrators have to be capable to bear their own weight plus the weight of protective coatings used against the impact of the environment (sealing glasses or clear plastics).

The microconcentrators and PV cells as RR are used to assemble CPV modules with the dimensions of traditional planar PV modules on the basis of crystal silicon and may be employed instead of them without material changes in the design and equipment of PV systems.

Experiments were carried out with two-zone, three-zone, and five-zone mirror microconcentrators with $2r = 10$ mm and $K = 10$. The construction height of the experimental samples ranged from 26.47 mm to 36.41 mm depending on the increase in the portion of concentration accumulated in the middle zone. With that, the average incidence angle cosine varied from 0.7307 to 0.8104.

The uniformity of the achieved concentrated radiation distribution was 98.3–98.5% all over the focal region of the samples except for the boundaries along the perimeter, with the expected concentration ratio of 10^+. On the periphery of the focal region (approx. 2% to 4% inward of the perimeter), a light distribution inhomogeneity rate of 3% at the edge was observed. The estimated radiation distribution uniformity was 99.7%. The fact that the experimental values of the radiation distribution uniformity were lower than the estimated ones can be explained by the immaturity of the technique used to manufacture both the microconcentrators and CPV modules as a whole (i.e., the imperfect mutual adjustment of the concentrator and RR).

Furthermore, the dependence of the microconcentrator's optical efficiency on the solar ray incidence angle was not observed in the case of a planar PV cell, and this efficiency reached 0.9 with an angle of 3 degrees. The average cosine of the ray incidence angle on the focal plane, within the above limits, was greater than that for a two-zone microconcentrator, and its value was over 0.67. Such redistribution of concentrations by zones led to an increase in the eventual angles of microconcentrator disorientation, but simultaneously, the average value of the cosine of angles of ray arrival at the focal plane decreased.

The CPV modules were manufactured from five-zone samples (Figure 6b) and high-voltage planar PV cells described in Section 2.2 with a special grid (see Figure 7).

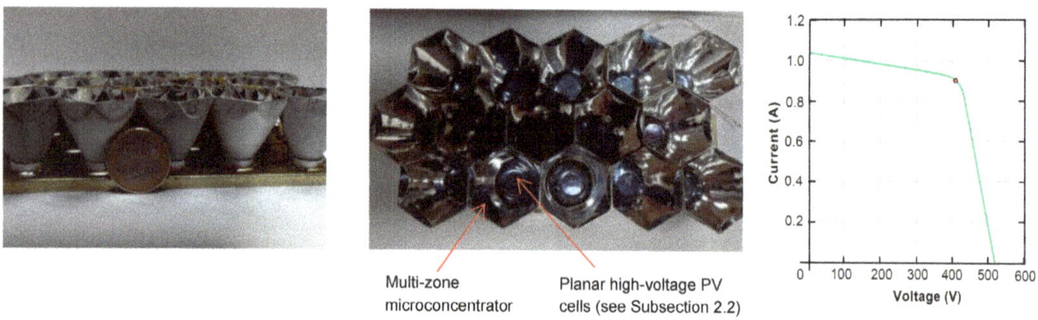

(a) (b)

Figure 7. (**a**) CPV module in the assembly process; (**b**) *I–V* curve of CPV modules at CSTC.

On the concentrator RS, above the upper zone, a supplementary ring was used forming a hexagon at the ray entrance into the concentrator. This ensured that there would be no voids between the microconcentrators in the plane of the active surface. The inscribed circle diameter of the hexagon was equal to $2R_5 = 31.6$ mm. The modules consisted of 29 parallel rows with 42 microconcentrators in each. The dimensions of the module's active surface were 1607.8 × 967.4 mm, and their height was 47.2 mm. Their *I–V* curve at concentrator standard test conditions (CSTC) (direct normal irradiance of 1000 W/m^2, cell temperature of 25 °C, and spectral irradiance distribution of direct normal AM1.5) is shown in Figure 7b. The efficiency change corresponds to the following equation [11]:

$$\eta_C = \eta_1(1 + \beta \ln K) \quad (7)$$

where $\beta = AkT/qV_{o.c}$ and β = 0.03–0.07.

3.2. One-Dimensional Double-Wing Concentrator with Low Fresnel Optical Loss

The body of the developed concentrator (Figure 8) was similar to that of a compound parabolic concentrator with a mirror internal RS operating on the principle of collecting the reflected rays into a rectangular focal region. The proposed principle of reducing Fresnel losses caused by radiation reflection from the RR's active surface involves minimizing the incidence angles for the reflected rays falling onto the concentrator focal region. For this purpose, RS has flat sections ($r \le x \le x_0$, $0 \le y \le y_0$) adjacent to the focal plane and smoothly passing into its curved surfaces ($x_0 \le x \le R$, $y_0 \le y$).

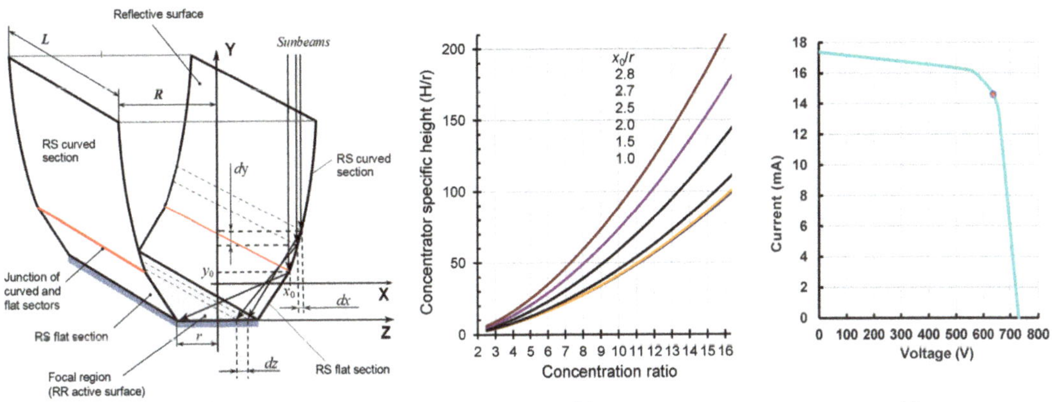

(a) (b) (c)

Figure 8. (**a**) Concentrator design (for the description of variables, see Equation (8)); (**b**) dependence of the concentrator specific height on radiation concentration; (**c**) *I–V* curve of the developed CPV module at CSTC.

The RS shape complying with the condition of uniform focal region illuminance with the maximum incidence angles value of rays falling onto the RR's active surface is described using the following set of equations:

$$\begin{cases} \left(\frac{dx}{dy}\right)^2 + \frac{2y}{x-z} \cdot \frac{dx}{dy} = 1;\ x_0 \le x \le R;\ y_0 \le y \\ \frac{dx}{dy} = \frac{x_0 - r}{y_0};\ r \le x \le x_0;\ 0 \le y \le y_0 \\ \frac{dx}{dz} = \frac{R - x_0}{2r};\ z = (x - x_0)\frac{2r}{r \cdot K - x_0} - r \\ tg\varphi = \frac{x_0 + r}{y_0} = \frac{\sqrt{(x_0 + r)(3r - x_0)}}{x_0 - r} \end{cases} \quad (8)$$

where, as shown in Figure 8a, x is the current position of the point of solar ray arrival at the RS on the X axis that is parallel with the focal plane and perpendicular to the concentrator symmetry plane ($x \ge r$); $y(x)$ is the point position on the concentrator's longitudinal symmetry plane in relation to the focal plane ($y \ge 0$); z is the position of the point of the reflected ray arrival at the focal region in relation to the concentrator's symmetry plane; x_0 and y_0 are the coordinates of the point on the junction line between the curved area ($x \ge x_0$, $y \ge y_0$) and the flat area ($r \le x \le x_0$, $0 \le y \le y_0$) of the RS ($r \le x_0 \le 3r$); K is the concentration ratio in the focal region; $2r$ is the width of the concentrator focal region

$-r \leq z \leq r$; $2R$ is the concentrator aperture size in cross-section; and $0 \leq x \leq R$ and φ is the ray incidence angle.

The shape of the concentrator's RS generatrix forms a flux of reflected radiation uniformly distributed over the rectangular focal region, thus reducing the current-spreading resistance in the illuminated layer and increasing the efficiency of solar radiation conversion into electric energy. Including flat sections in the concentrator made it possible to achieve efficient steep-wise incidence of the rays onto the focal plane at small incidence angles, resulting in a decrease in the coefficient of ray reflection from the RR's active surface. The primary important optical and energetic peculiarities of the developed product involved the symmetric property of the concentrator design, which considerably eased the requirements for the orientation towards the sun and increased the possibility of achieving high optical efficiency.

The maximum angle of the ray arrival at the RR's active surface decreased with the size of the flat horizontal section equal to $x_0 - r$. For the minimum value of $x_0 = r$, there was no flat section, and rays fell from the lower part of the concentrator at angles close to 90°, with a large reflection coefficient, while for the maximum value of $x_0 = 3r$, the incidence angle became equal to zero, and Fresnel losses were minimal. However, in this case, the size of the concentrator's flat section became unacceptably large, in a vertical direction.

The size and inclination of the flat section were chosen based on the conditions ensuring small angles of ray incidence onto the focal plane for acceptable values of the concentrator's entire height and weight. Varying this size and inclination angle enabled us to obtain a wide spectrum of concentrators ensuring target values of concentration ratios and relevant angles of radiation incidence on the focal plane.

Experiments were carried out with the concentrator option having the following parameters: $r = 12$ mm, $x_0 = 2r = 24$ mm, $R = 37.64$ mm, and $K = 10$. The concentrator height was $H = 75.42$ mm. Modules composed of vertical high-voltage PV cells (Figure 2b) were used as RRs. For 10-fold radiation concentration, the required area of the RR's active surface decreased 10 times, and concurrently, RR efficiency increased by relatively 15–18 percent. These effects together led to a considerable reduction in cost. The chosen value of parameter $x_0 = 2r$ approximately corresponded to the optimal concentrator design in terms of the coordination of Fresnel optical losses and concentrator height, i.e., its materials content and weight. The maximum angle of ray arrival at the RR surface was $\varphi = 60°$. The concentrator ensured a uniform radiation distribution of 98.9% over the RR. At the same time, as in Section 3.1, unevenness was localized on the periphery of the RR. These areas were not above the cells without substantial losses of the RR's active surface thanks to the form of the vertical PV cells.

The height values H increased with the concentration ratio K, leading to the decrease in the angle of ray incidence on the focal region and thus reducing Fresnel optical losses. The dependence of the concentrator height related to the width of the PV module's active surface (concentrator focal region) on the concentration ratio, for concentrators having different values of the design parameter x_0/r, is shown in Figure 8b. Additionally, in Figure 8c, we can see the averaged I–V curve of one section of the developed CPV modules at CSTC. The dimensions of the RR's active surface, for one module section, are 10 × 1020 mm. The efficiency change corresponds to Equation (7).

4. Increasing Energy Production by Increasing Systematization Level—Optimization of the Power Supply Process as a Whole

It should be noted that in many recent studies on normative instruments (e.g., [54–58]) the approach to the construction concept of PV systems for individual facilities is close to the one we investigated. In [58], the consuming equipment (consumption circuit) is supposed to be a part of the PV system. Additionally, in the event that the consuming equipment would not be included in the PV system supply contract, such equipment shall be detailed in the technical documentation for the PV system. Nevertheless, the evaluation of the system was limited to that of its PV part. In [57], efficiency was assessed through a conventional parameter called "service ratio", which rather schematically describes the relationships between the generated energy and satisfied needs. This methodology is only fit for a theoretical comparison of PV systems using this conventional parameter but not for the design process of PV systems and the efficiency assessment of real functioning systems.

The final objective of PV system operation, similar to any other power-generating equipment, is to provide the consumer with the required kind of energy in the specified and required quality.

The analysis of the process of power supply to individual facilities shows that the total losses caused by inefficient operation of power supply systems are 28–73% (see Figure 9). The average losses caused by improperly arranged power supply amount to 26% of the aggregate power losses. The losses caused by the unreliable performance of such systems as a result of their design features are 28% [6,17,59].

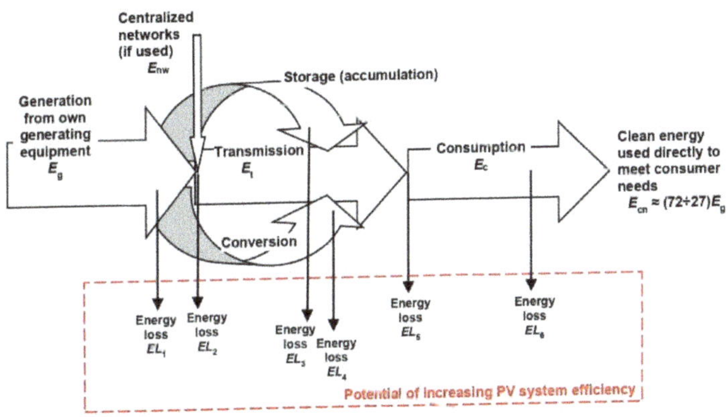

Figure 9. Diagram of the process of power supply to individual facilities.

Therefore, the actual efficiency of the power-generating equipment is determined not only by the characteristics of the generating equipment themselves, e.g., a PV system (E_g and EL_1 or $E_g + E_{nw}$ and $EL_1 + EL_2$ shown in Figure 9) but also by the peculiarities of the equipment, structures, and factors influencing the energy generation, transmission, and use processes (EL_3–EL_6). Moreover, due to the specificities of PV system functioning, even an inconsiderable change in losses (in conjunction with the time of day and of year, weather, load conditions, etc.) may substantially influence the PV system's operation and the quantity of energy delivered to the consumption points. The assembly of the above-mentioned equipment, structures, and factors, into a system and the optimization of such a system will allow for the minimization of losses, thus substantially increasing the amount of energy used by a consumer with the same output power of the PV system (both as stand-alone generating equipment and in combination with any other generating equipment and/or a network). Hence, by taking into account the final objective mentioned above, the operation of the generating equipment can become as efficient as possible.

At present, the efficiency of power supply with the use of PV systems is evaluated on the basis of their characteristics at the consuming equipment input. As seen in Figure 9, from the point of power input to the consuming equipment until the moment of meeting consumer needs, significant energy losses are possible. Accordingly, to estimate the efficiency of the PV system realistically, it should be determined based on the final result, i.e., the degree to which consumer needs are met.

This means that an increase in the PV system's efficiency can be achieved by expanding the boundaries of the system, which is mainly the new and higher level of generalization and systematization characterized by the following features:

- For power supply, a uniform complex system is created that includes all the systems, processes, and structures, involved in or influencing the power supply in any way;
- The system efficiency index is the degree to which consumer need(s) are met;
- The power supply facility (e.g., a building) is a part of the system;
- The principle of integrating functions is applied to the maximum possible extent: Each structure, technical means or subsystem shall perform as many functions as possible. When a structure (technical means, subsystem) can be built to perform the functions of several structures (technical means, subsystems), it has to be designed to combine maximum of these functions (e.g., BIPV modules);
- The development of a system begins at the time when the facility creation begins and includes a synchronized parallel interrelated development of all of the system components.

The block diagram of such a highly efficient PV system is shown in Figure 10.

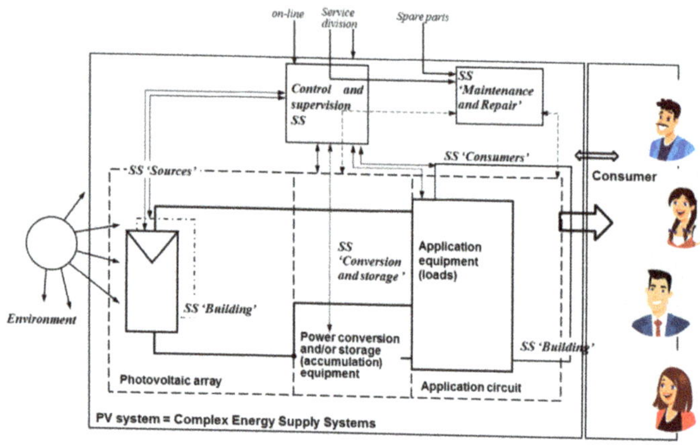

Figure 10. Design of highly efficient PV system for individual facilities created as a complex system of energy supply to the facility (SS means "subsystem").

Unlike ordinary network systems, the problem of the interconnection of all processes, from power production to power consumption, is critically important for PV systems. If a system is improperly organized or its structure is not optimized, there will be an energy shortage, or even an absence of energy supply, on the consumer side. It is obvious that in particular cases, the assertions of inapplicability (inoperability) of a PV system or its lower efficiency compared with the expected values of parameters, are, first of all, associated with improper organization and structure of the PV system. When a PV system is used for power supply to a facility, its maximum efficiency may only be achieved when it is designed in accordance with the foregoing principles.

The efficiency criterion of highly efficient PV systems is formulated as follows: "maximum satisfaction of needs with minimum costs", and the objective function, generally, is as follows:

$$W = \begin{cases} \mathrm{S} \to \max \\ \mathrm{C_E} \to \min \\ HE \to 0 \end{cases} \quad (9)$$

Provided that

$$(\mathrm{C_E} \to \min) \equiv \begin{Bmatrix} \mathrm{E_{cn}} \to \min \\ EL_\Sigma = EL_1 + EL_2 + EL_3 + EL_4 + EL_5 + EL_6 \to \min \end{Bmatrix} \quad (10)$$

$$(\mathrm{E_{cn}} \to \min) \equiv \begin{Bmatrix} \mathrm{E_d} \to \min \\ EL_6 \to \min \end{Bmatrix} \quad (11)$$

$$(E_d \to \min) \supset \begin{Bmatrix} \text{Need optimization/correct assessment of needs} \\ \text{Consumption equipment optimization} \end{Bmatrix} \quad (12)$$

where S is the degree of need satisfaction (if S is expressed as a percentage, then $S \to 100\%$); C_E is the energy expenditure; HE is the environmental damage; EL_d is the energy demand; and EL_Σ is the aggregate energy losses. E_{cn}, EL_1, EL_2, EL_3, EL_4, EL_5, and EL_6 are shown in Figure 9.

With that, the cost minimization is assumed to be, first and foremost, the minimization of energy expenditures and, therefore, the minimization of losses. Additionally, the maximum satisfaction of needs is understood to be the maximum number of needs met and the maximum satisfaction of each need. Methods for numeric evaluation at the level of individual facilities do not yet exist for all the indices through which the needs are expressed. In this case, one proceeds from the affirmation that a greater observance of this condition corresponds to greater satisfaction of needs and is always preferable, e.g., more ecologically sound solutions are always preferable.

The correct assessment and formalization of needs is the most difficult task in the creation of PV systems in accordance with the proposed efficiency criterion in Equation (9). This occurs in the following manner:

The consumer is described by a set of needs N for the satisfaction of which electrical energy (or electrical energy from a PV system) may be required. Each need n_i is described from 1 to J by parameters p_{ij}. In a general case, the number of parameters characteristic of different needs is different. Each of the parameters is characterized by a value (or permissible value range) of v_{ij} corresponding to the state of "need is satisfied", or a number of values/ranges and the relevant degrees of need satisfaction are established for each parameter. The general outline of consumer need formalization may be presented through the following identical relations:

$$\text{Consumer} \equiv \begin{Bmatrix} \textit{Set} \\ \textit{of needs} \\ \text{N} \\ \text{N} = \{n_1, n_2, \ldots, n_i\} \end{Bmatrix} \Leftrightarrow \begin{Bmatrix} \textit{Set} \\ \textit{of the parameters} \\ \textit{characterizing} \\ \textit{needs} \\ \text{NP} \\ \text{NP} = \begin{Bmatrix} \{\text{NP}_1\} \\ \{\text{NP}_2\} \\ \ldots \\ \{\text{NP}_i\} \end{Bmatrix} \end{Bmatrix} = \begin{vmatrix} \text{NP}_1 = \{p_{11}, p_{12}, \ldots, p_{1j}\} \\ \text{NP}_2 = \{p_{21}, p_{12}, \ldots, p_{2j}\} \\ \ldots \\ \text{NP}_i = \{p_{i1}, p_{i2}, \ldots, p_{ij}\} \end{vmatrix} \Leftrightarrow \begin{Bmatrix} \textit{Set} \\ \textit{of the parameter} \\ \textit{values} \\ \textit{characterizing} \\ \textit{needs} \\ \text{NPV} \\ \text{NPV} = \begin{Bmatrix} \{\text{NPV}_1\} \\ \{\text{NPV}_2\} \\ \ldots \\ \{\text{NPV}_i\} \end{Bmatrix} \end{Bmatrix} \equiv \begin{vmatrix} \text{NPV}_1 = \{v_{11}, v_{12}, \ldots, v_{1j}\} \\ \text{NPV}_2 = \{v_{21}, v_{12}, \ldots, v_{2j}\} \\ \ldots \\ \text{NPV}_i = \{v_{i1}, v_{i2}, \ldots, v_{ij}\} \end{vmatrix} \quad (13)$$

where NP is the set of parameters characterizing all consumer needs that require electrical energy to be met, NP_i is the set of the parameters characterizing the i-th need, NPV is the set of the parameter values characterizing all needs, and NPV_i is the set of parameter values characterizing the i-th need.

The task of creating a PV system is to ensure that the energy generated by it meets each need. Formally, this means that the value of each parameter from the set of parameters in Equation (13) is equal to the desired value (i.e., it is within the desired range of values).

The process that is fully applied today is shown in Figure 11. The system efficiency evaluation is carried out in reverse order.

Additionally, the efficiency of such PV systems is as follows:

$$\eta_{syst} = \frac{E_{cn} - \Delta_E}{E_{in}} \cdot 100\% \quad (14)$$

where E_{in} is the amount of energy supplied to the system (the product of the aggregate irradiance of PV module active surfaces or CPV module aperture by the aggregate area of active surfaces or aperture). In the case of power supply with the use of PV systems, these are only $E_{in} = E_{cn} + (EL_{PV} + EL_3 + EL_4 + EL_5 + EL_6)$; $EL_{PV} = EL_1$ is the losses in the generation process; and Δ_E is the difference between the energy required for consumer needs to be fully met and the clean energy directly used to meet consumer needs.

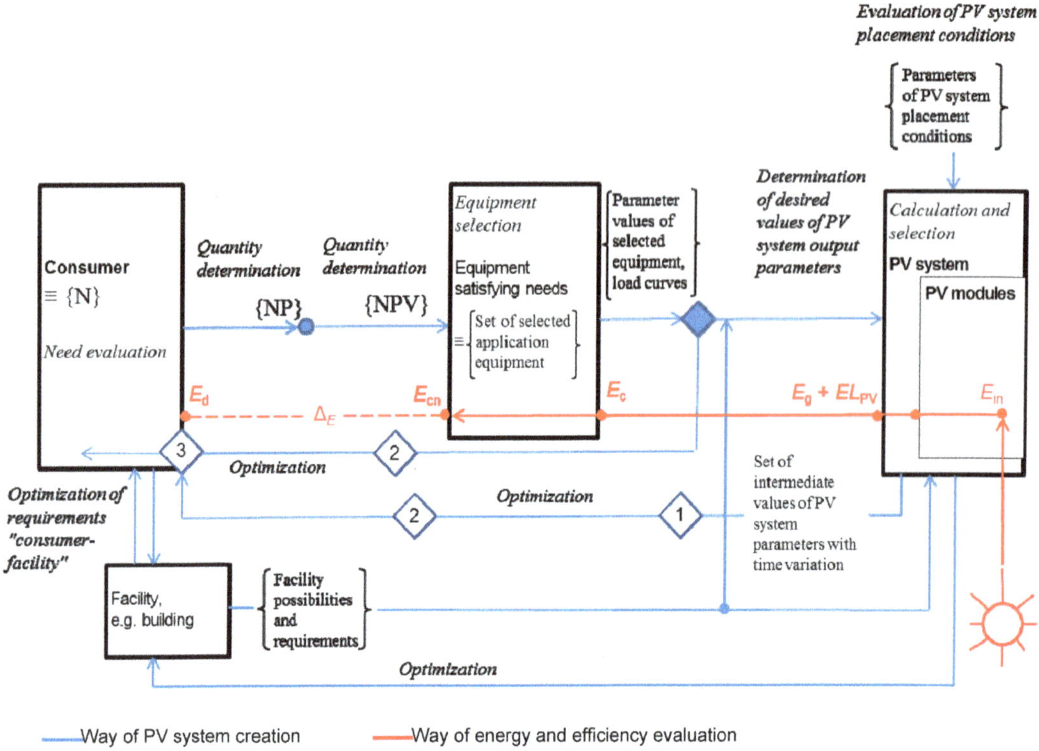

1 – evaluation of conformity with load diagrams; 2 – evaluation of value differences; 3 – evaluation of the possibility to adjust needs – need adjustment or transition to parameter adjustment or transition to parameter value adjustment

Figure 11. Diagram of PV system implementation and its efficiency evaluation.

Instead of energy, power may be used to simplify the evaluation process. In this case, to assess the efficiency, instead of Δ_E, it is necessary to determine Δ_P, which is the total power of all needs that are not fully met.

The efficiency evaluation according to the proposed criterion and the creation of a PV system in accordance with Figure 10 indispensably include the possibility of amendments, adjustments, and modifications as a normal functional condition of the process, from changing settings to the replacement of equipment, software logic, etc. This includes the possibility of modifications due to the inaccurate assessment of the state of need satisfaction resulting from (1) PV system's imperfection due to any errors in design and inaccurate determination of the consumer needs from the outset as well as (2) a change in needs.

The described approach may be quite properly combined with modern software possibilities, which allows for the implementation of the diagram in Figure 11 in practice.

5. Approbation of Proposed Solutions

The final characteristics of the developed PV/CPV modules, as well as typical average characteristics of conventional mono-Si modules, are shown in Table 3. As can be seen from the *I–V* curve of the developed modules (Figures 2c, 5b, 7b and 8c), it is possible to further increase the PV/CPV modules' efficiency, especially the modules based on planar high-voltage PV cells.

Table 3. Characteristics of the newly developed PV/CPV modules and typical characteristics of conventional crystalline Si PV modules.

Module Type	Active Surface or Aperture (mm)	Height (mm)	$V_{o.c}$ (V)	$J_{s.c}$ (mA/sm^2)	V_{mpp} (V)	J_{mpp} (mA/sm^2)	P_{max} (W/sm^2)
Conventional crystalline Si PV modules (60 single *p*–*n* junction mono-Si PV cells with size 156.75 × 156.75 mm)	1671.0 × 1002.0	35.0	44.0	0.567	36.3	0.526	0.019
PV module-based vertical high-voltage PV cells	60.0 × 10.0	0.4	640.0 [1]	0.559 [1]	548.5 [1]	0.456 [1]	0.250 [1]
PV module-based planar high-voltage PV cells	1578.5 × 945.5	37.1	439.7	0.0625	382.4	0.061	0.023
CPV module-based one-dimensional double-wing concentrator and vertical high-voltage PV cells (one section)	37.64 × 1020.0	75.4	730.0	0.0453	637	0.038	0.024
CPV module-based multi-zone microconcentrator and planar high-voltage PV cells	1607.0 × 967.4	47.2	512.9	0.068	405.0	0.058	0.023

[1] Under 10 suns.

In the period from 2011 to 2022, the concept considered above was used for designing PV systems in various climate conditions and at facilities with different needs [43,60–64]. The functioning of those systems confirms the correctness of the proposed approach. The composed systems can satisfy 30–50% more needs than PV systems with a conventional systematization level simultaneously installed at similar facilities, i.e., they are equivalent to PV systems with a conventional systematization level but generate more energy by 30–50%. The points of PV systems' installation were in various regions of rural territories with different climates, so the results may be deemed quite representative.

The research results presented in Sections 2–4 were combined in the base experimental project of an ecological complex (Lat 58.01 N, Lon 43.34 E) consisting of two main buildings and free-standing identical cottages whose number grew from year to year. The PV system was implemented from the outset as a comprehensive power supply system with conventional mono-Si PV modules located on the roof and integrated into the balcony rail and flexible a-Si panels on the roof with an arched form [60]. As new PV/CPV modules were developed, these new devices would be partially installed for conventional planar modules. The power supply to one of the cottages, in comparison, was ensured from the very beginning using a conventional PV system. The rest were designed with a different higher level of systematization. Depending on the particular cottage, the systematization level increased with the extension options by incorporating into the system other components involved in the power supply process (see Figures 8 and 9). Additionally, the level of systematization of parts of the PV system of the main buildings was gradually adjusted.

Thanks to the installation of the developed PV and CPV modules, the total maximum power at STC (CSTC) was 1120.86 kW. That said, thanks to a higher output voltage and other specific features of the modules, the losses inside the system decreased by 12 to 15%, and also maintenance losses decreased. In the case of PV systems of the cottages built simultaneously with the cottage equipped with a traditional PV system, the power directly used to meet consumer needs was 21% to 23% higher. As the proposed approach was improved, in the construction of new cottages, losses were reduced, and E_{cn} was increased by 12% to 16%.

The reduction values of the maximum power at STC/CSTC, and, consequently, those of the energy production of the developed PV modules, were from 0.2% to 0.24% per year under wet atmospheres with high concentrations of dissolved ammonia (see Figure 12). These values of the developed PV modules were from 0.22% to 0.37% per year. Measurements were carried out on a livestock complex experimental site and using an accelerated aging method in laboratory conditions. For comparison reasons, Figure 12 shows the deterioration range for conventional mono-Si PV modules, in a normal environment. The average annual deterioration of their performance may attain 5% to 7%, in a wet ammonia environment.

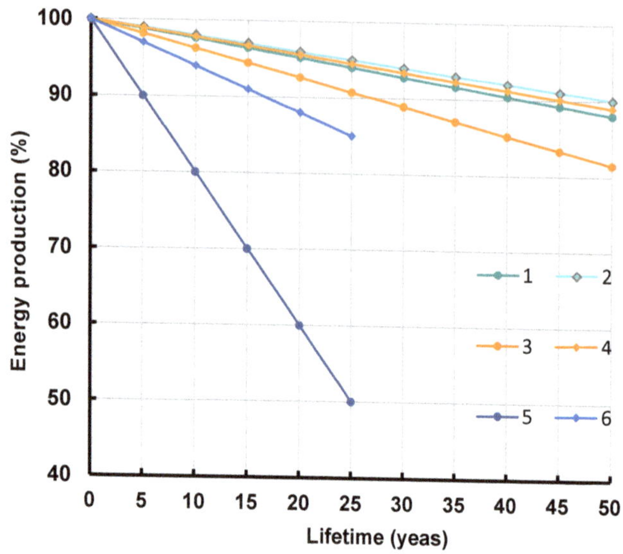

Figure 12. Reduction in the energy production of modules during their operation lifespan: 1, 2—newly developed PV modules with high-voltage PV cells in wet ammonia environment (1—maximum, 2—minimum); 3, 4—newly developed CPV modules in wet ammonia environment (3—maximum, 4—minimum); 5, 6—conventional mono-Si PV modules in normal operation conditions (5—maximum, 6—minimum).

6. Conclusions

The use of PV equipment is a real efficient method for fighting against climate change, and the protection and restoration of the environment. For the mass implementation of PV systems in agriculture in rural areas, their attractiveness to consumers should be increased, for which it is necessary to improve the energy output of PV devices, as well as their reliability and environmental resistance, to improve their aesthetic quality, as well as enhance the diversity of efficient PV equipment for any operational conditions ensuring the versatility of PV systems use as an energy source.

High-voltage vertical Si PV cells for terrestrial applications and high-voltage multi-junction planar Si PV cells were proposed and developed. A significant increase in the output voltage of the PV modules and PV arrays was attained with the design of PV cells that enabled a reduction in power losses, material consumption, required space, and cost. At the same time, the design of planar multijunction PV cells allowed for adherence to the conventional mass production technology for PV modules.

The modules based on high-voltage vertical PV cells composed of 40 cells had a size of 60 × 10 mm, a maximum power point voltage of up to 1000 V, specific power of up to 0.245 ± 0.01 W/cm^2, and an efficiency rate of up to 25.3%, for radiation concentration ratios from 10 to 100. The samples of the second type of modules composed of 60 PV cells (156.75 × 156.75 mm) had $V_{o.c}$ = 439.7 V, $I_{s.c}$ = 0.933 A, and P_{max} of 348 W, at STC. The maximum power degradation at STC was estimated to be within 0.2% to 0.24% per year in a wet ammonia environment.

New designs of concentrators with uniform radiation distribution in the focal region were developed, including a parabolic cylindrical concentrator with minimized Fresnel optical losses and a multi-zone parabolotoric microconcentrator whose dimensions enabled CPV modules similar in size to conventional planar modules to be manufactured. Two types of CPV modules were designed: (1) modules consisting of a parabolic cylindrical concentrator and an RR composed of vertical multi-junction cells, and (2) modules comprising multi-zone microconcentrators with a planar high-voltage multi-junction cell as the RR. The second-type CPV modules included 42 × 29 microconcentrators and had the dimensions of 1607.8 × 967.4 ×47.2 mm. These modules had the following average values of output parameters at CSTC: $V_{o.c}$ = 512.9 V, $I_{s.c}$ = 1.051 A, and P_{max} of 361.57 W. The degradation of maximum power at CSTC was estimated at 0.22% to 0.37% per year in a wet ammonia environment.

Through the use of PV and CPV modules based on the developed cells and concentrators, projects using the power supply on the basis of PV systems were designed. The projects implemented the principles of complex energy supply systems, which allowed for the minimization of the losses and requirements for the output power level of PV systems. The thus-created systems satisfied 30–50% more consumer needs than PV systems with a traditional systematization level simultaneously installed at similar facilities. Thanks to a higher output voltage and other specific features of the developed modules, the power loss inside the PV system decreased by 12% to 15%, along with a reduction in maintenance loss.

Author Contributions: Conceptualization, O.S.; methodology, O.S.; validation, O.S., A.I., Y.L. and A.D.; formal analysis, O.S.; investigation, O.S.; resources, A.I., Y.L. and A.D.; data curation, A.I., Y.L., and A.D.; writing—original draft preparation, writing—review and editing, O.S.; visualization, O.S. and A.D.; supervision, A.I., Y.L. and A.D.; project administration, A.I., Y.L. and A.D.; funding acquisition, A.I, Y.L. and A.D. All authors have read and agreed to the published version of the manuscript.

Funding: This research received no external funding.

Institutional Review Board Statement: Not applicable.

Data Availability Statement: Not applicable.

Conflicts of Interest: The authors declare no conflict of interest.

References

1. Konovalov, S.S. *Man and the Universe*; AST: Moscow, Russia, 2022.
2. Barron-Gafford, G.A.; Pavao-Zuckerman, M.A.; Minor, R.L.; Sutter, L.F.; Barnett-Moreno, I.; Blackett, D.T.; Thompson, M.; Dimond, K.; Gerlak, A.K.; Nabhan, G.P.; et al. Agrivoltaics provide mutual benefits across the food–energy–water nexus in drylands. *Nat. Sustain.* **2019**, *2*, 848–855. [CrossRef]
3. Weselek, A.; Bauerle, A.; Hartung, J.; Zikeli, S.; Lewandowski, I.; Högy, P. Agrivoltaic system impacts on microclimate and yield of different crops within an organic crop rotation in a temperate climate. *Agron. Sustain. Dev.* **2021**, *41*, 59. [CrossRef]
4. Ghosha, S.; Ranjana Yadav, R. Future of photovoltaic technologies: A comprehensive review. *Sustain. Energy Technol. Assess.* **2021**, *47*, 101410. [CrossRef]
5. Elahi, E.; Khalid, Z.; Zhang, Z. Understanding farmers' intention and willingness to install renewable energy technology: A solution to reduce the environmental emissions of agriculture. *Appl. Energy* **2022**, *309*, 118459. [CrossRef]
6. Izmailov, A.Y.; Lobachevsky, Y.P.; Shepovalova, O.V. Solar power systems implementation potential for energy supply in rural areas of Russia. *AIP Conf. Proc.* **2019**, *2123*, 020104. [CrossRef]
7. Ufa, R.A.; Malkova, Y.Y.; Rudnik, V.E.; Andreev, M.V.; Borisov, V.A. A review on distributed generation impacts on electric power system. *Int. J. Hydrogen Energy* **2022**, *47*, 20347–20361. [CrossRef]
8. Shepovalova, O.V. Specificity of the individual energy supply systems creation in Russia. In Proceedings of the 42nd ASES National Solar Conference SOLAR 2013, Including 42nd ASES Annual Conference and 38th National Passive Solar Conference, Baltimore, MD, USA, 16–20 April 2013.
9. Hafeez, M.; Yuan, C.; Shah, W.U.; Mahmood, M.T.; Li, X.; Iqbal, K. Evaluating the relationship among agriculture, energy demand, finance and environmental degradation in one belt and one road economies. *Carbon Manag.* **2020**, *11*, 139–154. [CrossRef]
10. Vasiliev, A.M.; Landsman, A.P. *Semiconductor Photoelectric Converters*; Soviet Radio Pupl.: Moscow, Russia, 1971.
11. Arbuzov, Y.D.; Evdokimov, V.M. *Fundamentals of Photovoltaics*; GNU VIESH: Moscow, Russia, 2008.
12. Dhilipan, J.; Vijayalakshmi, N.; Shanmugam, D.B.; Jai Ganesh, R.; Kodeeswaran, S.; Muralidharan, S. Performance and efficiency of different types of solar cell material—A review. *Mater. Today Proc.* **2022**, *66*, 1295–1302. [CrossRef]
13. Fraunhofer Institute for Solar Energy Systems (ISE) with Support of PSE Projects GmbH. *Photovoltaics Report*; Fraunhofer ISE FHG-SK—ISE-PUBLIC: Freiburg, Germany, 2023. Available online: https://www.ise.fraunhofer.de/content/dam/ise/de/documents/publications/studies/Photovoltaics-Report.pdf (accessed on 26 March 2023).

14. Strebkov, D.S.; Tver'yanovich, E.V. *Concentrator of Solar Radiation*; GNU VIESH: Moscow, Russia, 2007.
15. *Integrated Photovoltaics—Areas for the Energy Transformation*; Fraunhofer ISE: Freiburg, Germany, 2023. Available online: https://www.ise.fraunhofer.de/en/key-topics/integrated-photovoltaics.html (accessed on 26 March 2023).
16. Mamun, M.Z.; Dargusch, P.; Wadley, D.A.; Zulkarnain, N.A.; Aziz, A.A. A review of research on agrivoltaic systems. *Renew. Sustain. Energy Rev.* **2022**, *161*, 112351. [CrossRef]
17. Izmailov, A.Y.; Lobachevsky, Y.P.; Shepovalova, O.V. Complex energy supply systems for individual sites. *Energy Procedia* **2019**, *157*, 1445–1455. [CrossRef]
18. Shepovalova, O.V. Mandatory Characteristics and Parameters of Photoelectric Systems, Arrays and Modules and Methods of their Determining. *Energy Procedia* **2019**, *157*, 1434–1444. [CrossRef]
19. Green, M.A.; Dunlop, E.D.; Siefer, G.; Yoshita, M.; Kopidakis, N.; Bothe, K.; Hao, X. Solar cell efficiency tables (Version 61). *Prog. Photovolt. Res. Appl.* **2022**, *31*, 3–16. [CrossRef]
20. Bagher, A.M.; Vahid, M.M.A.; Mohsen, M. Types of solar cells and application. *Am. J. Opt. Photon.* **2015**, *3*, 94–113. [CrossRef]
21. Würfel, P.; Würfel, U. *Physics of Solar Cells: From Principles to New Concepts*, 3rd ed.; Wiley-VCH: Weinheim, Germany, 2009.
22. Morocha, A.K. On the theory of love waves in heterostructures with hexagonal symmetry. *Semiconductors* **2010**, *44*, 1625–1630. [CrossRef]
23. Strebkov, D.S. *Matrix Solar Cells*; GNU VIESH: Moscow, Russia, 2010; Volume I–III.
24. Xing, Y.; Han, P.; Wang, S.; Fan, Y.; Liang, P.; Ye, Z.; Li, X.; Hu, S.; Lou, S.; Zhao, C.; et al. Performance analysis of vertical multi-junction solar cell with front surface diffusion for high concentration. *Sol. Energy* **2013**, *94*, 8–18. [CrossRef]
25. Arbuzov, Y.D.; Evdokimov, V.M.; Strebkov, D.S.; Shepovalova, O.V. Semiconductor Photoelectric Generator (Versions) and Method of Making Said Generator (Versions). R.F. Patent 2371811, 27 October 2009.
26. Arbuzov, Y.D.; Evdokimov, V.M.; Strebkov, D.S.; Shepovalova, O.V. Semiconductor Photoelectric Generator and Method of Making Said Generator. R.F. Patent No. 2373607, 20 November 2009.
27. Arbuzov, Y.D.; Evdokimov, V.M.; Shepovalova, O.V. Cascade photoconverters on the basis of the homogeneous semiconductor tunnel structures. In Proceedings of the World Renewable Energy Forum, WREF 2012, Including World Renewable Energy Congress XII and Colorado Renewable Energy Society (CRES) Annual Conference, Denver, CO, USA, 13–19 May 2012; Volume 1, pp. 147–153.
28. Arbuzov, Y.D.; Evdokimov, V.M.; Shepovalova, O.V. Spectral characteristics of cascade photoelectric converters on the base of idealized tunnel homogeneous semiconductor structures. *AIP Conf. Proc.* **2018**, *1968*, 030007. [CrossRef]
29. Arbuzov, Y.D.; Evdokimov, V.M.; Majorov, V.A.; Saginov, L.D.; Shepovalova, O.V. Silicon PV Cell Design and Solar Intensity Radiation Optimization for CPV Systems. *Energy Procedia* **2015**, *74*, 1543–1550. [CrossRef]
30. Arbuzov, Y.D.; Evdokimov, V.M.; Shepovalova, O.V. New Photoelectric System on the Basis of Cascade Homogeneous Photoconverters and Solar Radiation Concentrators. *Energy Procedia* **2015**, *74*, 1533–1542. [CrossRef]
31. Vasiliev, A.M.; Evdokimov, V.M.; Landsman, A.P.; Milovanov, A.F. Operation of photoconverters under conditions of strong illumination. *Geliotekhnika* **1975**, *11*, 72–87.
32. Khamooshi, M.; Salati, H.; Egelioğlu, F.; Faghiri, A.H.; Tarabishi, J.; Babadi, S. A Review of Solar Photovoltaic Concentrators. *Int. J. Photoenergy* **2014**, *2014*, 1–17. [CrossRef]
33. Wang, L.; Yuan, Z.; Zhao, Y.; Guo, Z. Review on Development of Small Point-Focusing Solar Concentrators. *J. Therm. Sci.* **2019**, *28*, 929–947. [CrossRef]
34. Almonacid, G.; Lique, A.; Moledo, A.G. Photovoltaic static concentrator analysis. *Sol. Cells* **1984**, *1*, 163–178. [CrossRef]
35. Li, G.; Xuan, Q.; Akram, M.W.; Akhlaghi, Y.G.; Liu, H.; Shittu, S. Building integrated solar concentrating systems: A review. *Appl. Energy* **2020**, *260*, 114288. [CrossRef]
36. Shanks, K.; Senthilarasu, S.; Mallick, T.K. Optics for concentrating photovoltaics: Trends, limits and opportunities for materials and design. *Renew. Sustain. Energy Rev.* **2016**, *60*, 394–407. [CrossRef]
37. Zahidov, R.A.; Umarov, G.Y.; Vainer, A.A. *Theory and Calculating Helio-Technical Concentrating Systems*; FAN: Tashkent, Russia, 1977.
38. Paul, D.I. Application of compound parabolic concentrators to solar photovoltaic conversion: A comprehensive review. *Int. J. Energy Res.* **2019**, *43*, 4003–4050. [CrossRef]
39. Li, G.; Xuan, Q.; Pei, G.; Su, Y.; Ji, J. Effect of non-uniform illumination and temperature distribution on concentrating solar cell—A review. *Energy* **2018**, *144*, 1119–1136. [CrossRef]
40. Alves, M.; Pérez-Rodríguez, A.; Dale, P.J.; Domínguez, C.; Sadewasser, S. Thin-film micro-concentrator solar cells. *J. Phys. Energy* **2020**, *2*, 012001. [CrossRef]
41. AL-Hamadany, A.H.; Al-Abideen, F.S.Z.; Ali, J.H. Effect of Angle Orientation of Flat Mirror Concentrator on Solar Panel System Output. *IOSR J. Comput. Eng.* **2016**, *18*, 16–23.
42. Vu, N.H.; Pham, T.T.; Shin, S. Large Scale Spectral Splitting Concentrator Photovoltaic System Based on Double Flat Waveguides. *Energies* **2020**, *13*, 2360. [CrossRef]
43. Shepovalova, O.V.; Izmailov, A.Y.; Lobachevsky, Y.P.; Dorokhov, A.S.; Chirkov, S.V. Solar energy toroidal concentrators. *Energy Rep.* **2021**, *7*, 328–342. [CrossRef]
44. Shepovalova, O.V. Fabrication process study for matrix silicon solar cells. *AIP Conf. Proc.* **2017**, *1814*, 020077. [CrossRef]
45. Strebkov, D.S.; Shepovalova, O.V.; Bobovnikov, N.I. Investigation of High-Voltage Silicon Solar Modules. *AIP Conf. Proc.* **2019**, *2123*, 020103. [CrossRef]

46. Strebkov, D.S.; Shepovalova, O.V.; Zadde, V.V. Semiconductor Photoelectric Generator. R.F. Patent 2336596, 20 October 2008.
47. Wheeler, N.R.; Bruckman, L.S.; Ma, J.; Wang, E.; Wang, C.K.; Chou, I.; Sun, J.; French, R.H. Degradation pathway models for photovoltaics module lifetime performance. In Proceedings of the IEEE 39th Photovoltaic Specialists Conference (PVSC), Tampa, FL, USA, 16–21 June 2013; pp. 3185–3190. [CrossRef]
48. Poulek, V.; Strebkov, D.S.; Persic, I.S.; Libra, M. Towards 50 years lifetime of PV panels laminated with silicone gel technology. *Sol. Energy* **2012**, *86*, 3103–3108. [CrossRef]
49. *IEC 61215-2:2021*; Terrestrial Photovoltaic (PV) Modules—Design Qualification and Type Approval—Part 2: Test Procedures. International Electrotechnical Commission: Geneva, Switzerland, 2021.
50. *IEC/TS 63126:2020*; Guidelines for Qualifying PV Modules, Components and Materials for Operation at High Temperatures. International Electrotechnical Commission: Geneva, Switzerland, 2020.
51. *IEC 62716:2013*; Photovoltaic (PV) Modules—Ammonia Corrosion Testing. International Electrotechnical Commission: Geneva, Switzerland, 2013.
52. *IEC 61701:2020*; Salt Mist Corrosion Testing of Photovoltaic (PV) Modules. International Electrotechnical Commission: Geneva, Switzerland, 2020.
53. Arbuzov, Y.D.; Evdokimov, V.M.; Shepovalova, O.V. Multi-zone parabolotoric micro-concentrator of solar energy. *AIP Conf. Proc.* **2019**, *2190*, 020093. [CrossRef]
54. Tabunschikov, Y.A.; Brodach, M.M.; Shilkin, N.V. *Energy-Efficient Buildings*; AVOK–PRESS: Moscow, Russia, 2013.
55. Iturriagaa, E.; Aldasoro, U.; Campos-Celadorc, A.; Sala, J.M. A general model for the optimization of energy supply systems of buildings. *Energy* **2017**, *138*, 954–966. [CrossRef]
56. Petri, I.; Kubicki, S.; Rezgui, Y.; Guerriero, A.; Li, H. Optimizing Energy Efficiency in Operating Built Environment Assets through Building Information Modeling: A Case Study. *Energies* **2017**, *10*, 1167. [CrossRef]
57. *IEC/TS 62257-9-6:2019*; Recommendations for Small Renewable Energy and Hybrid Systems for Rural Electrification—Part 9–6: Integrated System—Selection of Photovoltaic Individual Electrification Systems (PV-IES). International Electrotechnical Commission: Geneva, Switzerland, 2019.
58. *IEC 62548:2016*; Photovoltaic (PV) Arrays—Design Requirements. International Electrotechnical Commission: Geneva, Switzerland, 2016.
59. Izmailov, A.Y.; Lobachevsky, Y.P.; Shepovalova, O.V. Comparison and selection of off-grid PV systems. *AIP Conf. Proc.* **2018**, *1968*, 030001. [CrossRef]
60. Strbkov, D.S.; Shepovalova, O.V.; Dunichkin, I.V. Energetically independent buildings of the resort-improving and educational-recreational complex in ecological settlement "GENOM". In Proceedings of the World Renewable Energy Forum, WREF 2012, Including World Renewable Energy Congress XII and Colorado Renewable Energy Society (CRES) Annual Conference, Denver, CO, USA, 13–19 May 2012; pp. 3767–3772.
61. Dunichkin, I. Aspects of designing construction of balcony parapets with photoelectric cells. *MATEC Web Conf.* **2016**, *86*, 02014. [CrossRef]
62. Dorokhov, A.S.; Shepovalova, O.V. Solar PV systems integrated into hardscape and sculptures. *AIP Conf. Proc.* **2019**, *2190*, 020094. [CrossRef]
63. Shepovalova, O.V.; Kondratenko, P.N. Complex power supply systems for agricultural facilities based on PV systems. In Proceedings of the International Scientific Conference "Prospects for the Development of Agriculture Energy Supply", Tomsk, Russia, 26–30 March 2020; pp. 25–37.
64. Noborov, P.R.; Barentzov, Y.I.; Kondratenko, P.N. PV systems integrated into barn buildings. In Proceedings of the Scientific Conference "Agriculture and Future", Krasnodar, Russia, 8–9 June 2021; pp. 118–127.

MDPI
St. Alban-Anlage 66
4052 Basel
Switzerland
Tel. +41 61 683 77 34
Fax +41 61 302 89 18
www.mdpi.com

Agriculture Editorial Office
E-mail: agriculture@mdpi.com
www.mdpi.com/journal/agriculture

www.ingramcontent.com/pod-product-compliance
Lightning Source LLC
Chambersburg PA
CBHW061558220926
51850CB00005B/363
9783036581606